Student's Solutions Manual

Essential Mathematics

KEEDY/BITTINGER/RUDOLPH

Student's Solutions Manual

Essential Mathematics

KEEDY/BITTINGER/RUDOLPH

6TH EDITION

Judith A. Penna

ADDISON-WESLEY PUBLISHING COMPANY
Reading, Massachusetts • Menlo Park, California • New York
Donn Mills, Ontario • Wokingham, England • Amsterdam • Bonn
Sydney • Singapore • Tokyo • Madrid • San Juan • Milan • Paris

Reproduced by Addison-Wesley from camera-ready copy supplied by the author.

ISBN 0-201-57883-2

2 3 4 5 6 7 8 9 10-AL-95949392

TABLE OF CONTENTS

Special thanks are extended to Patsy Hammond for her excellent typing and to Pam Smith for her careful proofreading. Their patience, efficiency, and good humor made the author's work much easier.

Student's Solutions Manual

Essential Mathematics

KEEDY/BITTINGER/RUDOLPH

Exercise Set 1.1

1. 5742 = 5 thousands + 7 hundreds + 4 tens + 2 ones

3. 27,342 = 2 ten thousands + 7 thousands + 3 hundreds + 4 tens + 2 ones

5. 9010 = 9 thousands + 1 ten

7. 2300 = 2 thousands + 3 hundreds

9. 2 thousands + 4 hundreds + 7 tens + 5 ones = 2475

11. 6 ten thousands + 8 thousands + 9 hundreds + 3 tens + 9 ones = 68,939

13. 7 thousands + 3 hundreds + 4 ones = 7304

15. 1 thousand + 9 ones = 1009

17. 77 = seventy-seven

19. 8 8, 0 0 0

Eighty-eight thousand

21. 1 2 3, 7 6 5

One hundred twenty-three thousand, seven hundred sixty-five

23.

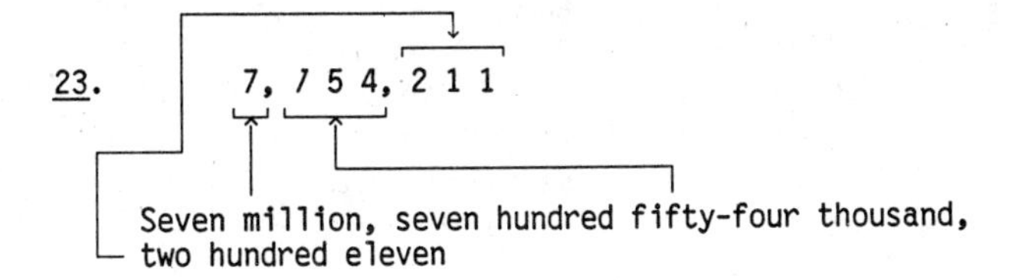

7, 7 5 4, 2 1 1
Seven million, seven hundred fifty-four thousand, two hundred eleven

25. 2 4 4, 8 3 9, 7 7 2

Two hundred forty-four million, eight hundred thirty-nine thousand, seven hundred seventy-two

27.

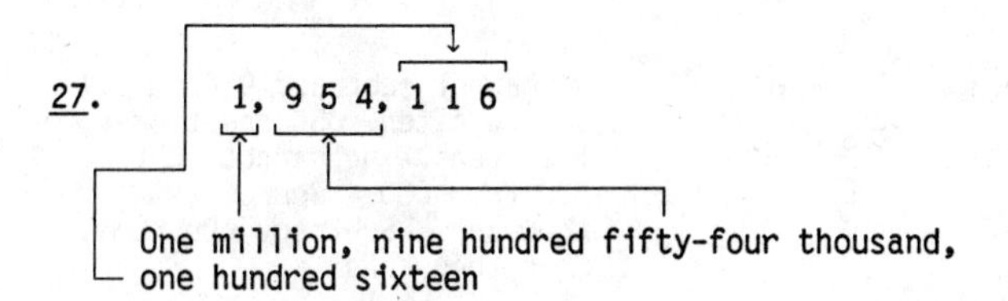

1, 9 5 4, 1 1 6
One million, nine hundred fifty-four thousand, one hundred sixteen

29.

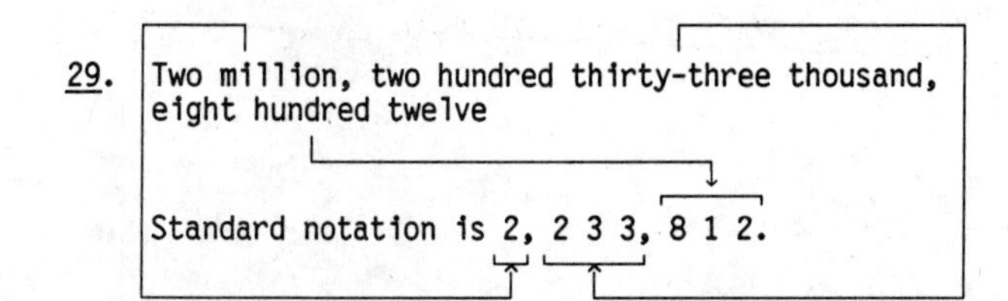

Two million, two hundred thirty-three thousand, eight hundred twelve
Standard notation is 2, 2 3 3, 8 1 2.

31. Eight billion

Standard notation is 8, 0 0 0, 0 0 0, 0 0 0.

33. Two hundred seventeen thousand, five hundred three

Standard notation is 2 1 7, 5 0 3.

35. Two million, one hundred seventy-three thousand, six hundred thirty-eight

Standard notation is 2, 1 7 3, 6 3 8

37. Two hundred six million, six hundred fifty-eight thousand

Standard notation is 2 0 6, 6 5 8, 0 0 0.

39. 23[5],888

The digit 5 means 5 thousands.

41. 488,[5]26

The digit 5 means 5 hundreds.

43. 89,[3]02

The digit 3 tells the number of hundreds.

45. 89,3[0]2

The digit 0 tells the number of tens.

47. All digits are 9's. Answers may vary. For an 8-digit readout, for example, it would be 99,999,999.

Exercise Set 1.2

1.

3 pins knocked down		6 pins knocked down		In all, 9 pins knocked down
3	+	6	=	9

3.

Earns \$23 one day		Earns \$31 the next day		The total earned is \$54
\$23	+	\$31	=	\$54

5.
```
  3 6 4
+   2 3
  3 8 7
```
Add ones, add tens, then add hundreds.

7.
```
  1 7 1 6
+ 3 2 8 2
---------
  4 9 9 8
```
Add ones, add tens, add hundreds, then add thousands

9.
```
   1 1
   9 9 9
+  1 1 1
--------
 1 1 1 0
```
Add ones: We get 10. Write 0 in the ones column and 1 above the tens. Add tens: We get 11. Write 1 in the tens column and 1 above the hundreds. Add hundreds: We get 11.

11.
```
     1
   9 0 9
+  1 0 1
--------
 1 0 1 0
```
Add ones: We get 10. Write 0 in the ones column and 1 above the tens. Add tens: We get 1. Add hundreds: We get 10.

13.
```
   1 1 1
   9 9 9 9
+  6 7 8 5
----------
 1 6,7 8 4
```
Add ones: We get 14. Write 4 in the ones column and 1 above the tens. Add tens: We get 18. Write 8 in the tens column and 1 above the hundreds. Add hundreds: We get 17. Write 7 in the hundreds column and 1 above the thousands. Add thousands: We get 16.

15.
```
     1 1 1
   2 3,4 4 3
+  1 0,9 8 9
------------
   3 4,4 3 2
```
Add ones: We get 12. Write 2 in the ones column and 1 above the tens. Add tens: We get 13. Write 3 in the tens column and 1 above the hundreds. Add hundreds: We get 14. Write 4 in the hundreds column and 1 above the thousands. Add thousands: We get 4. Add ten thousands: We get 3.

17.
```
    2 6
    8 2
+   6 1
-------
  1 6 9
```
Add ones, then add tens.

19.
```
   1 1 1
   2 0 3 7
   4 9 2 3
   3 4 7 1
+  1 2 4 8
----------
 1 1,6 7 9
```
Add ones: We get 19. Write 9 in the ones column and 1 above the tens. Add tens: We get 17. Write 7 in the tens column and 1 above the hundreds. Add hundreds: We get 16. Write 6 in the hundreds column and 1 above the thousands. Add thousands: We get 11.

21.

(Amount to begin with)		(Amount of check)		(Amount left)
\$650	-	\$100	=	☐

23. 6 + 9 = 15 (9 moves under the arrow)

The number 9 gets subtracted (moved).

6 = 15 - 9

6 + 9 = 15 (6 moves under the arrow)

The number 6 gets subtracted (moved).

9 = 15 - 6

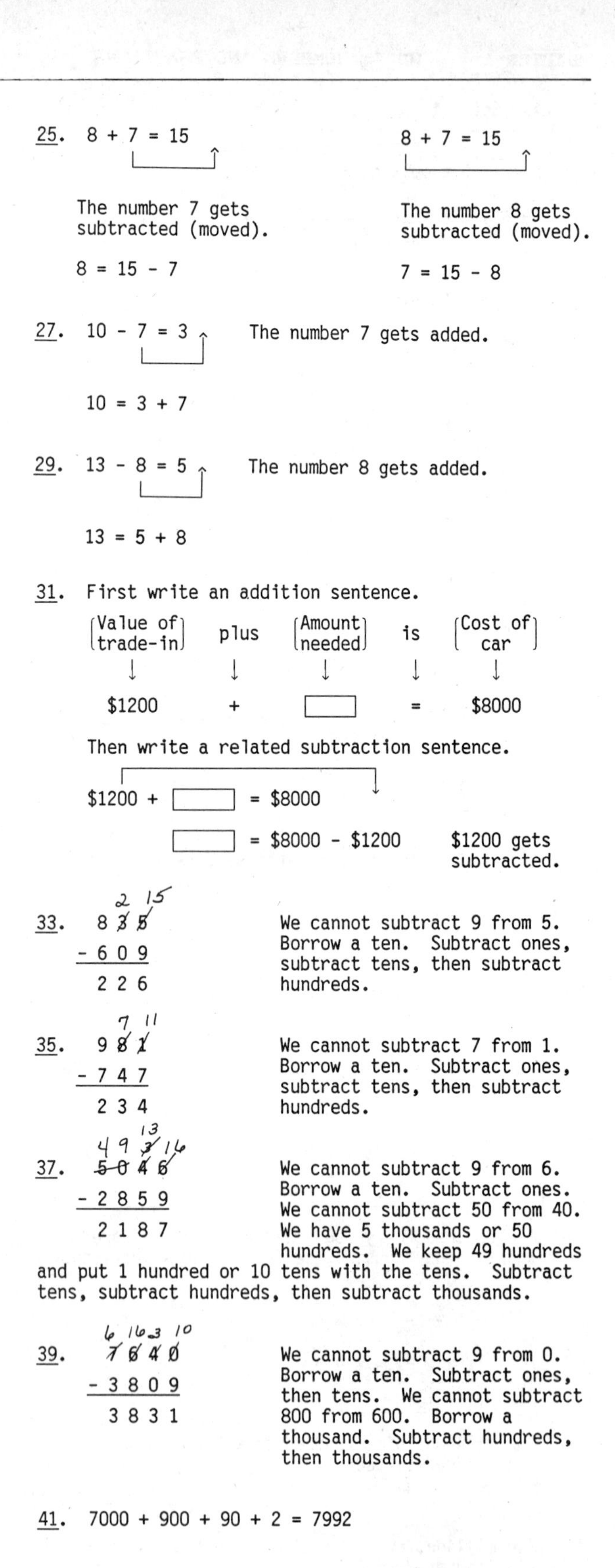

25. 8 + 7 = 15

The number 7 gets subtracted (moved).

8 = 15 - 7

8 + 7 = 15

The number 8 gets subtracted (moved).

7 = 15 - 8

27. 10 - 7 = 3 The number 7 gets added.

10 = 3 + 7

29. 13 - 8 = 5 The number 8 gets added.

13 = 5 + 8

31. First write an addition sentence.

(Value of trade-in)	plus	(Amount needed)	is	(Cost of car)
↓	↓	↓	↓	↓
\$1200	+	☐	=	\$8000

Then write a related subtraction sentence.

\$1200 + ☐ = \$8000

☐ = \$8000 - \$1200 \$1200 gets subtracted.

33.
```
    2 15
  8 3 5
- 6 0 9
-------
  2 2 6
```
We cannot subtract 9 from 5. Borrow a ten. Subtract ones, subtract tens, then subtract hundreds.

35.
```
    7 11
  9 8 1
- 7 4 7
-------
  2 3 4
```
We cannot subtract 7 from 1. Borrow a ten. Subtract ones, subtract tens, then subtract hundreds.

37.
```
        13
  4 9  3 16
  5 0  4 6
- 2 8  5 9
----------
  2 1  8 7
```
We cannot subtract 9 from 6. Borrow a ten. Subtract ones. We cannot subtract 50 from 40. We have 5 thousands or 50 hundreds. We keep 49 hundreds and put 1 hundred or 10 tens with the tens. Subtract tens, subtract hundreds, then subtract thousands.

39.
```
  6 16 3 10
  7  6 4  0
- 3  8 0  9
-----------
  3  8 3  1
```
We cannot subtract 9 from 0. Borrow a ten. Subtract ones, then tens. We cannot subtract 800 from 600. Borrow a thousand. Subtract hundreds, then thousands.

41. 7000 + 900 + 90 + 2 = 7992

43. $1 + 100 = 101$, $2 + 99 = 101$,
$3 + 98 = 101$, . . . $50 + 51 = 101$.

There are 50, or $\frac{100}{2}$, such sums, each equal to 101, or $100 + 1$. Then the sum of the numbers from 1 to 100 inclusive is $50 \times 101 = 5050$ or $\frac{100}{2} \times (100 + 1) = 5050$.

45. Perform the operations of addition and subtraction in order from left to right.

$3217 + 598 - 1349 = 3815 - 1349 = 2466$

Exercise Set 1.3

1. Round 48 to the nearest ten.

4[8]
↑

The digit 4 is in the tens place. Consider the next digit to the right. Since the digit, 8, is 5 or higher, round 4 tens up to 5 tens. Then change the digit to the right of the tens digit to zero.

The answer is 50.

3. Round 67 to the nearest ten.

6[7]
↑

The digit 6 is in the tens place. Consider the next digit to the right. Since the digit, 7, is 5 or higher, round 6 tens up to 7 tens. Then change the digit to the right of the tens digit to zero.

The answer is 70.

5. Round 731 to the nearest ten.

73[1]
↑

The digit 3 is in the tens place. Consider the next digit to the right. Since the digit, 1, is 4 or lower, round down, meaning that 3 tens stays as 3 tens. Then change the digit to the right of the tens digit to zero.

The answer is 730.

7. Round 895 to the nearest ten.

89[5]
↑

The digit 9 is in the tens place. Consider the next digit to the right. Since the digit, 5, is 5 or higher, we round up. The 89 tens become 90 tens. Then change the digit to the right of the tens digit to zero.

The answer is 900.

9. Round 146 to the nearest hundred.

1[4]6
↑

The digit 1 is in the hundreds place. Consider the next digit to the right. Since the digit, 4, is 4 or lower, round down, meaning that 1 hundred stays as 1 hundred. Then change all digits to the right of the hundreds digit to zeros.

The answer is 100.

11. Round 957 to the nearest hundred.

9[5]7
↑

The digit 9 is in the hundreds place. Consider the next digit to the right. Since the digit, 5, is 5 or higher, round up. The 9 hundreds become 10 hundreds. Then change all digits to the right of the hundreds digit to zeros.

The answer is 1000.

13. Round 3583 to the nearest hundred.

35[8]3
↑

The digit 5 is in the hundreds place. Consider the next digit to the right. Since the digit, 8, is 5 or higher, round 5 hundreds up to 6 hundreds. Then change all digits to the right of the hundreds digit to zeros.

The answer is 3600.

15. Round 2850 to the nearest hundred.

28[5]0
↑

The digit 8 is in the hundreds place. Consider the next digit to the right. Since the digit, 5, is 5 or higher, round 8 hundreds up to 9 hundreds. Then change all digits to the right of the hundreds digit to zeros.

The answer is 2900.

17. Round 5932 to the nearest thousand.

5[9]32
↑

The digit 5 is in the thousands place. Consider the next digit to the right. Since the digit, 9, is 5 or higher, round 5 thousands up to 6 thousands. Then change all digits to the right of the thousands digit to zeros.

The answer is 6000.

19. Round 7500 to the nearest thousand.

7[5]00
↑

The digit 7 is in the thousands place. Consider the next digit to the right. Since the digit, 5, is 5 or higher, round 7 thousands up to 8 thousands. Then change all digits to the right of the thousands digit to zeros.

The answer is 8000.

21. Round 45,340 to the nearest thousand.

45,[3]40
↑

The digit 5 is in the thousands place. Consider the next digit to the right. Since the digit, 3, is 4 or lower, round down, meaning that 5 thousands stays as 5 thousands. Then change all digits to the right of the thousands digit to zeros.

The answer is 45,000.

23. Round 373,405 to the nearest thousand.

373,[4]05
↑

The digit 3 is in the thousands place. Consider the next digit to the right. Since the digit, 4, is 4 or lower, round down, meaning that 3 thousands stays as 3 thousands. Then change all digits to the right of the thousands digit to zeros.

The answer is 373,000.

25.

	Rounded to the nearest ten
7848	7850
+ 9747	+ 9750
	17,600 ← Estimated answer

27.

	Rounded to the nearest ten
6882	6880
− 1748	− 1750
	5130 ← Estimated answer

29.

	Rounded to the nearest ten
45	50
77	80
25	30
+ 56	+ 60
343	220 ← Estimated answer

The sum 343 seems to be incorrect since 220 is not close to 343.

31.

	Rounded to the nearest ten
622	620
78	80
81	80
+ 111	+ 110
932	890 ← Estimated answer

The sum 932 seens to be incorrect since 890 is not close to 932.

33.

	Rounded to the nearest hundred
7848	7800
+ 9747	+ 9700
	17,500 ← Estimated answer

35.

	Rounded to the nearest hundred
6852	6900
− 1748	− 1700
	5200 ← Estimated answer

37.

	Rounded to the nearest hundred
216	200
84	100
745	700
+ 595	+ 600
1640	1600 ← Estimated answer

The sum 1640 seems to be correct since 1600 is close to 1640.

39.

	Rounded to the nearest hundred
750	800
428	400
63	100
+ 205	+ 200
1446	1500 ← Estimated answer

The sum 1446 seems to be correct since 1500 is close to 1446.

41.

	Rounded to the nearest thousand
9643	10,000
4821	5000
8943	9000
+ 7004	+ 7000
	31,000 ← Estimated answer

43.

	Rounded to the nearest thousand
92,149	92,000
− 22,255	− 22,000
	70,000 ← Estimated answer

45.

```
  1 1 1 1
  6 7,7 8 9
+ 1 8,9 6 5
  8 6,7 5 4
```

Add ones: We get 14. Write 4 in the ones column and 1 above the tens. Add tens: We get 15. Write 5 in the tens column and 1 above the hundreds. Add hundreds: We get 17. Write 7 in the hundreds column and 1 above the thousands. Add thousands: We get 16. Write 6 in the thousands column and 1 above the ten thousands. Add ten thousands: We get 8.

47. Using a calculator, we find that the sum is 30,411.

49. Using a calculator, we find that the difference is 69,894.

Exercise Set 1.4

1. Repeated addition fits best in this case.

[$10] [$10] [$10] · · · [$10]

32 addends

32 × $10 = $320

3. We have a rectangular array.

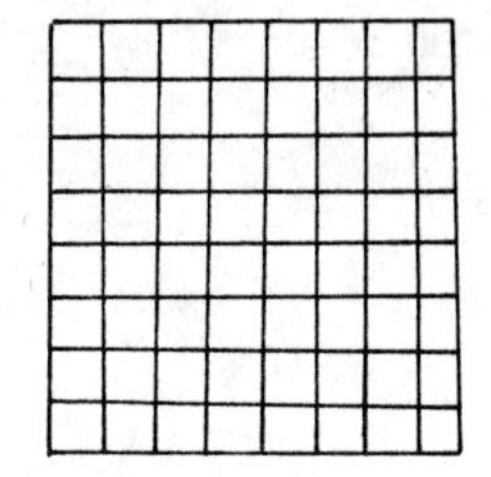

8 rows

8 squares

$8 \times 8 = 64$

5.

2 (carry over 5)	
5 3	
× 9 0	Multiplying by 9 tens (We write 0 and then multiply 53 by 9.)
4 7 7 0	

7.

2, 4 (carries over 6)	
6 5	
× 4 8	
5 2 0	Multiplying by 8
2 6 0 0	Multiplying by 40
3 1 2 0	Adding

9.

2 (carry over 6)	
6 4 0	
× 7 2	
1 2 8 0	Multiplying by 2
4 4 8 0 0	Multiplying by 70
4 6,0 8 0	Adding

11.

1 1	
4 4 4	
× 3 3	
1 3 3 2	Multiplying by 3
1 3 3 2 0	Multiplying by 30
1 4,6 5 2	Adding

13.

2, 3 (over 4); 2, 3 (over 8)	
4 8 9	
× 3 4 0	
1 9 5 6 0	Multiplying by 4 tens (We write 0 and then multiply 489 by 4.)
1 4 6,7 0 0	Multiplying by 3 hundreds (We write 00 and then multiply 489 by 3.)
1 6 6,2 6 0	Adding

15.

2, 3, 1 (over 4); 1, 4, 7, 3 (over 3); 1, 4, 7, 3 (over 7)	
4 3 7 8	
× 2 6 9 4	
1 7 5 1 2	Multiplying by 4
3 9 4 0 2 0	Multiplying by 90
2 6 2 6 8 0 0	Multiplying by 600
8 7 5 6 0 0 0	Multiplying by 2000
1 1,7 9 4,3 3 2	Adding

17.

2, 3, 3 (over 8); 1, 2, 3 (over 7)	
8 7 6	
× 3 4 5	
4 3 8 0	Multiplying by 5
3 5 0 4 0	Multiplying by 40
2 6 2 8 0 0	Multiplying by 300
3 0 2,2 2 0	Adding

19.

5, 1, 1, 3 (over 7); 5, 1, 1, 3 (over 8); 5, 1, 1, 3 (over 8)	
7 8 8 9	
× 6 2 2 4	
3 1 5 5 6	Multiplying by 4
1 5 7 7 8 0	Multiplying by 20
1 5 7 7 8 0 0	Multiplying by 200
4 7 3 3 4 0 0 0	Multiplying by 6000
4 9,1 0 1,1 3 6	Adding

21.

2, 2 (over 5); 2, 2 (over 5)	
5 5 5	
× 5 5	
2 7 7 5	Multiplying by 5
2 7 7 5 0	Multiplying by 50
3 0,5 2 5	Adding

23.

1, 2 (over 7); 1, 2 (over 3)	
7 3 4	
× 4 0 7	
5 1 3 8	Multiplying by 7
2 9 3 6 0 0	Multiplying by 4 hundreds (We write 00 and then multiply 734 by 4.)
2 9 8,7 3 8	Adding

25.

	Rounded to the nearest ten
4 5	5 0
× 6 7	× 7 0
	3 5 0 0 <— Estimated answer

27.

	Rounded to the nearest ten
3 4	3 0
× 2 9	× 3 0
	9 0 0 <— Estimated answer

29.

	Rounded to the nearest hundred
8 7 6	9 0 0
× 3 4 5	× 3 0 0
	2 7 0,0 0 0 <— Estimated answer

31.

	Rounded to the nearest hundred
4 3 2	4 0 0
× 1 9 9	× 2 0 0
	8 0,0 0 0 <— Estimated answer

33. Think of an array with 4 lb in each row.

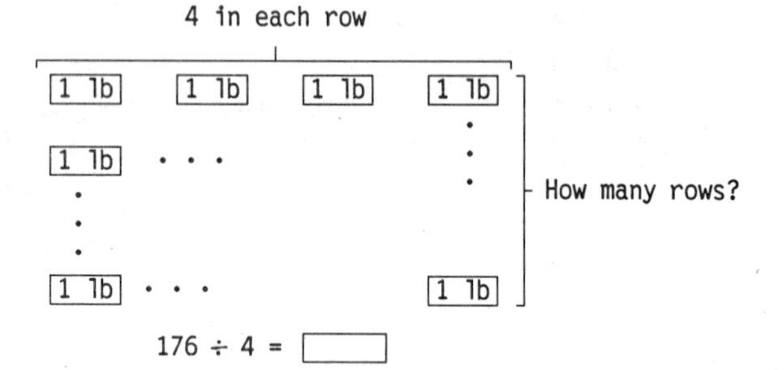

$176 \div 4 = \square$

35. Think of an array with \$23,000 in each row.

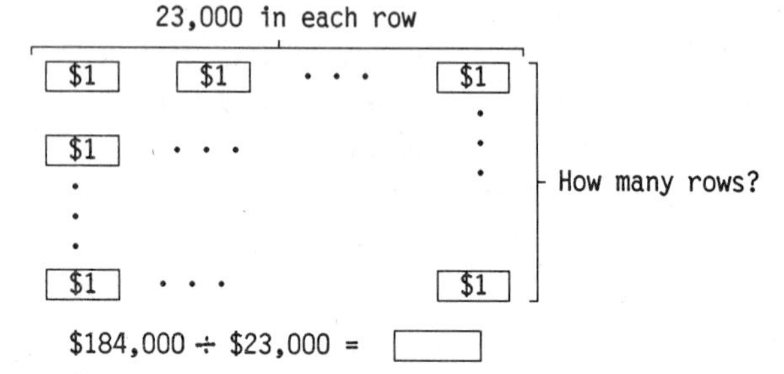

$\$184,000 \div \$23,000 = \square$

37. $24 \div 8 = 3$ The 8 moves to the right. A related multiplication sentence is $24 = 3 \cdot 8$.

39. $9 \times 5 = 45$

Move a factor to the other side and then write a division.

$9 \times 5 = 45$ $\qquad$ $9 \times 5 = 45$

$9 = 45 \div 5$ $\qquad$ $5 = 45 \div 9$

41. Any number divided by 1 is that same number.

$14 \div 1 = 14$

43. Zero divided by any number greater than zero is zero.

$0 \div 5 = 0$

45. Any number greater than zero divided by itself is 1.

$45 \div 45 = 1$

47. Any number greater than zero divided by itself is 1.

$37 \div 37 = 1$

49.

$$\begin{array}{r} 22 \\ 46\overline{)1058} \\ \underline{920} \\ 138 \\ \underline{92} \\ 46 \end{array}$$

Round 46 to 50.

Think: 105 tens ÷ 50. Estimate 2 tens.

Think: 138 ones ÷ 50. Estimate 2 ones.

Since 46 is not smaller than the divisor, 46, the estimate is too low.

49. (continued)

$$\begin{array}{r} 23 \\ 46\overline{)1058} \\ \underline{920} \\ 138 \\ \underline{138} \\ 0 \end{array}$$

Think: 138 ones ÷ 50. Estimate 3 ones.

The answer is 23.

51.

$$\begin{array}{r} 2904 \\ 306\overline{)888,888} \\ \underline{612000} \\ 276888 \\ \underline{275400} \\ 1488 \\ \underline{0} \\ 1488 \\ \underline{1224} \\ 264 \end{array}$$

Round 306 to 300.

Think: 888 thousands ÷ 300. Estimate 2 thousands.

Think: 2768 hundreds ÷ 300. Estimate 9 hundreds.

Think: 148 tens ÷ 300. Estimate 0 tens.

Think 1488 ones ÷ 300. Estimate 4 ones.

The answers is 2904 R 264.

53.

$$\begin{array}{r} \overset{5}{\cancel{6}}\,\overset{9}{\cancel{0}}\,\overset{9}{\cancel{0}}\,\overset{13}{\cancel{3}} \\ \underline{-\ 2\ 8\ 9\ 4} \\ 3\ 1\ 0\ 9 \end{array}$$

55. First we divide to find how many new cigarettes can be made from 29 cigarette butts:

$$\begin{array}{r} 7 \\ 4\overline{)29} \\ \underline{28} \\ 1 \end{array}$$

The person can make 7 new cigarettes, and there is 1 butt left over.

We divide again to find how many new cigarettes can be made by reusing the butts from the first seven cigarettes and the butt left over when the first seven were made. That is, the person now has a total of 8 butts to use:

$$\begin{array}{r} 2 \\ 4\overline{)8} \\ \underline{8} \\ 0 \end{array}$$

The person can now make 2 more cigarettes, so 7 + 2, or 9, cigarettes were made in all. After the last 2 cigarettes are smoked, 2 butts will be left over.

Exercise Set 1.5

1. $x + 0 = 14$

We replace x by different numbers until we get a true equation. If we replace x by 14, we get a true equation: $14 + 0 = 14$. No other replacement makes the equation true, so the solution is 14.

3. $y \cdot 17 = 0$

We replace y by different numbers until we get a true equation. If we replace y by 0 we get a true equation: $0 \cdot 17 = 0$. No other replacement makes the equation true, so the solution is 0.

5. $13 + x = 42$

$13 + x - 13 = 42 - 13$ Subtracting 13 on both sides

$x = 42 - 13$ 13 plus x minus 13 is x

$x = 29$ Doing the subtraction

7. $12 = 12 + m$

$12 - 12 = 12 + m - 12$ Subtracting 12 on both sides

$12 - 12 = m$ 12 plus m minus 12 is m

$0 = m$ Doing the subtraction

9. $3 \cdot x = 24$

$\frac{3 \cdot x}{3} = \frac{24}{3}$ Dividing by 3 on both sides

$x = 8$ Doing the division

11. $112 = n \cdot 8$

$\frac{112}{8} = \frac{n \cdot 8}{8}$ Dividing by 8 on both sides

$14 = n$ Doing the division

13. $45 \times 23 = x$

To solve the equation we carry out the calculation.

$$\begin{array}{r} 45 \\ \times\ 23 \\ \hline 135 \\ 90 \\ \hline 1035 \end{array}$$

The solution is 1035.

15. $t = 125 \div 5$

To solve the equation we carry out the calculation.

$$\begin{array}{r} 25 \\ 5\overline{)125} \\ \underline{100} \\ 25 \\ \underline{25} \\ 0 \end{array}$$

The solution is 25.

17. $p = 908 - 458$

To solve the equation we carry out the calculation.

$$\begin{array}{r} 908 \\ -\ 458 \\ \hline 450 \end{array}$$

The solution is 450.

19. $x = 12{,}345 + 78{,}555$

To solve the equation we carry out the calculation.

$$\begin{array}{r} 12{,}345 \\ +\ 78{,}555 \\ \hline 90{,}900 \end{array}$$

The solution is 90,900.

21. $3 \cdot m = 96$

$\frac{3 \cdot m}{3} = \frac{96}{3}$ Dividing by 3 on both sides

$m = 32$ Doing the division

23. $715 = 5 \cdot z$

$\frac{715}{5} = \frac{5 \cdot z}{5}$ Dividing by 5 on both sides

$143 = z$ Doing the division

25. $357 = 7 \cdot s$

$\frac{357}{7} = \frac{7 \cdot s}{7}$

$51 = s$

27. $20 + x = 57$

$20 + x - 20 = 57 - 20$

$x = 37$

29. $53 = 17 + w$

$53 - 17 = 17 + w - 17$

$36 = w$

31. $4 \cdot w = 3404$

$\frac{4 \cdot w}{4} = \frac{3404}{4}$

$w = 851$

33. $9 \cdot x = 1269$

$\frac{9 \cdot x}{9} = \frac{1269}{9}$

$x = 141$

35. $56 + p = 92$

$56 + p - 56 = 92 - 56$

$p = 36$

37. $z + 67 = 133$

$z + 67 - 67 = 133 - 67$

$z = 66$

39. $660 = 12 \cdot n$

$\frac{660}{12} = \frac{12 \cdot n}{12}$

$55 = n$

41. $784 = y \cdot 16$

$\frac{784}{16} = \frac{y \cdot 16}{16}$

$49 = y$

43. $x + 221 = 333$

$x + 221 - 221 = 333 - 221$

$x = 112$

45. $438 + x = 807$

$438 + x - 438 = 807 - 438$

$x = 369$

47. $19 \cdot x = 6080$

$\frac{19 \cdot x}{19} = \frac{6080}{19}$

$x = 320$

49. $20 \cdot x = 1500$

$\frac{20 \cdot x}{20} = \frac{1500}{20}$

$x = 75$

51. $9281 = 8322 + t$

$9281 - 8322 = 8322 + t - 8322$

$959 = t$

53. $9281 - 8322 = y$

$959 = y$

55. $10{,}534 \div 458 = q$

$23 = q$

57. $233 \cdot x = 22{,}135$

$\frac{233 \cdot x}{233} = \frac{22{,}135}{233}$

$x = 95$

59. 4[5],6 7 8,2 3 1

The digit 5 names millions.

61.

```
          6 5
  8 4)5 4 6 0     Round 84 to 80.
      5 0 4 0     Think:  546 tens ÷ 80
        4 2 0     Estimate 6 tens.
        4 2 0     Think:  420 ones ÷ 80
            0     Estimate 5 ones.
```

The answer is 65.

63. $29 \cdot y - 38 = 1325$

$29 \cdot y - 38 + 38 = 1325 + 38$ Adding 38 on both sides

$29 \cdot y = 1363$

$\frac{29 \cdot y}{29} = \frac{1363}{29}$ Dividing by 29 on both sides

$y = 47$

Exercise Set 1.6

1. Familiarize. We visualize the situation.

Ty Cobb's singles + Stan Musial's singles = Total singles

3052 singles   2641 singles   n singles

Translate. We write a number sentence that corresponds to the situation:

$3052 + 2641 = n$

Solve. We carry out the addition.

```
  3 0 5 2
+ 2 6 4 1
  5 6 9 3
```

Check. We can repeat the calculation. We can also find an estimated answer by rounding: $3052 + 2641 \approx 3000 + 2600 = 5600 \approx 5693$. Since the estimated answer is close to the calculation, our answer seems correct.

State. Ty Cobb and Stan Musial hit 5693 singles together.

3. Familiarize. We first make a drawing.

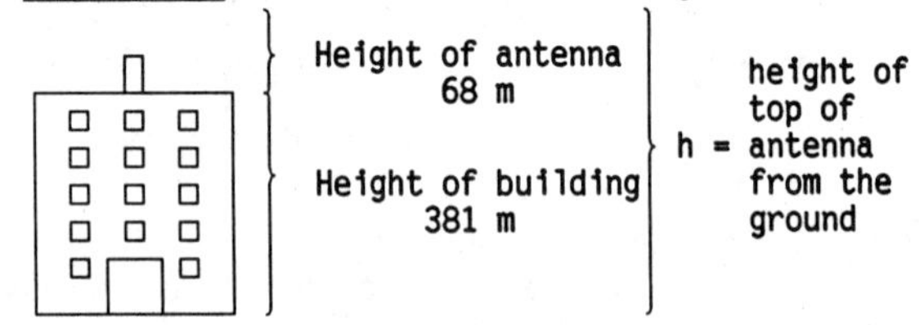

Since we are combining lengths, addition can be used. We let h = the height of the top of the antenna from the ground.

Translate. We translate to the following addition sentence:

$381 + 68 = h$

Solve. To solve we carry out the addition.

```
  3 8 1
+   6 8
  4 4 9
```

Check. We can repeat the calculation. We can also find an estimated answer by rounding: $381 + 68 \approx 380 + 70 = 450 \approx 449$. The answer checks.

State. The top of the antenna is 449 m from the ground.

5. Familiarize. We first make a drawing.

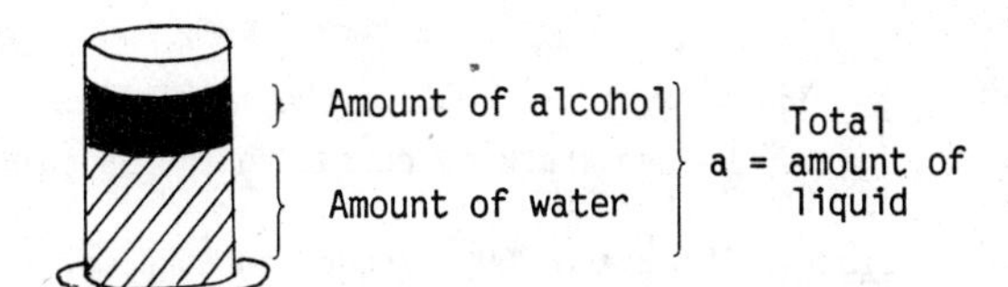

Since we are combining amounts, addition can be used. We let a = the total amount of liquid.

Translate. We translate to the following addition sentence:

$$2340 + 655 = a$$

Solve. To solve we carry out the addition.

$$\begin{array}{r} 2\,3\,4\,0 \\ +\ \ 6\,5\,5 \\ \hline 2\,9\,9\,5 \end{array}$$

Check. We can repeat the calculation. We can also find an estimated answer by rounding: $2340 + 655 \approx 2300 + 700 = 3000 \approx 2995$. The answer checks.

State. A total of 2995 cubic centimeters of liquid was poured.

7. Familiarize. We visualize the situation. We let x = the amount by which the water loss from the skin exceeds the water loss from the lungs.

Water loss from lungs 400 cubic centimeters 400 cc	Excess loss from skin x
Water loss from skin 500 cubic centimeters	

Translate. This is a "how-much-more" situation. Translate to an equation.

Water lost from lungs	plus	Additional amount of water	is	Water lost from skin
400	+	x	=	500

Solve. Solve the equation.

$$400 + x = 500$$

$$400 + x - 400 = 500 - 400 \quad \text{Subtracting 400 on both sides}$$

$$x = 100$$

Check. We can add the result to 400: $400 + 100 = 500$. The answer checks.

State. 100 cubic centimeters more water is lost from the skin than from the lungs.

9. Familiarize. We visualize the situation. We let p = the number by which the population of the Tokyo-Yokohama area exceeds the population of the New York-northeastern New Jersey area.

NY - NJ population 17,013,000	Excess p
Tokyo-Yokohama population 17,317,000	

Translate. This is a "how-much-more" situation. Translate to an equation.

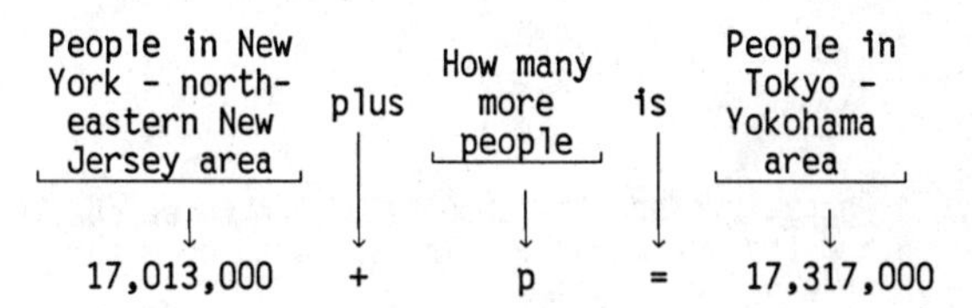

Solve. We solve the equation.

$$17{,}013{,}000 + p = 17{,}317{,}000$$

$$17{,}013{,}000 + p - 17{,}013{,}000 = 17{,}317{,}000 - 17{,}013{,}000$$

Subtracting 17,013,000 on both sides

$$p = 304{,}000$$

Check. We can add the result to 17,013,000: $17{,}013{,}000 + 304{,}000 = 17{,}317{,}000$. We can also estimate: $17{,}317{,}000 - 17{,}013{,}000 \approx 17{,}300{,}000 - 17{,}000{,}000 = 300{,}000 \approx 304{,}000$. The answer checks.

State. There are 304,000 more people in the Tokyo-Yokohama area.

11. Familiarize. This is a multistep problem.

To find the new balance, find the total amount of the three checks. Then take that amount away from the original balance. We visualize the situation. We let t = the total amount of the three checks and n = the new balance.

$246			
$45	$78	$32	n
t			n

Translate. To find the total amount of the three checks, write an addition sentence.

Amount of first check	plus	Amount of second check	plus	Amount of third check	is	Total Amount
$45	+	$78	+	$32	=	t

Solve. To solve the equation we carry out the addition.

$$\begin{array}{r} \$\ \ 4\,5 \\ 7\,8 \\ +\ \ 3\,2 \\ \hline \$1\,5\,5 \end{array}$$

11. (continued)

The total amount of the three checks is \$155.

To find the new balance, we have a "take-away" situation.

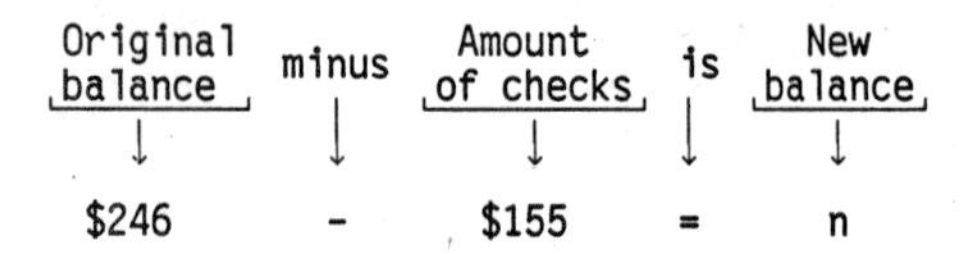

To solve we carry out the subtraction.

$$\begin{array}{r} \$\ 2\ 4\ 6 \\ -\ 1\ 5\ 5 \\ \hline \$\ \ \ 9\ 1 \end{array}$$

Check. We add the amounts of the three checks and the new balance. The sum should be the original balance.

\$45 + \$78 + \$32 + \$91 = \$246

The answer checks.

State. The new balance is \$91.

13. Familiarize. We first make a drawing. Repeated addition works well here. We let n = the number of calories burned in 5 hours.

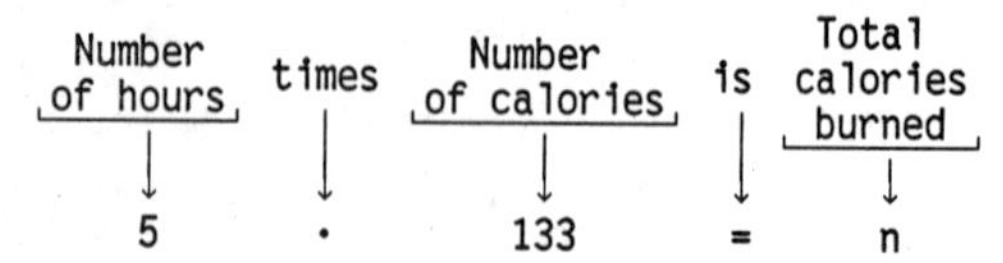

Translate. Translate to a number sentence.

Number of hours | times | Number of calories | is | Total calories burned
5 · 133 = n

Solve. To solve the equation, we carry out the multiplication.

$5 \cdot 133 = n$

$665 = n$ Doing the multiplication

Check. We can repeat the calculation. We can also estimate: $5 \cdot 133 \approx 5 \cdot 130 = 650 \approx 665$. The answer checks.

State. In 5 hours, 665 calories would be burned.

15. Familiarize. We first make a drawing. Repeated addition works well here. We let x = the number of seconds in an hour.

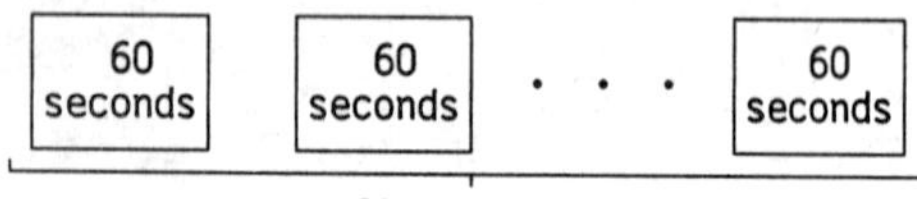

Translate. We translate to an equation.

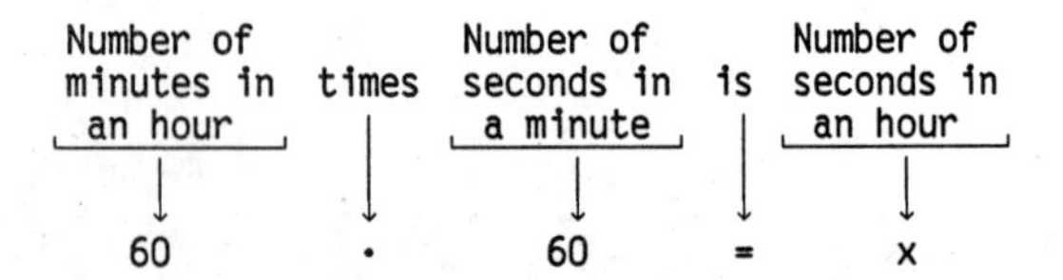

17. (continued)

Solve. We carry out the multiplication.

$60 \cdot 60 = x$

$3600 = x$ Doing the multiplication

Check. We can check by repeating the calculation. The answer checks.

State. There are 3600 seconds in an hour.

17. We first make a drawing.

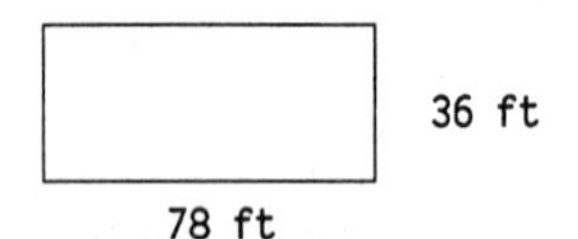

Translate. Using the formula for area, we have $A = \ell \cdot w = 78 \cdot 36$.

Solve. Carry out the multiplication.

$$\begin{array}{r} 7\ 8 \\ \times\ 3\ 6 \\ \hline 4\ 6\ 8 \\ 2\ 3\ 4\ 0 \\ \hline 2\ 8\ 0\ 8 \end{array}$$

Thus, A = 2808.

Check. We repeat the calculation. The answer checks.

State. The area is 2808 sq ft.

19. Familiarize. We visualize the situation. We let d = the diameter of the earth.

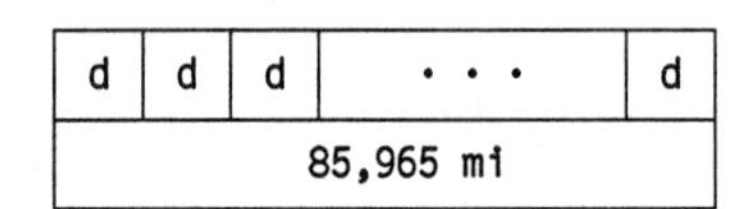

Translate. Repeated addition applies here. The following multiplication corresponds to the situation.

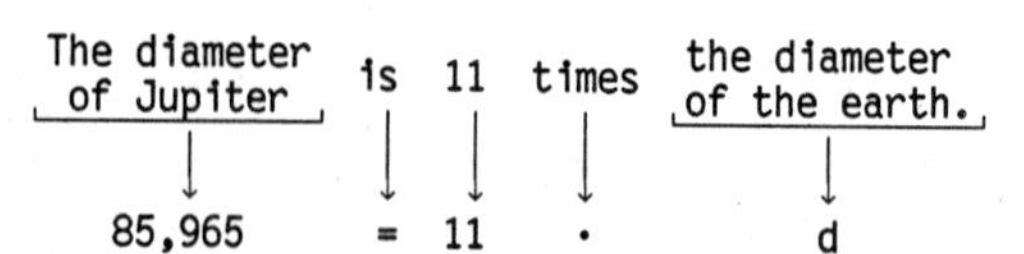

Solve. To solve the equation, we divide by 11 on both sides.

$85,965 = 11 \cdot d$

$\frac{85,965}{11} = \frac{11 \cdot d}{11}$

$7815 = d$

Check. To check we multiply 7815 by 11:

$11 \cdot 7815 = 85,965$

This checks.

State. The diameter of the earth is 7815 mi.

21. Familiarize. This is a multistep problem.

We must find the total cost of the 8 suits and the total cost of the 3 shirts. The amount spent is the sum of these two totals.

Repeated addition works well in finding the total cost of the 8 suits and the total cost of the 3 shirts. We let x = the total cost of the suits and y = the total cost of the shirts.

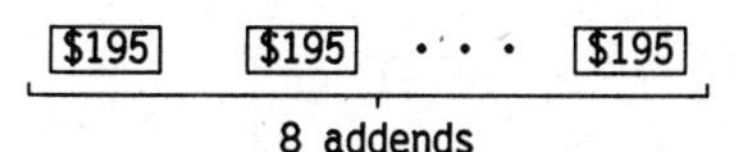

$26 $26 $26

3 addends

Translate. We translate to two equations.

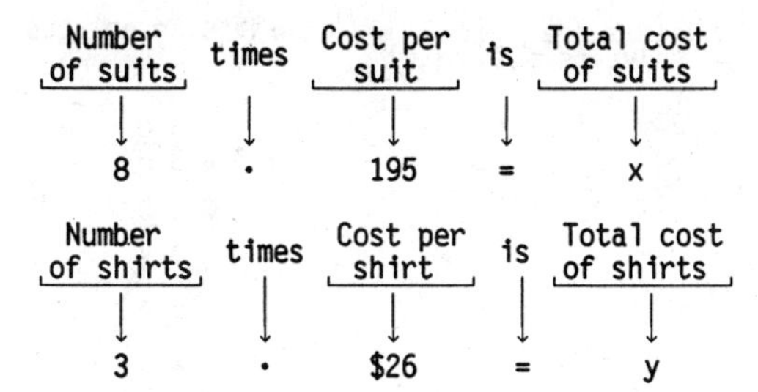

Solve. To solve the equations, we carry out the multiplications.

$$\begin{array}{r} 195 \\ \times\ \ \ 8 \\ \hline 1560 \end{array}$$

Thus, x = $1560.

$$\begin{array}{r} 26 \\ \times\ 3 \\ \hline 78 \end{array}$$

Thus, y = $78.

We let a = the total amount spent.

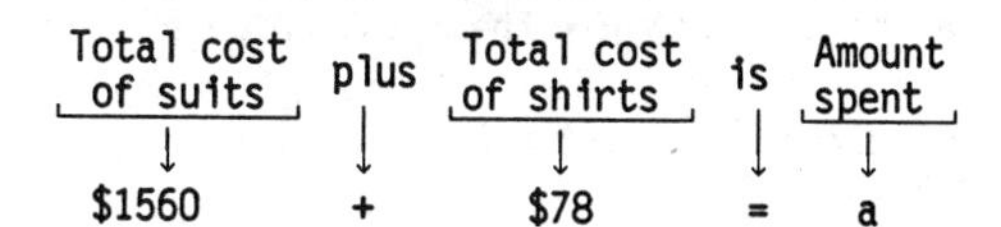

To solve the equation carry out the addition.

$$\begin{array}{r} \$1560 \\ +\ \ \ \ 78 \\ \hline \$1638 \end{array}$$

Check. We repeat the calculations. The answer checks.

State. The amount spent is $1638.

23. Familiarize. This is a multistep problem.

We must find the area of the lot and the area of the garden. Then we take the area of the garden away from the area of the lot. We let A = the area of the lot and G = the area of the garden.

First, make a drawing of the lot with the garden. The area left over is shaded.

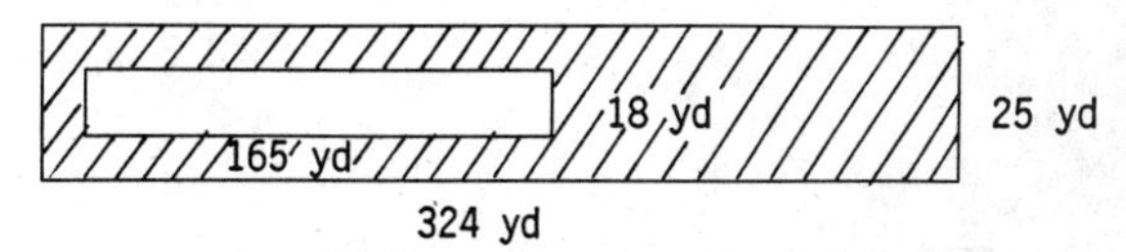

Translate. We use the formula for area twice.

$A = \ell \cdot w = 324 \cdot 25$

$G = \ell \cdot w = 185 \cdot 18$

Solve. We carry out the multiplications.

$$\begin{array}{r} 324 \\ \times\ \ 25 \\ \hline 1620 \\ 6480 \\ \hline 8100 \end{array}$$

A = 8100 sq yd.

$$\begin{array}{r} 165 \\ \times\ \ 18 \\ \hline 1320 \\ 1650 \\ \hline 2970 \end{array}$$

G = 2970 sq yd.

To find the area left over we have a "take-away" situation. We let a = the area left over.

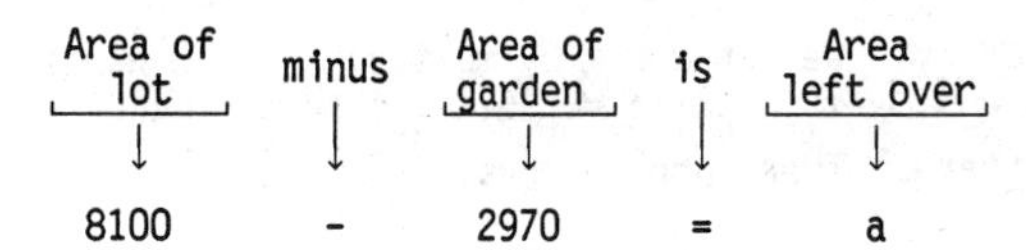

To solve, carry out the subtraction.

$$\begin{array}{r} 8100 \\ -\ 2970 \\ \hline 5130 \end{array}$$

Check. We repeat the calculations. The answer checks.

State. The area left over is 5130 sq yd.

25. Familiarize. We first draw a picture. We let n = the number of bottles to be filled.

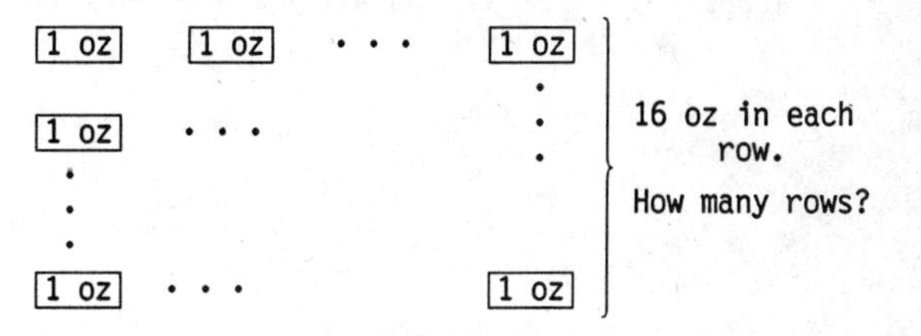

25. (continued)

Translate and Solve. We translate to an equation and solve as follows:

$608 \div 16 = n$

$$\begin{array}{r} 38 \\ 16\overline{)608} \\ \underline{480} \\ 128 \\ \underline{128} \\ 0 \end{array}$$

Check. We can check by multiplying the number of bottles by 16: $16 \cdot 38 = 608$. The answer checks.

State. 38 sixteen-oz bottles can be filled.

27. Familiarize. We draw a picture. We let y = the number of rows.

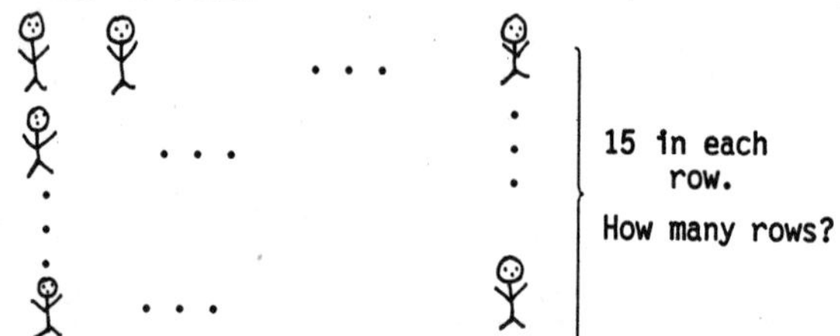

Translate and Solve. We translate to an equation and solve as follows:

$225 \div 15 = y$

$$\begin{array}{r} 15 \\ 15\overline{)225} \\ \underline{150} \\ 75 \\ \underline{75} \\ 0 \end{array}$$

Check. We can check by multiplying the number of rows by 15: $15 \cdot 15 = 225$. The answer checks.

State. There are 15 rows.

29. Familiarize. We first draw a picture. We let x = the amount of each payment.

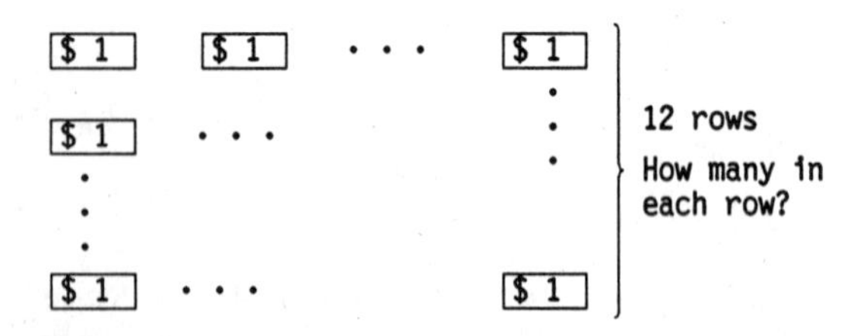

Translate and Solve. We translate to an equation and solve as follows:

$324 \div 12 = x$

$$\begin{array}{r} 27 \\ 12\overline{)324} \\ \underline{240} \\ 84 \\ \underline{84} \\ 0 \end{array}$$

29. (continued)

Check. We can check by multiplying 27 by 12: $12 \cdot 27 = 324$. the answer checks.

State. Each payment is $27.

31. Familiarize. We draw a picture. We let n = the number of bags to be filled.

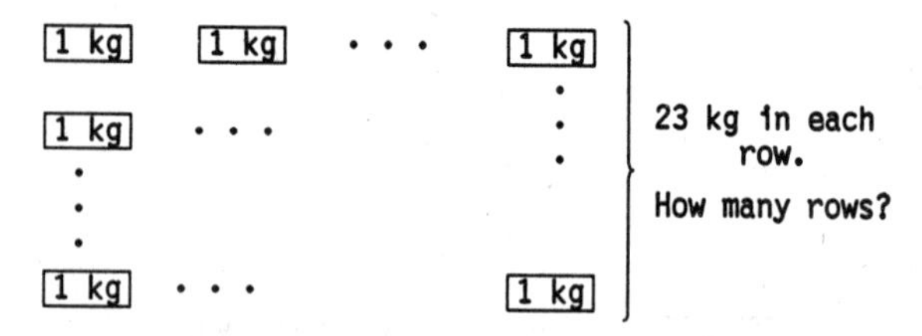

Translate and Solve. We translate to an equation and solve as follows:

$885 \div 23 = n$

$$\begin{array}{r} 38 \\ 23\overline{)885} \\ \underline{690} \\ 195 \\ \underline{184} \\ 11 \end{array}$$

Check. We can check by multiplying the number of bags by 23 and adding the remainder of 11:

$23 \cdot 38 = 874$

$874 + 11 = 885$

The answer checks.

State. 38 twenty-three-kg bags can be filled. There will be 11 kg of sand left over.

33. Familiarize. This is a multistep problem.

We must find the total price of the 5 coats. Then we must find how many 20's there are in the total price. Let p = the total price of the coats.

To find the total price of the 5 coats we can use repeated addition.

$64 $64 $64 $64 $64

5 addends

Translate.

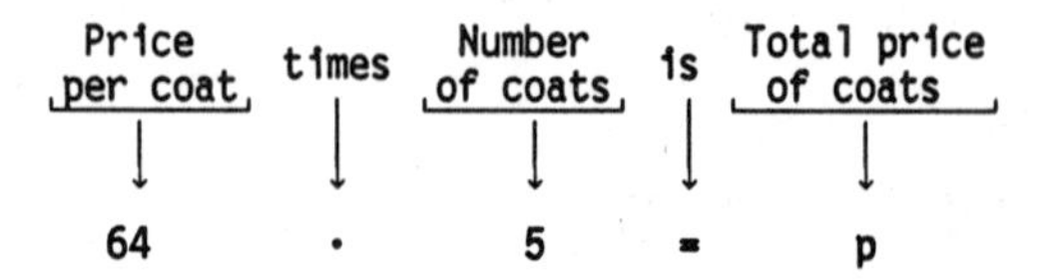

Solve. First we carry out the multiplication.

$64 \cdot 5 = p$

$320 = p$

The total price of the 5 coats is $320. Repeated addition can be used again to find how many 20's in $320. We let x = the number of $20 bills required.

33. (continued)

$320			
$20	$20	. . .	$20

Translate to an equation and solve.

$20 \cdot x = 320$

$\frac{20 \cdot x}{20} = \frac{320}{20}$ Dividing on both sides by 20

$x = 16$

Check. We repeat the calculations. The answer checks.

Solve. It took 16 twenty dollar bills.

35. Familiarize. To find how far apart on the map the cities are we must find how many 55's there are in 605. Repeated addition applies here. We let d = the distance between the cities on the map.

605 mi			
55 mi	55 mi	. . .	55 mi

Translate.

55 mi times How many 55's is 605 mi?

$55 \cdot d = 605$

Solve. We divide on both sides by 55.

$55 \cdot d = 605$

$\frac{55 \cdot d}{55} = \frac{605}{55}$ Dividing on both sides by 55

$d = 11$

Check. We multiply the number of inches by 55: $11 \cdot 55 = 605$. The answer checks.

State. The cities are 11 inches apart on the map.

Familiarize. When two cities are 14 in. apart on the map, we can use repeated addition to find how far apart they are in reality. We let n = the distance between the cities in reality.

Since 1 in. represents 55 mi, we can draw the following picture:

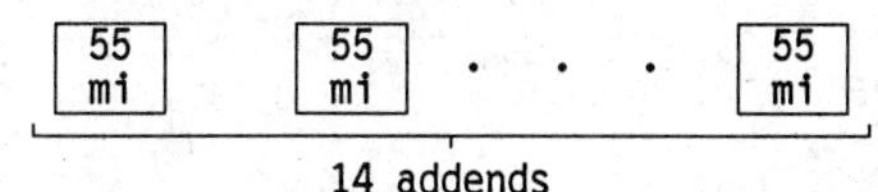

Translate.

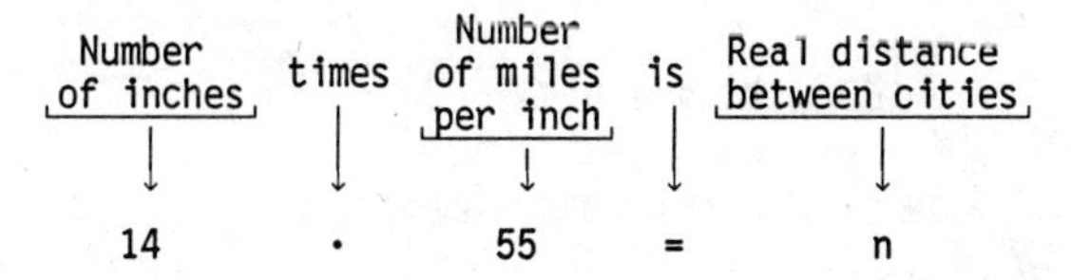

35. (continued)

Solve. We carry out the multiplication.

$14 \times 55 = n$

$770 = n$ Doing the multiplication

Check. We repeat the calculation. The answer checks.

State. The cities are 770 mi apart in reality.

37. Familiarize. This is a multistep problem.

We must find how many 100's there are in 3500. Then we must find that number times 15.

We first draw a picture.

One pound			
3500 calories			
100 cal	100 cal	. . .	100 cal
15 min	15 min	. . .	15 min

In Example 11 it was determined that there are 35 100's in 3500. We let t = the time you have to bicycle to lose a pound.

Translate and Solve. We know that bicycling at 9 mph for 15 min burns off 100 calories, so we need to bicycle for 35 times 15 min in order to burn off one pound. Translate to an equation and solve. We let t = the time required to lose one pound by bicycling.

$35 \times 15 = t$

$525 = t$

Check. Suppose you bicycle for 525 minutes. If we divide 525 by 15, we get 35, and 35 times 100 is 3500, the number of calories that must be burned off to lose one pound. The answer checks.

State. You must bicycle at 9 mph for 525 min, or 8 hr 45 min, to lose one pound.

39.

	Rounded to the nearest thousand
45,763	46,000
+ 38,427	+ 38,000
	84,000 ← Estimated answer

41. Familiarize. We visualize the situation. Let d = the distance light travels in 1 sec.

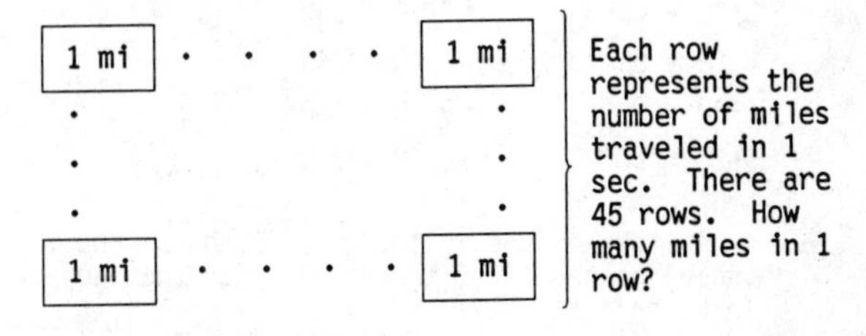

41. (continued)

Translate and Solve. We translate to an equation and solve as follows:

$8{,}370{,}000 \div 45 = d$

```
         1 8 6,0 0 0
  4 5 ) 8,3 7 0,0 0 0
        4,5 0 0,0 0 0
        3,8 7 0,0 0 0
        3,6 0 0,0 0 0
          2 7 0,0 0 0
          2 7 0,0 0 0
                    0
```

Check. We can check by multiplying the number of miles traveled in 1 sec by 45 sec:

$45 \cdot 186{,}000 = 8{,}370{,}000$

The answer checks.

State. Light travels 186,000 mi in 1 sec.

Exercise Set 1.7

1. We first find some factorizations:

$16 = 1 \cdot 16$ $\quad 16 = 8 \cdot 2$

$16 = 2 \cdot 8$ $\quad 16 = 16 \cdot 1$

$16 = 4 \cdot 4$

Factors: 1, 2, 4, 8, 16

3. We first find some factorizations:

$54 = 1 \cdot 54$ $\quad 54 = 9 \cdot 6$

$54 = 2 \cdot 27$ $\quad 54 = 18 \cdot 3$

$54 = 3 \cdot 18$ $\quad 54 = 27 \cdot 2$

$54 = 6 \cdot 9$ $\quad 54 = 54 \cdot 1$

Factors: 1, 2, 3, 6, 9, 18, 27, 54

5. $1 \cdot 4 = 4$ $\quad 6 \cdot 4 = 24$

$2 \cdot 4 = 8$ $\quad 7 \cdot 4 = 28$

$3 \cdot 4 = 12$ $\quad 8 \cdot 4 = 32$

$4 \cdot 4 = 16$ $\quad 9 \cdot 4 = 36$

$5 \cdot 4 = 20$ $\quad 10 \cdot 4 = 40$

7. $1 \cdot 20 = 20$ $\quad 6 \cdot 20 = 120$

$2 \cdot 20 = 40$ $\quad 7 \cdot 20 = 140$

$3 \cdot 20 = 60$ $\quad 8 \cdot 20 = 160$

$4 \cdot 20 = 80$ $\quad 9 \cdot 20 = 180$

$5 \cdot 20 = 100$ $\quad 10 \cdot 20 = 200$

9. Make lists of the multiples of 3 and 7, and underline the first three multiples that appear in both lists.

3, 6, 9, 12, 15, 18, 21, 24, 27, 30, 33, 36, 39, 42, 45, 48, 51, 54, 57, 60, 63, 66, 69, . . .

7, 14, 21, 28, 35, 42, 49, 56, 63, 70, 77, 84, . . .

9. (continued)

The first three common multiples of 3 and 7 are 21, 42, and 63.

11. Make lists of the multiples of 5, 8, and 9, and underline the first three multiples that appear in each list.

5, 10, 15, 20, 25, 30, 35, 40, 45, 50, 55, 60, 65, 70, 75, 80, 85, 90, 95, 100, 105, 110, 115, 120, 125, 130, 135, 140, 145, 150, 155, 160, 165, 170, 175, 180, 185, 190, 195, 200, 205, 210, 215, 220, 225, 230, 235, 240, 245, 250, 255, 260, 265, 270, 275, 280, 285, 290, 295, 300, 305, 310, 315, 320, 325, 330, 335, 340, 345, 350, 355, 360, 365, 370, 375, 380, 385, 390, 395, 400, 405, 410, 415, 420, 425, 430, 435, 440, 445, 450, 455, 460, 465, 470, 475, 480, 485, 490, 495, 500, 505, 510, 515, 520, 525, 530, 535, 540, 545, 550, 555, 560, 565, 570, 575, 580, 585, 590, 595, 600, 605, 610, 615, 620, 625, 630, 635, 640, 645, 650, 655, 660, 665, 670, 675, 680, 685, 690, 695, 700, 705, 710, 715, 720, 725, 730, 735, 740, 745, 750, 755, 760, 765, 770, 775, 780, 785, 790, 795, 800, 805, 810, 815, 820, 825, 830, 835, 840, 845, 850, 855, 860, 865, 870, 875, 880, 885, 890, 895, 900, 905, 910, 915, 920, 925, 930, 935, 940, 945, 950, 955, 960, 965, 970, 975, 980, 985, 990, 995, 1000, 1005, 1010, 1015, 1020, 1025, 1030, 1035, 1040, 1045, 1050, 1055, 1060, 1065, 1070, 1075, 1080, 1085, . . .

8, 16, 24, 32, 40, 48, 56, 64, 72, 80, 88, 96, 104, 112, 120, 128, 136, 144, 152, 160, 168, 176, 184, 192, 200, 208, 216, 224, 232, 240, 248, 256, 264, 272, 280, 288, 296, 304, 312, 320, 328, 336, 344, 352, 360, 368, 376, 384, 392, 400, 408, 416, 424, 432, 440, 448, 456, 464, 472, 480, 488, 496, 504, 512, 520, 528, 536, 544, 552, 560, 568, 576, 584, 592, 600, 608, 616, 624, 632, 640, 648, 656, 664, 672, 680, 688, 696, 704, 712, 720, 728, 736, 744, 752, 760, 768, 776, 784, 792, 800, 808, 816, 824, 832, 840, 848, 856, 864, 872, 880, 888, 896, 904, 912, 920, 928, 936, 944, 952, 960, 968, 976, 984, 992, 1000, 1008, 1016, 1024, 1032, 1040, 1048, 1056, 1064, 1072, 1080, 1088, 1096, . . .

9, 18, 27, 36, 45, 54, 63, 72, 81, 90, 99, 108, 117, 126, 135, 144, 153, 162, 171, 180, 189, 198, 207, 216, 225, 234, 243, 252, 261, 270, 279, 288, 297, 306, 315, 324, 333, 342, 351, 360, 369, 378, 387, 396, 405, 414, 423, 432, 441, 450, 459, 468, 477, 486, 495, 504, 513, 522, 531, 540, 549, 558, 567, 576, 585, 594, 603, 612, 621, 630, 639, 648, 657, 666, 675, 684, 693, 702, 711, 720, 729, 738, 747, 756, 765, 774, 783, 792, 801, 810, 819, 828, 837, 846, 855, 864, 873, 882, 891, 900, 909, 918, 927, 936, 945, 954, 963, 972, 981, 990, 999, 1008, 1017, 1026, 1035, 1044, 1053, 1062, 1071, 1080, 1089, . . .

The first three common multiples of 5, 8, and 9 are 360, 720, and 1080.

13. 1 is neither prime nor composite.

15. The number 9 has factors 1, 3, and 9.

Since 9 is not 1 and not prime, it is composite.

17. The number 11 is prime. It has only the factors 1 and 11.

19. The number 29 is prime. It has only the factors 1 and 29.

21.
```
      2 <—  2 is prime
    2|4
   2|8
  2|16
 2|32
```

$32 = 2\cdot2\cdot2\cdot2\cdot2$

23.
```
      5 <—  5 is prime
   2|10
  2|20
 2|40
```

$40 = 2\cdot2\cdot2\cdot5$

25.
```
     31 <—  31 is prime
   2|62
```

$62 = 2\cdot31$

27.
```
      7 <—  7 is prime
   5|35     (35 is not divisible by 2 or 3.
  2|70       We move to 5.)
 2|140
```

$140 = 2\cdot2\cdot5\cdot7$

29.
```
     11 <—  11 is prime
   5|55     (55 is not divisible by 2 or 3.
  2|110      We move to 5.)
```

$110 = 2\cdot5\cdot11$

31.
```
      7 <—  7 is prime
   5|35     (35 is not divisible by 2 or 3.
  2|70       We move to 5.)
```

$70 = 2\cdot5\cdot7$

33.
```
     43 <—  43 is prime
   2|86
```

$86 = 2\cdot43$

35.
```
     11 <—  11 is prime
   3|33     (33 is not divisible by 2.
  3|99       We move to 3.)
```

$99 = 3\cdot3\cdot11$

37. a) Find the prime factorization of each number.

Since 3 is prime it has no prime factorization. We still need it as a factor of the LCM.

$3 = 3$

$10 = 2 \times 5$

b) We write 2 as a factor 1 time (the greatest number of times it occurs in any one factorization). We write 3 as a factor 1 time (the greatest number of times it occurs in any one factorization). We write 5 as a factor 1 time (the greatest number of times it occurs in any one factorization).

The LCM is $2 \times 3 \times 5$, or 30.

39. a) Find the prime factorization of each number.

Since 3 is prime it has no prime factorization. We still need it as a factor of the LCM.

$3 = 3$

$15 = 3 \times 15$

b) We write 3 as a factor 1 time (the greatest number of times it occurs in any one factorization). We write 5 as a factor 1 time (the greatest number of times it occurs in any one factorization).

The LCM of 3 and 15 is 3×5, or 15.

41. a) Find the prime factorization of each number.

$12 = 2 \times 2 \times 3$

$18 = 2 \times 3 \times 3$

b) We write 2 as a factor 2 times (the greatest number of times it occurs in any one factorization). We write 3 as a factor 2 times (the greatest number of times it occurs in any one factorization).

The LCM of 12 and 18 is $2 \times 2 \times 3 \times 3$, or 36.

43. a) Find the prime factorization of each number.

$35 = 5 \times 7$

$45 = 5 \times 9$

b) We write 5 as a factor 1 time, 7 as a factor 1 time, and 9 as a factor 1 time (the greatest number of times that each occurs in any one factorization).

The LCM of 35 and 45 is $5 \times 7 \times 9$, or 315.

45. a) Find the prime factorization of each number.

$8 = 2 \times 2 \times 2$

$16 = 2 \times 2 \times 2 \times 2$

$22 = 2 \times 11$

b) We write 2 as a factor 4 times and 11 as a factor 1 time (the greatest number of times that each occurs in any one factorization).

The LCM of 8, 16, and 22 is $2\times2\times2\times2\times11$, or 176.

47. a) Find the prime factorization of each number.

$$12 = 2 \times 2 \times 3$$
$$18 = 2 \times 3 \times 3$$
$$40 = 2 \times 2 \times 2 \times 5$$

b) We write 2 as a factor 3 times, 3 as a factor 2 times, and 5 as a factor 1 time (the greatest number of times that each occurs in any one factorization).

The LCM of 12, 18, and 40 is $2 \times 2 \times 2 \times 3 \times 3 \times 5$, or 360.

49. a) Find the prime factorization of each number.

Since 7 and 3 are prime, they have no prime factorizations. We still need each as a factor of the LCM.

$$7 = 7$$
$$18 = 2 \times 3 \times 3$$
$$3 = 3$$

b) We write 2 as a factor 1 time, 3 as a factor 2 times, and 7 as a factor 1 time (the greatest number of times that each occurs in any one factorization).

The LCM of 7, 18, and 3 is $2 \times 3 \times 3 \times 7$, or 126.

51. a) Find the prime factorization of each number.

$$6 = 2 \times 3$$
$$12 = 2 \times 2 \times 2$$
$$18 = 2 \times 3 \times 3$$

b) We write 2 as a factor 2 times and 3 as a factor 2 times (the greatest number of times that each occurs in any one factorization).

The LCM of 6, 12, and 18 is $2 \times 2 \times 3 \times 3$, or 36.

53. Familiarize. We draw a picture. Repeated addition applies here.

$3250			
$13	$13	. . .	$13

Translate. We must determine how many 13's there are in 3250. This number is the number of seats in the auditorium. We let x = the number of seats in the auditorium.

$13 times number of seats is $3250.

$$13 \cdot x = 3250$$

Solve. To solve the equation, we divide on both sides by 13.

$$13 \cdot x = 3250$$
$$\frac{13 \cdot x}{13} = \frac{3250}{13}$$
$$x = 250$$

53. (continued)

Check. If 250 seats are sold at $13 each, the total receipts are 250·$13, or $3250. The result checks.

State. The auditorium contains 250 seats.

55. The length of the carton must be a multiple of both 6 and 8. The shortest length carton will be the least common multiple of 6 and 8.

$$6 = 2 \cdot 3$$
$$8 = 2 \cdot 2 \cdot 2$$

LCM is $2 \cdot 2 \cdot 2 \cdot 3$, or 24.

The shortest carton is 24 in. long.

Exercise Set 1.8

1. The top number is the numerator, and the bottom number is the denominator.

$\frac{3}{4}$ 3 ⟵ Numerator, 4 ⟵ Denominator

3. 11 ⟵ Numerator, 20 ⟵ Denominator

5. Since $2 \cdot 5 = 10$, we multiply by $\frac{5}{5}$.

$$\frac{1}{2} = \frac{1}{2} \cdot \frac{5}{5} = \frac{1 \cdot 5}{2 \cdot 5} = \frac{5}{10}$$

7. Since $4 \cdot 12 = 48$, we multiply by $\frac{12}{12}$.

$$\frac{3}{4} = \frac{3}{4} \cdot \frac{12}{12} = \frac{3 \cdot 12}{4 \cdot 12} = \frac{36}{48}$$

9. Since $10 \cdot 3 = 30$, we multiply by $\frac{3}{3}$.

$$\frac{9}{10} = \frac{9}{10} \cdot \frac{3}{3} = \frac{9 \cdot 3}{10 \cdot 3} = \frac{27}{30}$$

11. Since $8 \cdot 4 = 32$, we multiply by $\frac{4}{4}$.

$$\frac{7}{8} = \frac{7}{8} \cdot \frac{4}{4} = \frac{7 \cdot 4}{8 \cdot 4} = \frac{28}{32}$$

13. Since $3 \cdot 15 = 45$, we multiply by $\frac{15}{15}$.

$$\frac{5}{3} = \frac{5}{3} \cdot \frac{15}{15} = \frac{5 \cdot 15}{3 \cdot 15} = \frac{75}{45}$$

15. Since $22 \cdot 6 = 132$, we multiply by $\frac{6}{6}$.

$$\frac{7}{22} = \frac{7}{22} \cdot \frac{6}{6} = \frac{7 \cdot 6}{22 \cdot 6} = \frac{42}{132}$$

17. $\frac{2}{4} = \frac{1\cdot 2}{2\cdot 2}$ ⟵ Factor the numerator / ⟵ Factor the denominator

$= \frac{1}{2}\cdot\frac{2}{2}$ ⟵ Factor the fraction

$= \frac{1}{2}\cdot 1$

$= \frac{1}{2}$ ⟵ Removing a factor of 1

19. $\frac{6}{8} = \frac{3\cdot 2}{4\cdot 2}$ ⟵ Factor the numerator / ⟵ Factor the denominator

$= \frac{3}{4}\cdot\frac{2}{2}$ ⟵ Factor the fraction

$= \frac{3}{4}\cdot 1$

$= \frac{3}{4}$ ⟵ Removing a factor of 1

21. $\frac{3}{15} = \frac{1\cdot 3}{5\cdot 3}$ ⟵ Factor the numerator / ⟵ Factor the denominator

$= \frac{1}{5}\cdot\frac{3}{3}$ ⟵ Factor the fraction

$= \frac{1}{5}\cdot 1$

$= \frac{1}{5}$ ⟵ Removing a factor of 1

23. $\frac{24}{8} = \frac{3\cdot 8}{1\cdot 8} = \frac{3}{1}\cdot\frac{8}{8} = \frac{3}{1}\cdot 1 = \frac{3}{1} = 3$

25. $\frac{18}{24} = \frac{3\cdot 6}{4\cdot 6} = \frac{3}{4}\cdot\frac{6}{6} = \frac{3}{4}\cdot 1 = \frac{3}{4}$

27. $\frac{14}{16} = \frac{7\cdot 2}{8\cdot 2} = \frac{7}{8}\cdot\frac{2}{2} = \frac{7}{8}\cdot 1 = \frac{7}{8}$

29. $\frac{12}{10} = \frac{6\cdot 2}{5\cdot 2} = \frac{6}{5}\cdot\frac{2}{2} = \frac{6}{5}\cdot 1 = \frac{6}{5}$

31. $\frac{16}{48} = \frac{1\cdot 16}{3\cdot 16} = \frac{1}{3}\cdot\frac{16}{16} = \frac{1}{3}\cdot 1 = \frac{1}{3}$

33. $\frac{150}{25} = \frac{6\cdot 25}{1\cdot 25} = \frac{6}{1}\cdot\frac{25}{25} = \frac{6}{1}\cdot 1 = \frac{6}{1} = 6$

We could also simplify $\frac{150}{25}$ by doing the division $150 \div 25$. That is, $\frac{150}{25} = 150 \div 25 = 6$.

35. $\frac{17}{51} = \frac{1\cdot 17}{3\cdot 17} = \frac{1}{3}\cdot\frac{17}{17} = \frac{1}{3}\cdot 1 = \frac{1}{3}$

37. We multiply these two numbers: $3\cdot 12 = 36$ — We multiply these two numbers: $4\cdot 9 = 36$

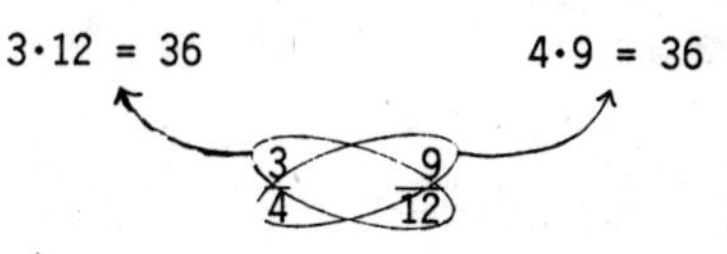

Since $36 = 36$, $\frac{3}{4} = \frac{9}{12}$.

39. We multiply these two numbers: $1\cdot 9 = 9$ — We multiply these two numbers: $5\cdot 2 = 10$

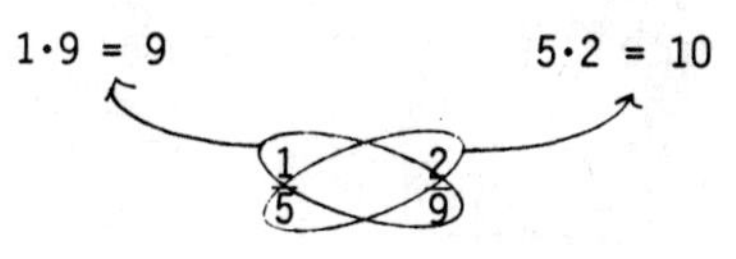

Since $9 \neq 10$, $\frac{1}{5} \neq \frac{2}{9}$.

41. We multiply these two numbers: $3\cdot 16 = 48$ — We multiply these two numbers: $8\cdot 6 = 48$

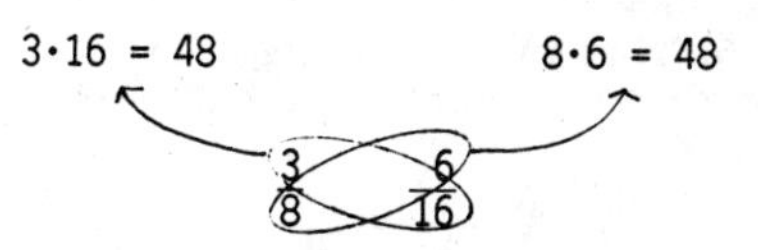

Since $48 = 48$, $\frac{3}{8} = \frac{6}{16}$.

43. We multiply these two numbers: $2\cdot 7 = 14$ — We multiply these two numbers: $5\cdot 3 = 15$

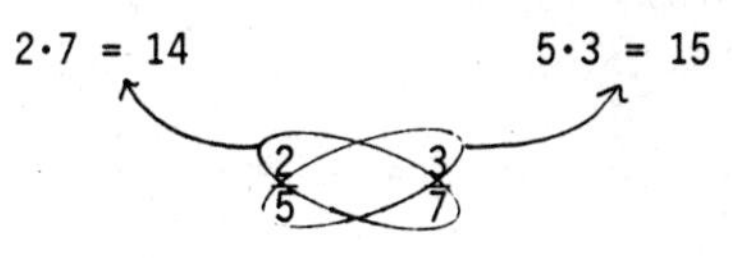

Since $14 \neq 15$, $\frac{2}{5} \neq \frac{3}{7}$.

45. We multiply these two numbers: $12\cdot 6 = 72$ — We multiply these two numbers: $9\cdot 8 = 72$

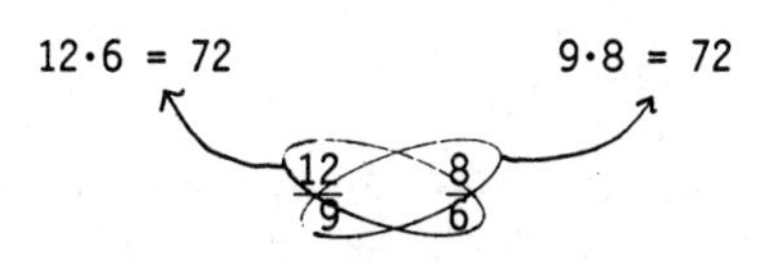

Since $72 = 72$, $\frac{12}{9} = \frac{8}{6}$.

47. We multiply these two numbers: $5\cdot 7 = 35$ — We multiply these two numbers: $2\cdot 17 = 34$

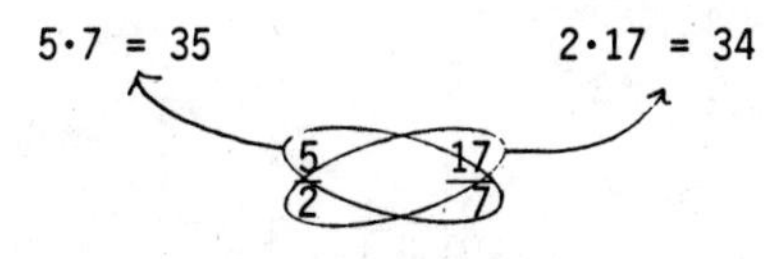

Since $35 \neq 34$, $\frac{5}{2} \neq \frac{17}{7}$.

49. We multiply these two numbers: $3 \cdot 100 = 300$

We multiply these two numbers: $10 \cdot 30 = 300$

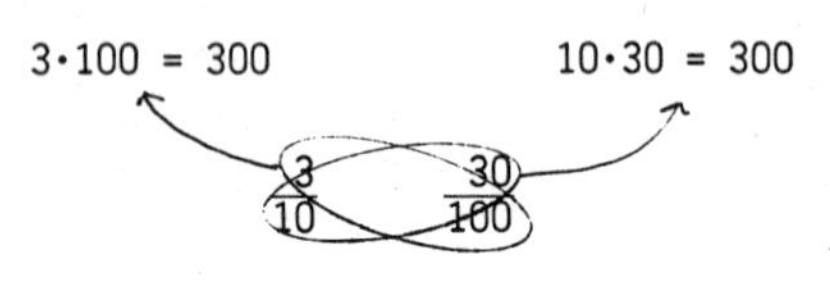

Since $300 = 300$, $\frac{3}{10} = \frac{30}{100}$.

51. We multiply these two numbers: $5 \cdot 1000 = 5000$

We multiply these two numbers: $10 \cdot 520 = 5200$

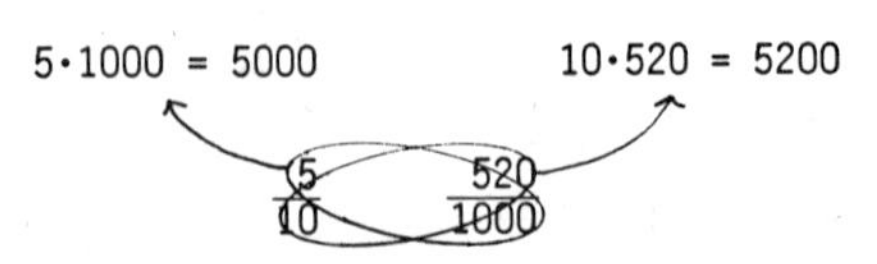

Since $5000 \neq 5200$, $\frac{5}{10} \neq \frac{520}{1000}$.

53. Round 34,56[2] to the nearest ten.

The digit 6 is in the tens place. Consider the next digit to the right. Since 2 is 4 or lower, we round down.

The answer is 34,560.

55. Round 34,[5]62 to the nearest thousand.

The digit 4 is in the thousands place. Consider the next digit to the right. Since 5 is 5 or higher, we round up.

The answer is 35,000.

57. Familiarize. We visualize the situation.

62 words	62 words	. . .	62 words
12,462 words			

Translate. We let t = the time it will take to type 12,462 words.

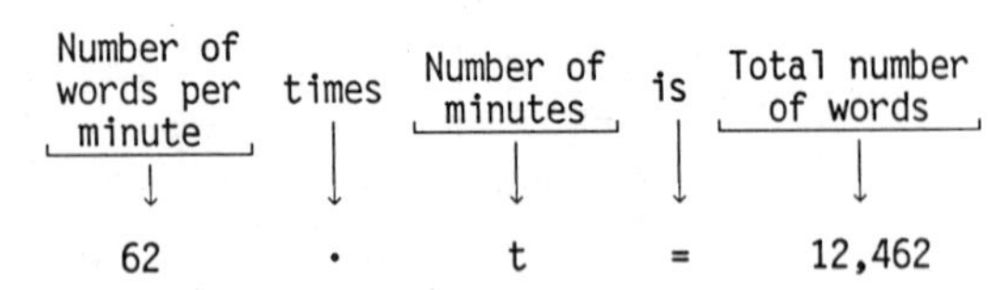

Solve. We solve the equation.

$$62 \cdot t = 12,462$$

$$\frac{62 \cdot t}{62} = \frac{12,462}{62}$$

$$t = 201$$

```
        2 0 1
  6 2)1 2,4 6 2
      1 2,4 0 0
            6 2
            6 2
              0
```

57. (continued)

Check. We can multiply 62 by the number of minutes: $201 \cdot 62 = 12,462$. The answer checks.

State. The answer is 201 minutes.

59. Think of dividing \$2700 into 2700 parts of the same size. (Each part is \$1.)

a) Think of \$1200 as 1200 of the 2700 parts. Thus, we have $\frac{1200}{2700} = \frac{4 \cdot 300}{9 \cdot 300} = \frac{4}{9} \cdot \frac{300}{300}$

$$= \frac{4}{9} \cdot 1 = \frac{4}{9}.$$

b) Think of \$540 as 540 of the 2700 parts. Thus, we have $\frac{540}{2700} = \frac{1 \cdot 540}{5 \cdot 540} = \frac{1}{5} \cdot \frac{540}{540} = \frac{1}{5} \cdot 1 = \frac{1}{5}$.

c) We have $\frac{360}{2700} = \frac{2 \cdot 180}{15 \cdot 180} = \frac{2}{15} \cdot \frac{180}{180} = \frac{2}{15} \cdot 1$

$$= \frac{2}{15}.$$

d) First, find the amount of miscellaneous expenses. To do this we must find the total of the amounts spent for tuition, rent, and food and then take that amount away from the amount earned.

To find the total spent for tuition, rent, and food translate to an equation and solve.

$$1200 + 540 + 360 = x$$

$$2100 = x \quad \text{Doing the addition}$$

The total is \$2100.

Now we have a "take-away" situation:

$$2700 - 2100 = m$$

$$600 = m \quad \text{Doing the subtraction}$$

Thus, \$600 went for miscellaneous expenses.

Think of \$600 as 600 of the 2700 parts. Thus, we have $\frac{600}{2700} = \frac{2 \cdot 300}{9 \cdot 300} = \frac{2}{9} \cdot \frac{300}{300} = \frac{2}{9} \cdot 1 = \frac{2}{9}$.

Exercise Set 1.9

1. $3 \cdot \frac{1}{5} = \frac{3 \cdot 1}{5} = \frac{3}{5}$

3. $5 \times \frac{1}{6} = \frac{5 \times 1}{6} = \frac{5}{6}$

5. $\frac{2}{5} \cdot 1 = \frac{2 \cdot 1}{5} = \frac{2}{5}$

7. $\frac{2}{5} \cdot 3 = \frac{2 \cdot 3}{5} = \frac{6}{5}$

9. $\frac{1}{8} \cdot \frac{4}{5} = \frac{1 \cdot 4}{8 \cdot 5} = \frac{1 \cdot 4}{2 \cdot 4 \cdot 5} = \frac{4}{4} \cdot \frac{1}{2 \cdot 5} = \frac{1}{2 \cdot 5} = \frac{1}{10}$

11. $\frac{1}{4} \cdot \frac{2}{3} = \frac{1 \cdot 2}{4 \cdot 3} = \frac{1 \cdot 2}{2 \cdot 2 \cdot 3} = \frac{2}{2} \cdot \frac{1}{2 \cdot 3} = \frac{1}{2 \cdot 3} = \frac{1}{6}$

13. $\frac{12}{5} \cdot \frac{9}{8} = \frac{12\cdot 9}{5\cdot 8} = \frac{4\cdot 3\cdot 9}{5\cdot 2\cdot 4} = \frac{4}{4} \cdot \frac{3\cdot 9}{5\cdot 2} = \frac{3\cdot 9}{5\cdot 2} = \frac{27}{10}$

15. $\frac{10}{9} \cdot \frac{7}{5} = \frac{10\cdot 7}{9\cdot 5} = \frac{5\cdot 2\cdot 7}{9\cdot 5} = \frac{5}{5} \cdot \frac{2\cdot 7}{9} = \frac{2\cdot 7}{9} = \frac{14}{9}$

17. $\frac{11}{24} \cdot \frac{3}{5} = \frac{11\cdot 3}{24\cdot 5} = \frac{11\cdot 3}{3\cdot 8\cdot 5} = \frac{3}{3} \cdot \frac{11}{8\cdot 5} = \frac{11}{8\cdot 5} = \frac{11}{40}$

19. $\frac{10}{21} \cdot \frac{3}{4} = \frac{10\cdot 3}{21\cdot 4} = \frac{2\cdot 5\cdot 3}{3\cdot 7\cdot 2\cdot 2} = \frac{2\cdot 3}{2\cdot 3} \cdot \frac{5}{7\cdot 2} = \frac{5}{7\cdot 2} = \frac{5}{14}$

21. $\frac{5}{6} \rightarrow \frac{6}{5}$ Interchange the numerator and denominator

The reciprocal of $\frac{5}{6}$ is $\frac{6}{5}$.

23. Think of 6 as $\frac{6}{1}$.

$\frac{6}{1} \rightarrow \frac{1}{6}$

The reciprocal of 6 is $\frac{1}{6}$.

25. $\frac{1}{6} \rightarrow \frac{6}{1}$ $\left(\frac{6}{1} = 6\right)$

The reciprocal of $\frac{1}{6}$ is 6.

27. $\frac{10}{3} \rightarrow \frac{3}{10}$

The reciprocal of $\frac{10}{3}$ is $\frac{3}{10}$.

29. $\frac{3}{5} \div \frac{3}{4} = \frac{3}{5} \cdot \frac{4}{3}$ Multiplying the dividend $\left(\frac{3}{5}\right)$ by the reciprocal of the divisor $\left(\text{The reciprocal of } \frac{3}{4} \text{ is } \frac{4}{3}.\right)$

$= \frac{3\cdot 4}{5\cdot 3}$ Multiplying numerators and denominators

$= \frac{3}{3} \cdot \frac{4}{5} = \frac{4}{5}$ Simplifying

31. $\frac{3}{5} \div \frac{9}{4} = \frac{3}{5} \cdot \frac{4}{9}$ Multiplying the dividend $\left(\frac{3}{5}\right)$ by the reciprocal of the divisor $\left(\text{The reciprocal of } \frac{9}{4} \text{ is } \frac{4}{9}.\right)$

$= \frac{3\cdot 4}{5\cdot 9}$ Multiplying numerators and denominators

$= \frac{3\cdot 4}{5\cdot 3\cdot 3} = \frac{3}{3} \cdot \frac{4}{5\cdot 3}$ Simplifying

$= \frac{4}{5\cdot 3} = \frac{4}{15}$

33. $\frac{4}{3} \div \frac{1}{3} = \frac{4}{3} \cdot 3 = \frac{4\cdot 3}{3} = \frac{3}{3} \cdot 4 = 4$

35. $\frac{1}{3} \div \frac{1}{6} = \frac{1}{3} \cdot 6 = \frac{1\cdot 6}{3} = \frac{1\cdot 2\cdot 3}{1\cdot 3} = \frac{1\cdot 3}{1\cdot 3} \cdot 2 = 2$

37. $\frac{3}{8} \div 3 = \frac{3}{8} \cdot \frac{1}{3} = \frac{3\cdot 1}{8\cdot 3} = \frac{3}{3} \cdot \frac{1}{8} = \frac{1}{8}$

39. $\frac{12}{7} \div 4 = \frac{12}{7} \cdot \frac{1}{4} = \frac{12\cdot 1}{7\cdot 4} = \frac{4\cdot 3\cdot 1}{7\cdot 4} = \frac{4}{4} \cdot \frac{3\cdot 1}{7} = \frac{3\cdot 1}{7} = \frac{3}{7}$

41. $12 \div \frac{3}{2} = 12 \cdot \frac{2}{3} = \frac{12\cdot 2}{3} = \frac{3\cdot 4\cdot 2}{3\cdot 1} = \frac{3}{3} \cdot \frac{4\cdot 2}{1} = \frac{4\cdot 2}{1}$

$= \frac{8}{1} = 8$

43. $28 \div \frac{4}{5} = 28 \cdot \frac{5}{4} = \frac{28\cdot 5}{4} = \frac{4\cdot 7\cdot 5}{4\cdot 1} = \frac{4}{4} \cdot \frac{7\cdot 5}{1} = \frac{7\cdot 5}{1} = 35$

45. $\frac{5}{8} \div \frac{5}{8} = \frac{5}{8} \cdot \frac{8}{5} = \frac{5\cdot 8}{8\cdot 5} = \frac{5\cdot 8}{5\cdot 8} = 1$

47. $\frac{8}{15} \div \frac{4}{5} = \frac{8}{15} \cdot \frac{5}{4} = \frac{8\cdot 5}{15\cdot 4} = \frac{2\cdot 4\cdot 5}{3\cdot 5\cdot 4} = \frac{4\cdot 5}{4\cdot 5} \cdot \frac{2}{3} = \frac{2}{3}$

49. Familiarize. We let n = the number of shy people.

Translate.

Two fifths	of	People interviewed	is	Number of shy people
↓	↓	↓	↓	↓
$\frac{2}{5}$	$\cdot$	650	=	n

Solve. We carry out the multiplication and simplify:

$\frac{2}{5} \cdot 650 = \frac{2\cdot 650}{5} = \frac{2\cdot 5\cdot 130}{1\cdot 5} = \frac{5}{5} \cdot \frac{2\cdot 130}{1}$

$= \frac{2\cdot 130}{1} = \frac{260}{1} = 260$

Check. We check by repeating the calculation.

State. 260 of the people interviewed might be shy.

51. Familiarize. We visualize the situation. Let a = the amount of the loan.

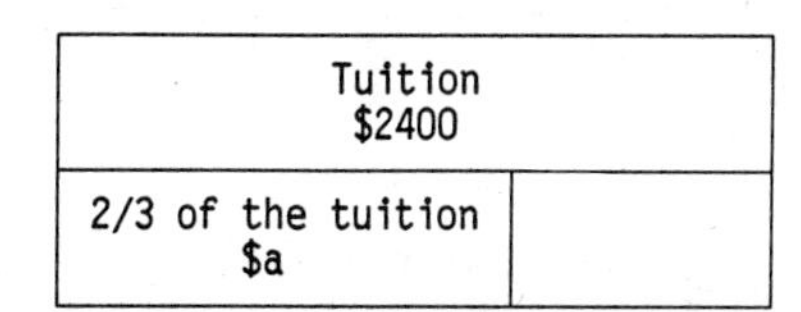

Translate. We write an equation.

Amount of loan	is	$\frac{2}{3}$	of	the tuition
↓	↓	↓	↓	↓
a	=	$\frac{2}{3}$	$\cdot$	2400

Solve. We carry out the multiplication.

$a = \frac{2}{3} \cdot 2400 = \frac{2\cdot 2400}{3}$

$= \frac{2\cdot 3\cdot 800}{3\cdot 1} = \frac{3}{3} \cdot \frac{2\cdot 800}{1}$

$= \frac{1600}{1} = 1600$

Check. We can repeat the calculation. We can also determine that the answer seems reasonable since we multiplied 2400 by a number less than 1 and the result is less than 2400. The answer checks.

State. The loan was \$1600.

53. Familiarize. We draw a picture. We let n = the amount the bucket could hold.

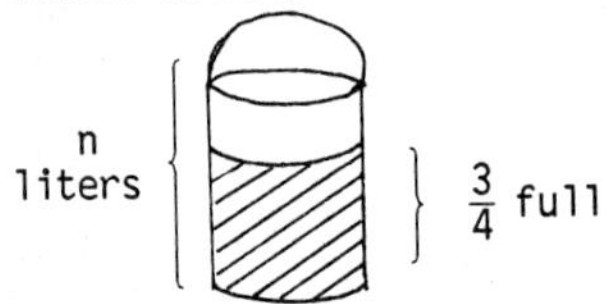

Translate. We write a multiplication sentence:

$$\frac{3}{4} \cdot n = 12$$

Solve. Solve the equation as follows:

$$\frac{3}{4} \cdot n = 12$$

$$n = 12 \div \frac{3}{4} = 12 \cdot \frac{4}{3} = \frac{12 \cdot 4}{3} = \frac{3 \cdot 4 \cdot 4}{3 \cdot 1}$$

$$= \frac{3}{3} \cdot \frac{4 \cdot 4}{1} = \frac{4 \cdot 4}{1} = 16$$

Check. We repeat the calculation. The answer checks.

State. The bucket could hold 16 L.

55.
```
      2 0 4
3 5)7 1 4 0
    7 0 0 0
        1 4 0
        1 4 0
            0
```

The answer is 204.

57.
```
    3 0 0 1
9)2 7,0 0 9
  2 7 0 0 0
          9
          9
          0
```

The answer is 3001.

59. Divide in the numerator and multiply in the denominator to get a single fraction in each place.

$$\frac{\frac{2}{5} \div \frac{3}{4}}{\frac{5}{6} \cdot \frac{1}{4}} = \frac{\frac{2}{5} \cdot \frac{4}{3}}{\frac{5}{6} \cdot \frac{1}{4}} = \frac{\frac{2 \cdot 4}{5 \cdot 3}}{\frac{5 \cdot 1}{6 \cdot 4}} = \frac{\frac{8}{15}}{\frac{5}{24}}$$

Now divide the fraction in the numerator by the fraction in the denominator.

$$\frac{8}{15} \div \frac{5}{24} = \frac{8}{15} \cdot \frac{24}{5} = \frac{8 \cdot 24}{15 \cdot 5} = \frac{8 \cdot 3 \cdot 8}{3 \cdot 5 \cdot 5} = \frac{3}{3} \cdot \frac{8 \cdot 8}{5 \cdot 5} = \frac{64}{25}$$

Exercise Set 1.10

1. $\frac{2}{3} + \frac{5}{6}$ — 3 is a factor of 6, so the LCM is 6.

$= \frac{2}{3} \cdot \frac{2}{2} + \frac{5}{6}$ ⟵ This fraction already has the LCM as denominator.

Think: 3 × ☐ = 6. The answer is 2, so we multiply by 1, using $\frac{2}{2}$.

$= \frac{4}{6} + \frac{5}{6} = \frac{9}{6}$

$= \frac{3}{2}$ — Simplifying

3. $\frac{1}{8} + \frac{1}{6}$ — $8 = 2 \cdot 2 \cdot 2$ and $6 = 2 \cdot 3$, so the LCM is $2 \cdot 2 \cdot 2 \cdot 3$, or 24.

$= \frac{1}{8} \cdot \frac{3}{3} + \frac{1}{6} \cdot \frac{4}{4}$

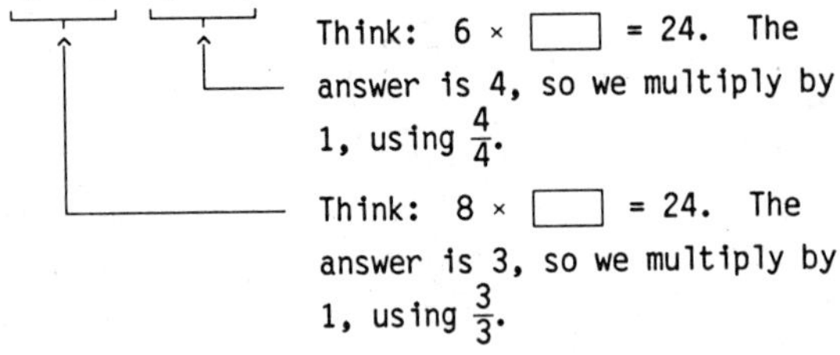

Think: 6 × ☐ = 24. The answer is 4, so we multiply by 1, using $\frac{4}{4}$.

Think: 8 × ☐ = 24. The answer is 3, so we multiply by 1, using $\frac{3}{3}$.

$= \frac{3}{24} + \frac{4}{24}$

$= \frac{7}{24}$

5. $\frac{4}{5} + \frac{7}{10}$ — 5 is a factor of 10, so the LCM is 10.

$= \frac{4}{5} \cdot \frac{2}{2} + \frac{7}{10}$ ⟵ This fraction already has the LCM as denominator.

Think: 5 × ☐ = 10. The answer is 2, so we multiply by 1, using $\frac{2}{2}$.

$= \frac{8}{10} + \frac{7}{10} = \frac{15}{10}$

$= \frac{3}{2}$ — Simplifying

7. $\frac{5}{12} + \frac{3}{8}$ — $12 = 2 \cdot 2 \cdot 3$ and $8 = 2 \cdot 2 \cdot 2$, so the LCM is $2 \cdot 2 \cdot 2 \cdot 3$, or 24

$= \frac{5}{12} \cdot \frac{2}{2} + \frac{3}{8} \cdot \frac{3}{3}$

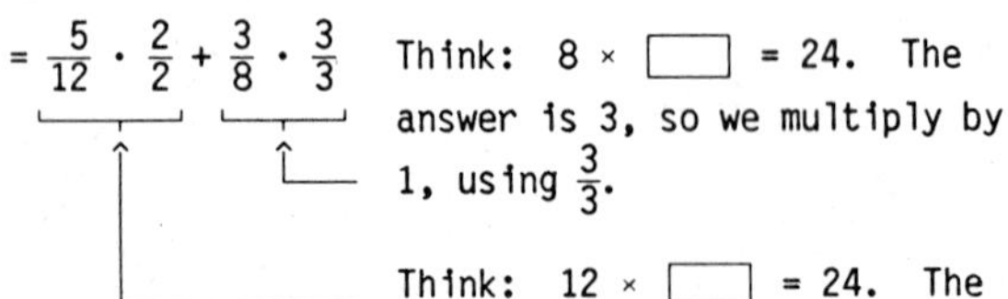

Think: 8 × ☐ = 24. The answer is 3, so we multiply by 1, using $\frac{3}{3}$.

Think: 12 × ☐ = 24. The answer is 2, so we multiply by 1, using $\frac{2}{2}$.

$= \frac{10}{24} + \frac{9}{24} = \frac{19}{24}$

9. $\frac{3}{20} + \frac{3}{4}$ 4 is a factor of 20, so the LCM is 20.

$= \frac{3}{20} + \frac{3}{4} \cdot \frac{5}{5}$ Multiplying by 1

$= \frac{3}{20} + \frac{15}{20} = \frac{18}{20} = \frac{9}{10}$

11. $\frac{5}{6} + \frac{7}{9}$ $6 = 2 \cdot 3$ and $9 = 3 \cdot 3$, so the LCM is $2 \cdot 3 \cdot 3$, or 18

$= \frac{5}{6} \cdot \frac{3}{3} + \frac{7}{9} \cdot \frac{2}{2}$ Multiplying by 1

$= \frac{15}{18} + \frac{14}{18} = \frac{29}{18}$

13. $\frac{3}{10} + \frac{1}{100}$ 10 is a factor of 100, so the LCM is 100.

$= \frac{3}{10} \cdot \frac{10}{10} + \frac{1}{100}$

$= \frac{30}{100} + \frac{1}{100} = \frac{31}{100}$

15. $\frac{5}{12} + \frac{4}{15}$ $12 = 2 \cdot 2 \cdot 3$ and $15 = 3 \cdot 5$, so the LCM is $2 \cdot 2 \cdot 3 \cdot 5$, or 60.

$= \frac{5}{12} \cdot \frac{5}{5} + \frac{4}{15} \cdot \frac{4}{4}$

$= \frac{25}{60} + \frac{16}{60} = \frac{41}{60}$

17. $\frac{1}{2} + \frac{3}{8} + \frac{1}{4}$

$= \frac{1}{2} + \frac{3}{2 \cdot 2 \cdot 2} + \frac{1}{2 \cdot 2}$ Factoring the denominators

The LCM is $2 \cdot 2 \cdot 2$, or 8.

$= \frac{1}{2} \cdot \frac{2 \cdot 2}{2 \cdot 2} + \frac{3}{2 \cdot 2 \cdot 2} + \frac{1}{2 \cdot 2} \cdot \frac{2}{2}$

In each case we multiply by 1 to get the LCM in the denominator.

$= \frac{1 \cdot 2 \cdot 2}{2 \cdot 2 \cdot 2} + \frac{3}{2 \cdot 2 \cdot 2} + \frac{1 \cdot 2}{2 \cdot 2 \cdot 2}$

$= \frac{4}{8} + \frac{3}{8} + \frac{2}{8} = \frac{9}{8}$

19. $\frac{5}{7} + \frac{25}{52} + \frac{7}{4}$

$= \frac{5}{7} + \frac{25}{2 \cdot 2 \cdot 13} + \frac{7}{2 \cdot 2}$ Factoring the denominators (7 is prime)

The LCM is $2 \cdot 2 \cdot 7 \cdot 13$, or 364.

$= \frac{5}{7} \cdot \frac{2 \cdot 2 \cdot 13}{2 \cdot 2 \cdot 13} + \frac{25}{2 \cdot 2 \cdot 13} \cdot \frac{7}{7} + \frac{7}{2 \cdot 2} \cdot \frac{7 \cdot 13}{7 \cdot 13}$

19. (continued)

In each case we multiply by 1 to get the LCM in the denominator.

$= \frac{5 \cdot 2 \cdot 2 \cdot 13}{7 \cdot 2 \cdot 2 \cdot 13} + \frac{25 \cdot 7}{2 \cdot 2 \cdot 13 \cdot 7} + \frac{7 \cdot 7 \cdot 13}{2 \cdot 2 \cdot 7 \cdot 13}$

$= \frac{260}{364} + \frac{175}{364} + \frac{637}{364} = \frac{1072}{364}$

$= \frac{268 \cdot 4}{91 \cdot 4} = \frac{268}{91} \cdot \frac{4}{4} = \frac{268}{91} \cdot 1 = \frac{268}{91}$

21. When denominators are the same, subtract the numerators and keep the denominator.

$\frac{7}{5} - \frac{2}{5} = \frac{7 - 2}{5} = \frac{5}{5} = 1$

23. When denominators are the same, subtract the numerators and keep the denominator.

$\frac{15}{16} - \frac{11}{16} = \frac{15 - 11}{16} = \frac{4}{16} = \frac{1}{4}$

25. The LCM of 3 and 9 is 9.

$\frac{2}{3} - \frac{1}{9} = \frac{2}{3} \cdot \frac{3}{3} - \frac{1}{9}$ ⟵ This fraction already has the LCM as the denominator.

Think: $3 \times \square = 9$. The answer is 3, so we multiply by 1, using $\frac{3}{3}$.

$= \frac{6}{9} - \frac{1}{9} = \frac{5}{9}$

27. The LCM of 6 and 8 is 24.

$\frac{1}{6} - \frac{1}{8} = \frac{1}{6} \cdot \frac{4}{4} - \frac{1}{8} \cdot \frac{3}{3}$

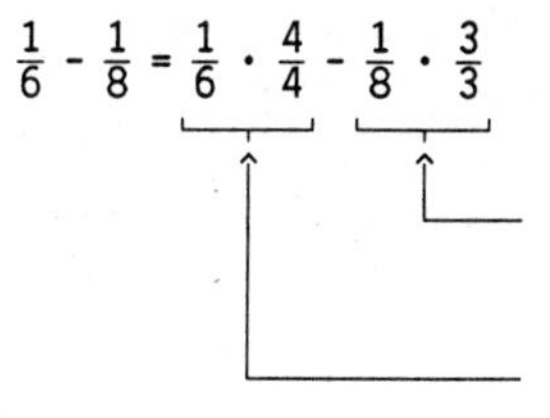

Think: $8 \times \square = 24$. The answer is 3, so we multiply by 1, using $\frac{3}{3}$.

Think: $6 \times \square = 24$. The answer is 4, so we multiply by 1, using $\frac{4}{4}$.

$= \frac{4}{24} - \frac{3}{24} = \frac{1}{24}$

29. The LCM of 8 and 16 is 16.

$\frac{7}{8} - \frac{1}{16} = \frac{7}{8} \cdot \frac{2}{2} - \frac{1}{16}$

$= \frac{14}{16} - \frac{1}{16}$

$= \frac{13}{16}$

31. The LCM of 5 and 15 is 15.

$\frac{2}{5} - \frac{2}{15} = \frac{2}{5} \cdot \frac{3}{3} - \frac{2}{15}$

$= \frac{6}{15} - \frac{2}{15}$

$= \frac{4}{15}$

33. The LCM of 10 and 16 is 80.

$$\frac{9}{10} - \frac{11}{16} = \frac{9}{10} \cdot \frac{8}{8} - \frac{11}{16} \cdot \frac{5}{5}$$
$$= \frac{72}{80} - \frac{55}{80}$$
$$= \frac{17}{80}$$

35. The LCM of 10 and 100 is 100.

$$\frac{9}{10} - \frac{3}{100} = \frac{9}{10} \cdot \frac{10}{10} - \frac{3}{100}$$
$$= \frac{90}{100} - \frac{3}{100}$$
$$= \frac{87}{100}$$

37. The LCM of 25 and 35 is 175.

$$\frac{18}{25} - \frac{4}{35} = \frac{18}{25} \cdot \frac{7}{7} - \frac{4}{35} \cdot \frac{5}{5}$$
$$= \frac{126}{175} - \frac{20}{175}$$
$$= \frac{106}{175}$$

39. The LCM of 100 and 20 is 100.

$$\frac{78}{100} - \frac{11}{20} = \frac{78}{100} - \frac{11}{20} \cdot \frac{5}{5}$$
$$= \frac{78}{100} - \frac{55}{100}$$
$$= \frac{23}{100}$$

41. Familiarize. We draw a picture. We let D = the total distance walked.

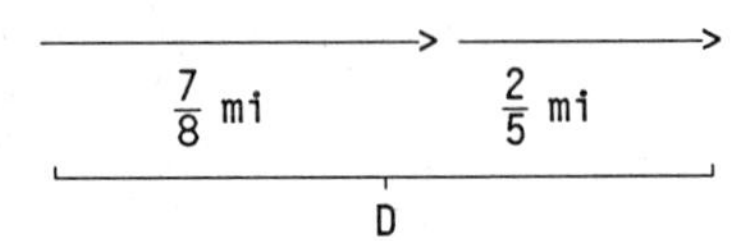

Translate. An addition sentence corresponds to this situation.

Distance to friend's house	plus	Distance to class	is	Total distance
↓	↓	↓	↓	↓
$\frac{7}{8}$	+	$\frac{2}{5}$	=	D

Solve. To solve the equation, carry out the addition. The LCM of the denominators is 8·5, or 40.

$$\frac{7}{8} \cdot \frac{5}{5} + \frac{2}{5} \cdot \frac{8}{8} = D$$
$$\frac{35}{40} + \frac{16}{40} = D$$
$$\frac{51}{40} = D$$

Check. We repeat the calculation. We also note that the sum is larger than either of the original distances, so the answer seems reasonable.

State. The student walked $\frac{51}{40}$ mi.

43. Familiarize. We let w = the total weight of the cubic meter of concrete mix. First we will find w. Then we will find what part of the total is made up by each component of the mix.

Translate. An addition sentence corresponds to this situation.

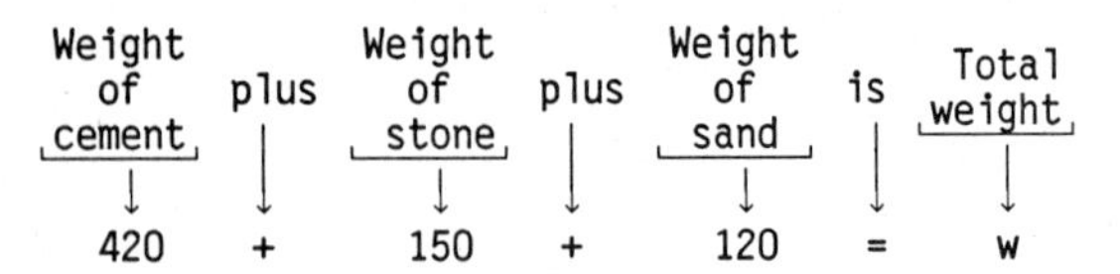

Solve. To solve the equation carry out the addition.

$$\begin{array}{r} 420 \\ 150 \\ +\,120 \\ \hline 690 \end{array}$$

The total weight is 690 kg.

Think of dividing the 690 kg into 690 parts of the same size. (Each part is 1 kg.) Then the 420 kg of cement are 420 of the 690 parts, or

$\frac{420}{690} = \frac{14}{23}$.

The 150 kg of stone are 150 of the 690 parts, or

$\frac{150}{690} = \frac{5}{23}$.

The 120 kg of sand are 120 of the 690 parts, or

$\frac{120}{690} = \frac{4}{23}$.

Add these amounts: $\frac{14}{23} + \frac{5}{23} + \frac{4}{23} = \frac{14 + 5 + 4}{23}$

$$= \frac{23}{23} = 1.$$

The result is equal to 1.

Check. We repeat the calculations.

State. The total weight is 690 kg. Of this, $\frac{14}{23}$ is cement, $\frac{5}{23}$ is stone, and $\frac{4}{23}$ is sand. The total of these fractional amounts is 1.

45. Familiarize. Let x = the fourth child's portion.

Translate. Write an addition sentence.

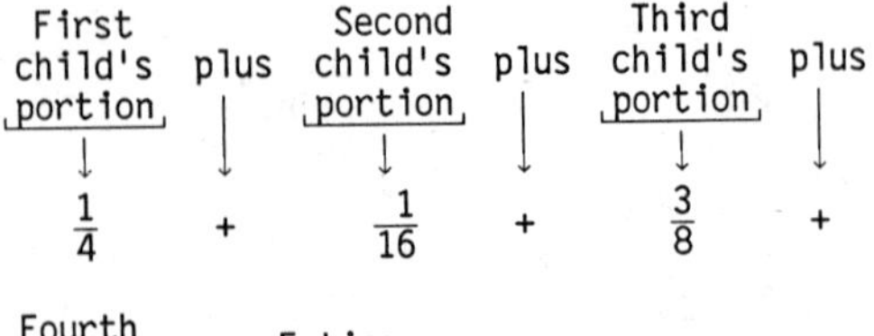

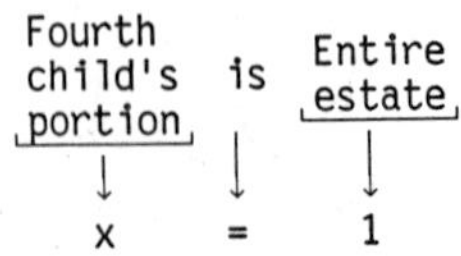

45. (continued)

Solve. We solve the equation.

$\frac{1}{4} + \frac{1}{16} + \frac{3}{8} + x = 1$

$$x = 1 - \frac{1}{4} - \frac{1}{16} - \frac{3}{8}$$

$$= 1 \cdot \frac{16}{16} - \frac{1}{4} \cdot \frac{4}{4} - \frac{1}{16} - \frac{3}{8} \cdot \frac{2}{2}$$

$$= \frac{16}{16} - \frac{4}{16} - \frac{1}{16} - \frac{6}{16}$$

$$= \frac{5}{16}$$

Check. We repeat the calculation.

State. The fourth child got $\frac{5}{16}$ of the estate.

47. Familiarize. Let t = the total number of cars recalled.

Translate. Write an addition sentence.

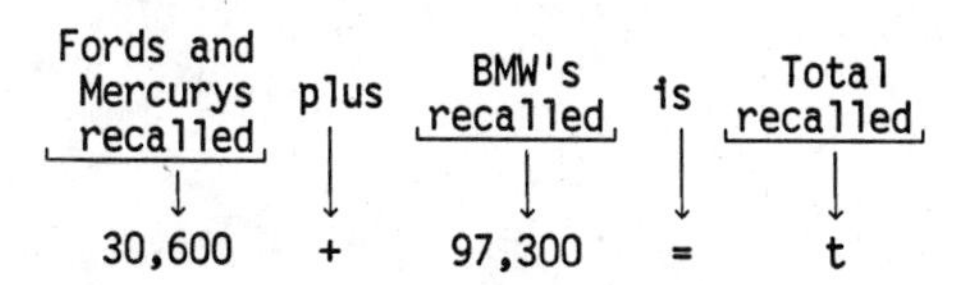

Fords and Mercurys recalled
plus
BMW's recalled
is
Total recalled
30,600
+
97,300
=
t

Solve. To solve the equation carry out the addition.

$$\begin{array}{r} 30{,}600 \\ +\ 97{,}300 \\ \hline 127{,}900 \end{array}$$

Check. We repeat the calculation.

State. In all, 127,900 cars were recalled.

49. Familiarize. This is a multistep problem. We let x and y represent the portion of the tape used at the 4-hr and 2-hr speeds, respectively. Then we let p = the unused portion of the tape and t = the time left on the tape at the 6-hr speed.

Translate and Solve.

First, we solve an equation to find the portion of the tape used at the 4-hr speed.

Total time	times	Portion used	is	Time used
↓	↓	↓	↓	↓
4	·	x	=	$\frac{1}{2}$

$$x = \frac{1}{2} \div 4$$

$$= \frac{1}{2} \cdot \frac{1}{4}$$

$$= \frac{1 \cdot 1}{2 \cdot 4}$$

$$= \frac{1}{8}$$

Thus, $\frac{1}{8}$ of the tape was used at the 4-hr speed.

49. (continued)

Now we solve an equation to find the portion of the tape used at the 2-hr speed.

Total time	times	Portion used	is	Time used
↓	↓	↓	↓	↓
2	·	y	=	$\frac{3}{4}$

$$y = \frac{3}{4} \div 2$$

$$= \frac{3}{4} \cdot \frac{1}{2}$$

$$= \frac{3 \cdot 1}{4 \cdot 2}$$

$$= \frac{3}{8}$$

Thus, $\frac{3}{8}$ of the tape is used at the 2-hr speed.

Next, we solve an equation to find the portion of the tape that is unused.

Portion used at 4-hr speed	plus	Portion used at 2-hr speed	plus	Unused portion	is	Entire tape
↓	↓	↓	↓	↓	↓	↓
$\frac{1}{8}$	+	$\frac{3}{8}$	+	p	=	1

$$p = 1 - \frac{1}{8} - \frac{3}{8}$$

$$= 1 \cdot \frac{8}{8} - \frac{1}{8} - \frac{3}{8}$$

$$= \frac{8}{8} - \frac{1}{8} - \frac{3}{8}$$

$$= \frac{4}{8}$$

$$= \frac{1}{2}$$

One-half of the tape is unused.

To find how much time is left on the tape at the 6-hr speed, we solve an equation.

Unused portion	of	Total time	is	Time left
↓	↓	↓	↓	↓
$\frac{1}{2}$	·	6	=	t

$$\frac{1 \cdot 6}{2} = t$$

$$\frac{6}{2} = t$$

$$\frac{2 \cdot 3}{2 \cdot 1} = t$$

$$\frac{2}{2} \cdot \frac{3}{1} = t$$

49. (continued)

$\frac{3}{1} = t$

$3 = t$

Check. We repeat the calculations.

State. There are three hours left at the 6-hr speed.

Exercise Set 1.11

1. $5\,\frac{2}{3} = \frac{17}{3}$

[1] Multiply: $5 \cdot 3 = 15$

[2] Add: $15 + 2 = 17$

[3] Keep the denominator

3. $6\,\frac{1}{4} = \frac{25}{4}$

[1] Multiply: $6 \cdot 4 = 24$

[2] Add: $24 + 1 = 25$

[3] Keep the denominator

5. $10\,\frac{1}{8} = \frac{81}{8}$ $(10 \cdot 8 + 1 = 81)$

7. $5\,\frac{1}{10} = \frac{51}{10}$ $(5 \cdot 10 + 1 = 51)$

9. $20\,\frac{3}{5} = \frac{103}{5}$ $(20 \cdot 5 + 3 = 103)$

11. $9\,\frac{5}{6} = \frac{59}{6}$ $(9 \cdot 6 + 5 = 59)$

13. $7\,\frac{3}{10} = \frac{73}{10}$ $(7 \cdot 10 + 3 = 73)$

15. $1\,\frac{5}{8} = \frac{13}{8}$ $(1 \cdot 8 + 5 = 13)$

17. $12\,\frac{3}{4} = \frac{51}{4}$ $(12 \cdot 4 + 3 = 51)$

19. $4\,\frac{3}{10} = \frac{43}{10}$ $(4 \cdot 10 + 3 = 43)$

21. $2\,\frac{3}{100} = \frac{203}{100}$ $(2 \cdot 100 + 3 = 203)$

23. $66\,\frac{2}{3} = \frac{200}{3}$ $(66 \cdot 3 + 2 = 200)$

25. $5\,\frac{29}{50} = \frac{279}{50}$ $(5 \cdot 50 + 29 = 279)$

27. $14\,\frac{5}{8} = \frac{117}{8}$ $(14 \cdot 8 + 5 = 117)$

29. To convert $\frac{8}{5}$ to a mixed numeral, we divide.

$$5\overline{)8} = 1,\ \text{remainder } 8 - 5 = 3 \qquad \frac{8}{5} = 1\,\frac{3}{5}$$

31. To convert $\frac{14}{3}$ to a mixed numeral, we divide.

$$3\overline{)14} = 4,\ \text{remainder } 14 - 12 = 2 \qquad \frac{14}{3} = 4\,\frac{2}{3}$$

33. $$6\overline{)27} = 4,\ \text{remainder } 27 - 24 = 3 \qquad \frac{27}{6} = 4\,\frac{3}{6} = 4\,\frac{1}{2}$$

35. $$10\overline{)57} = 5,\ \text{remainder } 57 - 50 = 7 \qquad \frac{57}{10} = 5\,\frac{7}{10}$$

37. $$7\overline{)53} = 7,\ \text{remainder } 53 - 49 = 4 \qquad \frac{53}{7} = 7\,\frac{4}{7}$$

39. $$6\overline{)45} = 7,\ \text{remainder } 45 - 42 = 3 \qquad \frac{45}{6} = 7\,\frac{3}{6} = 7\,\frac{1}{2}$$

41. $$4\overline{)46} = 11;\ 46 - 40 = 6,\ 6 - 4 = 2 \qquad \frac{46}{4} = 11\,\frac{2}{4} = 11\,\frac{1}{2}$$

43. $$8\overline{)12} = 1,\ \text{remainder } 12 - 8 = 4 \qquad \frac{12}{8} = 1\,\frac{4}{8} = 1\,\frac{1}{2}$$

45. $$100\overline{)757} = 7,\ \text{remainder } 757 - 700 = 57 \qquad \frac{757}{100} = 7\,\frac{57}{100}$$

47. $$8\overline{)345} = 43;\ 345 - 320 = 25,\ 25 - 24 = 1 \qquad \frac{345}{8} = 43\,\frac{1}{8}$$

49. We first divide as usual.

$$\begin{array}{r} 108 \\ 8\overline{)869} \\ \underline{800} \\ 69 \\ \underline{64} \\ 5 \end{array}$$

The answer is 108 R 5. We write a mixed numeral for the quotient as follows: $108\,\frac{5}{8}$

51. We first divide as usual.

$$\begin{array}{r} 906 \\ 7\overline{)6345} \\ \underline{6300} \\ 45 \\ \underline{42} \\ 3 \end{array}$$

The answer is 906 R 3. We write a mixed numeral for the quotient as follows: $906\,\frac{3}{7}$

53.

$$\begin{array}{r} 40 \\ 21\overline{)852} \\ \underline{840} \\ 12 \end{array}$$

We get $40\,\frac{12}{21}$. This simplifies as $40\,\frac{4}{7}$.

55.

$$\begin{array}{r} 55 \\ 102\overline{)5612} \\ \underline{5100} \\ 512 \\ \underline{510} \\ 2 \end{array}$$

We get $55\,\frac{2}{102}$. This simplifies as $55\,\frac{1}{51}$.

57. $\frac{6}{5} \cdot 15 = \frac{6 \cdot 15}{5} = \frac{6 \cdot 3 \cdot 5}{5 \cdot 1} = \frac{5}{5} \cdot \frac{6 \cdot 3}{1} = \frac{6 \cdot 3}{1} = \frac{18}{1} = 18$

59. $\frac{7}{10} \cdot \frac{5}{14} = \frac{7 \cdot 5}{10 \cdot 14} = \frac{7 \cdot 5 \cdot 1}{2 \cdot 5 \cdot 2 \cdot 7} = \frac{7 \cdot 5}{7 \cdot 5} \cdot \frac{1}{2 \cdot 2} = \frac{1}{2 \cdot 2} = \frac{1}{4}$

61. $\frac{56}{7} + \frac{2}{3} = 8 + \frac{2}{3}$ $\quad (56 \div 7 = 8)$

$= 8\,\frac{2}{3}$

63.

$$\begin{array}{r} 52 \\ 7\overline{)366} \\ \underline{350} \\ 16 \\ \underline{14} \\ 2 \end{array}$$

$\frac{366}{7} = 52\,\frac{2}{7}$

Exercise Set 1.12

1.

$$\begin{array}{r} 2\,\frac{7}{8} \\ +\ 3\,\frac{5}{8} \\ \hline 5\,\frac{12}{8} \end{array} = 5 + \frac{12}{8}$$

To find a mixed numeral for $\frac{12}{8}$ we divide:

$$\begin{array}{r} 1 \\ 8\overline{)12} \\ \underline{8} \\ 4 \end{array} \qquad \frac{12}{8} = 1\,\frac{4}{8} = 1\,\frac{1}{2}$$

$5\,\frac{12}{8} = 5 + 1\,\frac{1}{2} = 6\,\frac{1}{2}$

3. The LCM is 12.

$$\begin{array}{rcr} 1\boxed{\frac{1}{4} \cdot \frac{3}{3}} & = & 1\,\frac{3}{12} \\ +\ 1\boxed{\frac{2}{3} \cdot \frac{4}{4}} & = & +\ 1\,\frac{8}{12} \\ \hline & & 2\,\frac{11}{12} \end{array}$$

5. The LCM is 8.

$$\begin{array}{rcr} 14\,\frac{5}{8} & = & 14\,\frac{5}{8} \\ +\ 13\boxed{\frac{1}{4} \cdot \frac{2}{2}} & = & +\ 13\,\frac{2}{8} \\ \hline & & 27\,\frac{7}{8} \end{array}$$

7. The LCM is 24.

$$\begin{array}{rcr} 7\boxed{\frac{1}{8} \cdot \frac{3}{3}} & = & 7\,\frac{3}{24} \\ 9\boxed{\frac{2}{3} \cdot \frac{8}{8}} & = & 9\,\frac{16}{24} \\ +\ 10\boxed{\frac{3}{4} \cdot \frac{6}{6}} & = & +\ 10\,\frac{18}{24} \\ \hline & & 26\,\frac{37}{24} \end{array}$$

$26\,\frac{37}{24} = 26 + \frac{37}{24}$

$= 26 + 1\,\frac{13}{24}$

$= 27\,\frac{13}{24}$

9.

$$\begin{array}{rcr} 4\,\frac{1}{5} & = & 3\,\frac{6}{5} \\ -\ 2\,\frac{3}{5} & = & -\ 2\,\frac{3}{5} \\ \hline & & 1\,\frac{3}{5} \end{array}$$

Since $\frac{1}{5}$ is smaller than $\frac{3}{5}$, we cannot subtract until we borrow:

$4\,\frac{1}{5} = 3 + \frac{5}{5} + \frac{1}{5} = 3 + \frac{6}{5} = 3\,\frac{6}{5}$

11. The LCM is 10.

$$\begin{array}{rcr} 6\boxed{\frac{3}{5} \cdot \frac{2}{2}} & = & 6\,\frac{6}{10} \\ -\ 2\boxed{\frac{1}{2} \cdot \frac{5}{5}} & = & -\ 2\,\frac{5}{10} \\ \hline & & 4\,\frac{1}{10} \end{array}$$

13. The LCM is 24.

$$34\boxed{\frac{1}{3}\cdot\frac{8}{8}} = 34\,\frac{8}{24} = 33\,\frac{32}{24}$$

$$-\,12\boxed{\frac{5}{8}\cdot\frac{3}{3}} = -\,12\,\frac{15}{24} = -\,12\,\frac{15}{24}$$

$$21\,\frac{17}{24}$$

Since $\frac{8}{24}$ is smaller than $\frac{15}{24}$, we cannot subtract until we borrow:

$$34\,\frac{8}{24} = 33 + \frac{24}{24} + \frac{8}{24} = 33 + \frac{32}{24} = 33\,\frac{32}{24}$$

15. $21 = 20\,\frac{4}{4}$ $\quad$ $\left[21 = 20 + 1 = 20 + \frac{4}{4} = 20\frac{4}{4}\right]$

$$-\,8\,\frac{3}{4} = -\,8\,\frac{3}{4}$$

$$12\,\frac{1}{4}$$

17. $8 \times 2\,\frac{5}{6}$

$= \frac{8}{1} \times \frac{17}{6}$ Writing fractional notation

$= \frac{8 \times 17}{1 \times 6} = \frac{2 \times 4 \times 17}{1 \times 2 \times 3} = \frac{2}{2} \times \frac{4 \times 17}{1 \times 3} = \frac{68}{3} = 22\,\frac{2}{3}$

19. $3\,\frac{5}{8} \cdot \frac{2}{3}$

$= \frac{29}{8} \cdot \frac{2}{3}$ Writing fractional notation

$= \frac{29\cdot 2}{8\cdot 3} = \frac{29\cdot 2}{2\cdot 4\cdot 3} = \frac{2}{2} \cdot \frac{29}{4\cdot 3} = \frac{29}{12} = 2\,\frac{5}{12}$

21. $3\,\frac{1}{2} \cdot 2\,\frac{1}{3} = \frac{7}{2} \cdot \frac{7}{3} = \frac{49}{6} = 8\,\frac{1}{6}$

23. $3\,\frac{2}{5} \cdot 2\,\frac{7}{8} = \frac{17}{5} \cdot \frac{23}{8} = \frac{391}{40} = 9\,\frac{31}{40}$

25. $20 \div 3\,\frac{1}{5}$

$= 20 \div \frac{16}{5}$ Writing fractional notation

$= 20 \cdot \frac{5}{16}$ Multiplying by the reciprocal

$= \frac{20\cdot 5}{16} = \frac{4\cdot 5\cdot 5}{4\cdot 4} = \frac{4}{4} \cdot \frac{5\cdot 5}{4} = \frac{25}{4} = 6\,\frac{1}{4}$

27. $8\,\frac{2}{5} \div 7$

$= \frac{42}{5} \div 7$ Writing fractional notation

$= \frac{42}{5} \cdot \frac{1}{7}$ Multiplying by the reciprocal

$= \frac{42\cdot 1}{5\cdot 7} = \frac{6\cdot 7}{5\cdot 7} = \frac{7}{7} \cdot \frac{6}{5} = \frac{6}{5} = 1\,\frac{1}{5}$

29. $4\,\frac{3}{4} \div 1\,\frac{1}{3} = \frac{19}{4} \div \frac{4}{3} = \frac{19}{4} \cdot \frac{3}{4} = \frac{19\cdot 3}{4\cdot 4} = \frac{57}{16} = 3\,\frac{9}{16}$

31. $1\,\frac{7}{8} \div 1\,\frac{2}{3} = \frac{15}{8} \div \frac{5}{3} = \frac{15}{8} \cdot \frac{3}{5} = \frac{15\cdot 3}{8\cdot 5} = \frac{5\cdot 3\cdot 3}{8\cdot 5}$

$= \frac{5}{5} \cdot \frac{3\cdot 3}{8} = \frac{3\cdot 3}{8} = \frac{9}{8} = 1\,\frac{1}{8}$

33. Familiarize. We let w = the total weight of the meat.

Translate. We write an equation.

Weight of one package	plus	Weight of second package	is	Total weight
↓	↓	↓		↓
$1\frac{1}{3}$	+	$4\frac{3}{5}$	=	w

Solve. To solve the equation we carry out the addition. The LCD is 15.

$$1\boxed{\frac{1}{3}\cdot\frac{5}{5}} = 1\,\frac{5}{15}$$

$$+\,4\boxed{\frac{3}{5}\cdot\frac{3}{3}} = +\,4\,\frac{9}{15}$$

$$5\,\frac{14}{15}$$

Check. We repeat the calculation. We also note that the answer is larger than either of the individual weights, so the answer seems reasonable.

State. The total weight of the meat was $5\frac{14}{15}$ lb.

35. Familiarize. We let y = the length of the pencil.

Translate. We write an addition sentence.

Length of wood	+	Length of eraser	=	Total length
↓	↓	↓	↓	↓
$16\frac{9}{10}$	+	$1\frac{9}{10}$	=	y

Solve. To solve the equation we carry out the addition.

$$16\,\frac{9}{10}$$

$$+\;1\,\frac{9}{10}$$

$$17\,\frac{18}{10} = 17 + \frac{18}{10} = 17 + 1\frac{8}{10} = 18\frac{8}{10} = 18\frac{4}{5}$$

Check. We repeat the calculation. We also note that the total length is larger than either of the individual lengths, so the answer seems reasonable.

State. The length of the standard pencil is $18\frac{4}{5}$ cm.

37. Familiarize. We let c = the closing price.

Translate. We write an equation.

Amount at opening	$-$	Amount dropped	$=$	Amount at closing
↓	↓	↓	↓	↓
$104\frac{5}{8}$	$-$	$1\frac{1}{4}$	$=$	c

Solve. To solve we carry out the subtraction. The LCD is 8.

$$\begin{array}{rcr} 104\,\frac{5}{8} & = & 104\,\frac{5}{8} \\ -\;1\boxed{\frac{1}{4}\cdot\frac{2}{2}} & = & -\;\;1\,\frac{2}{8} \\ \hline & & 103\,\frac{3}{8} \end{array}$$

Check. We add the amount the stock dropped to the closing price:

$$1\frac{1}{4} + 103\frac{3}{8} = 1\frac{2}{8} + 103\frac{3}{8} = 104\frac{5}{8}$$

This checks.

State. The closing price was $\$103\frac{3}{8}$.

39. We see that d and the two smallest distances combined are the same as the largest distance. We translate and solve.

$$2\frac{3}{4} + d + 2\frac{3}{4} = 12\frac{7}{8}$$

$$d = 12\frac{7}{8} - 2\frac{3}{4} - 2\frac{3}{4}$$

$$= 10\frac{1}{8} - 2\frac{3}{4} \quad \text{Subtracting } 2\frac{3}{4} \text{ from } 12\frac{7}{8}$$

$$= 7\frac{3}{8} \quad \text{Subtracting } 2\frac{3}{4} \text{ from } 10\frac{1}{8}$$

The length of d is $7\frac{3}{8}$ ft.

41. Familiarize. We let w = the weight of $5\frac{1}{2}$ cubic feet of water.

Translate. We write an equation.

Weight per cubic foot	$\cdot$	Number of cubic feet	$=$	Total weight
↓	↓	↓	↓	↓
$62\frac{1}{2}$	$\cdot$	$5\frac{1}{2}$	$=$	w

Solve. To solve the equation, we carry out the multiplication.

$$w = 62\frac{1}{2} \cdot 5\frac{1}{2}$$

$$= \frac{125}{2} \cdot \frac{11}{2} = \frac{125\cdot 11}{2\cdot 2}$$

$$= \frac{1375}{4} = 343\,\frac{3}{4}$$

41. (continued)

Check. We repeat the calculation. We also note that $62\frac{1}{2} \approx 60$ and $5\frac{1}{2} \approx 5$. Then the product is about 300. Our answer seems reasonable.

State. The weight of $5\frac{1}{2}$ cubic feet of water is $343\frac{3}{4}$ lb.

43. Familiarize, Translate, and Solve. To find the ingredients for $\frac{1}{2}$ the recipe, we multiply each ingredient by $\frac{1}{2}$.

$$2\frac{1}{2} \cdot \frac{1}{2} = \frac{5}{2} \cdot \frac{1}{2} = \frac{5\cdot 1}{2\cdot 2} = \frac{5}{4} = 1\frac{1}{4}$$

$$1\frac{1}{3} \cdot \frac{1}{2} = \frac{4}{3} \cdot \frac{1}{2} = \frac{2\cdot 2\cdot 1}{3\cdot 2} = \frac{2}{2} \cdot \frac{2\cdot 1}{3} = \frac{2}{3}$$

$$\frac{2}{3} \cdot \frac{1}{2} = \frac{2\cdot 1}{3\cdot 2} = \frac{2}{2} \cdot \frac{1}{3} = \frac{1}{3}$$

$$\frac{1}{2} \cdot \frac{1}{2} = \frac{1\cdot 1}{2\cdot 2} = \frac{1}{4}$$

$$4 \cdot \frac{1}{2} = \frac{4\cdot 1}{2} = \frac{4}{2} = 2$$

$$2 \cdot \frac{1}{2} = \frac{2\cdot 1}{2} = \frac{2}{2} = 1$$

Check. We repeat the calculations.

State. The ingredients for $\frac{1}{2}$ the recipe are $1\frac{1}{4}$ lb opossum meat, $1\frac{1}{4}$ teaspoons salt, $\frac{2}{3}$ teaspoon black pepper, $\frac{1}{3}$ cup flour, $\frac{1}{4}$ cup water, 2 medium sweet potatoes, and 1 tablespoon sugar.

Familiarize, Translate, and Solve. To find the ingredients for 3 recipes, we multiply each ingredient by 3.

$$2\frac{1}{2} \cdot 3 = \frac{5}{2} \cdot 3 = \frac{5\cdot 3}{2} = \frac{15}{2} = 7\frac{1}{2}$$

$$1\frac{1}{3} \cdot 3 = \frac{4}{3} \cdot 3 = \frac{4\cdot 3}{3} = \frac{12}{3} = 4$$

$$\frac{2}{3} \cdot 3 = \frac{2\cdot 3}{3} = \frac{6}{3} = 2$$

$$\frac{1}{2} \cdot 3 = \frac{1\cdot 3}{2} = \frac{3}{2} = 1\,\frac{1}{2}$$

$$4\cdot 3 = 12$$

$$2\cdot 3 = 6$$

Check. We repeat the calculations.

State. The ingredients for 3 recipes are $7\frac{1}{2}$ lb opossum meat, $7\frac{1}{2}$ teaspoons salt, 4 teaspoons black pepper, 2 cups flour, $1\frac{1}{2}$ cups water, 12 medium sweet potatoes, and 6 tablespoons sugar.

45. Familiarize. We let m = the number of miles per gallon the car got.

Translate. We write an equation.

Total number of miles traveled	÷	Number of gallons of gas used	=	Miles per gallon
↓	↓	↓	↓	↓
213	÷	$14\frac{2}{10}$	=	m

Solve. To solve the equation we carry out the division.

$$m = 213 \div 14\frac{2}{10} = 213 \div \frac{142}{10}$$

$$= 213 \cdot \frac{10}{142} = \frac{3 \cdot 71 \cdot 2 \cdot 5}{2 \cdot 71 \cdot 1}$$

$$= \frac{2 \cdot 71}{2 \cdot 71} \cdot \frac{3 \cdot 5}{1} = 15$$

Check. We repeat the calculation.

State. Thus, the car got 15 miles per gallon of gas.

47. Familiarize. We let n = the number of cubic feet occupied by 250 lb of water.

Translate. We write an equation.

Total weight	÷	Weight per cubic foot	=	Number of cubic feet
↓	↓	↓	↓	↓
250	÷	$62\frac{1}{2}$	=	n

Solve. To solve the equation we carry out the division.

$$n = 250 \div 62\frac{1}{2} = 250 \div \frac{125}{2}$$

$$= 250 \cdot \frac{2}{125} = \frac{2 \cdot 125 \cdot 2}{125 \cdot 1}$$

$$= \frac{125}{125} \cdot \frac{2 \cdot 2}{1} = 4$$

Check. We repeat the calculation.

State. 4 cubic feet would be occupied.

49. Familiarize. We let p = the number of pounds of turkey needed for 32 servings.

Translate. We write an equation.

Total number of servings	÷	Servings per pound	=	Number of pounds needed
↓	↓	↓	↓	↓
32	÷	$1\frac{1}{3}$	=	p

49. (continued)

Solve. To solve the equation we carry out the division.

$$p = 32 \div 1\frac{1}{3} = 32 \div \frac{4}{3}$$

$$= 32 \cdot \frac{3}{4} = \frac{4 \cdot 8 \cdot 3}{4 \cdot 1}$$

$$= \frac{4}{4} \cdot \frac{8 \cdot 3}{1} = 24$$

Check. We repeat the calculation.

State. 24 pounds would be needed.

51. Familiarize. We draw a picture. We let n = the number of 16-oz bottles to be filled.

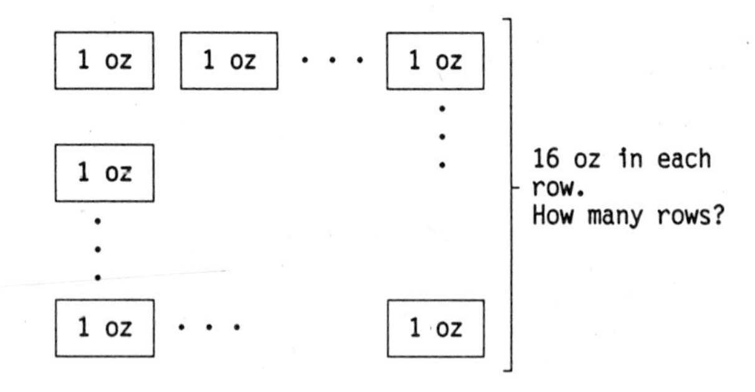

Translate and Solve. We translate to an equation and solve as follows:

$$4578 \div 16 = n$$

```
       2 8 6
 1 6 ) 4 5 7 8
       3 2 0 0
       1 3 7 8
       1 2 8 0
           9 8
           9 6
             2
```

Check. We can check by multiplying the number of cartons by 6 and adding the remainder of 2:

$$16 \cdot 286 = 4576, \quad 4576 + 2 = 4578$$

State. 286 16-oz bottles can be filled. There will be 2 oz of milk left over.

53. Familiarize. We draw a picture.

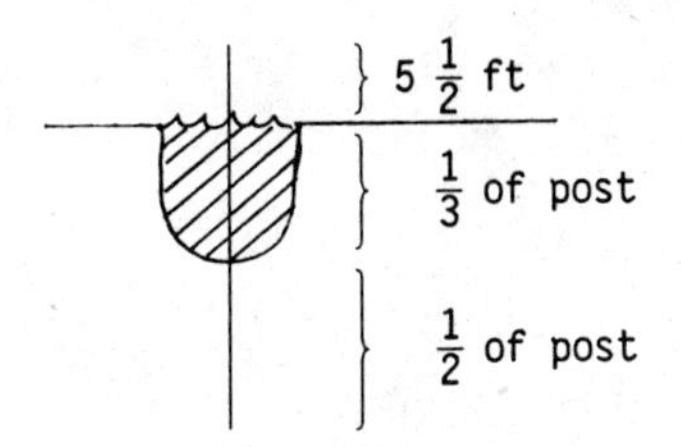

Translate and Solve. This is a two-step problem. First we find f, the fraction of the post above water. We translate and solve.

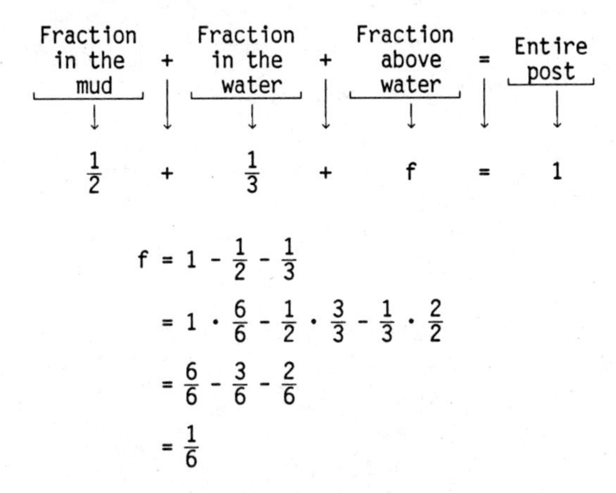

$$\frac{1}{2} + \frac{1}{3} + f = 1$$

$$\begin{aligned} f &= 1 - \frac{1}{2} - \frac{1}{3} \\ &= 1 \cdot \frac{6}{6} - \frac{1}{2} \cdot \frac{3}{3} - \frac{1}{3} \cdot \frac{2}{2} \\ &= \frac{6}{6} - \frac{3}{6} - \frac{2}{6} \\ &= \frac{1}{6} \end{aligned}$$

Thus, $\frac{1}{6}$ of the post is above water.

Next we find p, the length of the post. We translate to a multiplication sentence and solve.

Fraction above water	of	Entire length of post	is	Length above water
↓	↓	↓	↓	↓
$\frac{1}{6}$	$\cdot$	p	=	$5\frac{1}{2}$

$$\begin{aligned} p &= 5\frac{1}{2} \div \frac{1}{6} \\ &= 5\frac{1}{2} \cdot 6 \\ &= \frac{11}{2} \cdot 6 \\ &= \frac{11 \cdot 6}{2} = \frac{11 \cdot 2 \cdot 3}{2 \cdot 1} \\ &= \frac{2}{2} \cdot \frac{11 \cdot 3}{1} = \frac{11 \cdot 3}{1} = \frac{33}{1} \\ &= 33 \end{aligned}$$

Check. We repeat the calculations.

State. The post is 33 ft long.

Exercise Set 2.1

1. a) Write a word name for the whole number. **Twenty-three**

   b) Write "and" for the decimal point. Twenty-three **and**

   c) Write a word name for the number to the right of the decimal point, followed by the place value of the last digit. Twenty-three and **two tenths**

   A word name for 23.2 is twenty-three and two tenths.

3. 3 4 . 8 9 1

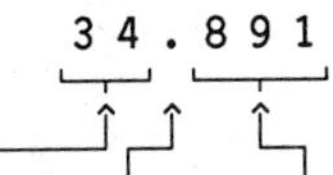

   Thirty-four and eight hundred ninety-one thousandths

5. Write "and 48 cents" as "and $\frac{48}{100}$ dollars." A word name for \$326.48 is three hundred twenty-six and $\frac{48}{100}$ dollars.

7. Write "67 cents" as "$\frac{67}{100}$ dollars." A word name for \$0.67 is $\frac{67}{100}$ dollars.

9. 6.8   6.8. $\frac{68}{10}$

   1 place   Move 1 place   1 zero

   $6.8 = \frac{68}{10}$

11. 0.17   0.17. $\frac{17}{100}$

    2 places   Move 2 places   2 zeros

    $0.17 = \frac{17}{100}$

13. 1.46   1.46. $\frac{146}{100}$

    2 places   Move 2 places   2 zeros

    $1.46 = \frac{146}{100}$

15. 204.6   204.6. $\frac{2046}{10}$

    1 place   Move 1 place   1 zero

    $204.6 = \frac{2046}{10}$

17. 3.142   3.142. $\frac{3142}{1000}$

    3 places   Move 3 places   3 zeros

    $3.142 = \frac{3142}{1000}$

19. 46.03   46.03. $\frac{4603}{100}$

    2 places   Move 2 places   2 zeros

    $46.03 = \frac{4603}{100}$

21. 1.0008   1.0008. $\frac{10008}{10000}$

    4 places   Move 4 places   4 zeros

    $1.0008 = \frac{10,008}{10,000}$

23. 4567.2   4567.2. $\frac{45672}{10}$

    1 place   Move 1 place   1 zero

    $4567.2 = \frac{45,672}{10}$

25. $\frac{8}{10}$ 0.8.

    1 zero   Move 1 place

    $\frac{8}{10} = 0.8$

27. $\frac{92}{100}$ 0.92.

    2 zeros   Move 2 places

    $\frac{92}{100} = 0.92$

29. $\frac{93}{10}$ 9.3.

    1 zero   Move 1 place

    $\frac{93}{10} = 9.3$

31. $\frac{889}{100}$ 8.89.

    2 places   Move 2 places

    $\frac{889}{100} = 8.89$

33. $\frac{2508}{10}$ 250.8.

1 place Move 1 place

$\frac{2508}{10} = 250.8$

35. $\frac{3798}{1000}$ 3.798.

3 places Move 3 places

$\frac{3798}{1000} = 3.798$

37. $\frac{78}{10{,}000}$ 0.0078.

4 zeros Move 4 places

$\frac{78}{10{,}000} = 0.0078$

39. $\frac{56{,}788}{100{,}000}$ 0.56788.

5 zeros Move 5 places

$\frac{56{,}788}{100{,}000} = 0.56788$

41. $\frac{2173}{100}$ 21.73.

2 zeros Move 2 places

$\frac{2173}{100} = 21.73$

43. $\frac{66}{100}$ 0.66.

2 zeros Move 2 places

$\frac{66}{100} = 0.66$

45. $\frac{3417}{100}$ 34.17.

2 zeros Move 2 places

$\frac{3417}{100} = 34.17$

47. $\frac{376{,}193}{1{,}000{,}000}$ 0.376193.

6 zeros Move 6 places

$\frac{376{,}193}{1{,}000{,}000} = 0.376193$

49. Round 617[2] to the nearest ten.

The digit 7 is in the tens place. Since the next digit to the right (2) is 4 or lower, we round down. The answer is 6170.

51. Round 6[1]72 to the nearest thousand.

Since 1 is 4 or lower we round down. The answer is 6000.

53. $4\frac{909}{1000} = \frac{4909}{1000}$ 4.909.

3 zeros Move 3 places

$4\frac{909}{1000} = 4.909$

Exercise Set 2.2

1.
```
      1
    3 1 6.2 5
  +   1 8.1 2
    3 3 4.3 7
```
Add hundredths.
Add tenths.
Write a decimal point in the answer.
Add ones.
Add tens.
Add hundreds.

3.
```
      1 1
    6 5 9.4 0 3
  + 9 1 6.8 1 2
  1 5 7 6.2 1 5
```
Add thousandths.
Add hundredths.
Add tenths.
Write a decimal point in the answer.
Add ones.
Add tens.
Add hundreds.

5.
```
    1       1
        9.1 0 4
  + 1 2 3.4 5 6
    1 3 2.5 6 0
```

7. Line up the decimal points.
```
        1
  6 1.0 0 6
+   3.4 0 7
  6 4.4 1 3
```
Adding

9. Line up the decimal points.
```
      2.0 2
  + 3 0.1 0  ←
    3 2.1 2
```
An extra 0 can be written here if desired.
Adding

11. Line up the decimal points.
```
    0.8 3 0  ←
  + 0.0 0 5
    0.8 3 5
```
An extra 0 can be written here if desired.
Adding

13. Line up the decimal points.

```
    0.3 4 0    Writing an extra 0
    3.5 0 0    Writing 2 extra 0's
    0.1 2 7
+ 7 6 8.0 0 0  Writing in the decimal point and 3 extra 0's
  7 7 1.9 6 7  Adding
```

15.

```
  1 7.0 0 0 0  Writing in the decimal point.
    3.2 4 0 0  You may find it helpful to
    0.2 5 6 0  write extra 0's.
+   0.3 6 8 9
  2 0.8 6 4 9
```

17. Line up the decimal points.

```
     2.7 0 3
+ 7 8.3 3 0   Writing an extra 0
  8 1.0 3 3   Adding
```

19.

```
   5.2   Borrow ones to subtract tenths. Subtract tenths.
 - 3.9   Write a decimal point in the answer.
   1.3   Subtract ones.
```

21.

```
  5 1.3 1   Borrow tenths to subtract hundredths. Subtract hundredths.
-   2.2 9   Subtract tenths.
  4 9.0 2   Write a decimal point in the answer.
            Borrow tens to subtract ones. Subtract ones.
            Subtract tens.
```

23.

```
  4 8.7 6
-   3.1 5
  4 5.6 1
```

25.

```
  9 2.3 4 1
-   6.4 2
  8 5.9 2 1
```

27. Line up the decimal points.

```
  2.5 0 0 0   Writing 3 extra 0's
- 0.0 0 2 5
  2.4 9 7 5
```

29. Line up the decimal points.

```
  3.4 0 0   Writing 2 extra 0's
- 0.0 0 3
  3.3 9 7
```

31. Line up the decimal points. Write an extra 0 if desired.

```
  2 8.2 0
- 1 9.3 5
    8.8 5
```

33.

```
  3 4.0 7
- 3 0.7
    3.3 7
```

35.

```
  8.4 5 0
- 7.4 0 5
  1.0 4 5
```

37.

```
  6.0 0 3
- 2.3
  3.7 0 3
```

39.

```
  6 2 4.0 0   Writing in the decimal point and 2 extra 0's
-   1 8.7 9
  6 0 5.2 1
```

41.

```
  2 5 4 8.9 8 0
-       2.0 0 7
  2 5 4 6.9 7 3
```

43.

```
  4 5.0 0 0
-   0.9 9 9
  4 4.0 0 1
```

45. 34,[4]96 Hundreds digit is less than 5. Round down.

34,000

Exercise Set 2.3

1.

```
     6 3   (0 decimal places)
× 0 .0 4   (2 decimal places)
   2.5 2   (2 decimal places)
```

3.

```
       8 7   (0 decimal places)
× 0.0 0 6   (3 decimal places)
  0.5 2 2   (3 decimal places)
```

5.
```
     3 2.6   (1 decimal place)
   ×   1 6   (0 decimal places)
   1 9 5 6
   3 2 6 0
   5 2 1.6   (1 decimal place)
```

7.
```
   0.9 8 4   (3 decimal places)
   ×   3.3   (1 decimal place)
   2 9 5 2
 2 9 5 2 0
 3.2 4 7 2   (4 decimal places)
```

9.
```
     3 7 4   (0 decimal places)
   ×   2.4   (1 decimal place)
   1 4 9 6
   7 4 8 0
   8 9 7.6   (1 decimal place)
```

11.
```
     7 4 9   (0 decimal places)
   × 0.4 3   (2 decimal places)
   2 2 4 7
 2 9 9 6 0
 3 2 2.0 7   (2 decimal places)
```

13.
```
       8 1.7   (1 decimal place)
   × 0.6 1 2   (3 decimal places)
     1 6 3 4
     8 1 7 0
 4 9 0 2 0 0
 5 0.0 0 0 4   (4 decimal places)
```

15.
```
     1 8.0 0 0   (3 decimal places)
     ×   2.0 2   (2 decimal places)
     3 6 0 0 0
 3 6 0 0 0 0 0
 3 6.3 6 0 0 0   (5 decimal places)
```

Since the last 3 decimal places are 0's, we could also write this answer as 36.36.

17.
```
     8.4 3   (2 decimal places)
   × 1 0 0   (0 decimal places)
 8 4 3.0 0   (2 decimal places)
```

We could also write this as 843.

19.
```
   4 6.5 0   (2 decimal places)
   ×   7 5   (0 decimal places)
 2 3 2 5 0
 3 2 5 5 0 0
 3 4 8 7.5 0   (2 decimal places)
```

Since the last decimal place is 0, we could also write this answer as 3487.5.

21.
```
   3 2.4   (1 decimal place)
   × 2.8   (1 decimal place)
 2 5 9 2
 6 4 8
 9 0.7 2   (2 decimal places)
```

23.
```
     0.0 0 3 4 2   (5 decimal places)
   ×       0.8 4   (2 decimal places)
         1 3 6 8
       2 7 3 6 0
 0.0 0 2 8 7 2 8   (7 decimal places)
```

25.
```
         1 6.3 4   (2 decimal places)
   × 0.0 0 0 5 1 2   (6 decimal places)
           3 2 6 8
         1 6 3 4 0
       8 1 7 0 0 0
 0.0 0 8 3 6 6 0 8   (8 decimal places)
```

27.
```
      3.2 4
   8)2 5.9 2
     2 4 0 0
       1 9 2
       1 6 0
         3 2
         3 2
           0
```

Divide as though dividing whole numbers. Place the decimal point directly above the decimal point in the dividend.

29.
```
        1.0 9
   2 1)2 2.8 9
       2 1 0 0
         1 8 9
         1 8 9
             0
```

Divide as though dividing whole numbers. Place the decimal point directly above the decimal point in the dividend.

31.
```
   3.1
 3)9.3
   9 0
     3
     3
     0
```

33.
```
   0.0 7
 5)0.3 5
     3 5
       0
```

35.
```
              7 0.
0.1 2^ )8.4 0^
         8 4 0
             0
```
Multiply the divisor by 100 (move the decimal point 2 places). Multiply the same way in the dividend (move 2 places).

Then divide. Place the decimal point in the answer above the new decimal point in the dividend.

37.
```
           2 0.
3.4^ )6 8.0^
        6 8 0
            0
            0
            0
```
Put a decimal point at the end of the whole number. Multiply the divisor by 10 (move the decimal point 1 place). Multiply the same way in the dividend (move 1 place), adding an extra 0.

Then divide. Place the decimal point in the answer above the new decimal point in the dividend.

39.
```
      0.4
1 5)6.0
    6 0
      0
```
Put a decimal point at the end of the whole number.

Write an extra 0 to the right of the decimal point.

Then divide and place the decimal point in the answer above the decimal point in the dividend.

41.
```
      0.4 1
3 6)1 4.7 6
    1 4 4 0
        3 6
        3 6
          0
```

43.
```
              2 2 5.
0.2 8^ )6 3.0 0^
          5 6 0 0
            7 0 0
            5 6 0
            1 4 0
            1 4 0
                0
```
Put a decimal point at the end of the whole number. Multiply the divisor by 100 (move the decimal point 2 places). Multiply the same way in the dividend (move 2 places, adding 2 extra 0's).

Then divide. Place the decimal point in the answer above the new decimal point in the dividend.

45.
```
                9 3.
0.0 1 7^ )1.5 8 1^
          1 5 3 0
              5 1
              5 1
                0
```

47.
```
       .0 0 4 3
3 4)0.1 4 6 2
      1 3 6 0
        1 0 2
        1 0 2
            0
```

49.

$$\begin{array}{rcrcr} 10\frac{1}{2} & = & 10\frac{4}{8} & = & 9\frac{12}{8} \\ -\ 4\frac{5}{8} & = & -\ 4\frac{5}{8} & = & -\ 4\frac{5}{8} \\ \hline & & & & 5\frac{7}{8} \end{array}$$

51.
```
      3   <--  3 is prime
    3| 9
   3|2 7
  3| 8 1       81 is not divisible by 2. We
 2|1 6 2       go to 3.
```

$162 = 2 \cdot 3 \cdot 3 \cdot 3 \cdot 3$

53. Multiply and divide, in order, from left to right.

$78.66 \div 5.7 \times 0.45$

$= 13.8 \times 0.45$ $(78.66 \div 5.7 = 13.8)$

$= 6.21$ Multiplying

Exercise Set 2.4

1. $\frac{3}{5} = 3 \div 5$
```
  0.6
5)3.0
  3 0
    0
```
$\frac{3}{5} = 0.6$

3. $\frac{13}{40} = 13 \div 40$
```
    0.3 2 5
4 0)1 3.0 0 0
    1 2 0
      1 0 0
        8 0
        2 0 0
        2 0 0
            0
```
$\frac{13}{40} = 0.325$

5. $\frac{1}{5} = 1 \div 5$

$$\begin{array}{r} 0.2 \\ 5\overline{)1.0} \\ \underline{1\,0} \\ 0 \end{array}$$

$\frac{1}{5} = 0.2$

7. $\frac{17}{20} = 17 \div 20$

$$\begin{array}{r} 0.85 \\ 20\overline{)17.00} \\ \underline{160} \\ 100 \\ \underline{100} \\ 0 \end{array}$$

$\frac{17}{20} = 0.85$

9. $\frac{19}{40} = 19 \div 40$

$$\begin{array}{r} 0.475 \\ 40\overline{)19.000} \\ \underline{160} \\ 300 \\ \underline{280} \\ 200 \\ \underline{200} \\ 0 \end{array}$$

$\frac{19}{40} = 0.475$

11. $\frac{39}{40} = 39 \div 40$

$$\begin{array}{r} 0.975 \\ 40\overline{)39.000} \\ \underline{360} \\ 300 \\ \underline{280} \\ 200 \\ \underline{200} \\ 0 \end{array}$$

$\frac{39}{40} = 0.975$

13. $\frac{13}{25} = 13 \div 25$

$$\begin{array}{r} 0.52 \\ 25\overline{)13.00} \\ \underline{125} \\ 50 \\ \underline{50} \\ 0 \end{array}$$

$\frac{13}{25} = 0.52$

15. $\frac{2502}{125} = 2502 \div 125$

$$\begin{array}{r} 20.016 \\ 125\overline{)2502.000} \\ \underline{2500} \\ 200 \\ \underline{125} \\ 750 \\ \underline{750} \\ 0 \end{array}$$

$\frac{2502}{125} = 20.016$

17. $\frac{1}{4} = 1 \div 4$

$$\begin{array}{r} 0.25 \\ 4\overline{)1.00} \\ \underline{8} \\ 20 \\ \underline{20} \\ 0 \end{array}$$

$\frac{1}{4} = 0.25$

19. $\frac{23}{40} = 23 \div 40$

$$\begin{array}{r} 0.575 \\ 40\overline{)23.000} \\ \underline{200} \\ 300 \\ \underline{280} \\ 200 \\ \underline{200} \\ 0 \end{array}$$

$\frac{23}{40} = 0.575$

21. $\frac{4}{7} = 4 \div 7$

$$\begin{array}{r} 0.5714285 \\ 7\overline{)4.0000000} \\ \underline{35} \\ 50 \\ \underline{49} \\ 10 \\ \underline{7} \\ 30 \\ \underline{28} \\ 20 \\ \underline{14} \\ 60 \\ \underline{56} \\ 40 \\ \underline{35} \\ 5 \end{array}$$

21. (continued)

Since 5 reappears as a remainder, the sequence repeats and

$\frac{4}{7} = 0.571428571428 \ldots$ or $0.\overline{571428}$

23. $\frac{7}{6} = 7 \div 6$

$$\begin{array}{r} 1.1\,6\,6 \\ 6\overline{)7.0\,0\,0} \\ \underline{6} \\ 1\,0 \\ \underline{6} \\ 4\,0 \\ \underline{3\,6} \\ 4\,0 \\ \underline{3\,6} \\ 4 \end{array}$$

Since 4 keeps reappearing as a remainder, the digits repeat and

$\frac{7}{6} = 1.166 \ldots$ or $1.1\overline{6}$

25. $\frac{4}{15} = 4 \div 15$

$$\begin{array}{r} .2\,6\,6 \\ 15\overline{)4.0\,0\,0} \\ \underline{3\,0} \\ 1\,0\,0 \\ \underline{9\,0} \\ 1\,0\,0 \\ \underline{9\,0} \\ 1\,0 \end{array}$$

Since 10 keeps reappearing as a remainder, the digits repeat and

$\frac{4}{15} = 0.266 \ldots$ or $0.2\overline{6}$

27. $\frac{11}{12} = 11 \div 12$

$$\begin{array}{r} 0.9\,1\,6\,6 \\ 12\overline{)1\,1.0\,0\,0\,0} \\ \underline{1\,0\,8} \\ 2\,0 \\ \underline{1\,2} \\ 8\,0 \\ \underline{7\,2} \\ 8\,0 \\ \underline{7\,2} \\ 8\,0 \end{array}$$

Since 8 keeps reappearing as a remainder, the digits repeat and

$\frac{11}{12} = 0.91666 \ldots$ or $0.91\overline{6}$

29. $0.\underline{1}\boxed{1} \rightarrow 0.1$ Hundredths digit is less than 5. Round down.

31. $0.\underline{1}\boxed{6} \rightarrow 0.2$ Hundredths digit is 5 or higher. Round up.

33. $0.\underline{5}\boxed{7}94 \rightarrow 0.6$ Hundredths digit is 5 or higher. Round up.

35. $2.\underline{7}\boxed{4}49 \rightarrow 2.7$ Hundredths digit is less than 5. Round down.

37. $0.8\underline{9}\boxed{3} \rightarrow 0.89$ Thousandths digit is less than 5. Round down.

39. $0.6\underline{6}\boxed{6} \rightarrow 0.67$ Thousandths digit is 5 or higher. Round up.

41. $0.4\underline{2}\boxed{4}6 \rightarrow 0.42$ Thousandths digit is less than 5. Round down.

43. $1.4\underline{3}\boxed{5} \rightarrow 1.44$ Thousandths digit is 5 or higher. Round up.

45. $0.32\underline{4}\boxed{6} \rightarrow 0.325$ Ten-thousandths digit is 5 or higher. Round up.

47. $0.66\underline{6}\boxed{6} \rightarrow 0.667$ Ten-thousandths digit is 5 or higher. Round up.

49. $17.00\underline{1}\boxed{5} \rightarrow 17.002$ Ten-thousandths digit is 5 or higher. Round up.

51. $0.00\underline{0}\boxed{9} \rightarrow 0.001$ Ten-thousandths digit is 5 or higher. Round up.

53. 2[8]3.1359 Tens digit is 5 or higher. Round up.
↓
300

55. 283.135[9] Ten-thousandths digit is 5 or higher. Round up.
↓
283.136

57. 283.[1]359 Tenths digit is less than 5. Round down.
↓
283

59. $17 \cdot x = 408$

$x = 408 \div 17$ Dividing by 17 on both sides

$x = 24$

61. Familiarize. We first make a drawing. We let d = the distance the car can travel on a full tank.

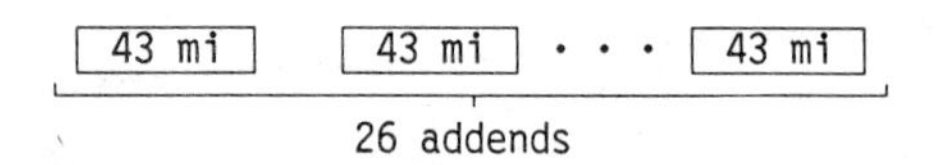

Translate. Repeated addition applies here. Thus the following multiplication corresponds to the situation.

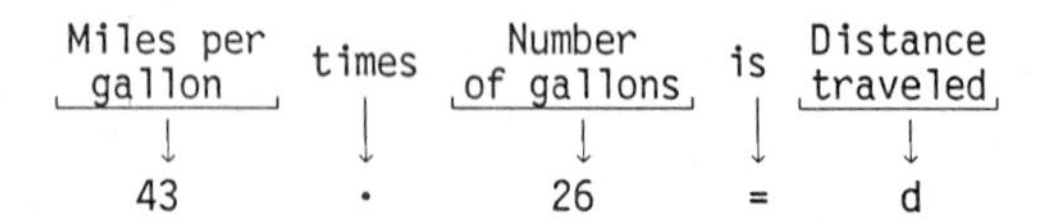

Solve. To solve the equation we carry out the multiplication.

$43 \cdot 26 = d$

$1118 = d$

```
    2 6
  × 4 3
    7 8
1 0 4 0
1 1 1 8
```

Check. We repeat the calculation.

State. The car can go 1118 mi on a full tank of 26 gal.

63.

```
       4 0 8
3 6)1 4,6 9 7
    1 4 4 0 0
        2 9 7
        2 8 8
            9
```

The answer is 408 R 9. We can write a mixed numeral for the quotient as follows:

$$408\frac{9}{36}, \text{ or } 408\frac{1}{4}$$

65. We can express 1 hit in 3 "at bats" as the fraction $\frac{1}{3}$. Then we convert to decimal notation by dividing.

```
   .3 3
3)1.0 0
    9
    1 0
      9
      1
```

$\frac{1}{3} = 0.\overline{3}$

Rounding 0.3333 . . . to the nearest thousandth, we get 0.333.

Exercise Set 2.5

1. Familiarize. We visualize the situation. We let g = the number of gallons purchases.

23.6 gal	17.7 gal	20.8 gal	17.2 gal	25.4 gal	13.8 gal
g					

Translate. Amounts are being combined. We translate to an equation.

First plus second plus third plus fourth plus fifth plus sixth is total
$23.6 + 17.7 + 20.8 + 17.2 + 25.4 + 13.8 = g$

Solve. To solve the equation we carry out the addition.

```
  ² ³
  2 3.6
  1 7.7
  2 0.8
  1 7.2
  2 5.4
+ 1 3.8
1 1 8.5
```

Thus, $g = 118.5$.

Check. We can check by repeating the addition. We also note that the answer seems reasonable since it is larger than any of the numbers being added. We can also check by rounding:

$23.6 + 17.7 + 20.8 + 17.2 + 25.4 + 13.8 \approx$

$24 + 18 + 21 + 17 + 25 + 14 = 119 \approx 118.5$

State. 118.5 gallons of gasoline were purchased.

3. Familiarize. We visualize the situation. We let m = the odometer reading at the end of the trip.

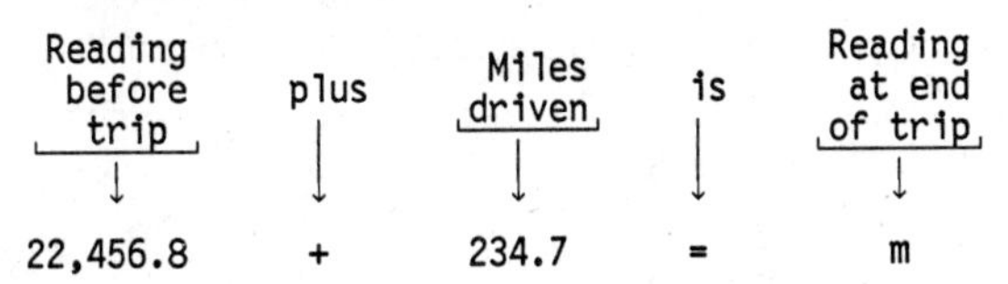

Translate. We are combining amounts.

Reading before trip plus Miles driven is Reading at end of trip

$$22{,}456.8 + 234.7 = m$$

Solve. To solve we carry out the addition.

```
  2 2,4 5 6.8
+     2 3 4.7
  2 2,6 9 1.5
```

Thus m = 22,691.5.

Check. We can check by repeating the addition. We can also check by rounding:

$22{,}456.8 + 234.7 \approx 22{,}460 + 230 = 22{,}690 \approx 22{,}691.5$

State. The odometer reading at the end of the trip was 22,691.5.

5. Familiarize. We vizualize the situation. We let s = the number of miles per hour by which Bobby Rahal's speed exceeded Ray Harroun's speed.

74.59 mph	s
170.722 mph	

Translate. This is a "how-much-more" situation.

Harroun's speed plus Additional speed is Rahal's speed

$$74.59 + s = 170.722$$

Solve. We subtract 74.59 on both sides on the equation.

$s = 170.722 - 74.59$

$s = 96.132$

```
  1 7 0.7 2 2
-   7 4.5 9 0
    9 6.1 3 2
```

Check. We check by adding 96.132 to 74.59 to get 170.722. This checks.

State. Rahal was 96.132 mph faster.

7. Familiarize. We visualize the situation. We let y = the number of years by which the average age in 1984 exceeded the average age in 1961.

19.8	y
23.2	

7. (continued)

Translate. This is a "how-much-more" situation.

Average age in 1961 + Additional years in 1984 is Average age in 1984

$$19.8 + y = 23.2$$

Solve. To solve the equation we subtract 19.8 on both sides.

$y = 23.2 - 19.8$

$y = 3.4$

```
  2 3.2
- 1 9.8
    3.4
```

Check. We check by adding 3.4 to 19.8 to get 23.2. This checks.

State. The average bride was 3.4 years older in 1984 than in 1961.

9. Familiarize. We visualize the situation. We let t = the combined time.

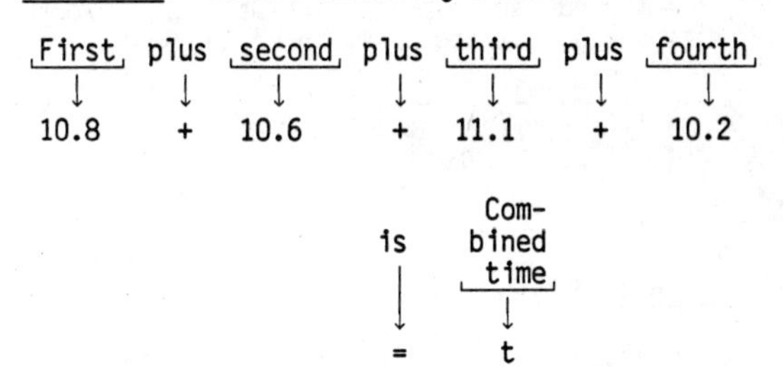

Translate. We are combining amounts.

First plus second plus third plus fourth is Combined time

$$10.8 + 10.6 + 11.1 + 10.2 = t$$

Solve. To solve we carry out the addition.

```
  1 0.8
  1 0.6
  1 1.1
+ 1 0.2
  4 2.7
```

Check. We can check by repeating the addition. We can also check by rounding:

$10.8 + 10.6 + 11.1 + 10.2 \approx 11 + 11 + 11 + 10 = 43 \approx 42.7$

State. Their combined time was 42.7 sec.

11. Familiarize. We visualize the situation. We let p = the number by which the O'Hare passengers exceed the Kennedy passengers, in millions.

45.7	
29.9	p

Translate. We have a "how-much-more" situation.

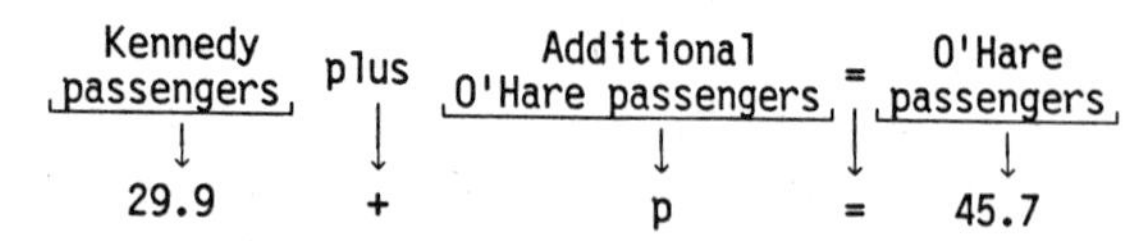

Solve. To solve we subtract 29.9 on both sides of the equation.

$p = 45.7 - 29.9$

$p = 15.8$

$$\begin{array}{r} 4\ 5.7 \\ -\ 2\ 9.9 \\ \hline 1\ 5.8 \end{array}$$

Check. We add 15.8 to 29.9 to get 45.7. This checks.

State. O'Hare handles 15.8 million more passengers than Kennedy.

13. Familiarize. We visualize the situation. We let t = the number of passengers Dallas/Ft. Worth and Kennedy handle together, in millions.

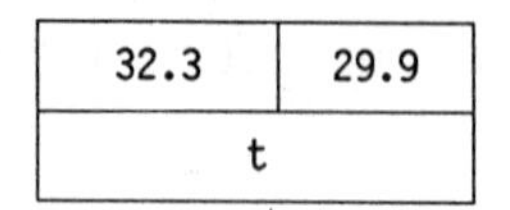

Translate. We are combining amounts.

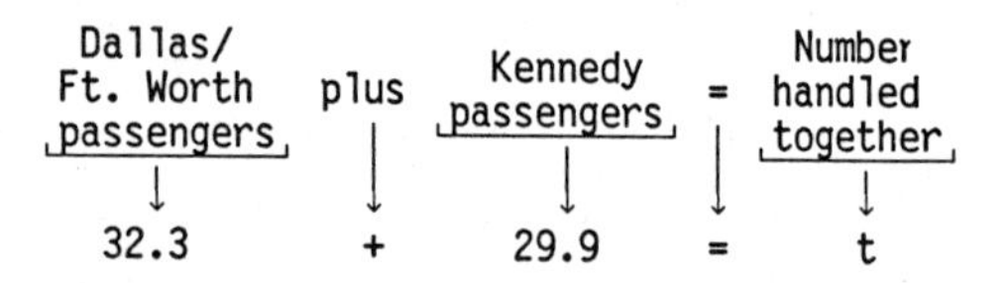

Solve. To solve we carry out the addition.

$$\begin{array}{r} {}^{1}3\ {}^{1}2.3 \\ 2\ 9.9 \\ \hline 6\ 2.2 \end{array}$$

Thus, $t = 62.2$.

Check. To check we can repeat the addition. We can also check by rounding:

$32.3 + 29.9 \approx 30 + 30 = 60 \approx 62.2$

State. Dallas/Ft. Worth and Kennedy handle 62.2 million passengers together.

15. Familiarize. We make a drawing. We let t = the total amount of chocolate consumed in one year, in millions of pounds.

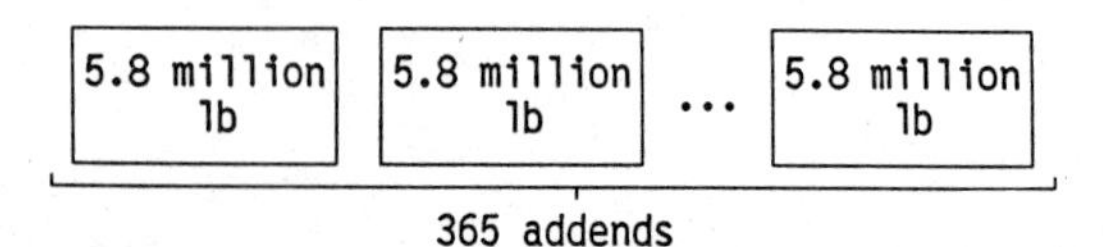

15. (continued)

Translate. Repeated addition fits this situation.

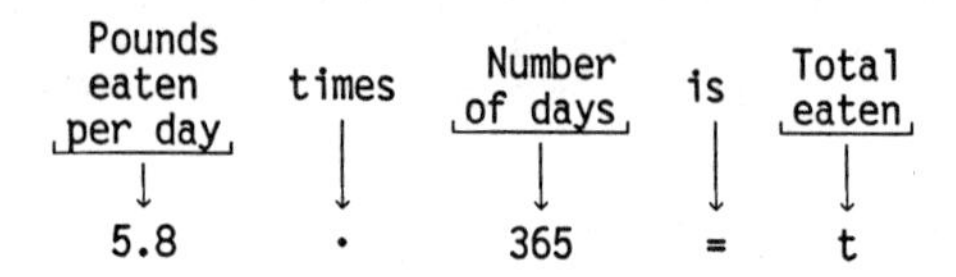

Solve. To solve the equation we carry out the multiplication.

$$\begin{array}{r} 3\ 6\ 5 \\ \times\ \ 5.8 \\ \hline 2\ 9\ 2\ 0 \\ 1\ 8\ 2\ 5\ 0 \\ \hline 2\ 1\ 1\ 7.0 \end{array}$$

Check. We repeat the calculation.

State. We consume 2117 million lb of chocolate in one year.

17. Familiarize. This is a two-step problem. First, we find the number of games that can be played in one hour. Think of an array containing 60 minutes (1 hour = 60 minutes) with 1.5 minutes in each row. We want to find how many rows there are. We let g represent the number.

Translate and Solve. We think (Number of minutes) ÷ (Number of minutes per game) = (Number of games).

$60 \div 1.5 = g$

To solve the equation we carry out the division.

$$\begin{array}{r} 4\ 0. \\ 1.5_{\wedge}\overline{)6\ 0.0_{\wedge}} \\ \underline{6\ 0\ 0} \\ 0 \\ \underline{0} \\ 0 \end{array}$$

Thus, $g = 40$. Second, we find the cost t of playing 40 video games. Repeated addition fits this situation. (We express 25¢ as \$0.25.)

Cost of one game	times	Number of games played	is	Total cost
↓	↓	↓	↓	↓
0.25	×	40	=	t

To solve the equation we carry out the multiplication.

$$\begin{array}{r} 0.2\ 5 \\ \times\ \ \ 4\ 0 \\ \hline 1\ 0.0\ 0 \end{array}$$

Thus, $t = 10$.

Check. To check, we first divide the total cost by the cost per game to find the number of games played:

$10 \div 0.25 = 40$

Then we multiply 40 by 1.5 to find the total time:

$1.5 \times 40 = 60$

17. (continued)

The number 10 checks.

State. It costs $10 to play video games for one hour.

19. $24\frac{1}{4} - 10\frac{2}{3} = 24\frac{3}{12} - 10\frac{8}{12}$

$= 23\frac{15}{12} - 10\frac{8}{12} = 13\frac{7}{12}$

21. $24\frac{1}{4} + 10\frac{2}{3} = 24\frac{3}{12} + 10\frac{8}{12}$

$= 34\frac{11}{12}$

23. Familiarize. This is a multistep problem. We begin by letting f = the taxi fare, excluding the tip.

Translate and Solve.

Fare	plus	Tip	is	Total
↓	↓	↓	↓	↓
f	+	5.00	=	35.25

To solve the equation we subtract 5.00 on both sides.

$f = 35.25 - 5.00$
$= 30.25$

```
  3 5.2 5
-   5.0 0
  3 0.2 5
```

The fare, excluding the tip, is $30.25.

Next find the amount that was charged for mileage after the charge for the first mile.

Fare for first mile	plus	Fare for additional miles	is	Total fare, excluding tip
↓	↓	↓	↓	↓
1.50	+	a	=	30.25

To solve the equation we subtract 1.50 on both sides.

$a = 30.25 - 1.50$
$= 28.75$

```
  2 9 12
  3 0.2 5
-   1.5 0
  2 8.7 5
```

The charge for the mileage after the first mile was $28.75.

Now we want to determine how many $0.25 charges are in the $28.75 charge. Think of an array containing $28.75 with $0.25 in each row. We want to find how many rows there are. This can be translated to the following equation:

$28.75 \div 0.25 = c$

To solve the equation we carry out the division.

23. (continued)

```
              1 1 5.
0.2 5^ ) 2 8.7 5^
         2 5 0 0
           3 7 5
           2 5 0
           1 2 5
           1 2 5
               0
```

There are 115 charges of $0.25 in $28.75. Each one represents $\frac{1}{5}$ mile. To determine how many miles were driven after the first mile we use repeated addition.

Number of $0.25 charges	times	Miles per charge	is	Miles after first mile
↓	↓	↓	↓	↓
115	·	$\frac{1}{5}$	=	m

To solve the equation we carry out the multiplication.

$115 \cdot \frac{1}{5} = \frac{115}{5} = \frac{5 \cdot 23}{5 \cdot 1} = \frac{23}{1} = 23$

After the first mile, 23 miles were driven.

We combine amounts to find the total distance from the airport to Addy's business.

First mile	plus	Miles after first mile	is	Total distance
↓	↓	↓	↓	↓
1	+	23	=	D

To solve the equation we carry out the addition.

```
   1
+ 2 3
  2 4
```

Check. We repeat the calculations.

State. It is 24 mi from the airport to Addy's business.

25. Add and subtract, in order, from left to right.

$451.63 - 21.7 + 13.7$

$= 429.93 + 13.7$ $(451.63 - 21.7 = 429.93)$

$= 443.63$ Adding

Exercise Set 2.6

1. The ratio of 4 to 5 can be written $\frac{4}{5}$.

3. The ratio of 0.4 to 12 can be written $\frac{0.4}{12}$.

5. The ratio of milk to flour can be written $\frac{2}{12}$. Simplifying, we get $\frac{2}{12} = \frac{1}{6}$.

7. The ratio of \$2.7 billion to \$13.1 billion can be written $\frac{2.7}{13.1}$.

The ratio of \$13.1 billion to \$2.7 billion can be written $\frac{13.1}{2.7}$.

9. $\frac{18}{4} = \frac{x}{10}$

$18 \cdot 10 = 4 \cdot x$ Finding cross products

$\frac{18 \cdot 10}{4} = x$ Dividing by 4

$\frac{180}{4} = x$ Multiplying

$45 = x$ Dividing

11. $\frac{x}{8} = \frac{9}{6}$

$6 \cdot x = 8 \cdot 9$ Finding cross products

$x = \frac{8 \cdot 9}{6}$ Dividing by 6

$x = \frac{72}{6}$ Multiplying

$x = 12$ Dividing

13. $\frac{t}{12} = \frac{5}{6}$

$6 \cdot t = 12 \cdot 5$

$t = \frac{12 \cdot 5}{6}$

$t = \frac{60}{6}$

$t = 10$

15. $\frac{2}{5} = \frac{8}{n}$

$2 \cdot n = 5 \cdot 8$

$n = \frac{5 \cdot 8}{2}$

$n = \frac{40}{2}$

$n = 20$

17. $\frac{n}{15} = \frac{10}{30}$

$30 \cdot n = 15 \cdot 10$

$n = \frac{15 \cdot 10}{30}$

$n = \frac{150}{30}$

$n = 5$

19. $\frac{16}{12} = \frac{24}{x}$

$16 \cdot x = 12 \cdot 24$

$x = \frac{12 \cdot 24}{16}$

$x = \frac{288}{16}$

$x = 18$

21. $\frac{t}{0.16} = \frac{0.15}{0.40}$

$0.40 \times t = 0.16 \times 0.15$

$t = \frac{0.16 \times 0.15}{0.40}$

$t = \frac{0.024}{0.40}$

$t = 0.06$

23. $\frac{25}{100} = \frac{n}{20}$

$25 \cdot 20 = 100 \cdot n$

$\frac{25 \cdot 20}{100} = n$

$\frac{500}{100} = n$

$5 = n$

25. $\frac{186{,}000 \text{ mi}}{1 \text{ sec}} = 186{,}000 \frac{\text{mi}}{\text{sec}}$

27. $\frac{4.6 \text{ km}}{2 \text{ hr}} = 2.3 \frac{\text{km}}{\text{hr}}$

29. Let A represent the earned run average.

Earned runs ⟶ $\frac{A}{9} = \frac{71}{179}$ ⟵ Earned runs
Innings ⟶ ⟵ Innings

Solve: $179 \cdot A = 9 \cdot 71$

$A = \frac{9 \cdot 71}{179}$

$A = \frac{639}{179}$

$A \approx 3.57$

Tom Browning's earned run average was about 3.57.

31. Let D represent the number of deer in the game preserve.

Deer tagged originally ⟶ $\frac{318}{D} = \frac{56}{168}$ ⟵ Tagged deer caught later

Deer in game preserve ⟶ D; 168 ⟵ Deer caught later

Solve: $318 \cdot 168 = 56 \cdot D$

$$\frac{318 \cdot 168}{56} = D$$

$$954 = D$$

We estimate that there are 954 deer in the game preserve.

33. We find the cross products:

$12 \cdot 4 = 48$ $\qquad$ $8 \cdot 6 = 48$

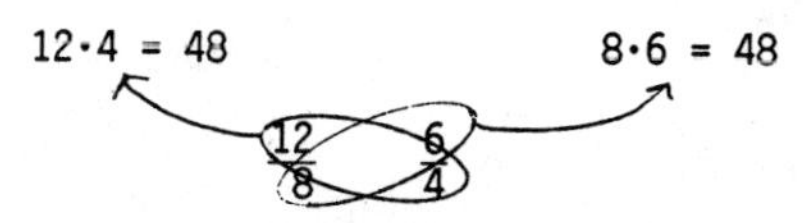

Since the cross products are equal,

$$\frac{12}{8} = \frac{6}{4}.$$

35. $200 \div 4 = 50$

$$\begin{array}{r} 50 \\ 4\overline{)200} \\ \underline{200} \\ 0 \\ \underline{0} \\ 0 \end{array}$$

37. $232 \div 16 = 14.5$

$$\begin{array}{r} 14.5 \\ 16\overline{)232.0} \\ \underline{160} \\ 72 \\ \underline{64} \\ 80 \\ \underline{80} \\ 0 \end{array}$$

39. Use a calculator.

The ratio of people to sheep is

$\frac{13{,}339{,}000}{145{,}304{,}000}$ or about 0.09.

The ratio of sheep to people is

$\frac{145{,}304{,}000}{13{,}339{,}000}$ or about 10.89.

Exercise Set 2.7

1. $90\% = \frac{90}{100}$ — A ratio of 90 to 100

$90\% = 90 \times \frac{1}{100}$ — Replacing % by $\times \frac{1}{100}$

$90\% = 90 \times 0.01$ — Replacing % by $\times$ 0.01

3. $12.5\% = \frac{12.5}{100}$ — A ratio of 12.5 to 100

$12.5\% = 12.5 \times \frac{1}{100}$ — Replacing % by $\times \frac{1}{100}$

$12.5\% = 12.5 \times 0.01$ — Replacing % by $\times$ 0.01

5. 59.01% → 59.01 (Drop percent symbol) → .59.01 (Move decimal point 2 places to the left) → 59.01% = 0.5901

7. 10% → 10 (Drop percent symbol) → .10. (Move decimal point 2 places to the left) → 10% = 0.10

9. 0.1% → 0.00.1 → 0.1% = 0.001

11. 0.09% → 0.00.09 → 0.09% = 0.0009

13. 90% → 0.90. → 90% = 0.9

15. 10.8% → .10.8 → 10.8% = 0.108

17. 1.00 → 1.00. (Move decimal point 2 places to the right) → 100% (Write a % symbol) → 1.00 = 100%

19. 0.334 → 0.33.4 (Move decimal point 2 places to the right) → 33.4% (Write a % symbol) → 0.334 = 33.4%

21. 0.00.6 → 0.6% → 0.006 = 0.6%

23. 0.01.7 → 1.7% → 0.017 = 1.7%

25. 0.02.5 → 2.5% → 0.025 = 2.5%

27. We multiply by 1 to get 100 in the denominator.

$$\frac{3}{10} = \frac{3}{10} \cdot \frac{10}{10} = \frac{30}{100} = 30\%$$

29. We multiply by 1 to get 100 in the denominator.

$$\frac{1}{2} = \frac{1}{2} \cdot \frac{50}{50} = \frac{50}{100} = 50\%$$

31. Find decimal notation by division.

$$\begin{array}{r} 0.625 \\ 8\overline{)5.000} \\ \underline{48} \\ 20 \\ \underline{16} \\ 40 \\ \underline{40} \\ 0 \end{array}$$

Convert to percent notation.

0.62.5 $\qquad \frac{5}{8} = 62.5\%$ or $62\frac{1}{2}\%$

33. $\frac{2}{5} = \frac{2}{5} \cdot \frac{20}{20} = \frac{40}{100} = 40\%$

35. Find decimal notation by division.

$$\begin{array}{r} 0.666 \\ 3\overline{)2.000} \\ \underline{18} \\ 20 \\ \underline{18} \\ 20 \\ \underline{18} \\ 2 \end{array}$$

We get a repeating decimal: $0.06\overline{6}$

Convert to percent notation.

$0.66.\overline{6}$ $\qquad \frac{2}{3} = 66.\overline{6}\%$ or $66\frac{2}{3}\%$

37. $$\begin{array}{r} 0.166 \\ 6\overline{)1.000} \\ \underline{6} \\ 40 \\ \underline{36} \\ 40 \\ \underline{36} \\ 4 \end{array}$$

We get a repeating decimal: $0.16\overline{6}$

$0.16.\overline{6}$ $\qquad \frac{1}{6} = 16.\overline{6}\%$ or $16\frac{2}{3}\%$

39. $\frac{9}{25} = \frac{9}{25} \cdot \frac{4}{4} = \frac{36}{100} = 36\%$

41. $80\% = \frac{80}{100}$ Definition of percent

$= \frac{4}{5} \cdot \frac{20}{20}$
$= \frac{4}{5}$ Simplifying

43. $62.5\% = \frac{62.5}{100}$ Definition of percent

$= \frac{62.5}{100} \cdot \frac{10}{10}$ Multiplying by 1 to get rid of the decimal point in the numerator

$= \frac{625}{1000}$

$= \frac{5}{8} \cdot \frac{125}{125}$
$= \frac{5}{8}$ Simplifying

45. $33\frac{1}{3}\% = \frac{100}{3}\%$ Converting from mixed numeral to fractional notation

$= \frac{100}{3} \times \frac{1}{100}$ Definition of percent

$= \frac{100}{300}$ Multiplying

$= \frac{1}{3} \cdot \frac{100}{100}$
$= \frac{1}{3}$ Simplifying

47. $16.\overline{6}\% = 16\frac{2}{3}\%$ $\quad \left[16.\overline{6} = 16\frac{2}{3}\right]$

$= \frac{50}{3}\%$ Converting from mixed numeral to fractional notation

$= \frac{50}{3} \times \frac{1}{100}$ Definition of percent

$= \frac{50}{300}$ Multiplying

$= \frac{1}{6} \cdot \frac{50}{50}$
$= \frac{1}{6}$ Simplifying

49. $35\% = \frac{35}{100} = \frac{7}{20} \cdot \frac{5}{5} = \frac{7}{20}$

51. Solve using an equation.

Restate: What is 17% of 160?

Translate: $a = 17\% \times 160$

This tells us what to do. We convert 17% to decimal notation and multiply.

$$\begin{array}{r} 160 \\ \times\ 0.17 \\ \hline 1120 \\ 1600 \\ \hline 27.20 \end{array} \qquad (17\% = 0.17)$$

You would expect 27 bowlers to be left-handed.

53. 

$18,600	$?
100%	5%

First, find the increase. We ask:

What is 5% of $18,600?

$a = 5\% \times 18{,}600$ Translating

We convert to decimal notation and multiply:

$$\begin{array}{r} 18{,}600 \\ \times\ 0.05 \\ \hline 930.00 \end{array}$$

The increase is $930.

The new salary is $18,600 + $930 = $19,530.

55. 

$12,000	
	$?
100%	
70%	30%

First, find the decrease. We ask:

What is 30% of $12,000?

$a = 30\% \times \$12{,}000$ Translating

Convert 30% to decimal notation and multiply.

$$\begin{array}{r} 12{,}000 \\ \times\ 0.3 \\ \hline 3600.0 \end{array}$$

55. (continued)

The decrease is $3600.

The value 1 year later is $12,000 - $3600 = $8400.

57. Solve using an equation.

Restate: 2190 is what percent of 8760?

Translate: $2190 = n\ \% \times 8760$

$2190 = n \times 0.01 \times 8760$

$2190 = n \times 87.60$

$2190 \div 87.60 = n$

$25 = n$

$$\begin{array}{r} 25. \\ 87.6_\wedge \overline{)\,2190.0_\wedge} \\ 1752\,0 \\ \hline 438\,0 \\ 438\,0 \\ \hline 0 \end{array}$$

Thus, 2190 is 25% of 8760.

59.

$200	
$200	$16
100%	
100%	?%

First find the increase by subtracting.

$$\begin{array}{rl} 216 & \text{New amount} \\ -\ 200 & \text{Original amount} \\ \hline 16 & \text{Increase} \end{array}$$

The increase is $16. Now we ask:

$16 is what percent of $200 (the original amount)?

Solve using an equation.

$16 is what percent of $200?

$16 = n\ \% \times 200$ Translating

$16 = n \times 0.01 \times 200$ Using the definition of percent

$16 = n \times 2$

$16 \div 2 = n$ Dividing on both sides by 2

$8 = n$

The percent of increase was 8%.

61.

$70	
$56	$14
100%	
	?%

First find the decrease by subtracting.

$$\begin{array}{r} 70 \\ -\ 56 \\ \hline 14 \end{array}$$

The decrease is $14. Now we ask:

$14 is what percent of $70 (the original price)?

Solve using an equation.

$14 is what percent of $70?

↓ ↓ ↓ ↓ ↓ ↓

$14 = n \; \% \times 70$

$14 = n \times 0.01 \times 70$ Using the definition of percent

$14 = n \times 0.7$

$14 \div 0.7 = n$ Dividing on both sides by 0.7

$20 = n$

$$\begin{array}{r} 2\,0. \\ 0.7_{\wedge}\overline{)1\,4.0_{\wedge}} \\ \underline{1\,4\,0} \\ 0 \\ \underline{0} \\ 0 \end{array}$$

The percent of decrease was 20%.

63. Commission = Commission rate × Sales

C = 20% × 18,450

This tells us what to do. We multiply.

$$\begin{array}{r} 1\,8{,}4\,5\,0 \\ \times\ \ 0.2 \\ \hline 3\,6\,9\,0.0 \end{array}$$ (20% = 0.20 = 0.2)

The commission is $3690.

65. Commission = Commission rate × Sales

C = 7% × 38,000

This tells us what to do. We multiply:

$$\begin{array}{r} 3\,8{,}0\,0\,0 \\ \times\ 0.0\,7 \\ \hline 2\,6\,6\,0.0\,0 \end{array}$$

The commission is $2660.

67. $\frac{5}{9} = 5 \div 9$

$$\begin{array}{r} 0.5\,5\,5 \\ 9\overline{)5.0\,0\,0} \\ \underline{4\,5} \\ 5\,0 \\ \underline{4\,5} \\ 5\,0 \\ \underline{4\,5} \\ 5 \end{array}$$

We get a repeating decimal.

$\frac{5}{9} = 0.\overline{5}$

69. $\frac{11}{12} = 11 \div 12$

$$\begin{array}{r} 0.9\,1\,6\,6 \\ 1\,2\overline{)1\,1.0\,0\,0\,0} \\ \underline{1\,0\,8} \\ 2\,0 \\ \underline{1\,2} \\ 8\,0 \\ \underline{7\,2} \\ 8\,0 \\ \underline{7\,2} \\ 8 \end{array}$$

We get a repeating decimal.

$\frac{11}{12} = 0.91\overline{6}$

71. The new income is 103.1% of the previous income (100% of the previous income plus the increase of 3.1% of the previous income). We ask:

103.1% of what is $47,200?

↓ ↓ ↓ ↓ ↓

$103.1\% \times b = 47{,}200$

We convert to decimal notation and solve the equation.

$1.031 \times b = 47{,}200$

$b = 47{,}200 \div 1.031$

$b = 45{,}780.80$

$$\begin{array}{r} 4\,5\,7\,8\,0\,.7\,9\,5 \\ 1.0\,3\,1_{\wedge}\overline{)4\,7{,}2\,0\,0.0\,0\,0_{\wedge}0\,0\,0} \\ \underline{4\,1\,2\,4\,0\,0\,0\,0} \\ 5\,9\,6\,0\,0\,0\,0 \\ \underline{5\,1\,5\,5\,0\,0\,0} \\ 8\,0\,5\,0\,0\,0 \\ \underline{7\,2\,1\,7\,0\,0} \\ 8\,3\,3\,0\,0 \\ \underline{8\,2\,4\,8\,0} \\ 8\,2\,0\,0 \\ \underline{7\,2\,1\,7} \\ 9\,8\,3\,0 \\ \underline{9\,2\,7\,9} \\ 5\,5\,1\,0 \\ \underline{5\,1\,5\,5} \\ 3\,5\,5 \end{array}$$

71. (continued)

The average annual income in the previous year was $45,780.80.

Exercise Set 2.8

---

1. $\underbrace{3 \times 3 \times 3 \times 3}_{\text{4 factors}} = 3^4$ ← 4 is the exponent; 3 is the base

3. $\underbrace{5 \times 5}_{\text{2 factors}} = 5^2$ ← 2 is the exponent; 5 is the base

5. $\underbrace{7 \cdot 7 \cdot 7 \cdot 7 \cdot 7}_{\text{5 factors}} = 7^5$ ← 5 is the exponent; 7 is the base

7. $5^2 = \underbrace{5 \times 5}_{\text{2 factors}} = 25$

9. $9^5 = \underbrace{9 \times 9 \times 9 \times 9 \times 9}_{\text{5 factors}} = 59{,}049$

11. $10^2 = \underbrace{10 \times 10}_{\text{2 factors}} = 100$

13. $1^4 = \underbrace{1 \times 1 \times 1 \times 1}_{\text{4 factors}} = 1$

15. $(2.3)^2 = \underbrace{2.3 \times 2.3}_{\text{2 factors}} = 5.29$

```
  2.3
× 2.3
  6 9
4 6 0
5.2 9
```

17. $(0.2)^3 = \underbrace{0.2 \times 0.2 \times 0.2}_{\text{3 factors}} = 0.008$

```
  0.2        0.0 4
× 0.2      ×   0.2
0.0 4      0.0 0 8
```

19. $(20.4)^2 = \underbrace{20.4 \times 20.4}_{\text{2 factors}} = 416.16$

```
    2 0.4
  × 2 0.4
    8 1 6
4 0 8 0 0
4 1 6.1 6
```

21. $\left(\frac{3}{8}\right)^2 = \underbrace{\frac{3}{8} \times \frac{3}{8}}_{\text{2 factors}} = \frac{3 \times 3}{8 \times 8} = \frac{9}{64}$

23. Remember: $a = a^1$, for any base a.

$6 = 6^1$

25. Remember: $a = a^1$, for any base a.

$33 = 33^1$

27. Remember: $a^1 = a$, for any base a.

$4^1 = 4$

29. Remember: $a^1 = a$, for any base a.

$15^1 = 15$

31. Remember: $a^0 = 1$, if a is not zero.

$43^0 = 1$

33. Remember: $a^0 = 1$, if a is not zero.

$\left(\frac{11}{12}\right)^0 = 1$

35. $22^2 = 22 \cdot 22 = 484$

37. $24^2 = 24 \cdot 24 = 576$

39. $\left(\frac{1}{5}\right)^2 = \frac{1}{5} \cdot \frac{1}{5} = \frac{1}{25}$

41. $\left(\frac{7}{6}\right)^2 = \frac{7}{6} \cdot \frac{7}{6} = \frac{49}{36}$

43. $\left(\frac{4}{5}\right)^2 = \frac{4}{5} \cdot \frac{4}{5} = \frac{16}{25}$

45. $\sqrt{25} = 5$

The square root of 25 is 5 because $5^2 = 25$.

47. $\sqrt{441} = 21$

The square root of 441 is 21 because $21^2 = 441$.

49. $\sqrt{\frac{512}{200}}$

$= \sqrt{\frac{64}{25}}$ Simplifying the fraction

$= \frac{8}{5}$ Taking the square root of the numerator and the denominator

51. $\sqrt{\frac{10}{40}}$

$= \sqrt{\frac{1}{4}}$ Simplifying the fraction

$= \frac{1}{2}$ Taking the square root of the numerator and the denominator

53. $\sqrt{\frac{10,000}{225}}$

$= \sqrt{\frac{400}{9}}$ Simplifying the fraction

$= \frac{20}{3}$ Taking the square root of the numerator and the denominator

55. $\sqrt{\frac{169}{196}} = \frac{13}{14}$ Taking the square root of the numerator and the denominator

57. Restate: What is 90% of 30,955?

Translate: $a = 90\% \times 30{,}955$

Convert 90% to decimal notation and multiply.

$$\begin{array}{r} 3\,0{,}9\,5\,5 \\ \times \quad 0.9 \\ \hline 2\,7\,8\,5\,9.5 \end{array}$$

27,859.5 sq mi of Maine are forest.

Exercise Set 2.9

1. To find the average, add the numbers. Then divide by the number of addends.

$$\frac{8 + 7 + 15 + 15 + 15 + 12}{6} = \frac{72}{6} = 12$$

The average is 12.

To find the median, first list the numbers in order.

7, 8, 12, 15, 15, 15

Median = 13.5

The median is halfway between 12 and 15.

We find it as follows:

$$\frac{12 + 15}{2} = \frac{27}{2} = 13.5$$

The median is 13.5.

Find the mode:

8, 7, 15, 15, 15, 12

The number that occurs most often is 15. Thus the mode is 15.

3. To find the average, add the numbers. Then divide by the number of addends.

$$\frac{5 + 10 + 15 + 20 + 25 + 30 + 35}{7} = \frac{140}{7} = 20$$

The average is 20.

3. (continued)

Find the median:

5, 10, 15, 20, 25, 30, 35

Median is 20

The middle number is the median. Thus, 20 is the median of the set of numbers.

Find the mode:

5, 10, 15, 20, 25, 30, 35

Each number occurs most often (exactly one time). Thus the modes are 5, 10, 15, 20, 25, 30, and 35.

5. Find the average:

$$\frac{1.2 + 4.3 + 5.7 + 7.4 + 7.4}{5} = \frac{26}{5} = 5.2$$

The average is 5.2.

Find the median.

1.2, 4.3, 5.7, 7.4, 7.4

Median is 5.7

The middle number is the median. Thus, 5.7 is the median of the set of numbers.

Find the mode:

1.2, 4.3, 5.7, 7.4, 7.4

The number that occurs most often is 7.4. Thus the mode is 7.4.

7. Find the average:

$$\frac{\$6.30 + \$7.70 + \$9.40 + \$9.40}{4} = \frac{\$32.80}{4} = \$8.20$$

The average is $8.20.

Find the median:

$6.30, $7.70, $9.40, $9.40

Median is $8.55

The median is halfway between $7.70 and $9.40. We find it as follows:

$$\frac{\$7.70 + \$9.40}{2} = \frac{\$17.10}{2} = \$8.55$$

Find the mode:

$6.30, $7.70, $9.40, $9.40

The number that occurs most often is $9.40. Thus the mode is $9.40.

9. Find the average:

$$\frac{1 + 2 + 3 + 4 + 5}{5} = \frac{15}{5} = 3$$

The average is 3.

Find the median:

1, 2, [3], 4, 5
Median is [3]

The middle number is the median. Thus the median is 3.

Find the mode.

1, 2, 3, 4, 5

Each number occurs most often (exactly one time). Thus the modes are 1, 2, 3, 4, and 5.

11. Find the average price per pound:

$$\frac{\$1.59 + \$1.49 + \$1.69 + \$1.79 + \$1.79}{5} = \frac{\$8.35}{5}$$

$$= \$1.67$$

The average price per pound of hamburger was $1.67.

13. To find the average number of miles per gallon we divide the total number of miles, 779, by the number of gallons, 19.

$$\frac{779}{19} = 41$$

The average was 41 miles per gallon.

15. Find the average:

$$\frac{56{,}668 + 56{,}691 + 55{,}992 + 55{,}995 + 55{,}955 + 56{,}407}{6}$$

$$= \frac{337{,}708}{6} = 56{,}284.\overline{6}$$

The average attendance was 56,285 (rounding to the nearest one).

Find the median:

55,955; 55,992; 55,995; 56,407; 56,668; 56,691
Median is [56,201]

The median is halfway between 55,995 and 56,407.

We find it as follows:

$$\frac{55{,}995 + 56{,}407}{2} = \frac{112{,}402}{2} = 56{,}201$$

The median is 56,201.

Find the mode:

56,668; 56,691; 55,992; 55,995; 55,955; 56,407

Each number occurs most often (exactly one time). Thus the modes are 56,668; 56,691; 55,992; 55,995; 55,955; and 56,407.

17. Find the average weight in kilograms:

$$\frac{113 + 116 + 118 + 118}{4} = \frac{465}{4} = 116.25$$

The average weight is 116.25 kg.

Find the median:

113, 116, 118, 118
Median is [117]

The median is halfway between 116 and 118. We find it as follows:

$$\frac{116 + 118}{2} = \frac{234}{2} = 117$$

The median weight is 117 kg.

19. Find the average:

$$\frac{74 + 72 + 75 + 69}{4} = \frac{290}{4} = 72.5$$

Her average score was 72.5.

Find the median:

69, 72, 74, 75
Median is [73]

The median is halfway between 72 and 74. We find it as follows:

$$\frac{72 + 74}{2} = \frac{146}{2} = 73$$

The median score was 73.

21. To find the GPA we first add the grade point values for each hour taken. This is done by first multiplying the grade point value by the number of hours in the course and then adding as follows:

A ——> 4.00 × 5 = 20
B ——> 3.00 × 4 = 12
B ——> 3.00 × 3 = 9
C ——> 2.00 × 5 = 10
51 (Total)

The total number of hours taken is

5 + 4 + 3 + 5, or 17. We divide 51 by 17.

$$\frac{51}{17} = 3.00$$

The student's grade point average is 3.00.

23. We can find the total of the five scores needed as follows:

90 + 90 + 90 + 90 + 90 = 450

The total of the scores on the first four tests is

90 + 91 + 81 + 92 = 354

Thus the student needs to get at least

450 - 354, or 96

to get an A. We can check this as follows:

$$\frac{90 + 91 + 81 + 92 + 96}{5} = \frac{450}{5} = 90$$

25. The shortest bar represents San Francisco. Thus San Francisco is the least expensive city.

27. We locate 1988 on the Year scale, and then move up until we reach the line. At that point we move to the left to the Estimated sales (millions) scale and read the information we are seeking. Estimate sales in 1988 are 17 million.

29. We are combining amounts.

Percent of soul	plus	Percent of pop/rock	is	Total
↓	↓	↓	↓	↓
12.0%	+	58.1%	=	t

To solve the equation we carry out the addition.

$$\begin{array}{r} 12.0 \\ +\ 58.1 \\ \hline 70.1 \end{array}$$

70.1% of all records sold are either soul or pop/rock.

31. We can see from the graph that the most was spent on gas purchased.

33. We can see from the graph that 4¢ of each dollar was spent on dividends.

35. We can use a proportion.

Snow ⟶ / Water ⟶ $\dfrac{1\frac{1}{2}}{2} = \dfrac{5\frac{1}{2}}{w}$ ⟵ Snow / ⟵ Water

$1\frac{1}{2} \cdot w = 2 \cdot 5\frac{1}{2}$ Cross multiplying

$\frac{3}{2} \cdot w = 2 \cdot \frac{11}{2}$ Converting to fractional notation

$\frac{3}{2} \cdot w = 11$ Multiplying on the right side

$w = 11 \div \frac{3}{2}$ Dividing by $\frac{3}{2}$ on both sides

$w = 11 \cdot \frac{2}{3}$

$w = \frac{22}{3}$

$w = 7\frac{1}{3}$

Thus, $5\frac{1}{2}$ ft of snow will melt to $7\frac{1}{3}$ in. of water.

37. Divide the total by the number of games. Use a calculator.

$$\frac{547}{3} = 182.\overline{3}$$

Drop the amount to the right of the decimal point.

182.[3]

This is the average (182) — Drop this amount (3)

The bowling average is 182.

Exercise Set 3.1

1. Substitute 29 for x: $29 - 6 = 23$
   Substitute 34 for x: $34 - 6 = 28$
   Substitute 47 for x: $47 - 6 = 41$

3. $bh = 6.5(15.4) = 100.1\text{ cm}^2$

5. $rt = 55(4) = 220$ mi

7. $6x = 6 \cdot 7 = 42$

9. $\frac{x}{y} = \frac{9}{3} = 3$

11. $\frac{3p}{q} = \frac{3 \cdot 2}{6} = \frac{6}{6} = 1$

13. $\frac{x + y}{5} = \frac{10 + 20}{5} = \frac{30}{5} = 6$

15. $\frac{x - y}{8} = \frac{20 - 4}{8} = \frac{16}{8} = 2$

17. $b + 6$, or $6 + b$

19. $c - 9$

21. $q + 6$, or $6 + q$

23. $a + b$, or $b + a$

25. $y - x$

27. $w + x$, or $x + w$

29. $n - m$

31. $r + s$, or $s + r$

33. $2x$

35. $5t$

37. Let x represent the number. Then we have 97%x, or 0.97x.

39. The amount is \$29.95 less than d dollars, or \$d - \$29.95.

41.
```
     3
  3⟌ 9
 3⟌2 7
2⟌ 5 4
```
The prime factorization is $2 \cdot 3 \cdot 3 \cdot 3$.

43. $6 = 2 \cdot 3$
    $18 = 2 \cdot 3 \cdot 3$
    The LCM is $2 \cdot 3 \cdot 3$, or 18.

45. $x + 3y$

47. $2x - 3$

Exercise Set 3.2

1. The integer 18 corresponds to 18° above zero, and the integer -2 corresponds to 2° below zero.

3. The integer 750 corresponds to a \$750 deposit, and the integer -125 corresponds to a \$125 withdrawal.

5. The integers 20, -150, and 300 correspond to the interception of the missile, the loss of the starship, and the capture of the base, respectively.

7. The number $\frac{10}{3}$ can be named $3\frac{1}{3}$, or $3.\overline{33}$. The graph is $\frac{1}{3}$ of the way from 3 to 4.

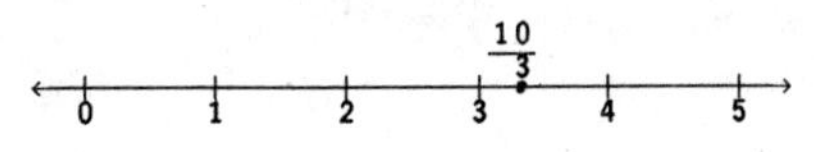

9. The graph of -4.3 is $\frac{3}{10}$ of the way from -4 to -5.

-4.3
-5 -4 -3 -2 -1 0

11. $x < -6$ has the same meaning as $-6 > x$.

13. $y \geqslant -10$ has the same meaning as $-10 \leqslant y$.

15. We first find decimal notation for $\frac{3}{8}$. Since $\frac{3}{8}$ means $3 \div 8$, we divide.
```
   .3 7 5
8⟌3.0 0 0
  2 4
    6 0
    5 6
      4 0
      4 0
        0
```
Thus, $\frac{3}{8} = 0.375$, so $-\frac{3}{8} = -0.375$.

17. $\frac{5}{3}$ means $5 \div 3$, so we divide.
```
  1.6 6. . .
3⟌5.0 0
  3
  2 0
  1 8
    2 0
    1 8
      2
```
We have $\frac{5}{3} = 1.\overline{6}$.

19. $\frac{7}{6}$ means $7 \div 6$, so we divide.

```
   1.1 6 6. . .
 6)7.0 0 0
   6
   1 0
     6
     4 0
     3 6
       4 0
       3 6
         4
```

We have $\frac{7}{6} = 1.1\overline{6}$.

21. $\frac{2}{3}$ means $2 \div 3$, so we divide.

```
   0.6 6 6. . .
 3)2.0 0 0
   1 8
     2 0
     1 8
       2 0
       1 8
         2
```

We have $\frac{2}{3} = 0.\overline{6}$

23. We first find decimal notation for $\frac{1}{2}$. Since $\frac{1}{2}$ means $1 \div 2$, we divide.

```
   0.5
 2)1.0
   1 0
     0
```

Thus, $\frac{1}{2} = 0.5$, so $-\frac{1}{2} = -0.5$.

25. $\frac{1}{10}$ means $1 \div 10$, so we divide.

```
    0.1
 10)1.0
    1 0
      0
```

We have $\frac{1}{10} = 0.1$

27. Since 5 is to the right of 0, we have $5 > 0$.

29. Since -9 is to the left of 5, we have $-9 < 5$.

31. Since -6 is to the left of 6, we have $-6 < 6$.

33. Since -8 is to the left of -5, we have $-8 < -5$.

35. Since -5 is to the right of -11, we have $-5 > -11$.

37. Since -6 is to the left of -5, we have $-6 < -5$.

39. Convert to decimal notation: $-\frac{14}{17} \approx -0.8235$ and $-\frac{27}{35} \approx -0.7714$. Since -0.8235 is to the left of -0.7714, we have $-\frac{14}{17} < -\frac{27}{35}$.

41. $5 \leqslant -5$ is false since neither $5 < -5$ nor $5 = -5$ is true.

43. $-5 \leqslant 7$ is true since $-5 < 7$ is true.

45. The solutions of $y < 0$ are those numbers less than 0. They are shown on the graph by shading all points to the left of 0. The open circle at 0 indicates that 0 is not part of the graph.

$y < 0$

47. In order to be a solution of the inequality $-5 \leqslant x < 2$, a number must be a solution of both $-5 \leqslant x$ and $x < 2$. The solution is graphed as follows:

$-5 \leqslant x < 2$

49. In order to be a solution of the inequality $-5 \leqslant x \leqslant 0$, a number must be a solution of both $-5 \leqslant x$ and $x \leqslant 0$. The solution is graphed as follows:

$-5 \leqslant x \leqslant 0$

51. The distance of -4 from 0 is 4, so $|-4| = 4$.

53. The distance of 325 from 0 is 325, so $|325| = 325$.

55. The distance of $-\frac{10}{7}$ from 0 is $\frac{10}{7}$, so $\left|-\frac{10}{7}\right| = \frac{10}{7}$.

57. The distance of 14.8 from 0 is 14.8, so $|14.8| = 14.8$.

59. Rewrite 7^1, -5, $|-6|$, 4, $|3|$, -100, 0, 1^7, $\frac{14}{4}$ as 7, -5, 6, 4, 3, -100, 0, 1, 3.5, respectively. Listing from the least to the greatest we have -100, -5, 0, 1^7, $|3|$, $\frac{14}{4}$, 4, $|-6|$, 7^1.

Exercise Set 3.3

1. $-9 + 2$ The absolute values are 9 and 2. The difference is 9 - 2, or 7. The negative number has the larger absolute value, so the answer is negative. $-9 + 2 = -7$

3. $-10 + 6$ The absolute values are 10 and 6. The difference is 10 - 6, or 4. The negative number has the larger absolute value, so the answer is negative. $-10 + 6 = -4$

5. $-8 + 8$ A positive and a negative number. The numbers have the same absolute value. The sum is 0. $-8 + 8 = 0$

7. $-3 + (-5)$ Two negatives. Add the absolute values, getting 8. Make the answer negative. $-3 + (-5) = -8$

9. $-7 + 0$ One number is 0. The answer is the other number. $-7 + 0 = -7$

11. $0 + (-27)$ One number is 0. The answer is the other number. $0 + (-27) = -27$

13. $17 + (-17)$ A positive and a negative number. The numbers have the same absolute value. The sum is 0. $17 + (-17) = 0$

15. $-17 + (-25)$ Two negatives. Add the absolute values, getting 42. Make the answer negative. $-17 + (-25) = -42$

17. $18 + (-18)$ A positive and a negative number. The numbers have the same absolute value. The sum is 0. $18 + (-18) = 0$

19. $-18 + 18$ A positive and a negative number. The numbers have the same absolute value. The sum is 0. $-18 + 18 = 0$

21. $8 + (-5)$ The absolute values are 8 and 5. The difference is $8 - 5$, or 3. The positive number has the larger absolute value, so the answer is positive. $8 + (-5) = 3$

23. $-4 + (-5)$ Two negatives. Add the absolute values, getting 9. Make the answer negative. $-4 + (-5) = -9$

25. $13 + (-6)$ The absolute values are 13 and 6. The difference is $13 - 6$, or 7. The positive number has the larger absolute value, so the answer is positive. $13 + (-6) = 7$

27. $-25 + 25$ A positive and a negative number. The numbers have the same absolute value. The sum is 0. $-25 + 25 = 0$

29. $63 + (-18)$ The absolute values are 63 and 18. The difference is $63 - 18$, or 45. The positive number has the larger absolute value, so the answer is positive. $63 + (-18) = 45$

31. $-6.5 + 4.7$ The absolute values are 6.5 and 4.7. The difference is $6.5 - 4.7$, or 1.8. The negative number has the larger absolute value, so the answer is negative. $-6.5 + 4.7 = -1.8$

33. $-2.8 + (-5.3)$ Two negatives. Add the absolute values, getting 8.1. Make the answer negative. $-2.8 + (-5.3) = -8.1$

35. $-\frac{3}{5} + \frac{2}{5}$ The absolute values are $\frac{3}{5}$ and $\frac{2}{5}$. The difference is $\frac{3}{5} - \frac{2}{5}$, or $\frac{1}{5}$. The negative number has the larger absolute value, so the answer is negative. $-\frac{3}{5} + \frac{2}{5} = -\frac{1}{5}$

37. $-\frac{3}{7} + \left(-\frac{5}{7}\right)$ Two negatives. Add the absolute values, getting $\frac{8}{7}$. Make the answer negative.

$-\frac{3}{7} + \left(-\frac{5}{7}\right) = -\frac{8}{7}$

39. $-\frac{5}{8} + \frac{1}{4}$ The absolute values are $\frac{5}{8}$ and $\frac{1}{4}$. The difference is $\frac{5}{8} - \frac{2}{8}$, or $\frac{3}{8}$. The negative number has the larger absolute value, so the answer is negative. $-\frac{5}{8} + \frac{1}{4} = -\frac{3}{8}$

41. $-\frac{3}{7} + \left(-\frac{2}{5}\right)$ Two negatives. Add the absolute values, getting $\frac{15}{35} + \frac{14}{35}$, or $\frac{29}{35}$. Make the answer negative. $-\frac{3}{7} + \left(-\frac{2}{5}\right) = -\frac{29}{35}$

43. $-\frac{7}{16} + \frac{7}{8}$ The absolute values are $\frac{7}{16}$ and $\frac{7}{8}$. The difference is $\frac{14}{16} - \frac{7}{16}$, or $\frac{7}{16}$. The positive number has the larger absolute value, so the answer is positive. $-\frac{7}{16} + \frac{7}{8} = \frac{7}{16}$

45. $75 + (-14) + (-17) + (-5)$

a) $-14 + (-17) + (-5) = -36$ Adding the negative numbers

b) $75 + (-36) = 39$ Adding the results

47. $-44 + \left(-\frac{3}{8}\right) + 95 + \left(-\frac{5}{8}\right)$

a) $-44 + \left(-\frac{3}{8}\right) + \left(-\frac{5}{8}\right) = -45$ Adding the negative numbers

b) $-45 + 95 = 50$ Adding the results

49. We add from left to right.

$$\begin{aligned} & 98 + (-54) + 113 + (-998) + 44 + (-612) \\ = {} & 44 + 113 + (-998) + 44 + (-612) \\ = {} & 157 + (-998) + 44 + (-612) \\ = {} & -841 + 44 + (-612) \\ = {} & -797 + (-612) \\ = {} & -1409 \end{aligned}$$

51. The additive inverse of 24 is -24 because $24 + (-24) = 0$.

53. The additive inverse of -26.9 is 26.9 because $-26.9 + 26.9 = 0$.

55. If $x = 9$, then $-x = -(9) = -9$. (The additive inverse of 9 is -9.)

57. If $x = -\frac{14}{3}$, then $-x = -\left(-\frac{14}{3}\right) = \frac{14}{3}$.

[The additive inverse of $-\frac{14}{3}$ is $\frac{14}{3}$.]

59. If $x = -65$, then $-(-x) = -[-(-65)] = -65$

(The opposite of the opposite of -65 is -65.)

61. If $x = \frac{5}{3}$, then $-(-x) = -\left(-\frac{5}{3}\right) = \frac{5}{3}$.

[The opposite of the opposite of $\frac{5}{3}$ is $\frac{5}{3}$.]

63. $-(-14) = 14$

65. $-(10) = -10$

67. When x is positive, the inverse of x, -x, is negative.

69. If n is positive, -n is negative. Thus -n + m, the sum of two negatives, is negative.

Exercise Set 3.4

1. $3 - 7 = 3 + (-7) = -4$

3. $0 - 7 = 0 + (-7) = -7$

5. $-8 - (-2) = -8 + 2 = -6$

7. $-10 - (-10) = -10 + 10 = 0$

9. $12 - 16 = 12 + (-16) = -4$

11. $20 - 27 = 20 + (-27) = -7$

13. $-9 - (-3) = -9 + 3 = -6$

15. $-40 - (-40) = -40 + 40 = 0$

17. $7 - 7 = 7 + (-7) = 0$

19. $7 - (-7) = 7 + 7 = 14$

21. $8 - (-3) = 8 + 3 = 11$

23. $-6 - 8 = -6 + (-8) = -14$

25. $-4 - (-9) = -4 + 9 = 5$

27. $2 - 9 = 2 + (-9) = -7$

29. $-6 - (-5) = -6 + 5 = -1$

31. $8 - (-10) = 8 + 10 = 18$

33. $0 - 5 = 0 + (-5) = -5$

35. $-5 - (-2) = -5 + 2 = -3$

37. $-7 - 14 = -7 + (-14) = -21$

39. $0 - (-5) = 0 + 5 = 5$

41. $-8 - 0 = -8 + 0 = -8$

43. $7 - (-5) = 7 + 5 = 12$

45. $2 - 25 = 2 + (-25) = -23$

47. $-42 - 26 = -42 + (-26) = -68$

49. $-71 - 2 = -71 + (-2) = -73$

51. $24 - (-92) = 24 + 92 = 116$

53. $-50 - (-50) = -50 + 50 = 0$

55. $\frac{3}{8} - \frac{5}{8} = \frac{3}{8} + \left(-\frac{5}{8}\right) = -\frac{2}{8} = -\frac{1}{4}$

57. $\frac{3}{4} - \frac{2}{3} = \frac{9}{12} - \frac{8}{12} = \frac{9}{12} + \left(-\frac{8}{12}\right) = \frac{1}{12}$

59. $-\frac{3}{4} - \frac{2}{3} = -\frac{9}{12} - \frac{8}{12} = -\frac{9}{12} + \left(-\frac{8}{12}\right) = -\frac{17}{12}$

61. $-\frac{5}{8} - \left(-\frac{3}{4}\right) = -\frac{5}{8} - \left(-\frac{6}{8}\right) = -\frac{5}{8} + \frac{6}{8} = \frac{1}{8}$

63. $6.1 - (-13.8) = 6.1 + 13.8 = 19.9$

65. $-3.2 - 5.8 = -3.2 + (-5.8) = -9$

67. $0.99 - 1 = 0.99 + (-1) = -0.01$

69. $-79 - 114 = -79 + (-114) = -193$

71. $0 - (-500) = 0 + 500 = 500$

73. $-2.8 - 0 = -2.8 + 0 = -2.8$

75. $7 - 10.53 = 7 + (-10.53) = -3.53$

77. $\frac{1}{6} - \frac{2}{3} = \frac{1}{6} - \frac{4}{6} = \frac{1}{6} + \left(-\frac{4}{6}\right) = -\frac{3}{6} = -\frac{1}{2}$

79. $-\frac{4}{7} - \left(-\frac{10}{7}\right) = -\frac{4}{7} + \frac{10}{7} = \frac{6}{7}$

81. $-\frac{7}{10} - \frac{10}{15} = -\frac{21}{30} - \frac{20}{30} = -\frac{21}{30} + \left(-\frac{20}{30}\right) = -\frac{41}{30}$

83. $\frac{1}{13} - \frac{1}{12} = \frac{12}{156} - \frac{13}{156} = \frac{12}{156} + \left(-\frac{13}{156}\right) = -\frac{1}{156}$

85. $18 - (-15) - 3 - (-5) + 2 =$
$18 + 15 + (-3) + 5 + 2 = 37$

87. $-31 + (-28) - (-14) - 17 =$
$(-31) + (-28) + 14 + (-17) = -62$

89. $-34 - 28 + (-33) - 44 =$
$(-34) + (-28) + (-33) + (-44) = -139$

91. $-93 - (-84) - 41 - (-56) =$
$(-93) + 84 + (-41) + 56 = 6$

93. $-5 - (-30) + 30 + 40 - (-12) =$
$(-5) + 30 + 30 + 40 + 12 = 107$

95. $132 - (-21) + 45 - (-21) = 132 + 21 + 45 + 21 =$
219

97. We subtract the amount borrowed from the original amount of assets:

$619.46 - $950 = $619.46 + (-$950) = -$330.54

Your total assets now are -$330.54.

99. The number -$215.50 represents the debt. We subtract to find the answer:

$y - (-$215.50) = $y + $215.50

You will need $y + $215.50.

101. We draw a picture of the situation.

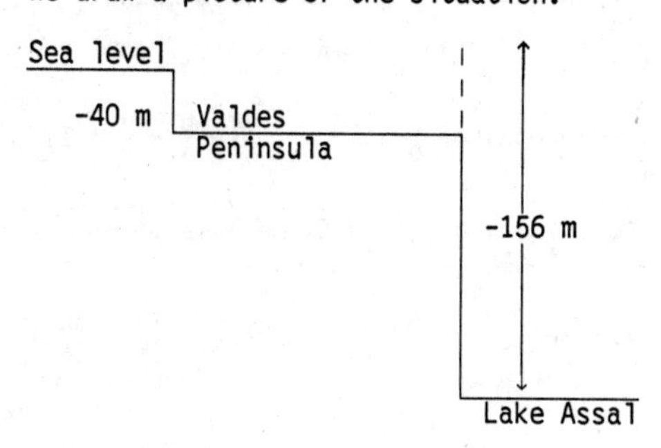

To find how much lower Lake Assal is, we subtract:

$-40 - (-156) = -40 + 156 = 116$

Lake Assal is 116 m lower than the Valdes Peninsula.

103. $5^3 = 5 \times 5 \times 5 = 125$

105. Use a calculator to do this exercise.

$123{,}907 - 433{,}789 = -309{,}882$

107. False. $3 - 0 = 3$, $0 - 3 = -3$, $3 - 0 \neq 0 - 3$

109. True

111. True by definition of additive inverses.

Exercise Set 3.5

---

1. -16

3. -42

5. -24

7. -72

9. 16

11. 42

13. -120

15. -238

17. 1200

19. 98

21. -72

23. -12.4

25. 24

27. 21.7

29. $\frac{2}{3} \cdot \left(-\frac{3}{5}\right) = -\left(\frac{2 \cdot 3}{3 \cdot 5}\right) = -\left(\frac{2}{5} \cdot \frac{3}{3}\right) = -\frac{2}{5}$

31. $-\frac{3}{8} \cdot \left(-\frac{2}{9}\right) = \frac{3 \cdot 2 \cdot 1}{4 \cdot 2 \cdot 3 \cdot 3} = \frac{1}{12}$

33. -17.01

35. $-\frac{5}{9} \cdot \frac{3}{4} = -\frac{5 \cdot 3}{3 \cdot 3 \cdot 4} = -\frac{5}{12}$

37. $7 \cdot (-4) \cdot (-3) \cdot 5 = 7 \cdot 12 \cdot 5 = 7 \cdot 60 = 420$

39. $-\frac{2}{3} \cdot \frac{1}{2} \cdot \left(-\frac{6}{7}\right) = -\frac{2}{6} \cdot \left(-\frac{6}{7}\right) = \frac{2 \cdot 6}{7 \cdot 6} = \frac{2}{7}$

41. $-3 \cdot (-4) \cdot (-5) = 12 \cdot (-5) = -60$

43. $-2 \cdot (-5) \cdot (-3) \cdot (-5) = 10 \cdot 15 = 150$

45. $-\frac{2}{45}$

47. $-7 \cdot (-21) \cdot 13 = 147 \cdot 13 = 1911$

49. $-4 \cdot (-1.8) \cdot 7 = (7.2) \cdot 7 = 50.4$

51. $-\frac{1}{9} \cdot \left(-\frac{2}{3}\right) \cdot \left(\frac{5}{7}\right) = \frac{2}{27} \cdot \frac{5}{7} = \frac{10}{189}$

53. $4 \cdot (-4) \cdot (-5) \cdot (-12) = -16 \cdot (60) = -960$

55. $0.07 \cdot (-7) \cdot 6 \cdot (-6) = 0.07 \cdot 6 \cdot (-7) \cdot (-6) = 0.42 \cdot (42) = 17.64$

57. $\left(-\frac{5}{6}\right)\left(\frac{1}{8}\right)\left(-\frac{3}{7}\right)\left(-\frac{1}{7}\right) = \left(-\frac{5}{48}\right)\left(\frac{3}{49}\right) = -\frac{5 \cdot 3}{16 \cdot 3 \cdot 49} = -\frac{5}{784}$

59. 0, The product of 0 and any real number is 0.

61. $(-8)(-9)(-10) = 72(-10) = -720$

63. $(-6)(-7)(-8)(-9)(-10) = 42 \cdot 72 \cdot (-10) = 3024 \cdot (-10) = -30{,}240$

65. $(-3x)^2 = (-3 \cdot 7)^2$ Substituting

$= (-21)^2$ Multiplying inside the parentheses

$= (-21)(-21)$ Evaluating the power

$= 441$

$-3x^2 = -3(7)^2$ Substituting

$= -3 \cdot 49$ Evaluating the power

$= -147$

67. When $x = 2$: $5x^2 = 5(2)^2$ Substituting
$= 5\cdot 4$ Evaluating the power
$= 20$

When $x = -2$: $5x^2 = 5(-2)^2$ Substituting
$= 5\cdot 4$ Evaluating the power
$= 20$

69. $-6[(-5) + (-7)] = -6[-12] = 72$

71. $-(3^5)\cdot[-(2^3)] = -243[-8] = 1944$

73. $|(-2)^3 + 4^2| - (2 - 7)^2 = |(-2)^3 + 4^2| - (-5)^2 =$
$|-8 + 16| - 25 = |8| - 25 = 8 - 25 = -17$

75. a) m and n have different signs;

b) either m or n is zero;

c) m and n have the same sign

Exercise Set 3.6

---

1. $36 \div (-6) = -6$ Check: $-6\cdot(-6) = 36$

3. $\frac{26}{-2} = -13$ Check: $-13\cdot(-2) = 26$

5. $\frac{-16}{8} = -2$ Check: $-2\cdot 8 = -16$

7. $\frac{-48}{-12} = 4$ Check: $4(-12) = -48$

9. $\frac{-72}{9} = -8$ Check: $-8\cdot 9 = -72$

11. $-100 \div (-50) = 2$ Check: $2(-50) = -100$

13. $-108 \div 9 = -12$ Check: $9(-12) = -108$

15. $\frac{200}{-25} = -8$ Check: $-8(-25) = 200$

17. Undefined

19. $\frac{88}{-9} = -\frac{88}{9}$ Check: $-\frac{88}{9}\cdot(-9) = 88$

21. The reciprocal of $\frac{15}{7}$ is $\frac{7}{15}$ because $\frac{15}{7}\cdot\frac{7}{15} = 1$.

23. The reciprocal of $-\frac{47}{13}$ is $-\frac{13}{47}$ because $\left(-\frac{47}{13}\right)\cdot\left(-\frac{13}{47}\right) = 1$.

25. The reciprocal of 13 is $\frac{1}{13}$ because $13\cdot\frac{1}{13} = 1$.

27. The reciprocal of 4.3 is $\frac{1}{4.3}$ because $4.3\cdot\frac{1}{4.3} = 1$.

29. The reciprocal of $-\frac{1}{7.1}$ is -7.1 because $\left(-\frac{1}{7.1}\right)(-7.1) = 1$.

31. The reciprocal of $\frac{p}{q}$ is $\frac{q}{p}$ because $\frac{p}{q}\cdot\frac{q}{p} = 1$.

33. The reciprocal of $\frac{1}{4y}$ is $4y$ because $\frac{1}{4y}\cdot 4y = 1$.

35. The reciprocal of $\frac{2a}{3b}$ is $\frac{3b}{2a}$ because $\frac{2a}{3b}\cdot\frac{3b}{2a} = 1$.

37. $3\cdot\frac{1}{19}$

39. $6\left(-\frac{1}{13}\right)$

41. $13.9\left(-\frac{1}{1.5}\right)$

43. $x\cdot y$

45. $(3x + 4)\frac{1}{5}$

47. $(5a - b)\left(\frac{1}{5a + b}\right)$

49. $\frac{3}{4} \div \left(-\frac{2}{3}\right) = \frac{3}{4}\cdot\left(-\frac{3}{2}\right) = -\frac{9}{8}$

51. $-\frac{5}{4} \div \left(-\frac{3}{4}\right) = -\frac{5}{4}\cdot\left(-\frac{4}{3}\right) = \frac{20}{12} = \frac{5\cdot 4}{3\cdot 4} = \frac{5}{3}$

53. $-\frac{2}{7} \div \left(-\frac{4}{9}\right) = -\frac{2}{7}\cdot\left(-\frac{9}{4}\right) = \frac{18}{28} = \frac{9\cdot 2}{14\cdot 2} = \frac{9}{14}$

55. $-\frac{3}{8} \div \left(-\frac{8}{3}\right) = -\frac{3}{8}\cdot\left(-\frac{3}{8}\right) = \frac{9}{64}$

57. $-6.6 \div 3.3 = -2$ Do the long division. Make the answer negative.

59. $\frac{-11}{-13} = \frac{11}{13}$ The opposite of a number divided by the opposite of another number is the quotient of the two numbers.

61. $\frac{48.6}{-3} = -16.2$ Do the long division. Make the answer negative.

63. $\frac{-9}{17 - 17} = \frac{-9}{0}$
Division by zero is undefined.

65. $\frac{264}{468} = \frac{4\cdot 66}{4\cdot 117} = \frac{\cancel{4}\cdot\cancel{3}\cdot 22}{\cancel{4}\cdot\cancel{3}\cdot 39} = \frac{22}{39}$

67. $2^3 - 5\cdot 3 + 8\cdot 10 \div 2$
$= 8 - 5\cdot 3 + 8\cdot 10 \div 2$ Evaluating the power
$= 8 - 15 + 80 \div 2$ Multiplying and dividing
$= 8 - 15 + 40$ in order from left to right
$= -7 + 40$ Adding and subtracting in
$= 33$ order from left to right

69. We find $\frac{1}{-10.5}$ using a calculator:
$\frac{1}{-10.5} = -\frac{1}{10.5}$, or $-0.\overline{095238}$

71. There are none. For $a \neq 0$, $-a$ and a have opposite signs but a and $\frac{1}{a}$ have the same sign. For $a = 0$, $-a = a$, but $\frac{1}{a}$ is undefined.

73. $-n$ is positive and m is negative, so $\frac{-n}{m}$ is the quotient of a positive and a negative number and, thus, is negative.

75. $\frac{-n}{m}$ is negative (see Exercise 73), so $-\left(\frac{-n}{m}\right)$ is the additive inverse of a negative number and, thus, is positive.

77. $-n$ and $-m$ are both positive, so $\frac{-n}{-m}$ is the quotient of two positive numbers and, thus, is positive. Then, $-\left(\frac{-n}{-m}\right)$ is the additive inverse of a positive number and, thus, is negative.

Exercise Set 3.7

1. Since $10 = 5\cdot 2$, we multiply by $\frac{2}{2}$:

$$\frac{3x}{5}\cdot\frac{2}{2} = \frac{6x}{10}$$

3. Since $4x = 4\cdot x$, we multiply by $\frac{x}{x}$:

$$\frac{3}{4}\cdot\frac{x}{x} = \frac{3x}{4x}$$

5. $\frac{25x}{15x} = \frac{5\cdot 5x}{3\cdot 5x}$ We look for the largest common factor of the numerator and denominator and factor each.

$= \frac{5}{3}\cdot\frac{5x}{5x}$ Factoring the expression

$= \frac{5}{3}\cdot 1$ $\left(\frac{5x}{5x} = 1\right)$

$= \frac{5}{3}$

7. $\frac{100}{25x} = \frac{4\cdot 25}{x\cdot 25}$ Factoring numerator and denominator

$= \frac{4}{x}\cdot\frac{25}{25}$ Factoring the expression

$= \frac{4}{x}\cdot 1$ $\left(\frac{25}{25} = 1\right)$

$= \frac{4}{x}$

9. $m + 9 = 9 + m$ Commutative law of addition

11. $pq = qp$ Commutative law of multiplication

13. $8 + ab = ab + 8$ Commutative law of addition

or

$8 + ab = ab + 8 = ba + 8$ Commutative laws of addition and multiplication

or

$8 + ab = 8 + ba$ Commutative law of multiplication

15. $xy + z = yx + z$ Commutative law of multiplication

or

$xy + z = yx + z = z + yx$ Commutative laws of multiplication and addition

or

$xy + z = z + xy$ Commutative law of addition

17. $m + (n + 2) = (m + n) + 2$ Associative law of addition

19. $(7\cdot x)\cdot y = 7\cdot(x\cdot y)$ Associative law of multiplication

21. $(a + b) + 8 = a + (b + 8)$ Associative law
$= a + (8 + b)$ Commutative law

$(a + b) + 8 = a + (b + 8)$ Associative law
$= a + (8 + b)$ Commutative law
$= (a + 8) + b$ Associative law

$(a + b) + 8 = (b + a) + 8$ Commutative law
$= b + (a + 8)$ Associative law

Other answers are possible.

23. $7\cdot(a\cdot b) = 7\cdot(b\cdot a)$ Commutative law
$= (7\cdot b)\cdot a$ Associative law

$7\cdot(a\cdot b) = (7\cdot a)\cdot b$ Associative law
$= b\cdot(7\cdot a)$ Commutative law
$= b\cdot(a\cdot 7)$ Commutative law

$7\cdot(a\cdot b) = 7\cdot(b\cdot a)$ Commutative law
$= (b\cdot a)\cdot 7$ Commutative law

25. Substitute -2 for x and 3 for y.
$xy + x = -2\cdot 3 + (-2) = -6 + (-2) = -8$

27. Substitute -2 for x, 3 for y, and -4 for z.
$(x + y)z = (-2 + 3)(-4) = 1(-4) = -4$

29. Substitute -2 for x, 3 for y, and -4 for z.
$xy - xz = -2\cdot 3 - (-2)(-4) = -6 - 8 = -14$

31. Substitute -2 for x.
$3x + 7 = 3(-2) + 7 = -6 + 7 = 1$

33. Substitute 120 for P, 6% for r, and 1 for t.
$P(1 + rt) = 120(1 + 6\%\cdot 1)$
$= 120(1 + 0.06)$
$= 120(1.06)$
$= 127.2$

The value of the account is $127.20.

35. $3(a + 1)$
$= 3 \cdot a + 3 \cdot 1$
$= 3a + 3$

37. $4(x - y)$
$= 4 \cdot x - 4 \cdot y$
$= 4x - 4y$

39. $-5(2a + 3b)$
$= -5 \cdot 2a + (-5) \cdot 3b$
$= -10a - 15b$

41. $2a(b - c + d)$
$= 2a \cdot b - 2a \cdot c + 2a \cdot d$
$= 2ab - 2ac + 2ad$

43. $2\pi r(h + 1)$
$= 2\pi r \cdot h + 2\pi r \cdot 1$
$= 2\pi rh + 2\pi r$

45. $\frac{1}{2} h(a + b)$
$= \frac{1}{2} h \cdot a + \frac{1}{2} h \cdot b$
$= \frac{1}{2} ha + \frac{1}{2} hb$

47. $4a - 5b + 6 = 4a + (-5b) + 6$
The terms are $4a$, $-5b$, and 6.

49. $2x - 3y - 2z = 2x + (-3y) + (-2z)$
The terms are $2x$, $-3y$, and $-2z$.

51. $18x + 18y$
$= 18 \cdot x + 18 \cdot y$
$= 18(x + y)$

53. $9p - 9$
$= 9 \cdot p - 9 \cdot 1$
$= 9(p - 1)$

55. $7x - 21$
$= 7 \cdot x - 7 \cdot 3$
$= 7(x - 3)$

57. $xy + x$
$= x \cdot y + x \cdot 1$
$= x(y + 1)$

59. $2x - 2y + 2z$
$= 2 \cdot x - 2 \cdot y + 2 \cdot z$
$= 2(x - y + z)$

61. $3x + 6y - 3$
$= 3 \cdot x + 3 \cdot 2y - 3 \cdot 1$
$= 3(x + 2y - 1)$

63. $ab + ac - ad$
$= a \cdot b + a \cdot c - a \cdot d$
$= a(b + c - d)$

65. $\frac{1}{4} \pi rr + \frac{1}{4} \pi rs$
$= \frac{1}{4} \pi r \cdot r + \frac{1}{4} \pi r \cdot s$
$= \frac{1}{4} \pi r(r + s)$

67. $4a + 5a$
$= (4 + 5)a$
$= 9a$

69. $8b - 11b$
$= (8 - 11)b$
$= -3b$

71. $14y + y$
$= 14y + 1y$
$= (14 + 1)y$
$= 15y$

73. $12a - a$
$= 12a - 1a$
$= (12 - 1)a$
$= 11a$

75. $t - 9t$
$= 1t - 9t$
$= (1 - 9)t$
$= -8t$

77. $5x - 3x + 8x$
$= (5 - 3 + 8)x$
$= 10x$

79. $5x - 8y + 3x$
$= (5 + 3)x - 8y$
$= 8x - 8y$

81. $7c + 8d - 5c + 2d$
$= (7 - 5)c + (8 + 2)d$
$= 2c + 10d$

83. $4x - 7 + 18x + 25$
$= (4 + 18)x + (-7 + 25)$
$= 22x + 18$

85. $1.3x + 1.4y - 0.11x - 0.47y$
$= (1.3 - 0.11)x + (1.4 - 0.47)y$
$= 1.19x + 0.93y$

87. $\frac{2}{3}a + \frac{5}{6}b - 27 - \frac{4}{5}a - \frac{7}{6}b$

$= (\frac{2}{3} - \frac{4}{5})a + (\frac{5}{6} - \frac{7}{6})b - 27$

$= (\frac{10}{15} - \frac{12}{15})a + (-\frac{2}{6})b - 27$

$= -\frac{2}{15}a - \frac{1}{3}b - 27$

89. $2\ell + 2w$

$= 2\cdot\ell + 2\cdot w$

$= 2(\ell + w)$

91. $-(-4b) = -1\cdot(-4b)$

$= [-1\cdot(-4)]b$

$= 4b$

93. $-(a + 2) = -1(a + 2)$

$= -1\cdot a + (-1)\cdot 2$

$= -a + (-2)$

$= -a - 2$

95. $-(b - 3) = -1(b - 3)$

$= -1\cdot b - (-1)\cdot 3$

$= -b + [-(-1)3]$

$= -b + 3$, or $3 - b$

97. $-(t - y) = -1(t - y)$

$= -1\cdot t - (-1)\cdot y$

$= -t + [-(-1)y]$

$= -t + y$, or $y - t$

99. $-(a + b + c)$

$= -a - b - c$ (Changing the sign of every term inside parentheses)

101. $-(8x - 6y + 13)$

$= -8x + 6y - 13$ (Changing the sign of every term inside parentheses)

103. $-(-2c + 5d - 3e + 4f)$

$= 2c - 5d + 3e - 4f$ (Changing the sign of every term inside parentheses)

105. $-(-1.2x + 56.7y - 34z - \frac{1}{4})$

$= 1.2x - 56.7y + 34z + \frac{1}{4}$ (Changing the sign of every term inside parentheses)

107. $a + (2a + 5)$

$= a + 2a + 5$

$= 3a + 5$

109. $4m - (3m - 1)$

$= 4m - 3m + 1$

$= m + 1$

111. $3d - 7 - (5 - 2d)$

$= 3d - 7 - 5 + 2d$

$= 5d - 12$

113. $-2(x + 3) - 5(x - 4)$

$= -2(x + 3) + [-5(x - 4)]$

$= -2x - 6 + [-5x + 20]$

$= -2x - 6 - 5x + 20$

$= -7x + 14$

115. $5x - 7(2x - 3) - 4$

$= 5x + [-7(2x - 3)] - 4$

$= 5x + [-14x + 21] - 4$

$= 5x - 14x + 21 - 4$

$= -9x + 17$

117. $8x - (-3y + 7) + (9x - 11)$

$= 8x + 3y - 7 + 9x - 11$

$= 17x + 3y - 18$

119. $\frac{1}{4}(24x - 8) - \frac{1}{2}(-8x + 6) - 14$

$= \frac{1}{4}(24x - 8) + [-\frac{1}{2}(-8x + 6)] - 14$

$= 6x - 2 + [4x - 3] - 14$

$= 6x - 2 + 4x - 3 - 14$

$= 10x - 19$

121. $-12 - (-19) = -12 + 19 = 7$

123. $-\frac{11}{5} - (-\frac{17}{10})$

$= -\frac{11}{5} + \frac{17}{10}$

$= -\frac{22}{10} + \frac{17}{10}$

$= -\frac{5}{10}$

$= -\frac{1}{2}$

125. The signs are different, so the product is negative.

$-45(20) = -900$

127. The signs are the same, so the product is positive.

$-45(-90) = 4050$

129. $\frac{1}{2}\cdot 0 = 0$, $\frac{0}{2} = 0$ Substituting 0

$\frac{1}{2}(-4) = -2$, $\frac{-4}{2} = -2$ Substituting -4

$\frac{1}{2}(12) = 6$, $\frac{12}{2} = 6$ Substituting 12

The expressions seem to be equivalent.

131. $\frac{5\cdot 0}{9} = 0$, $\frac{5}{9\cdot 0}$ is undefined

The expressions are not equivalent.

Exercise Set 3.8

1. $\underbrace{4\cdot 4\cdot 4\cdot 4\cdot 4}_{\text{5 factors}} = 4^5$

3. $\underbrace{5\cdot 5\cdot 5\cdot 5\cdot 5\cdot 5}_{\text{6 factors}} = 5^6$

5. $\underbrace{m\cdot m\cdot m\cdot m}_{\text{4 factors}} = m^4$

7. $\underbrace{3a\cdot 3a\cdot 3a\cdot 3a}_{\text{4 factors}} = (3a)^4$

9. There are 2 factors of 5, 3 factors of c, and 4 factors of d.

   $5\cdot 5\cdot c\cdot c\cdot c\cdot d\cdot d\cdot d\cdot d = 5^2c^3d^4$

11. $2^5 = 2\cdot 2\cdot 2\cdot 2\cdot 2$, or 32

13. $(-3)^4 = (-3)\cdot(-3)\cdot(-3)\cdot(-3)$, or 81

15. $x^4 = x\cdot x\cdot x\cdot x$, or xxxx

17. $(-4b)^3 = (-4b)(-4b)(-4b)$, or -64bbb

19. $(ab)^4 = ab\cdot ab\cdot ab\cdot ab$

21. $5^1 = 5$ (For any a, $a^1 = a$.)

23. $(3z)^0 = 1$ (For any nonzero a, $a^0 = 1$.)

25. $(\sqrt{8})^0 = 1$ (For any nonzero a, $a^0 = 1$.)

27. $(\frac{7}{8})^1 = \frac{7}{8}$ (For any a, $a^1 = a$.)

29. $x^{-3} = \frac{1}{x^3}$

31. $\frac{1}{a^{-2}} = a^2$

33. $(-11)^{-1} = \frac{1}{(-11)^1}$

35. $\frac{1}{3^4} = 3^{-4}$

37. $\frac{1}{b^3} = b^{-3}$

39. $\frac{1}{(-16)^2} = (-16)^{-2}$

41. $[10 - 3(6 - 1)]$
$= [10 - 3\cdot 5]$
$= [10 - 15]$
$= -5$

43. $9[8 - 7(5 - 2)]$
$= 9[8 - 7\cdot 3]$
$= 9[8 - 21]$
$= 9[-13]$
$= -117$

45. $[5(8 - 6) + 12] - [24 - (8 - 4)]$
$= [5\cdot 2 + 12] - [24 - 4]$
$= [10 + 12] - [24 - 4]$
$= 22 - 20$
$= 2$

47. $[64 \div (-4)] \div (-2)$
$= -16 \div (-2)$
$= 8$

49. $17(-24) + 50$
$= -408 + 50$
$= -358$

51. $(5 + 7)^2 = 12^2 = 144$
$5^2 + 7^2 = 25 + 49 = 74$

53. $2^3 + 2^4 - 20\cdot 30$
$= 8 + 16 - 600$
$= 24 - 600$
$= -576$

55. $5^3 + 36\cdot 72 - (18 + 25\cdot 4)$
$= 5^3 + 36\cdot 72 - (18 + 100)$
$= 5^3 + 36\cdot 72 - 118$
$= 125 + 36\cdot 72 - 118$
$= 125 + 2592 - 118$
$= 2717 - 118$
$= 2599$

57. $(13\cdot 2 - 8\cdot 4)^2$
$= (26 - 32)^2$
$= (-6)^2$
$= 36$

59. $4000\cdot(1 + 0.12)^3$
$= 4000(1.12)^3$
$= 4000(1.404928)$
$= 5619.712$

61. $(20 \cdot 4 + 13 \cdot 8)^2 - (39 \cdot 59)^3$
$= (80 + 104)^2 - (2301)^3$
$= 184^2 - 2301^3$
$\approx 33{,}856 - 12{,}182{,}876{,}900$
$(2301^3 \approx 12{,}182{,}876{,}900)$
$\approx -12{,}182{,}843{,}044$
(Answers may vary due to rounding.)

63. $18 - 2 \cdot 3 - 9$
$= 18 - 6 - 9$
$= 12 - 9$
$= 3$

65. $(18 - 2 \cdot 3) - 9$
$= (18 - 6) - 9$
$= 12 - 9$
$= 3$

67. $[24 \div (-3)] \div \left(-\frac{1}{2}\right)$
$= -8 \div \left(-\frac{1}{2}\right)$
$= -8 \cdot (-2)$
$= 16$

69. $15 \cdot (-24) + 50$
$= -360 + 50$
$= -310$

71. $4 \div (8 - 10)^2 + 1$
$= 4 \div (-2)^2 + 1$
$= 4 \div 4 + 1$
$= 1 + 1$
$= 2$

73. $6^3 + 25 \cdot 71 - (16 + 25 \cdot 4)$
$= 6^3 + 25 \cdot 71 - (16 + 100)$
$= 6^3 + 25 \cdot 71 - 116$
$= 216 + 25 \cdot 71 - 116$
$= 216 + 1775 - 116$
$= 1991 - 116$
$= 1875$

75. $5000 \cdot (1 + 0.16)^3$
$= 5000 \cdot (1.16)^3$
$= 5000(1.560896)$
$= 7804.48$

77. $4 \cdot 5 - 2 \cdot 6 + 4$
$= 20 - 12 + 4$
$= 8 + 4$
$= 12$

79. $4 \cdot (6 + 8)/(4 + 3)$
$= 4 \cdot 14/7$
$= 56/7$
$= 8$

81. $[2 \cdot (5 - 3)]^2$
$= [2 \cdot 2]^2$
$= 4^2$
$= 16$

83. $8(-7) + 6(-5)$
$= -56 - 30$
$= -86$

85. $19 - 5(-3) + 3$
$= 19 + 15 + 3$
$= 34 + 3$
$= 37$

87. $9 \div (-3) + 16 \div 8$
$= -3 + 2$
$= -1$

89. $7 + 10 - (-10 \div 2)$
$= 7 + 10 - (-5)$
$= 7 + 10 + 5$
$= 17 + 5$
$= 22$

91. $3^2 - 8^2$
$= 9 - 64$
$= -55$

93. $20 + 4^3 \div (-8)$
$= 20 + 64 \div (-8)$
$= 20 + (-8)$
$= 12$

95. $-7(3^4) + 18$
$= -7 \cdot 81 + 18$
$= -567 + 18$
$= -549$

97. $8[(6 - 13) - 11]$
$= 8[-7 - 11]$
$= 8[-18]$
$= -144$

99. $256 \div (-32) \div (-4)$
$= -8 \div (-4)$
$= 2$

101. $\dfrac{5^2 - |4^3 - 8|}{9^2 - 2^2 - 1^5}$

$= \dfrac{5^2 - |64 - 8|}{81 - 4 - 1}$

$= \dfrac{5^2 - |56|}{77 - 1}$

$= \dfrac{5^2 - 56}{77 - 1}$

$= \dfrac{25 - 56}{76}$

$= \dfrac{-31}{76}$

$= -\dfrac{31}{76}$

103. $\dfrac{30(8 - 3) - 4(10 - 3)}{10|2 - 6| - 2(5 + 2)}$

$= \dfrac{30\cdot 5 - 4\cdot 7}{10|-4| - 2\cdot 7}$

$= \dfrac{150 - 28}{10\cdot 4 - 2\cdot 7}$

$= \dfrac{122}{40 - 14}$

$= \dfrac{122}{26}$

$= \dfrac{61}{13}$

105. $9a - [7 - 5(7a - 3)]$
$= 9a - [7 - 35a + 15]$
$= 9a - [22 - 35a]$
$= 9a - 22 + 35a$
$= 44a - 22$

107. $5\{-2 + 3[4 - 2(3 + 5)]\}$
$= 5\{-2 + 3[4 - 2(8)]\}$
$= 5\{-2 + 3[4 - 16]\}$
$= 5\{-2 + 3[-12]\}$
$= 5\{-2 - 36\}$
$= 5\{-38\}$
$= -190$

109. $[10(x + 3) - 4] + [2(x - 1) + 6]$
$= [10x + 30 - 4] + [2x - 2 + 6]$
$= 10x + 26 + 2x + 4$
$= 12x + 30$

111. $[7(x + 5) - 19] - [4(x - 6) + 10]$
$= [7x + 35 - 19] - [4x - 24 + 10]$
$= [7x + 16] - [4x - 14]$
$= 7x + 16 - 4x + 14$
$= 3x + 30$

113. $3\{[7(x - 2) + 4] - [2(2x - 5) + 6]\}$
$= 3\{[7x - 14 + 4] - [4x - 10 + 6]\}$
$= 3\{[7x - 10] - [4x - 4]\}$
$= 3\{7x - 10 - 4x + 4\}$
$= 3\{3x - 6\}$
$= 9x - 18$

115. $4\{[5(x - 3) + 2^2] - 3[2(x + 5) - 9^2]\}$
$= 4\{[5(x - 3) + 4] - 3[2(x + 5) - 81]\}$
$= 4\{[5x - 15 + 4] - 3[2x + 10 - 81]\}$
$= 4\{[5x - 11] - 3[2x - 71]\}$
$= 4\{5x - 11 - 6x + 213\}$
$= 4\{-x + 202\}$
$= -4x + 808$

117. $2y + \{8[3(2y - 5) - (8y + 9)] + 6\}$
$= 2y + \{8[6y - 15 - 8y - 9] + 6\}$
$= 2y + \{8[-2y - 24] + 6\}$
$= 2y + \{-16y - 192 + 6\}$
$= 2y + \{-16y - 186\}$
$= 2y - 16y - 186$
$= -14y - 186$

119. $[8(x - 2) + 9x] - \{7[3(2y - 5) - (8y + 7)] + 9\}$
$= [8x - 16 + 9x] - \{7[6y - 15 - 8y - 7] + 9\}$
$= [17x - 16] - \{7[-2y - 22] + 9\}$
$= 17x - 16 - \{-14y - 154 + 9\}$
$= 17x - 16 - \{-14y - 145\}$
$= 17x - 16 + 14y + 145$
$= 17x + 14y + 129$

121. $-3[9(x - 4) + 5x] - 8\{3[5(3y + 4)] - 12\}$
$= -3[9x - 36 + 5x] - 8\{3[15y + 20] - 12\}$
$= -3[14x - 36] - 8\{45y + 60 - 12\}$
$= -42x + 108 - 8\{45y + 48\}$
$= -42x + 108 - 360y - 384$
$= -42x - 360y - 276$

123. Substitute 78 for ℓ and 36 for w.
$P = 2\ell + 2w$, or $2(\ell + w)$
$2\ell + 2w = 2\cdot 78 + 2\cdot 36 = 156 + 72 = 228$
$2(\ell + w) = 2(78 + 36) = 2\cdot 114 = 228$

The perimeter is 228 ft.

$A = \ell w$
$A = 78\cdot 36$, or 2808 ft^2.

125. $(-x)^2 = (-x)(-x) = (-1)x(-1)x = (-1)(-1)x\cdot x = x^2$
Thus, $(-x^2) = x^2$ is true for any real number x.

127. $z - \{2z + [3z - (4z + 5x) - 6z] + 7z\} - 8z$
$= z - \{2z + [3z - 4z - 5x - 6z] + 7z\} - 8z$
$= z - \{2z - 7z - 5x + 7z\} - 8z$
$= z - \{2z - 5x\} - 8z$
$= z - 2z + 5x - 8z$
$= -9z + 5x$

129. $x-\{x+1 - [x+2 - (x-3 - \{x+4 - [x-5 + (x-6)]\})]\}$
$= x-\{x+1 - [x+2 - (x-3 - \{x+4 -[2x-11]\})]\}$
$= x-\{x+1 - [x+2 - (x-3 - \{x+4 - 2x+11\})]\}$
$= x - \{x + 1 - [x + 2 - (x - 3 - \{-x + 15\})]\}$
$= x - \{x + 1 - [x + 2 - (x - 3 + x - 15)]\}$
$= x - \{x + 1 - [x + 2 - (2x - 18)]\}$
$= x - \{x + 1 - [x + 2 - 2x + 18]\}$
$= x - \{x + 1 - [-x + 20]\}$
$= x - \{x + 1 + x - 20\}$
$= x - \{2x - 19\}$
$= x - 2x + 19$
$= -x + 19$

Exercise Set 3.9

1. $5^6 \cdot 5^3 = 5^{6+3} = 5^9$

3. $8^{-6} \cdot 8^2 = 8^{-6+2} = 8^{-4} = \frac{1}{8^4}$

5. $8^{-2} \cdot 8^{-4} = 8^{-2+(-4)} = 8^{-6} = \frac{1}{8^6}$

7. $b^2 \cdot b^{-5} = b^{2+(-5)} = b^{-3} = \frac{1}{b^3}$

9. $a^{-3} \cdot a^4 \cdot a^2 = a^{-3+4+2} = a^3$

11. $(2x)^3(3x)^2$
$= 8x^3 \cdot 9x^2$
$= 8 \cdot 9 \cdot x^3 \cdot x^2$
$= 72x^{3+2}$
$= 72x^5$

13. $(14m^2n^3)(-2m^3n^2)$
$= 14 \cdot (-2) \cdot m^2 \cdot m^3 \cdot n^3 \cdot n^2$
$= -28m^{2+3}n^{3+2}$
$= -28m^5n^5$

15. $(-2x^{-3})(7x^{-8})$
$= -2 \cdot 7 \cdot x^{-3} \cdot x^{-8}$
$= -14x^{-3+(-8)}$
$= -14x^{-11} = -\frac{14}{x^{11}}$

17. $(15x^{4t})(7x^{-6t}) = 15 \cdot 7 \cdot x^{4t} \cdot x^{-6t}$
$= 105x^{4t+(-6t)}$
$= 105x^{-2t}$
$= \frac{105}{x^{2t}}$

19. $\frac{6^8}{6^3} = 6^{8-3} = 6^5$

21. $\frac{4^3}{4^{-2}} = 4^{3-(-2)} = 4^{3+2} = 4^5$

23. $\frac{10^{-3}}{10^6} = 10^{-3-6} = 10^{-3+(-6)} = 10^{-9} = \frac{1}{10^9}$

25. $\frac{9^{-4}}{9^{-6}} = 9^{-4-(-6)} = 9^{-4+6} = 9^2$

27. $\frac{x^{-4n}}{x^{6n}} = x^{-4n-6n} = x^{-10n} = \frac{1}{x^{10n}}$

29. $\frac{w^{-11q}}{w^{-6q}} = w^{-11q-(-6q)} = w^{-11q+6q} = w^{-5q} =$
$\frac{1}{w^{5q}}$

31. $\frac{a^3}{a^{-2}} = a^{3-(-2)} = a^{3+2} = a^5$

33. $\frac{9a^2}{(-3a)^2} = \frac{9a^2}{9a^2} = 1 \quad [(-3a)(-3a) = 9a^2]$

35. $\frac{-24x^6y^7}{18x^{-3}y^9} = \frac{-24}{18} x^{6-(-3)}y^{7-9}$
$= -\frac{24}{18} x^{6+3}y^{-2}$
$= -\frac{4}{3} x^9y^{-2} = -\frac{4x^9}{3y^2}$

37. $\frac{-18x^{-2}y^3}{-12x^{-5}y^5} = \frac{-18}{-12} x^{-2-(-5)}y^{3-5}$
$= \frac{18}{12} x^{-2+5}y^{-2}$
$= \frac{3}{2} x^3y^{-2} = \frac{3x^3}{2y^2}$

39. $(4^3)^2 = 4^{3 \cdot 2} = 4^6$

41. $(8^4)^{-3} = 8^{4(-3)} = 8^{-12} = \frac{1}{8^{12}}$

43. $(6^{-4})^{-3} = 6^{-4(-3)} = 6^{12}$

45. $(3x^2y^2)^3 = 3^3(x^2)^3(y^2)^3$
$= 27x^{2 \cdot 3}y^{2 \cdot 3}$
$= 27x^6y^6$

47. $(-2x^3y^{-4})^{-2} = (-2)^{-2}(x^3)^{-2}(y^{-4})^{-2}$
$= \frac{1}{(-2)^2} x^{3(-2)}y^{-4(-2)}$
$= \frac{1}{4} x^{-6}y^8 = \frac{y^8}{4x^6}$

49. $(-6a^{-2}b^3c)^{-2}$
$= (-6)^{-2}(a^{-2})^{-2}(b^3)^{-2}c^{-2}$
$= \frac{1}{(-6)^2} a^{-2(-2)}b^{3(-2)}c^{-2}$
$= \frac{1}{36} a^4b^{-6}c^{-2} = \frac{a^4}{36b^6c^2}$

51. $\left[\frac{4^{-3}}{3^4}\right]^3 = \frac{(4^{-3})^3}{(3^4)^3} = \frac{4^{-3\cdot3}}{3^{4\cdot3}} = \frac{4^{-9}}{3^{12}} = \frac{1}{4^9\cdot3^{12}}$

53. $\left[\frac{2x^3y^{-2}}{3y^{-3}}\right]^3 = \frac{(2x^3y^{-2})^3}{(3y^{-3})^3}$

$= \frac{2^3(x^3)^3(y^{-2})^3}{3^3(y^{-3})^3}$

$= \frac{8x^9y^{-6}}{27y^{-9}}$

$= \frac{8}{27}x^9y^{-6-(-9)}$

$= \frac{8}{27}x^9y^3$

55. $\left[\frac{125a^2b^{-3}}{5a^4b^{-2}}\right]^{-5} = \left[\frac{5a^4b^{-2}}{125a^2b^{-3}}\right]^5$

$= \left[\frac{a^{4-2}b^{-2-(-3)}}{25}\right]^5$

$= \left[\frac{a^2b}{25}\right]^5$

$= \frac{(a^2)^5(b)^5}{(25)^5} = \frac{a^{2\cdot5}b^{1\cdot5}}{25^{1\cdot5}}$

$= \frac{a^{10}b^5}{25^5}$, or $\frac{a^{10}b^5}{5^{10}}$

$[25^5 = (5^2)^5 = 5^{10}]$

57. $\left[\frac{-6^5y^4z^{-5}}{2^{-2}y^{-2}z^3}\right]^6 = (-1\cdot6^5\cdot2^2y^{4-(-2)}z^{-5-3})^6$

$= (-1\cdot6^5\cdot2^2y^6z^{-8})^6$

$= (-1)^6(6^5)^6(2^2)^6(y^6)^6(z^{-8})^6$

$= 1\cdot6^{5\cdot6}2^{2\cdot6}y^{6\cdot6}z^{-8\cdot6}$

$= 6^{30}2^{12}y^{36}z^{-48}$

$= \frac{6^{30}2^{12}y^{36}}{z^{48}}$

59. $[(-2x^{-4}y^{-2})^{-3}]^{-2} = [(-2)^{-3}(x^{-4})^{-3}(y^{-2})^{-3}]^{-2}$

$= [(-2)^{-3}x^{12}y^6]^{-2}$

$= [(-2)^{-3}]^{-2}(x^{12})^{-2}(y^6)^{-2}$

$= (-2)^6x^{-24}y^{-12}$

$= \frac{64}{x^{24}y^{12}}$

61. $\left[\frac{3a^{-2}b}{5a^{-7}b^5}\right]^{-7} = \left[\frac{5a^{-7}b^5}{3a^{-2}b}\right]^7$

$= \left[\frac{5b^{5-1}}{3a^{-2}a^7}\right]^7$

$= \left[\frac{5b^4}{3a^5}\right]^7$

$= \frac{5^7(b^4)^7}{3^7(a^5)^7}$

$= \frac{5^7b^{28}}{3^7a^{35}}$

63. $\frac{10^{2a+1}}{10^{a+1}} = 10^{2a+1-(a+1)} = 10^{2a+1-a-1} = 10^a$

65. $\frac{9a^{x-2}}{3a^{2x+2}} = \frac{9}{3}\cdot\frac{a^{x-2}}{a^{2x+2}} = 3a^{x-2-(2x+2)} =$

$3a^{x-2-2x-2} = 3a^{-x-4}$

67. $\frac{45x^{2a+4}y^{b+1}}{-9x^{a+3}y^{2+b}} = \frac{45}{-9}\cdot\frac{x^{2a+4}y^{b+1}}{x^{a+3}y^{2+b}} =$

$-5x^{2a+4-(a+3)}y^{b+1-(2+b)} = -5x^{2a+4-a-3}y^{b+1-2-b} =$

$-5x^{a+1}y^{-1}$

69. $(8^x)^{4y} = 8^{x\cdot4y} = 8^{4xy}$

71. $(12^{3-a})^{2b} = 12^{(3-a)(2b)} = 12^{6b-2ab}$

73. $(5x^{a-1}y^{b+1})^{2c} = 5^{2c}x^{(a-1)(2c)}y^{(b+1)(2c)} =$

$5^{2c}x^{2ac-2c}y^{2bc+2c}$, or $(5^2)^cx^{2ac-2c}y^{2bc+2c} =$

$25^cx^{2ac-2c}y^{2bc+2c}$

75. $\frac{4x^{2a+3}y^{2b-1}}{2x^{a+1}y^{b+1}} = \frac{4}{2}\cdot\frac{x^{2a+3}y^{2b-1}}{x^{a+1}y^{b+1}} =$

$2x^{2a+3-(a+1)}y^{2b-1-(b+1)} = 2x^{2a+3-a-1}y^{2b-1-b-1} =$

$2x^{a+2}y^{b-2}$

77. 4.7,000,000,000.

10 places

Large number, so the exponent is positive.

$47{,}000{,}000{,}000 = 4.7 \times 10^{10}$

79. \$9.32,000,000,000.

11 places

Large number, so the exponent is positive.

$\$932{,}000{,}000{,}000 = \$9.32 \times 10^{11}$

81. 0.00000001.6

8 places

Small number, so the exponent is negative.

$0.000000016 = 1.6 \times 10^{-8}$

83. 0.00000000007.

11 places

Small number, so the exponent is negative.

$0.00000000007 = 7 \times 10^{-11}$

85. 6.73000000.

8 places

Positive exponent so the number is large.

$6.73 \times 10^8 = 673{,}000{,}000$

87. 0.00006.6 cm

(decimal point moved 5 places)

Negative exponent, so the number is small.

6.6×10^{-5} cm = 0.000066 cm

89. 0.00000000004.8

(decimal point moved 11 places)

Negative exponent, so the number is small.

$4.8 \times 10^{-11} = 0.000000000048$

91. 0.0000000008.923

(decimal point moved 10 places)

Negative exponent, so the number is small.

$8.923 \times 10^{-10} = 0.0000000008923$

93.
$$(2.3 \times 10^6)(4.2 \times 10^{-11})$$
$$= (2.3 \times 4.2)(10^6 \times 10^{-11})$$
$$= 9.66 \times 10^{-5}$$

95.
$$(2.34 \times 10^{-8})(5.7 \times 10^{-4})$$
$$= (2.34 \times 5.7)(10^{-8} \times 10^{-4})$$
$$= 13.338 \times 10^{-12}$$
$$= (1.3338 \times 10^1) \times 10^{-12}$$
$$= 1.3338 \times (10^1 \times 10^{-12})$$
$$= 1.3338 \times 10^{-11}$$

97.
$$\frac{8.5 \times 10^8}{3.4 \times 10^5}$$
$$= \frac{8.5}{3.4} \times \frac{10^8}{10^5}$$
$$= 2.5 \times 10^3$$

99.
$$\frac{4.0 \times 10^{-6}}{8.0 \times 10^{-3}}$$
$$= \frac{4.0}{8.0} \times \frac{10^{-6}}{10^{-3}}$$
$$= 0.5 \times 10^{-3}$$
$$= (5 \times 10^{-1}) \times 10^{-3}$$
$$= 5 \times (10^{-1} \times 10^{-3})$$
$$= 5 \times 10^{-4}$$

101. Each day we purchase 25,000 new automobiles. There are 365 days in a year, so the total number of cars purchased in one year is
$$(365)(25{,}000)$$
$$= (3.65 \times 10^2)(2.5 \times 10^4)$$
$$= (3.65 \times 2.5)(10^2 \times 10^4)$$
$$= 9.125 \times 10^6.$$

Automobiles per person can be expressed as
$$\frac{9{,}125{,}000}{244{,}000{,}000} = \frac{9.125 \times 10^6}{2.44 \times 10^8}$$
$$= \frac{9.125}{2.44} \times \frac{10^6}{10^8}$$
$$\approx 3.74 \times 10^{-2}.$$

103.
$$\text{Distance} = \text{Speed} \times \text{Time}$$
$$= (300{,}000 \text{ km/sec})(4{,}500{,}000 \text{ sec})$$
$$= (3 \times 10^5 \text{ km/sec}) \times (4.5 \times 10^6 \text{ sec})$$
$$= (3 \times 4.5)(10^5 \times 10^6) \text{ km}$$
$$= 13.5 \times 10^{11} \text{ km}$$
$$= (1.35 \times 10^1) \times 10^{11} \text{ km}$$
$$= 1.35 \times 10^{12} \text{ km}$$

105. Each day Americans eat 6.5 million, or 6.5×10^6, gal of popcorn. There are 365 days in a year, so the amount of popcorn eaten in one year is
$$(6.5 \times 10^6 \text{ gal})(365)$$
$$= (6.5 \times 10^6 \text{ gal})(3.65 \times 10^2)$$
$$= (6.5 \times 3.65)(10^6 \times 10^2) \text{ gal}$$
$$= 23.725 \times 10^8 \text{ gal}$$
$$= (2.37525 \times 10^1) \times 10^8 \text{ gal}$$
$$= 2.3725 \times 10^9 \text{ gal}.$$

107. 1 hour = 60 minutes = 60(60 seconds) = 3600 seconds

The amount of water discharged in one hour is
$$(4{,}200{,}000 \text{ ft}^3\text{/sec}) \times (3600 \text{ sec})$$
$$= (4.2 \times 10^6 \text{ ft}^3\text{/sec}) \times (3.6 \times 10^3 \text{ sec})$$
$$= (4.2 \times 3.6)(10^6 \times 10^3) \text{ ft}^3$$
$$= 15.12 \times 10^9 \text{ ft}^3$$
$$= (1.512 \times 10^1) \times 10^9 \text{ ft}^3$$
$$= 1.512 \times 10^{10} \text{ ft}^3.$$

1 year = 365 days = 365(24 hours) = $365 \cdot 24 \cdot 3600$ seconds = 31,536,000 seconds

The amount of water discharged in one year is
$$(4{,}200{,}000 \text{ ft}^3\text{/sec}) \times (31{,}536{,}000 \text{ sec})$$
$$= (4.2 \times 10^6 \text{ ft}^3\text{/sec}) \times (3.1536 \times 10^7 \text{ sec})$$
$$= (4.2 \times 3.1536)(10^6 \times 10^7) \text{ ft}^3$$
$$= 13.24512 \times 10^{13} \text{ ft}^3$$
$$= (1.324512 \times 10^1) \times 10^{13} \text{ ft}^3$$
$$= 1.324512 \times 10^{14} \text{ ft}^3.$$

109.
$$9x - (-4y + 8) + (10x - 12)$$
$$= 9x + 4y - 8 + 10x - 12$$
$$= 19x + 4y - 20$$

111. $$\frac{(2^{-2})^{-4} \times (2^3)^{-2}}{(2^{-2})^2 \cdot (2^5)^{-3}} = \frac{2^8 \times 2^{-6}}{2^{-4} \cdot 2^{-15}}$$
$$= \frac{2^{8+(-6)}}{2^{-4+(-15)}}$$
$$= \frac{2^2}{2^{-19}}$$
$$= 2^{2-(-19)}$$
$$= 2^{21}$$

113. $$\left[\left(\frac{a^{-2}}{b^7}\right)^{-3} \cdot \left(\frac{a^4}{b^{-3}}\right)^2\right]^{-1} = \left[\frac{(a^{-2})^{-3}}{(b^7)^{-3}} \cdot \frac{(a^4)^2}{(b^{-3})^2}\right]^{-1}$$
$$= \left[\frac{a^6}{b^{-21}} \cdot \frac{a^8}{b^{-6}}\right]^{-1}$$
$$= \left[\frac{a^{6+8}}{b^{-21+(-6)}}\right]^{-1}$$
$$= \left[\frac{a^{14}}{b^{-27}}\right]^{-1}$$
$$= \frac{b^{-27}}{a^{14}}$$
$$= \frac{1}{a^{14}b^{27}}$$

115. $$\left[\frac{(2x^ay^b)^3}{(-2x^ay^b)^2}\right]^2$$
$$= \left[\frac{(2x^ay^b)^3}{(2x^ay^b)^2}\right]^2 \qquad [(-2x^ay^b)^2 = (2x^ay^b)^2]$$
$$= [(2x^ay^b)^{3-2}]^2$$
$$= (2x^ay^b)^2$$
$$= 2^2(x^a)^2(y^b)^2$$
$$= 4x^{2a}y^{2b}$$

Exercise Set 4.1

1. $$\begin{aligned} x + 5 &= 14 \\ x + 5 - 5 &= 14 - 5 \\ x + 0 &= 9 \\ x &= 9 \end{aligned}$$

Check:

$x + 5 = 14$	
$9 + 5$	14
14	TRUE

The solution is 9.

3. $$\begin{aligned} -22 &= x - 18 \\ -22 + 18 &= x - 18 + 18 \\ -4 &= x + 0 \\ -4 &= x \end{aligned}$$

Check:

$-22 = x - 18$	
-22	$-4 - 18$
TRUE	-22

The solution is -4.

5. $$\begin{aligned} -8 + y &= 15 \\ 8 + (-8) + y &= 8 + 15 \\ 0 + y &= 23 \\ y &= 23 \end{aligned}$$

Check:

$-8 + y = 15$	
$-8 + 23$	15
15	TRUE

The solution is 23.

7. $$\begin{aligned} -12 + z &= -51 \\ 12 + (-12) + z &= 12 + (-51) \\ 0 + z &= -39 \\ z &= -39 \end{aligned}$$

The number -39 checks, so it is the solution.

9. $$\begin{aligned} p - 2.96 &= 83.9 \\ p - 2.96 + 2.96 &= 83.9 + 2.96 \\ p + 0 &= 86.86 \\ p &= 86.86 \end{aligned}$$

The number 86.86 checks, so it is the solution.

11. $$\begin{aligned} -\frac{3}{8} + x &= -\frac{5}{24} \\ \frac{3}{8} + \left(-\frac{3}{8}\right) + x &= \frac{3}{8} + \left(-\frac{5}{24}\right) \\ 0 + x &= \frac{3}{8}\cdot\frac{3}{3} + \left(-\frac{5}{24}\right) \\ x &= \frac{9}{24} + \left(-\frac{5}{24}\right) \\ x &= \frac{4}{24} \\ x &= \frac{1}{6} \end{aligned}$$

The number $\frac{1}{6}$ checks, so it is the solution.

13. $$\begin{aligned} 5x &= 20 \\ \frac{5x}{5} &= \frac{20}{5} \\ 1\cdot x &= \frac{20}{5} \\ x &= 4 \end{aligned}$$

Check:

$5x = 20$	
$5\cdot 4$	20
20	TRUE

The solution is 4.

15. $$\begin{aligned} -4x &= 88 \\ \frac{-4x}{-4} &= \frac{88}{-4} \\ 1\cdot x &= -\frac{88}{4} \\ x &= -22 \end{aligned}$$

Check:

$-4x = 88$	
$-4(-22)$	88
88	TRUE

The solution is -22.

17. $$\begin{aligned} 4 &= 24t \\ \frac{4}{24} &= \frac{24t}{24} \\ \frac{4}{24} &= 1\cdot t \\ \frac{1}{6} &= t \end{aligned}$$

Check:

$4 = 24t$	
4	$24\cdot\frac{1}{6}$
TRUE	4

The solution is $\frac{1}{6}$.

19. $$\begin{aligned} -3z &= -96 \\ \frac{-3z}{-3} &= \frac{-96}{-3} \\ 1\cdot z &= 32 \\ z &= 32 \end{aligned}$$

The number 32 checks, so it is the solution.

21. $$\begin{aligned} 4.8y &= -28.8 \\ \frac{4.8y}{4.8} &= \frac{-28.8}{4.8} \\ 1\cdot y &= -\frac{28.8}{4.8} \\ y &= -6 \end{aligned}$$

The number -6 checks, so it is the solution.

23. $$\begin{aligned} \frac{3}{2}t &= -\frac{1}{4} \\ \frac{2}{3}\cdot\frac{3}{2}t &= \frac{2}{3}\cdot\left(-\frac{1}{4}\right) \\ 1\cdot t &= -\frac{2}{12} \\ t &= -\frac{1}{6} \end{aligned}$$

The number $-\frac{1}{6}$ checks, so it is the solution.

25. $$\begin{aligned} 4x - 12 &= 60 \\ 4x - 12 + 12 &= 60 + 12 \\ 4x &= 72 \\ \frac{4x}{4} &= \frac{72}{4} \\ x &= 18 \end{aligned}$$

Check:

$4x - 12 = 60$	
$4\cdot 18 - 12$	60
$72 - 12$	
60	TRUE

The solution is 18.

27. $5x - 10 = 45$
$5x - 10 + 10 = 45 + 10$
$5x = 55$
$\frac{5x}{5} = \frac{55}{5}$
$x = 11$

Check:

$5x - 10$	$= 45$
$5 \cdot 11 - 10$	45
$55 - 10$	
45	TRUE

The solution is 11.

29. $9t + 4 = -104$
$9t + 4 - 4 = -104 - 4$
$9t = -108$
$\frac{9t}{9} = \frac{-108}{9}$
$t = -12$

Check:

$9t + 4$	$= -104$
$9(-12) + 4$	-104
$-108 + 4$	
-104	TRUE

The solution is -12.

31. $-\frac{7}{3}x + \frac{2}{3} = -18$, LCM = 3
$3(-\frac{7}{3}x + \frac{2}{3}) = 3(-18)$ (Multiplying by 3 to clear fractions)
$-7x + 2 = -54$
$-7x = -56$ (Subtracting 2)
$x = \frac{-56}{-7}$ (Dividing by -7)
$x = 8$

The number 8 checks. It is the solution.

33. $\frac{6}{5}x + \frac{4}{10}x = \frac{32}{10}$, LCM = 10
$10(\frac{6}{5}x + \frac{4}{10}x) = 10 \cdot \frac{32}{10}$ (Multiplying by 10 to clear fractions)
$12x + 4x = 32$
$16x = 32$ (Collecting like terms)
$x = \frac{32}{16}$ (Dividing by 16)
$x = 2$

The number 2 checks. It is the solution.

35. $0.9y - 0.7y = 4.2$
$10(0.9y - 0.7y) = 10(4.2)$ (Multiplying by 10 to clear fractions)
$9y - 7y = 42$
$2y = 42$ (Collecting like terms)
$y = \frac{42}{2}$ (Dividing by 2)
$y = 21$

The number 21 checks, so it is the solution.

37. $8x + 48 = 3x - 12$
$5x + 48 = -12$ (Subtracting 3x)
$5x = -60$ (Subtracting 48)
$x = \frac{-60}{5}$ (Dividing by 5)
$x = -12$

The number -12 checks, so it is the solution.

39. $7y - 1 = 23 - 5y$
$12y - 1 = 23$ (Adding 5y)
$12y = 24$ (Adding 1)
$y = \frac{24}{12}$ (Dividing by 12)
$y = 2$

The number 2 checks. It is the solution.

41. $4x - 3 = 5 + 12x$
$-3 = 5 + 8x$ (Subtracting 4x)
$-8 = 8x$ (Subtracting 5)
$\frac{-8}{8} = x$ (Dividing by 8)
$-1 = x$

The number -1 checks, so it is the solution.

43. $5 - 4a = a - 13$
$5 = 5a - 13$ (Adding 4a)
$18 = 5a$ (Adding 13)
$\frac{18}{5} = a$ (Dividing by 5)

The number $\frac{18}{5}$ checks. It is the solution.

45. $3m - 7 = -7 - 4m - m$
$3m - 7 = -7 - 5m$ (Collecting like terms)
$3m = -5m$ (Adding 7)
$8m = 0$ (Adding 5m)
$m = \frac{0}{8}$ (Dividing by 8)
$m = 0$

The number 0 checks, so it is the solution.

47. $5x + 3 = 11 - 4x + x$
$5x + 3 = 11 - 3x$ (Collecting like terms)
$8x + 3 = 11$ (Adding 3x)
$8x = 8$ (Subtracting 3)
$x = \frac{8}{8}$ (Dividing by 8)
$x = 1$

The number 1 checks, so it is the solution.

49. $-7 + 9x = 9x - 7$

$-7 = -7$ (Subtracting 9x)

The equation $-7 = -7$ is true. Replacing x by any real number gives a true sentence. Thus, any real number is a solution.

51. $6y - 8 = 9 + 6y$

$-8 = 9$ (Subtracting 6y)

The equation $-8 = 9$ is false. No matter what number we try for x we get a false sentence. Thus, the equation has no solution.

53. $2(x + 6) = 8x$

$2x + 12 = 8x$

$12 = 6x$

$2 = x$

Check:

$2(x + 6)$	$= 8x$
$2(2 + 6)$	$8 \cdot 2$
$2 \cdot 8$	16
16	TRUE

The solution is 2.

55. $80 = 10(3t + 2)$

$80 = 30t + 20$

$60 = 30t$

$2 = t$

Check:

$80 =$	$10(3t + 2)$
80	$10(3 \cdot 2 + 2)$
	$10(6 + 2)$
	$10 \cdot 8$
TRUE	80

The solution is 2.

57. $180(n - 2) = 900$

$180n - 360 = 900$

$180n = 1260$

$n = 7$

Check:

$180(n - 2)$	$= 900$
$180(7 - 2)$	900
$180 \cdot 5$	
900	TRUE

The solution is 7.

59. $5y - (2y - 10) = 25$

$5y - 2y + 10 = 25$

$3y + 10 = 25$

$3y = 15$

$y = 5$

Check:

$5y - (2y - 10)$	$= 25$
$5 \cdot 5 - (2 \cdot 5 - 10)$	25
$25 - (10 - 10)$	
$25 - 0$	
25	TRUE

The solution is 5.

61. $7(3x + 6) = 11 - (x + 2)$

$21x + 42 = 11 - x - 2$

$21x + 42 = 9 - x$

$22x + 42 = 9$

$22x = -33$

$x = \frac{-33}{22}$

$x = -\frac{3}{2}$

The number $-\frac{3}{2}$ checks and is the solution.

63. $\frac{1}{8}(16y + 8) - 17 = -\frac{1}{4}(8y - 16)$

$2y + 1 - 17 = -2y + 4$

$2y - 16 = -2y + 4$

$4y - 16 = 4$

$4y = 20$

$y = 5$

The number 5 checks and is the solution.

65. $3[5 - 3(4 - t)] - 2 = 5[3(5t - 4) + 8] - 26$

$3[5 - 12 + 3t] - 2 = 5[15t - 12 + 8] - 26$

$3[-7 + 3t] - 2 = 5[15t - 4] - 26$

$-21 + 9t - 2 = 75t - 20 - 26$

$9t - 23 = 75t - 46$

$-23 = 66t - 46$

$23 = 66t$

$\frac{23}{66} = t$

The number $\frac{23}{66}$ checks and is the solution.

67. $\frac{2}{3}\left(\frac{7}{8} + 4x\right) - \frac{5}{8} = \frac{3}{8}$

$\frac{2}{3}\left(\frac{7}{8} + 4x\right) = 1$

$\frac{7}{12} + \frac{8}{3}x = 1$

$12\left(\frac{7}{12} + \frac{8}{3}x\right) = 12 \cdot 1$

$7 + 32x = 12$

$32x = 5$

$x = \frac{5}{32}$

The number $\frac{5}{32}$ checks and is the solution.

69. $(6x^5y^{-4})(-3x^{-3}y^{-7})$

$= 6 \cdot (-3) \cdot x^5 \cdot x^{-3} \cdot y^{-4} \cdot y^{-7}$

$= -18x^{5+(-3)}y^{-4+(-7)}$

$= -18x^2y^{-11} = -\frac{18x^2}{y^{11}}$

71. $-4(3x - 2y + z)$

$= -4 \cdot 3x - (-4) \cdot 2y + (-4) \cdot z$

$= -12x + 8y - 4z$

73. $\frac{3x}{2} + \frac{5x}{3} - \frac{13x}{6} - \frac{2}{3} = \frac{5}{6}$, LCM = 6

$6(\frac{3x}{2} + \frac{5x}{3} - \frac{13x}{6} - \frac{2}{3}) = 6 \cdot \frac{5}{6}$

(Multiplying by 6 to clear fractions)

$3 \cdot 3x + 2 \cdot 5x - 1 \cdot 13x - 2 \cdot 2 = 1 \cdot 5$

$9x + 10x - 13x - 4 = 5$

$6x - 4 = 5$ (Collecting like terms)

$6x = 9$ (Adding 4)

$x = \frac{9}{6}$ (Dividing by 6)

$x = \frac{3}{2}$

The number $\frac{3}{2}$ checks, so it is the solution.

75. $2x - 4 - (x+1) - 3(x-2) = 6(2x-3) - 3(6x-1) - 8$

$2x - 4 - x - 1 - 3x + 6 = 12x - 18 - 18x + 3 - 8$

$-2x + 1 = -6x - 23$

$4x + 1 = -23$

$4x = -24$

$x = -6$

The number -6 checks, so it is the solution.

Exercise Set 4.2

1. Familiarize. We first make a drawing.

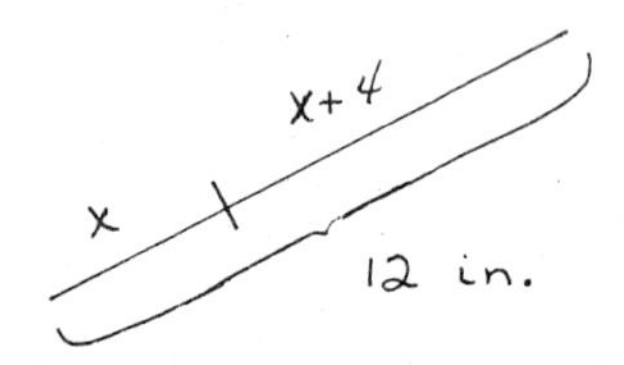

We let x represent the length of one piece and x + 4 the other.

Translate.

Length of one piece	plus	length of other	is	12.
↓	↓	↓	↓	↓
x	+	(x + 4)	=	12

Solve.

$x + x + 4 = 12$

$2x + 4 = 12$

$2x = 8$

$x = 4$

Check. If the length of one piece is 4 in., then the length of the other is 4 + 4, or 8 in. The total length is 4 + 8, or 12 in. The lengths check.

State. The lengths of the pieces are 4 in. and 8 in.

3. Familiarize. It is helpful to make a drawing.

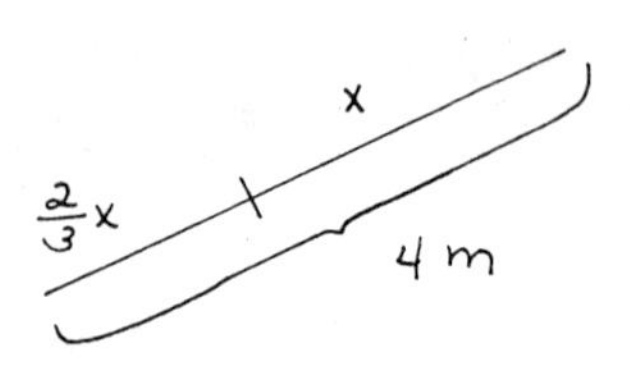

We let x represent the length of the longer piece and $\frac{2}{3}$ x represent the length of the shorter piece.

Translate.

Length of one piece	plus	length of other	is	4 m.
↓	↓	↓	↓	↓
x	+	$\frac{2}{3}x$	=	4

Solve.

$x + \frac{2}{3}x = 4$

$\frac{3}{3}x + \frac{2}{3}x = 4$

$\frac{5}{3}x = 4$

$\frac{3}{5} \cdot \frac{5}{3}x = \frac{3}{5} \cdot 4$

$x = \frac{12}{5}$, or $2\frac{2}{5}$

Check. If $x = \frac{12}{5}$, then $\frac{2}{3} \cdot \frac{12}{5} = \frac{8}{5}$, or $1\frac{3}{5}$. The sum of the lengths is $2\frac{2}{5} + 1\frac{3}{5}$, or 4 m. The lengths check.

State. The lengths of the pieces are $2\frac{2}{5}$ m and $1\frac{3}{5}$ m.

5. Familiarize. Let x represent the Burger King sales. Then x + 475,000 represents the McDonalds sales.

Translate.

Burger King sales	plus	McDonalds sales	are	$2,525,000.
↓	↓	↓	↓	↓
x	+	x + 475,000	=	2,525,000

Solve.

$x + x + 475,000 = 2,525,000$

$2x + 475,000 = 2,525,000$

$2x = 2,050,000$

$x = 1,025,000$

Check. If x = \$1,025,000, then x + \$475,000 = \$1,500,000. The total of these amounts is \$2,525,000. The answers check.

State. The average Burger King sales were \$1,025,000, and the average McDonalds sales were \$1,500,000.

7. Familiarize. Let x represent the original cost.

Translate.

2	%	of	the original cost	is	$480,000.
2	%	$\cdot$	x	=	480,000

Solve.

$$2\% \cdot x = 480{,}000$$
$$0.02x = 480{,}000 \quad (2\% = 0.02)$$
$$x = \frac{480{,}000}{0.02}$$
$$x = 24{,}000{,}000$$

Check. 2% of $24,000,000 is $480,000. The answer checks.

State. The original cost was $24,000,000.

9. Familiarize. Let m represent the number of miles driven.

Translate.

Daily rate	plus	$0.29	times	number of miles	is	$102.30.
39.95	+	0.29	$\cdot$	m	=	102.30

Solve.

$$39.95 + 0.29m = 102.30$$
$$3995 + 29m = 10{,}230 \quad \text{(Multiplying by 100)}$$
$$29m = 6235$$
$$m = 215$$

Check. $39.95 + $0.29(215) = $39.95 + $62.35 = $102.30. The answer checks.

State. The businesswoman drove 215 miles.

11. Familiarize. Let m represent the number of minutes the call lasted. Then m - 1 minutes are charged at the 26¢ rate.

Translate.

Cost of first minute	plus	$0.26	times	number of additional minutes	is	$11.74.
0.30	+	0.26	$\cdot$	(m - 1)	=	11.74

Solve.

$$0.30 + 0.26(m - 1) = 11.74$$
$$30 + 26(m - 1) = 1174 \quad \text{(Multiplying by 100)}$$
$$30 + 26m - 26 = 1174$$
$$26m + 4 = 1174$$
$$26m = 1170$$
$$m = 45$$

Check. If the call lasted 45 minutes, the charge was $0.30 + $0.26(44) = $0.30 + $11.44 = $11.74. The answer checks.

State. The call was 45 minutes long.

13. Familiarize. Let x = the number.

Translate.

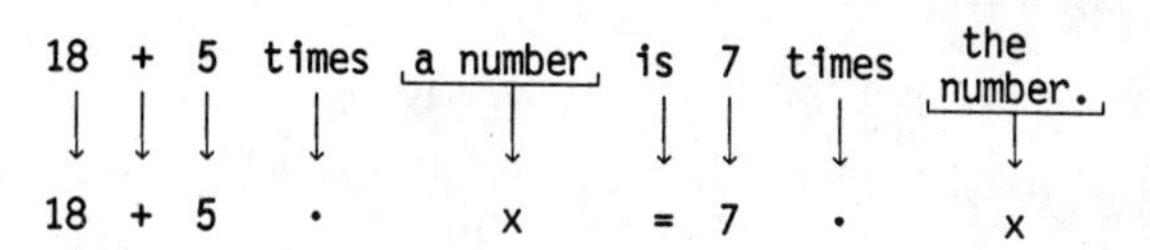

Solve.

$$18 + 5x = 7x$$
$$18 = 2x$$
$$9 = x$$

Check. Five times 9 is 45. If we add 18, we get 63. This is 7·9. The answer checks.

State. The number is 9.

15. Familiarize. Let n = the number.

Translate.

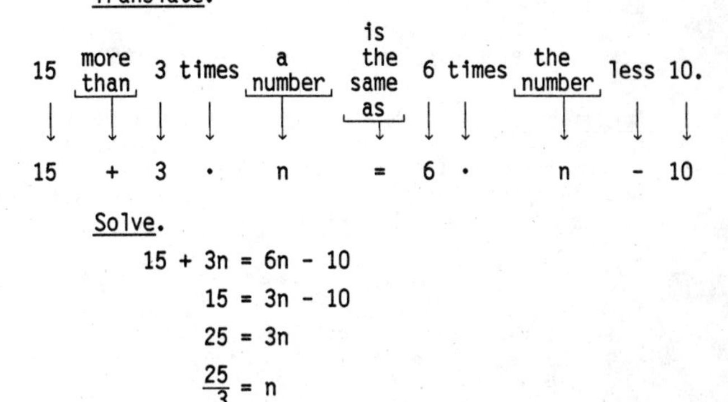

Solve.

$$15 + 3n = 6n - 10$$
$$15 = 3n - 10$$
$$25 = 3n$$
$$\frac{25}{3} = n$$

Check. Three times $\frac{25}{3}$ is 25. If we add 15, we get 40. Also, $6 \cdot \frac{25}{3} = 50$ and 10 less than 50 is 40. The answer checks.

State. The number is $\frac{25}{3}$.

17. Familiarize. We let x represent the former price.

Translate.

Former price	minus	24%	times	former price	is	sale price.
x	-	24%	$\cdot$	x	=	$78.66

Solve.

$$x - 24\%x = 78.66$$
$$1x - 0.24x = 78.66$$
$$0.76x = 78.66$$
$$x = \frac{78.66}{0.76}$$
$$x = 103.5$$

Check. If the former price is $103.50, then 24%·$103.50 is $24.84 and the sale price is $103.50 - $24.84, or $78.66. The value checks.

State. The former price is $103.50.

19. Familiarize. We let x represent the amount of the loan.

Translate.

Loan plus 9% times the loan is the payoff.

$x + 9\%x = \$708.50$

Solve.

$$x + 9\%x = 708.50$$
$$1x + 0.09x = 708.50$$
$$1.09x = 708.50$$
$$x = \frac{708.50}{1.09}$$
$$x = 650$$

Check. If \$650 was borrowed, then 9%·650, or \$58.50, is the simple interest owed for the loan. The payoff was 650 + 58.50, or \$708.50. The amount checks.

State. The original amount borrowed was \$650.

21. Familiarize. We draw a picture. We use x for the measure of the first angle. The second angle is three times the first, so its measure will be 3x. The third angle is 12° less than twice the first, so its measure will be 2x - 12.

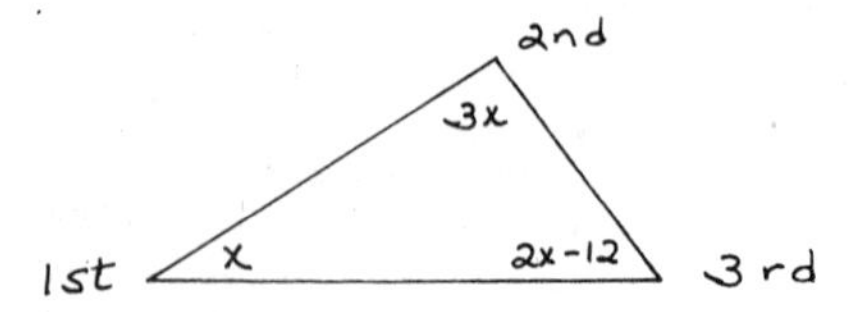

The measures of the angles of a triangle add up to 180°.

Translate.

Measure of 1st angle + Measure of 2nd angle + Measure of 3rd angle is 180°.

$x + 3x + (2x - 12) = 180$

Solve.

$$x + 3x + 2x - 12 = 180$$
$$6x - 12 = 180$$
$$6x = 192$$
$$x = 32$$

The angles will then have measures as follows:

1st angle: x = 32°
2nd angle: 3x = 3·32, or 96°
3rd angle: 2x - 12 = 2·32 - 12, or 52°

Check. The angle measures add up to 180° so they give an answer to the problem.

State. The angles are 32°, 96°, and 52°.

23. Familiarize. We first draw a picture.

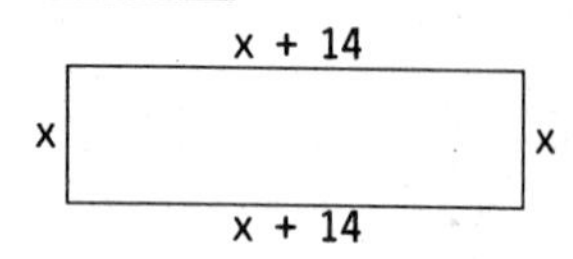

We let x represent the width. Then x + 14 represents the length. The perimeter is 96 m.

Translate.

Length + Length + Width + Width = 96

$(x + 14) + (x + 14) + x + x = 96$

Solve.

$$x + 14 + x + 14 + x + x = 96$$
$$4x + 28 = 96$$
$$4x = 68$$
$$x = 17$$

Check. If the width is 17 m, then the length is 17 + 4, or 31 m. The perimeter is 2·31 + 2·17, or 96 m. The answer checks.

State.

The dimensions are 31 m and 17 m.

25. Familiarize. Consecutive integers are next to each other like 7, 8, and 9, or -7, -6, and -5. We let x represent the first integer. Then the next two are x + 1 and x + 1 + 1, or x + 2.

Translate.

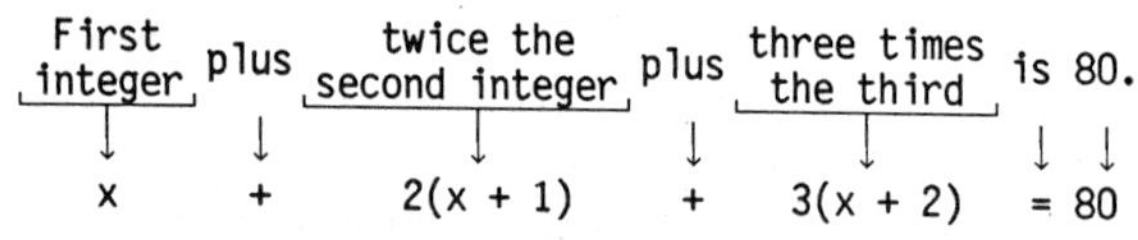

First integer plus twice the second integer plus three times the third is 80.
x + 2(x + 1) + 3(x + 2) = 80

Solve.

$$x + 2(x + 1) + 3(x + 2) = 80$$
$$x + 2x + 2 + 3x + 6 = 80$$
$$6x + 8 = 80$$
$$6x = 72$$
$$x = 12$$

Check. If x = 12, then x + 1 = 13 and x + 2 = 14. The sum of 12 plus 2·13 plus 3·14 is 12 + 26 + 42, or 80. The numbers check.

State. The consecutive integers are 12, 13, and 14.

27. Familiarize. We let x represent the old salary. Then 20%x represents the raise.

Translate.

Old salary plus raise is new salary.

$$x + 20\%x = 9600$$

Solve.

$$x + 20\%x = 9600$$
$$1x + 0.2x = 9600$$
$$1.2x = 9600$$
$$x = \frac{9600}{1.2}$$
$$x = 8000$$

Check. If the old salary is \$8000, the raise is 20%(\$8000), or \$1600. The new salary would be \$8000 + \$1600, or \$9600. The value checks.

State. The old salary was \$8000.

29. Familiarize. Let x represent the cost of room and board. Then x + 704 represents the tuition cost.

Translate.

Tuition plus room and board is \$2584.

$$(x + 704) + x = 2584$$

Solve.

$$x + 704 + x = 2584$$
$$2x + 704 = 2584$$
$$2x = 1880$$
$$x = 940$$

Check. When x = \$940, x + \$704 = \$940 + \$704, or \$1644. The total cost is \$940 + \$1644, or \$2584. The amounts check.

State. The tuition cost is \$1644.

31. Familiarize. Let x represent the number.

Translate.

Eleven less than seven times a number is five more than six times the number.

$$7x - 11 = 6x + 5$$

Solve.

$$7x - 11 = 6x + 5$$
$$x - 11 = 5$$
$$x = 16$$

Check. Eleven less than seven times 16 is $7 \cdot 16 - 11$, or 101. Five more than six times 16 is $6 \cdot 16 + 5$, or 101. The number checks.

State. The number is 16.

33. Familiarize. Consecutive odd integers are like 11 and 13, or -9 and -7. Let x represent the first odd integer, then x + 2 represents the next odd integer.

Translate.

First integer plus second integer is 137.

$$x + (x + 2) = 137$$

Solve.

$$x + x + 2 = 137$$
$$2x + 2 = 137$$
$$2x = 135$$
$$x = 67.5$$

Check and State. The number 67.5 is not an odd integer. There is no solution to the problem.

This problem could also be solved as follows. The sum of any two odd integers is always even. Thus, the sum can never be 137. There is no solution.

35. Familiarize. After the next test there will be six test scores. The average of the six scores is their sum divided by 6. We let x represent the next test score.

Translate.

Average of the six scores is 88.

$$\frac{93 + 89 + 72 + 80 + 96 + x}{6} = 88$$

Solve.

$$\frac{93 + 89 + 72 + 80 + 96 + x}{6} = 88$$

$$93 + 89 + 72 + 80 + 96 + x = 528$$

(Multiplying by 6)

$$430 + x = 528$$
$$x = 98$$

Check. If the next test score is 98%, the average will be $\frac{93 + 89 + 72 + 80 + 96 + 98}{6} = \frac{528}{6}$, or 88%. The answer checks.

State. The sixth score must be 98%.

37. Familiarize. Note that each odd integer is two more than the one preceding it. If we let n represent the first odd integer, then the second is 2 more than the first and the third is 2 more than the second, or 4 more than the first. We are told that the sum of the first, twice the second, and three times the third is 70.

Translate. The three odd integers are n, n + 2, and n + 4. Translate to an equation.

First plus two times second plus three times third is 70.

$$n + 2(n + 2) + 3(n + 4) = 70$$

37. (continued)

Solve.

$$n + 2(n + 2) + 3(n + 4) = 70$$
$$n + 2n + 4 + 3n + 12 = 70$$
$$6n + 16 = 70$$
$$6n = 54$$
$$n = 9$$

Check. The numbers are 9, 11, and 13. They are consecutive odd integers. Also, $9 + 2\cdot11 + 3\cdot13 = 9 + 22 + 39$, or 70. The numbers check.

State. The integers are 9, 11, and 13.

39. Familiarize. We first make drawings. We let x represent the length of the shorter piece of wire. Then the length of the longer piece is 100 - x.

x 100 - x

The squares have sides of length $\frac{x}{4}$ and $\frac{100 - x}{4}$.

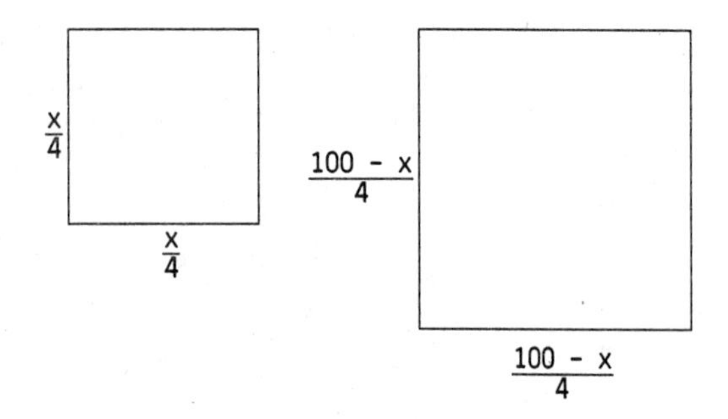

Translate.

Length of a side of the large square is 2 cm more than length of a side of the smaller square.

$$\frac{100 - x}{4} = 2 + \frac{x}{4}$$

Solve.

$$\frac{100 - x}{4} = 2 + \frac{x}{4}$$
$$100 - x = 8 + x \quad \text{(Multiplying by 4)}$$
$$100 = 8 + 2x$$
$$92 = 2x$$
$$46 = x$$

Check. If one piece of wire is 46 cm, the other piece is 100 - 46, or 54 cm. The side of one square is $\frac{46}{4}$, or $11\frac{1}{2}$ cm. The side of the other is $\frac{54}{4}$, or $13\frac{1}{2}$ cm, which is 2 cm greater than the side of the first square. The numbers check.

State. The wire should be cut into pieces of 46 cm and 54 cm.

41. Familiarize. Let p represent the average number of points scored in a game. Note that 48 minutes = 48(60 seconds) = 2880 seconds.

Translate.

Total seconds divided by average points is 24.

$$2880 \div p = 24$$

Solve.

$$2880 \div p = 24$$
$$\frac{2880}{p} = 24$$
$$2880 = 24p$$
$$120 = p$$

Check. Since $2880 \div 120 = 24$, the answer checks.

State. The average number of points scored in a game is 120.

43. Familiarize. If we let x represent the height, then the three sides are x + 1, x + 2, and x + 3, with x + 2 representing the base. The perimeter is 42 in.

Translate.

The perimeter is 42 in.

$$(x + 1) + (x + 2) + (x + 3) = 42$$

Solve.

$$x + 1 + x + 2 + x + 3 = 42$$
$$3x + 6 = 42$$
$$3x = 36$$
$$x = 12$$

If x = 12, then x + 1 = 12 + 1, or 13; x + 2 = 12 + 2, or 14; and x + 3 = 12 + 3, or 15.

Check. 12, 13, 14, and 15 are four consecutive integers. Also, 13 + 14 + 15 = 42. The numbers check. The height is 12 in., and the base is 14 in.

Now we find the area of the triangle.

$$\text{Area} = \frac{1}{2} \cdot \text{base} \cdot \text{height}$$
$$= \frac{1}{2} \cdot 14 \text{ in.} \cdot 12 \text{ in.}$$
$$= 84 \text{ in}^2$$

State. The area of the triangle is 84 in^2.

45. Familiarize. Let n represent the number of romance novels. Then 65%n, or 0.65n, represents the number of science fiction novels; 46%(0.65n), or (0.46)(0.65n), or 0.299n, represents the number of horror novels; and 17%(0.299n), or 0.17(0.299n), or 0.05083n, represents the number of mystery novels.

Translate.

Romance novels + Science fiction novels + Horror novels + Mystery novels is 400.

$$n + 0.65n + 0.299n + 0.05083n = 400$$

Solve.

$$n + 0.65n + 0.299n + 0.05083n = 400$$
$$1.99983n = 400$$
$$n \approx 200$$

Check. If the number of romance novels is 200, then the other novels are:

Science fiction: $0.65(200) = 130$

Horror: $0.299(200) \approx 60$

Mystery: $0.05083(200) \approx 10$

The total number of novels is 200 + 130 + 60 + 10, or 400. The numbers check.

State. The professor has 130 science fiction novels.

47. Familiarize. We add some labels to the figure in the text.

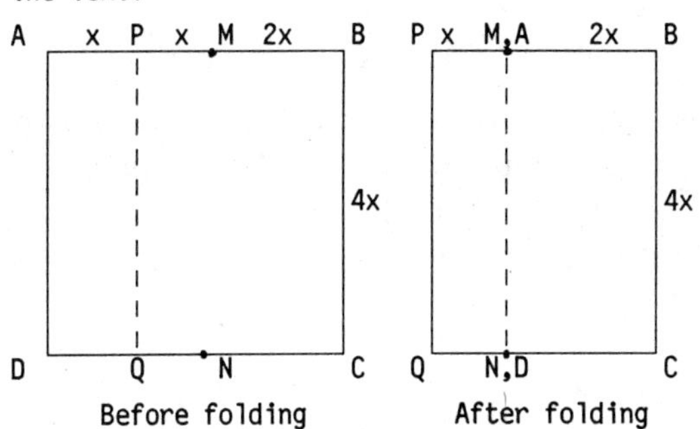

Before folding After folding

Let x represent the length of $\overline{AP}$. Then the length of a side of the square is 4x. The smaller figure has sides of length 3x and 4x.

Translate.

Perimeter of smaller figure is 25 in.

$$2(3x) + 2(4x) = 25$$

Solve.

$$2(3x) + 2(4x) = 25$$
$$6x + 8x = 25$$
$$14x = 25$$
$$x = \frac{25}{14}$$

47. (continued)

Check. If $x = \frac{25}{14}$ in., then the smaller figure has sides of $3\left(\frac{25}{14}\right)$, or $\frac{75}{14}$ in., and $4\left(\frac{25}{14}\right)$, or $\frac{50}{7}$ in., and the perimeter is $2\left(\frac{75}{14}\right) + 2\left(\frac{50}{7}\right) = \frac{75}{7} + \frac{100}{7} = \frac{175}{7} = 25$ in. The number checks.

Now we find the area of the square. The square has sides of length 4x, so $A = (4x)^2 = 16x^2 = 16\left(\frac{25}{14}\right)^2 = 16\left(\frac{625}{196}\right) = \frac{2500}{49} = 51\frac{1}{49}$ in^2.

State. The area of the square is $51\frac{1}{49}$ in^2.

Exercise Set 4.3

1. $A = \ell w$

$\frac{A}{w} = \ell$ (Dividing by w)

3. $W = EI$

$\frac{W}{E} = I$ (Dividing by E)

5. $d = rt$

$\frac{d}{r} = t$ (Dividing by r)

7. $I = Prt$

$\frac{I}{Pr} = t$ (Dividing by Pr)

9. $E = mc^2$

$\frac{E}{c^2} = m$ (Dividing by c^2)

11. $P = 2\ell + 2w$

$P - 2w = 2\ell$ (Subtracting 2w)

$\frac{P - 2w}{2} = \ell$ (Dividing by 2)

or $\frac{P}{2} - w = \ell$

13. $c^2 = a^2 + b^2$

$c^2 - b^2 = a^2$ (Subtracting b^2)

15. $Ax + By = C$

$Ax = C - By$

$x = \frac{C - By}{A}$

17. $A = \pi r^2$

$\frac{A}{\pi} = r^2$

19. $W = \frac{11}{2}(h - 40)$

$\frac{2}{11} W = h - 40$ $\left[\text{Multiplying by } \frac{2}{11}\right]$

$\frac{2}{11} W + 40 = h$ (Adding 40)

21. $V = \frac{4}{3} \pi r^3$

$\frac{3V}{4\pi} = r^3$ $\left[\text{Multiplying by } \frac{3}{4\pi}\right]$

23. $A = \frac{1}{2} h(c - d)$

$2A = h(c - d)$ (Multiplying by 2)

$2A = hc - hd$

$2A + hd = hc$ (Adding hd)

$\frac{2A + hd}{h} = c$ (Dividing by h)

or $\frac{2A}{h} + d = c$

25. $F = \frac{mv^2}{r}$

$F = m \cdot \frac{v^2}{r}$

$\frac{Fr}{v^2} = m$ $\left[\text{Multiplying by } \frac{r}{v^2}\right]$

27. $P = 5a - 3ab$

$P = a(5 - 3b)$ (Factoring)

$\frac{P}{5 - 3b} = a$

$P = 5a - 3ab$

$P - 5a = -3ab$

$\frac{P - 5a}{-3a} = b$, or

$\frac{5}{3} - \frac{P}{3a} = b$

29. $\frac{80}{-16} = -\frac{80}{16} = -5$

31. $-\frac{1}{2} \div \frac{1}{4} = -\frac{1}{2} \cdot \frac{4}{1} = -\frac{4}{2} = -2$

33. $s = v_1t + \frac{1}{2} at^2$

$s - v_1t = \frac{1}{2} at^2$

$2(s - v_1t) = at^2$

$\frac{2(s - v_1t)}{t^2} = a$

33. (continued)

$s = v_1t + \frac{1}{2}at^2$

$S - \frac{1}{2}at^2 = v_1t$

$\frac{s - \frac{1}{2}at^2}{t} = v_1$, or

$\frac{s}{t} - \frac{1}{2}at = v_1$, or

$\frac{2s - at^2}{2t} = v_1$

35. $\frac{P_1V_1}{T_1} = \frac{P_2V_2}{T_2}$

$V_1 = \frac{T_1P_2V_2}{P_1T_2}$ $\left[\text{Multiplying by } \frac{T_1}{P_1}\right]$

$\frac{P_1V_1}{T_1} = \frac{P_2V_2}{T_2}$

$\frac{P_1V_1T_2}{T_1V_2} = P_2$ $\left[\text{Multiplying by } \frac{T_2}{V_2}\right]$

37.

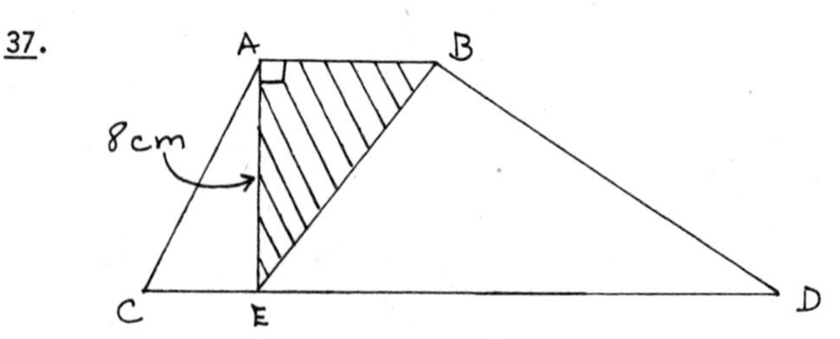

Triangle ABE is a right triangle. We first find the length of $\overline{AB}$. The formula for the area of a triangle is $\frac{1}{2}$ bh. We know the area is 20 cm² and the height is 8 cm. We can substitute to find the base, which is $\overline{AB}$.

$\text{Area} = \frac{1}{2} bh$

$20 = \frac{1}{2} \cdot b \cdot 8$ (Substituting)

$20 = 4b$

$5 = b$

The length of $\overline{AB}$ is 5 cm.

The formula for the area of a trapezoid is $\frac{1}{2} h(b_1 + b_2)$, where h is the height and b_1 and b_2 are the bases. We can substitute 8 for h, 5 for b_1, and 13 for b_2 and find the area.

$\text{Area} = \frac{1}{2} h(b_1 + b_2)$

$\text{Area} = \frac{1}{2} \cdot 8 \cdot (5 + 13)$ (Substituting)

$= 4(18)$

$= 72$

The area of the trapezoid is 72 cm².

Exercise Set 4.4

1. $x - 2 \geqslant 6$

   -4: We substitute and get $-4 - 2 \geqslant 6$, or $-6 \geqslant 6$, a false sentence. Therefore, -4 is not a solution.

   0: We substitute and get $0 - 2 \geqslant 6$, or $-2 \geqslant 6$, a false sentence. Therefore, 0 is not a solution.

   4: We substitute and get $4 - 2 \geqslant 6$, or $2 \geqslant 6$, a false sentence. Therefore, 4 is not a solution.

   8: We substitute and get $8 - 2 \geqslant 6$, or $6 \geqslant 6$, a true sentence. Therefore, 8 is a solution.

3. $t - 8 > 2t - 3$

   0: We substitute and get $0 - 8 > 2\cdot 0 - 3$, or $-8 > -3$, a false sentence. Therefore, 0 is not a solution.

   -8: We substitute and get $-8 - 8 > 2(-8) - 3$, or $-16 > -19$, a true sentence. Therefore, -8 is a solution.

   -9: We substitute and get $-9 - 8 > 2(-9) - 3$, or $-17 > -21$, a true sentence. Therefore, -9 is a solution.

   -3: We substitute and get $-3 - 8 > 2(-3) - 3$, or $-11 > -9$, a false sentence. Therefore, -3 is not a solution.

5. $x + 8 > 3$

   $x + 8 - 8 > 3 - 8$ Subtracting 8

   $x > -5$

   The solution set is $\{x|x > -5\}$.

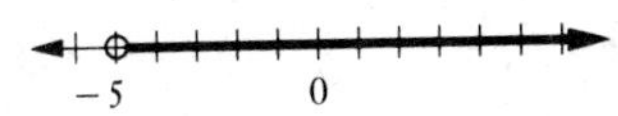

7. $y + 3 < 9$

   $y + 3 - 3 < 9 - 3$ Subtracting 3

   $y < 6$

   The solution set is $\{y|y < 6\}$.

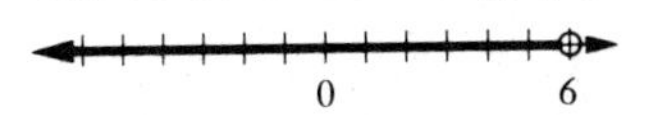

9. $a + 9 \leqslant -12$

   $a + 9 - 9 \leqslant -12 - 9$ Subtracting 9

   $a \leqslant -21$

   The solution set is $\{a|a \leqslant -21\}$.

   −21 0

11. $t + 14 \geqslant 9$

    $t + 14 - 14 \geqslant 9 - 14$ Subtracting 14

    $t \geqslant -5$

    The solution set is $\{t|t \geqslant -5\}$.

    −5 0

13. $y - 8 > -14$

    $y - 8 + 8 > -14 + 8$ Adding 8

    $y > -6$

    The solution set is $\{y|y > -6\}$.

    −6 0

15. $x - 11 \leqslant -2$

    $x - 11 + 11 \leqslant -2 + 11$ Adding 11

    $x \leqslant 9$

    The solution set is $\{x|x \leqslant 9\}$.

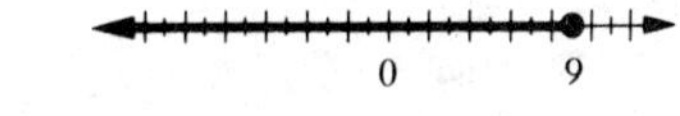

17. $8x \geqslant 24$

    $\frac{8x}{8} \geqslant \frac{24}{8}$ Dividing by 8

    $x \geqslant 3$

    The solution set is $\{x|x \geqslant 3\}$.

    0 3

19. $0.3x < -18$

    $\frac{0.3x}{0.3} < \frac{-18}{0.3}$ Dividing by 0.3

    $x < -60$

    The solution set is $\{x|x < -60\}$.

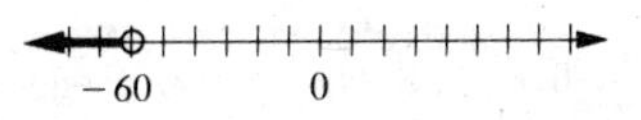

21. $-9x \geqslant -8.1$

    $\frac{-9x}{-9} \leqslant \frac{-8.1}{-9}$ Dividing by -9 and reversing the inequality sign

    $x \leqslant 0.9$

    The solution set is $\{x|x \leqslant 0.9\}$.

23. $-\frac{3}{4}x \geqslant -\frac{5}{8}$

    $-\frac{4}{3}\left(-\frac{3}{4}x\right) \leqslant -\frac{4}{3}\left(-\frac{5}{8}\right)$ Multiplying by $-\frac{4}{3}$ and reversing the inequality sign

    $x \leqslant \frac{20}{24}$

    $x \leqslant \frac{5}{6}$

    The solution set is $\left\{x\middle|x \leqslant \frac{5}{6}\right\}$.

25. $2x + 7 < 19$

    $2x + 7 - 7 < 19 - 7$ Subtracting 7

    $2x < 12$

    $\frac{2x}{2} < \frac{12}{2}$ Dividing by 2

    $x < 6$

    The solution set is $\{x|x < 6\}$.

    0 6

27. $5y + 2y \le -21$

$7y \le -21$ Collecting like terms

$\frac{7y}{7} \le \frac{-21}{7}$ Dividing by 7

$y \le -3$

The solution set is $\{y \mid y \le -3\}$.

$-3 \quad 0$

29. $2y - 7 < 5y - 9$

$-5y + 2y - 7 < -5y + 5y - 9$ Adding $-5y$

$-3y - 7 < -9$

$-3y - 7 + 7 < -9 + 7$ Adding 7

$-3y < -2$

$\frac{-3y}{-3} > \frac{-2}{-3}$ Dividing by -3 and reversing the inequality sign

$y > \frac{2}{3}$

The solution set is $\left\{y \middle| y > \frac{2}{3}\right\}$.

31. $0.4x + 5 \le 1.2x - 4$

$-1.2x + 0.4x + 5 \le -1.2x + 1.2x - 4$ Adding $-1.2x$

$-0.8x + 5 \le -4$

$-0.8x + 5 - 5 \le -4 - 5$ Subtracting 5

$-0.8x \le -9$

$\frac{-0.8x}{-0.8} \ge \frac{-9}{-0.8}$ Dividing by -0.8 and reversing the inequality sign

$x \ge 11.25$, or $x \ge \frac{45}{4}$

The solution set is $\left\{x \middle| x \ge \frac{45}{4}\right\}$.

33. $3x - \frac{1}{8} \le \frac{3}{8} + 2x$

$-2x + 3x - \frac{1}{8} \le -2x + \frac{3}{8} + 2x$ Adding $-2x$

$x - \frac{1}{8} \le \frac{3}{8}$

$x - \frac{1}{8} + \frac{1}{8} \le \frac{3}{8} + \frac{1}{8}$ Adding $\frac{1}{8}$

$x \le \frac{4}{8}$

$x \le \frac{1}{2}$

The solution set is $\left\{x \middle| x \le \frac{1}{2}\right\}$.

35. $4(3y - 2) \ge 9(2y + 5)$

$12y - 8 \ge 18y + 45$

$-6y - 8 \ge 45$

$-6y \ge 53$

$y \le -\frac{53}{6}$

The solution set is $\left\{y \middle| y \le -\frac{53}{6}\right\}$.

37. $3(2 - 5x) + 2x < 2(4 + 2x)$

$6 - 15x + 2x < 8 + 4x$

$6 - 13x < 8 + 4x$

$6 - 17x < 8$

$-17x < 2$

$x > -\frac{2}{17}$

The solution set is $\left\{x \middle| x > -\frac{2}{17}\right\}$.

39. $5[3m - (m + 4)] > -2(m - 4)$

$5(3m - m - 4) > -2(m - 4)$

$5(2m - 4) > -2(m - 4)$

$10m - 20 > -2m + 8$

$12m - 20 > 8$

$12m > 28$

$m > \frac{28}{12}$

$m > \frac{7}{3}$

The solution set is $\left\{m \middle| m > \frac{7}{3}\right\}$.

41. $3(r - 6) + 2 > 4(r + 2) - 21$

$3r - 18 + 2 > 4r + 8 - 21$

$3r - 16 > 4r - 13$

$-r - 16 > -13$

$-r > 3$

$r < -3$

The solution set is $\{r \mid r < -3\}$.

43. $19 - (2x + 3) \le 2(x + 3) + x$

$19 - 2x - 3 \le 2x + 6 + x$

$16 - 2x \le 3x + 6$

$16 - 5x \le 6$

$-5x \le -10$

$x \ge 2$

The solution set is $\{x \mid x \ge 2\}$.

45. $\frac{1}{4}(8y + 4) - 17 < -\frac{1}{2}(4y - 8)$

$2y + 1 - 17 < -2y + 4$

$2y - 16 < -2y + 4$

$4y - 16 < 4$

$4y < 20$

$y < 5$

The solution set is $\{y \mid y < 5\}$.

47. $2[4 - 2(3 - x)] - 1 \geqslant 4[2(4x - 3) + 7] - 25$

$2[4 - 6 + 2x] - 1 \geqslant 4[8x - 6 + 7] - 25$

$2[-2 + 2x] - 1 \geqslant 4[8x + 1] - 25$

$-4 + 4x - 1 \geqslant 32x + 4 - 25$

$4x - 5 \geqslant 32x - 21$

$-28x - 5 \geqslant -21$

$-28x \geqslant -16$

$x \leqslant \frac{-16}{-28}$, or $\frac{4}{7}$

The solution set is $\left\{x \middle| x \leqslant \frac{4}{7}\right\}$.

49. $\frac{2}{3}(2x - 1) > 10$

$3 \cdot \frac{2}{3}(2x - 1) > 3 \cdot 10$ Multiplying by 3 to clear the fraction

$2(2x - 1) > 30$

$4x - 2 > 30$

$4x > 32$

$x > 8$

The solution set is $\{x|x > 8\}$.

51. $\frac{3}{4}(3 + 2x) + 1 \geqslant 13$

$4\left[\frac{3}{4}(3 + 2x) + 1\right] \geqslant 4 \cdot 13$ Multiplying by 4 to clear the fraction

$3(3 + 2x) + 4 \geqslant 52$

$9 + 6x + 4 \geqslant 52$

$6x + 13 \geqslant 52$

$6x \geqslant 39$

$x \geqslant \frac{39}{6}$, or $\frac{13}{2}$

The solution set is $\left\{x \middle| x \geqslant \frac{13}{2}\right\}$.

53. $\frac{3}{4}\left(3x - \frac{1}{2}\right) - \frac{2}{3} < \frac{1}{3}$

$\frac{9x}{4} - \frac{3}{8} - \frac{2}{3} < \frac{1}{3}$

$24\left(\frac{9x}{4} - \frac{3}{8} - \frac{2}{3}\right) < 24 \cdot \frac{1}{3}$ Multiplying by 24 to clear fractions

$54x - 9 - 16 < 8$

$54x - 25 < 8$

$54x < 33$

$x < \frac{33}{54}$, or $\frac{11}{18}$

The solution set is $\left\{x \middle| x < \frac{11}{18}\right\}$.

55. $0.7(3x + 6) \geqslant 1.1 - (x + 2)$

$10[0.7(3x + 6)] \geqslant 10[1.1 - (x + 2)]$ Multiplying by 10 to clear decimals

$7(3x + 6) \geqslant 11 - 10(x + 2)$

$21x + 42 \geqslant 11 - 10x - 20$

$21x + 42 \geqslant -9 - 10x$

$31x + 42 \geqslant -9$

$31x \geqslant -51$

$x \geqslant -\frac{51}{31}$

The solution set is $\left\{x \middle| x \geqslant -\frac{51}{31}\right\}$.

57. $a + (a - 3) \leqslant (a + 2) - (a + 1)$

$a + a - 3 \leqslant a + 2 - a - 1$

$2a - 3 \leqslant 1$

$2a \leqslant 4$

$a \leqslant 2$

The solution set is $\{a|a \leqslant 2\}$.

59. A number is less than 8.

$n < 8$

61. The price of a movie ticket is greater than or equal to \$6.

$p \geqslant \$6$

63. The price of compact disks is at most \$17.95.

$d \leqslant \$17.95$

65. 24 minus 3 times a number is less than 16 plus the number.

$24 - 3x < 16 + x$

67. Fifteen times the sum of two numbers is at least 78.

$15(a + b) \geqslant 78$

69. Familiarize. We let x represent the number of miles traveled in a day. Then the total rental cost for a day is $30 + 0.20x$.

Translate. The total cost for a day must be less than or equal to \$96. This translates to the following inequality:

$30 + 0.20x \leqslant 96$

Solve.

$0.20x \leqslant 66$ Adding -30

$x \leqslant \frac{66}{0.20}$ Multiplying by $\frac{1}{0.20}$

$x \leqslant 330$

Check. If you travel 330 miles, the total cost is $30 + 0.20(330)$, or $30 + 66 = \$96$. Any mileage less than 330 will also stay within the budget.

State. Mileage less than or equal to 330 miles allows you to stay within budget.

71. Familiarize. List the information in a table. Let x represent the score on the fourth test.

Test	Score
Test 1	89
Test 2	92
Test 3	95
Test 4	x
Total	360 or more

Translate. We can easily get an inequality from the table.

$88 + 92 + 95 + x \geqslant 360$

Solve.

$276 + x \geqslant 360$ Collecting like terms

$x \geqslant 84$ Adding -276

Check. If you get 84 on the fourth test, your total score will be 89 + 92 + 95 + 84, or 360. Any higher score will also give you an A.

State. A score of 84 or better will give you an A.

73. Familiarize. We let x represent the ticket price. Then 300x represents the total receipts from the ticket sales assuming 300 people will attend. The first band will play for \$250 + 50%(300x). The second band will play for \$550.

Translate. For school profit to be greater when the first band plays, the amount the first band charges must be less than the amount the second band charges. We now have an inequality.

$250 + 50\%(300x) < 550$

Solve.

$250 + 0.5(300x) < 550$

$250 + 150x < 550$

$150x < 300$ Subtracting 250

$x < 2$ Dividing by 150

Check. For x = \$2, the total receipts are 300(2), or \$600. The first band charges 250 + 50%(600), or \$550. The second band also charges \$550. The school profit is the same using either band. For x = \$1.99, the total receipts are 300(1.99), or \$597. The first band charges 250 + 50%(597), or \$548.50. Using the first band, the school profit would be 597 - 548.50, or \$48.50. Using the second band, the profit would be 597 - 550, or \$47. Thus, the first band produces more profit. For x = \$2.01, the total receipts are 300(2.01), or \$603. The first band charges 250 + 50%(603), or \$551.50. Using the first band, the school profit would be 603 - 551.50, or \$51.50. Using the second band, the school profit would be 603 - 550, or \$53. Thus, the second band produces more profit. For these values, the inequality x < 2 gives correct results.

State. The ticket price must be less than \$2. The highest price, rounded to the nearest cent, less than \$2 is \$1.99.

75. Familiarize. We make a table listing the information. We let x represent the total medical bill.

	Plan A	Plan B
You pay the first	100	250
Insurance pays	80%(x - 100)	90%(x - 250)
You also pay	20%(x - 100)	10%(x - 250)
Total you pay	100+20%(x-100)	250+10%(x-250)

Translate. We write an inequality stating that the amount you pay is less when you choose Plan B. This gives us an inequality.

$250 + 10\%(x - 250) < 100 + 20\%(x - 100)$

Solve.

$250 + 0.1x - 25 < 100 + 0.2x - 20$

$225 + 0.1x < 80 + 0.2x$ Collecting like terms

$145 < 0.1x$ Subtracting 80 and 0.1x

$1450 < x$ Dividing by 0.1

or

$x > 1450$

Check. We calculate for x = \$1450 and also for some amount greater than \$1450 and some amount less than \$1450.

Plan A:	Plan B:
100 + 20%(1450 - 100)	250 + 10%(1450 - 250)
100 + 0.2(1350)	250 + 0.1(1200)
100 + 270	250 + 120
\$370	\$370

When x = \$1450, you pay the same with either plan.

Plan A:	Plan B:
100 + 20%(1500 - 100)	250 + 10%(1500 - 250)
100 + 0.2(1400)	250 + 0.1(1250)
100 + 280	250 + 125
\$380	\$375

When x = \$1500, you pay less with Plan B.

Plan A:	Plan B:
100 + 20%(1400 - 100)	250 + 10%(1400 - 100)
100 + 0.2(1300)	250 + 0.1(1300)
100 + 260	250 + 130
\$360	\$380

When x = \$1400, you pay more with Plan B.

For these values, the inequality x > \$1450 gives correct results.

State. For Plan B to save you money, the total of the medical bills must be greater than \$1450.

77. Familiarize. We make a table of information.

Plan A: Monthly Income	Plan B: Monthly Income
\$500 salary 4% of sales Total: 500 + 4% of sales	\$750 salary 5% of sales over \$8000 Total: 750 + 5% of sales over 8000

Translate. We write an inequality stating that the income from Plan B is greater than the income from Plan A. We let S represent gross sales. Then S - 8000 represents gross sales over 8000.

$$750 + 5\%(S - 8000) > 500 + 4\%S$$

Solve.

$$750 + 0.05S - 400 > 500 + 0.04S$$
$$350 + 0.05S > 500 + 0.04S$$
$$0.01S > 150$$
$$S > \frac{150}{0.01}$$
$$S > 15{,}000$$

Check. We calculate for x = \$15,000 and for some amount greater than \$15,000 and some amount less than \$15,000.

Plan A:	Plan B:
500 + 4%(15,000)	750 + 5%(15,000 - 8000)
500 + 0.04(15,000)	750 + 0.05(7000)
500 + 600	750 + 350
\$1100	\$1100

When x = \$15,000, income from Plan A is equal to the income from Plan B.

Plan A:	Plan B:
500 + 4%(16,000)	750 + 5%(16,000 - 8000)
500 + 0.04(16,000)	750 + 0.05(8000)
500 + 640	750 + 400
\$1140	\$1150

When x = \$16,000, income from Plan B is greater than the income from Plan A.

Plan A:	Plan B:
500 + 4%(14,000)	750 + 5%(14,000 - 8000)
500 + 0.04(14,000)	750 + 0.05(6000)
500 + 560	750 + 300
\$1060	\$1050

When x = \$14,000, income from Plan B is less than the income from Plan A.

State. Plan B is better than Plan A when gross sales are greater than \$15,000.

79. Familiarize. We let n represent the number of hours it will take the mason to complete the job. Plan A will pay the mason 500 + 5n, while Plan B will pay the mason 8n.

Translate. We write an inequality stating that the income from the job under Plan A is greater than the income from the job under Plan B.

$$500 + 5n \geq 8n$$

Solve.

$$500 \geq 3n$$
$$\frac{500}{3} \geq n$$

or

$$n \leq 166\tfrac{2}{3}$$

Check.

Plan A:	Plan B:
$500 + 5\left(\frac{500}{3}\right)$	$8\left(\frac{500}{3}\right)$
$\frac{1500}{3} + \frac{2500}{3}$	$\frac{4000}{3}$, or \$1333.33
$\frac{4000}{3}$, or \$1333.33	

When $n = 166\frac{2}{3}$ hours, Plan A = Plan B.

We also calculate the income when n = 166 and when n = 167.

Plan A:	Plan B:
500 + 5(166)	8(166)
500 + 830	\$1328
\$1330	

When n = 166, Plan A > Plan B.

Plan A:	Plan B:
500 + 5(167)	8(167)
500 + 835	\$1336
\$1335	

When n = 167, Plan A < Plan B.

State. Plan A is better than Plan B when $n < 166\frac{2}{3}$ hours.

81. Familiarize. We want to find the values of s for which I > 36.

Translate. $2(s + 10) > 36$

Solve.

$$2s + 20 > 36$$
$$2s > 16$$
$$s > 8$$

Check. For s = 8, I = f(8) = 2(8 + 10) = 2·18 = 36. Then any U.S. size larger than 8 will give a size larger than 36 in Italy.

State. For U.S. dress sizes larger than 8, dress sizes in Italy will be larger than 36.

83. a) Familiarize. We will use the formula

$N = 12{,}197.8t + 44{,}000$.

Translate. We substitute for t.

In 1971($t = 0$): $N = 12{,}197.8(0) + 44{,}000 = 44{,}000$.

In 1981($t = 10$): $N = 12{,}197.8(10) + 44{,}000 = 121{,}978 + 44{,}000 = 165{,}978$.

In 1990($t = 19$): $N = 12{,}197.8(19) + 44{,}000 \approx 231{,}758 + 44{,}000 = 275{,}758$.

Check. We go over our calculations.

State. In 1971 there were 44,000 women in the military. In 1981 there were 165,978, and in 1990 there were about 275,758.

b) Familiarize. We will use the formula

$N = 12{,}197.8t + 44{,}000$.

Translate. The number of women is to be at least 250,000. We have the inequality

$N \geqslant 250{,}000$.

Substituting for N, we have

$12{,}197.8t + 44{,}000 \geqslant 250{,}000$.

Solve.

$12{,}197.8t + 44{,}000 \geqslant 250{,}000$

$12{,}197.8t \geqslant 206{,}000$

$t \geqslant 17$ Rounding

$1971 + 17 = 1988$

Check. We can substitute a value for t less than 17. We try 16:

$12{,}197.8(16) + 44{,}000 \approx 239{,}165$

Our result seems correct.

State. The number of women will be at least 250,000 from 1988 on.

85. $-12a + 30b\ell = 6\cdot(-2a) + 6\cdot 5b\ell = 6(-2a + 5b\ell)$

87. $4(a - 2b) - 6(2a - 5b) = 4a - 8b - 12a + 30b$

$= 4a - 12a - 8b + 30b$

$= -8a + 22b$

89. a) Familiarize. We will use

$D = 2000 - 60p$ and $S = 460 + 94p$.

Translate. Demand is to exceed supply, so we have

$D > S$, or

$2000 - 60p > 460 + 94p$.

Solve.

$2000 - 60p > 460 + 94p$

$2000 > 460 + 154p$

$1540 > 154p$

$10 > p$

89. (continued)

Check. For $p = 10$, $D = 2000 - 60\cdot 10 = 1400$ and $S = 460 + 94\cdot 10 = 1400$.

We compute D and S for some value of p less than 10 and for some value of p greater than 10.

Suppose $p = 9$:

$D = 2000 - 60\cdot 9 = 1460$ and $S = 460 + 94\cdot 9 = 1306$. Thus $D > S$.

Suppose $p = 11$:

$D = 2000 - 60\cdot 11 = 1340$ and $S = 460 + 94\cdot 11 = 1494$. Thus $D < S$.

Our result seems correct.

State. Demand exceeds supply for values of p less than 10.

b) We have seen in part (a) that $D = S$ for $p = 10$ and $D > S$ for $p < 10$. Thus $D < S$ for $p > 10$.

91. False; $-3 < -2$, but $(-3)^2 > (-2)^2$.

93. $x + 5 \leqslant 5 + x$

$5 \leqslant 5$ Subtracting x

We get an inequality that is true for all real numbers x. Thus the solution is all real numbers.

95. $0^2 = 0$, $x^2 > 0$ for $x \neq 0$

The solution is all real numbers except 0.

Exercise Set 4.5

1. $\{5,6,7,8\} \cap \{4,6,8,10\}$

The numbers 6 and 8 are common to the two sets, so the intersection is $\{6,8\}$.

3. $\{2,4,6,8\} \cap \{1,3,5\}$

There are no numbers common to the two sets, so the intersection is the empty set, $\emptyset$.

5. $\{1,2,3,4\} \cap \{1,2,3,4\}$

The numbers 1, 2, 3, and 4 are common to the two sets, so the intersection is $\{1,2,3,4\}$.

7. $1 < x < 6$

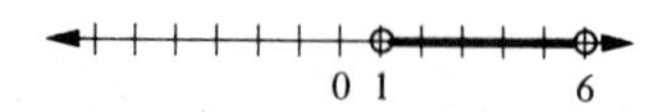

9. $6 > -x \geqslant -2$

$-6 < x \leqslant 2$ Multiplying by -1

−6 0 2

11. $-10 \leqslant 3x + 2$ and $3x + 2 < 17$

$-12 \leqslant 3x$ and $3x < 15$

$-4 \leqslant x$ and $x < 5$

The solution set is the intersection of the solution sets of the individual inequalities. The numbers common to both sets are those that are greater than or equal to -4 and less than 5. Thus the solution set is $\{x|-4 \leqslant x < 5\}$.

13. $3x + 7 \geqslant 4$ and $2x - 5 \geqslant -1$

$3x \geqslant -3$ and $2x \geqslant 4$

$x \geqslant -1$ and $x \geqslant 2$

$\{x|x \geqslant -1\} \cap \{x|x \geqslant 2\} = \{x|x \geqslant 2\}$

15. $4 - 3x \geqslant 10$ and $5x - 2 > 13$

$-3x \geqslant 6$ and $5x > 15$

$x \leqslant -2$ and $x > 3$

$\{x|x \leqslant -2\} \cap \{x|x > 3\} = \emptyset$

17. $-2 < x + 2 < 8$

$-2 - 2 < x + 2 - 2 < 8 - 2$ Subtracting 2

$-4 < x < 6$

The solution set is $\{x|-4 < x < 6\}$.

19. $1 < 2y + 5 \leqslant 9$

$1 - 5 < 2y + 5 - 5 \leqslant 9 - 5$ Subtracting 5

$-4 < 2y \leqslant 4$

$\frac{-4}{2} < \frac{2y}{2} \leqslant \frac{4}{2}$ Dividing by 2

$-2 < y \leqslant 2$

The solution set is $\{y|-2 < y \leqslant 2\}$.

21. $-10 \leqslant 3x - 5 \leqslant -1$

$-10 + 5 \leqslant 3x - 5 + 5 \leqslant -1 + 5$ Adding 5

$-5 \leqslant 3x \leqslant 4$

$\frac{-5}{3} \leqslant \frac{3x}{3} < \frac{4}{3}$ Dividing by 3

$-\frac{5}{3} \leqslant x \leqslant \frac{4}{3}$

The solution set is $\left\{x\middle|-\frac{5}{3} \leqslant x \leqslant \frac{4}{3}\right\}$.

23. $2 < x + 3 \leqslant 9$

$2 - 3 < x + 3 - 3 \leqslant 9 - 3$ Subtracting 3

$-1 < x \leqslant 6$

The solution set is $\{x|-1 < x \leqslant 6\}$.

25. $-6 \leqslant 2x - 3 < 6$

$-6 + 3 \leqslant 2x - 3 + 3 < 6 + 3$

$-3 \leqslant 2x < 9$

$\frac{-3}{2} \leqslant \frac{2x}{2} < \frac{9}{2}$

$-\frac{3}{2} \leqslant x < \frac{9}{2}$

The solution set is $\left\{x\middle|-\frac{3}{2} \leqslant x < \frac{9}{2}\right\}$.

27. $-\frac{1}{2} < \frac{1}{4}x - 3 \leqslant \frac{1}{2}$

$-\frac{1}{2} + 3 < \frac{1}{4}x - 3 + 3 \leqslant \frac{1}{2} + 3$

$\frac{5}{2} < \frac{1}{4}x \leqslant \frac{7}{2}$

$4 \cdot \frac{5}{2} < 4 \cdot \frac{1}{4}x \leqslant 4 \cdot \frac{7}{2}$

$10 < x \leqslant 14$

The solution set is $\{x|10 < x \leqslant 14\}$.

29. $-3 < \frac{2x - 5}{4} < 8$

$4(-3) < 4\left[\frac{2x - 5}{4}\right] < 4 \cdot 8$

$-12 < 2x - 5 < 32$

$-12 + 5 < 2x - 5 + 5 < 32 + 5$

$-7 < 2x < 37$

$\frac{1}{2}(-7) < \frac{1}{2} \cdot 2x < \frac{1}{2} \cdot 37$

$-\frac{7}{2} < x < \frac{37}{2}$

The solution set is $\left\{x\middle|-\frac{7}{2} < x < \frac{37}{2}\right\}$.

31. $\{4,5,6,7,8\} \cup \{1,4,6,11\}$

The numbers in either or both sets are 1, 4, 5, 6, 7, 8, and 11, so the union is $\{1,4,5,6,7,8,11\}$.

33. $\{2,4,6,8\} \cup \{1,3,5\}$

The numbers in either or both sets are 1, 2, 3, 4, 5, 6, and 8, so the union is $\{1,2,3,4,5,6,8\}$.

35. $\{4,8,11\} \cup \emptyset$

The numbers in either or both sets are 4, 8, and 11, so the union is $\{4,8,11\}$.

37. $x < -1$ or $x > 2$

−1 0 2

39. $x \leqslant -3$ or $x > 1$

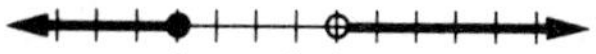

−3 0 1

41. $x + 7 < -2$ or $x + 7 > 2$

$x + 7 + (-7) < -2 + (-7)$ or $x + 7 + (-7) > 2 + (-7)$

$x < -9$ or $x > -5$

The solution set is $\{x|x < -9 \text{ or } x > -5\}$.

43. $2x - 8 \leqslant -3$ or $x - 8 \geqslant 3$

$2x - 8 + 8 \leqslant -3 + 8$ or $x - 8 + 8 \geqslant 3 + 8$

$2x \leqslant 5$ or $x \geqslant 11$

$\frac{2x}{2} \leqslant \frac{5}{2}$ or $x \geqslant 11$

$x \leqslant \frac{5}{2}$ or $x \geqslant 11$

The solution set is $\left\{x\middle|x \leqslant \frac{5}{2} \text{ or } x \geqslant 11\right\}$.

45. $7x + 4 \geqslant -17$ or $6x + 5 \geqslant -7$

$7x \geqslant -21$ or $6x \geqslant -12$

$x \geqslant -3$ or $x \geqslant -2$

The solution set is $\{x \mid x \geqslant -3\}$.

47. $7 > -4x + 5$ or $10 \leqslant -4x + 5$

$7 - 5 > -4x + 5 - 5$ or $10 - 5 \leqslant -4x + 5 - 5$

$2 > -4x$ or $5 \leqslant -4x$

$\frac{2}{-4} < \frac{-4x}{-4}$ or $\frac{5}{-4} \geqslant \frac{-4x}{-4}$

$-\frac{1}{2} < x$ or $-\frac{5}{4} \geqslant x$

The solution set is $\left\{x \middle| x \leqslant -\frac{5}{4} \text{ or } x > -\frac{1}{2}\right\}$.

49. $3x - 7 > -10$ or $5x + 2 \leqslant 22$

$3x > -3$ or $5x \leqslant 20$

$x > -1$ or $x \leqslant 4$

All real numbers are solutions.

51. $-2x - 2 < -6$ or $-2x - 2 > 6$

$-2x - 2 + 2 < -6 + 2$ or $-2x - 2 + 2 > 6 + 2$

$-2x < -4$ or $-2x > 8$

$\frac{-2x}{-2} > \frac{-4}{-2}$ or $\frac{-2x}{-2} < \frac{8}{-2}$

$x > 2$ or $x < -4$

The solution set is $\{x \mid x < -4 \text{ or } x > 2\}$.

53. $\frac{2}{3}x - 14 < -\frac{5}{6}$ or $\frac{2}{3}x - 14 > \frac{5}{6}$

$6\left(\frac{2}{3}x - 14\right) < 6\left(-\frac{5}{6}\right)$ or $6\left(\frac{2}{3}x - 14\right) > 6 \cdot \frac{5}{6}$

$4x - 84 < -5$ or $4x - 84 > 5$

$4x - 84 + 84 < -5 + 84$ or $4x - 84 + 84 > 5 + 84$

$4x < 79$ or $4x > 89$

$\frac{4x}{4} < \frac{79}{4}$ or $\frac{4x}{4} > \frac{89}{4}$

$x < \frac{79}{4}$ or $x > \frac{89}{4}$

The solution set is $\left\{x \middle| x < \frac{79}{4} \text{ or } x > \frac{89}{4}\right\}$.

55. $\frac{2x - 5}{6} \leqslant -3$ or $\frac{2x - 5}{6} \geqslant 4$

$6\left(\frac{2x - 5}{6}\right) \leqslant 6(-3)$ or $6\left(\frac{2x - 5}{6}\right) \geqslant 6 \cdot 4$

$2x - 5 \leqslant -18$ or $2x - 5 \geqslant 24$

$2x - 5 + 5 \leqslant -18 + 5$ or $2x - 5 + 5 \geqslant 24 + 5$

$2x \leqslant -13$ or $2x \geqslant 29$

$\frac{2x}{2} \leqslant \frac{-13}{2}$ or $\frac{2x}{2} \geqslant \frac{29}{2}$

$x \leqslant -\frac{13}{2}$ or $x \geqslant \frac{29}{2}$

The solution set is $\left\{x \middle| x \leqslant -\frac{13}{2} \text{ or } x \geqslant \frac{29}{2}\right\}$.

57. $(-2x^{-4}y^6)^5 = (-2)^5(x^{-4})^5(y^6)^5$

$= -32x^{-20}y^{30}$

$= -\frac{32y^{30}}{x^{20}}$

59. $\frac{-4a^5b^{-7}}{5a^{-12}b^8} = -\frac{4a^{5-(-12)}b^{-7-8}}{5}$

$= -\frac{4a^{17}b^{-15}}{5}$

$= -\frac{4a^{17}}{5b^{15}}$

61. a) Substitute $\frac{5}{9}(F - 32)$ for C in the given inequality.

$1063 \leqslant \frac{5}{9}(F - 32) < 2660$

$9 \cdot 1063 \leqslant 9 \cdot \frac{5}{9}(F - 32) < 9 \cdot 2660$

$9567 \leqslant 5(F - 32) < 23{,}940$

$9567 \leqslant 5F - 160 < 23{,}940$

$9727 \leqslant 5F < 24{,}100$

$1945.4 \leqslant F < 4820$

The inequality for Fahrenheit temperatures is $1945.4° \leqslant F < 4820°$.

b) Substitute $\frac{5}{9}(F - 32)$ for C in the given inequality.

$960.8 \leqslant \frac{5}{9}(F - 32) < 2180$

$9(960.8) \leqslant 9 \cdot \frac{5}{9}(F - 32) < 9 \cdot 2180$

$8647.2 \leqslant 5(F - 32) < 19{,}620$

$8647.2 \leqslant 5F - 160 < 19{,}620$

$8807.2 \leqslant 5F < 19{,}780$

$1761.44 \leqslant F < 3956$

The inequality for Fahrenheit temperatures is $1761.44° \leqslant F < 3956°$.

63. Find x such that $-8 < x - 3 < 8$:

$-8 < x - 3 < 8$

$-5 < x < 11$

The solution set is $\{x \mid -5 < x < 11\}$.

65. $x - 10 < 5x + 6 \leqslant x + 10$

$-10 < 4x + 6 \leqslant 10$

$-16 < 4x \leqslant 4$

$-4 < x \leqslant 1$

The solution set is $\{x \mid -4 < x \leqslant 1\}$.

67. $-\frac{2}{15} \leqslant \frac{2}{3}x - \frac{2}{5} \leqslant \frac{2}{15}$

$-\frac{2}{15} \leqslant \frac{2}{3}x - \frac{6}{15} \leqslant \frac{2}{15}$

$\frac{4}{15} \leqslant \frac{2}{3}x \leqslant \frac{8}{15}$

$\frac{3}{2} \cdot \frac{4}{15} \leqslant \frac{3}{2} \cdot \frac{2}{3}x \leqslant \frac{3}{2} \cdot \frac{8}{15}$

$\frac{2}{5} \leqslant x \leqslant \frac{4}{5}$

The solution set is $\left\{x\middle|\frac{2}{5} \leqslant x \leqslant \frac{4}{5}\right\}$.

69. $3x < 4 - 5x < 5 + 3x$

$0 < 4 - 8x < 5$

$-4 < -8x < 1$

$\frac{1}{2} > x > -\frac{1}{8}$

The solution set is $\left\{x\middle|-\frac{1}{8} < x < \frac{1}{2}\right\}$.

71. $x + 4 < 2x - 6 \leqslant x + 12$

$4 < x - 6 \leqslant 12$ Subtracting x

$10 < x \leqslant 18$

The solution set is $\{x|10 < x \leqslant 18\}$.

Exercise Set 4.6

1. $|3x| = |3|\cdot|x| = 3|x|$

3. $|9x^2| = |9|\cdot|x^2|$

$= 9|x^2|$

$= 9x^2$ x^2 is never negative

5. $|-4x^2| = |-4|\cdot|x^2| = 4|x^2| = 4x^2$

7. $|-8y| = |-8|\cdot|y| = 8|y|$

9. $\left|\frac{-4}{x}\right| = \frac{|-4|}{|x|} = \frac{4}{|x|}$

11. $\left|\frac{x^2}{-y}\right| = \frac{|x^2|}{|-y|}$

$= \frac{x^2}{|-y|}$

$= \frac{x^2}{|y|}$ The absolute value of the additive inverse of a number is the same as the absolute value of the number.

13. $\left|\frac{-8x^2}{2x}\right| = |-4x| = |-4|\cdot|x| = 4|x|$

15. $|-8 - (-42)| = |34| = 34$, or

$|-42 - (-8)| = |-34| = 34$

17. $|26 - 15| = |11| = 11$, or

$|15 - 26| = |-11| = 11$

19. $|-3.9 - 2.4| = |-6.3| = 6.3$, or

$|2.4 - (-3.9)| = |6.3| = 6.3$

21. $|-5 - 0| = |-5| = 5$, or

$|0 - (-5)| = |5| = 5$

23. $|x| = 3$

$x = -3$ or $x = 3$ Using the absolute-value principle

The solution set is $\{-3, 3\}$.

25. $|x| = -3$

The absolute value of a number is always nonnegative. Therefore, the solution set is $\emptyset$.

27. $|p| = 0$

The only number whose absolute value is 0 is 0. The solution set is $\{0\}$.

29. $|x - 3| = 12$

$x - 3 = -12$ or $x - 3 = 12$ Absolute-value principle

$x = -9$ or $x = 15$

The solution set is $\{-9, 15\}$.

31. $|2x - 3| = 4$

$2x - 3 = -4$ or $2x - 3 = 4$ Absolute-value principle

$2x = -1$ or $2x = 7$

$x = -\frac{1}{2}$ or $x = \frac{7}{2}$

The solution set is $\left\{-\frac{1}{2}, \frac{7}{2}\right\}$.

33. $|4x - 9| = 14$

$4x - 9 = -14$ or $4x - 9 = 14$

$4x = -5$ or $4x = 23$

$x = -\frac{5}{4}$ or $x = \frac{23}{4}$

The solution set $\left\{-\frac{5}{4}, \frac{23}{4}\right\}$.

35. $|x| + 7 = 18$

$|x| + 7 - 7 = 18 - 7$ Subtracting 7

$|x| = 11$

$x = -11$ or $x = 11$ Absolute-value principle

The solution set is $\{-11, 11\}$.

37. $678 = 289 + |t|$

$389 = |t|$ Subtracting 289

$t = -389$ or $t = 389$ Absolute-value principle

The solution set is $\{-389, 389\}$.

39. $|5x| = 40$

$5x = -40$ or $5x = 40$

$x = -8$ or $x = 8$

The solution set is $\{-8, 8\}$.

41. $|3x| - 4 = 17$

$|3x| = 21$ Adding 4

$3x = -21$ or $3x = 21$

$x = -7$ or $x = 7$

The solution set is $\{-7, 7\}$.

43. $5|q| - 2 = 9$

$5|q| = 11$ Adding 2

$|q| = \frac{11}{5}$ Dividing by 5

$q = -\frac{11}{5}$ or $q = \frac{11}{5}$

The solution set is $\left\{-\frac{11}{5}, \frac{11}{5}\right\}$.

45. $\left|\frac{2x - 1}{3}\right| = 5$

$\frac{2x - 1}{3} = -5$ or $\frac{2x - 1}{3} = 5$

$2x - 1 = -15$ or $2x - 1 = 15$

$2x = -14$ or $2x = 16$

$x = -7$ or $x = 8$

The solution set is $\{-7, 8\}$.

47. $|m + 5| + 9 = 16$

$|m + 5| = 7$ Subtracting 9

$m + 5 = -7$ or $m + 5 = 7$

$m = -12$ or $m = 2$

The solution set is $\{-12, 2\}$.

49. $10 - |2x - 1| = 4$

$-|2x - 1| = -6$ Subtracting 10

$|2x - 1| = 6$ Multiplying by -1

$2x - 1 = -6$ or $2x - 1 = 6$

$2x = -5$ or $2x = 7$

$x = -\frac{5}{2}$ or $x = \frac{7}{2}$

The solution set is $\left\{-\frac{5}{2}, \frac{7}{2}\right\}$.

51. $|3x - 4| = -2$

The absolute value of a number is always nonnegative. The solution set is $\emptyset$.

53. $\left|\frac{5}{9} + 3x\right| = \frac{1}{6}$

$\frac{5}{9} + 3x = -\frac{1}{6}$ or $\frac{5}{9} + 3x = \frac{1}{6}$

$3x = -\frac{13}{18}$ or $3x = -\frac{7}{18}$

$x = -\frac{13}{54}$ or $x = -\frac{7}{54}$

The solution set is $\left\{-\frac{13}{54}, -\frac{7}{54}\right\}$.

55. $|3x + 4| = |x - 7|$

$3x + 4 = x - 7$ or $3x + 4 = -(x - 7)$

$2x + 4 = -7$ or $3x + 4 = -x + 7$

$2x = -11$ or $4x + 4 = 7$

$x = -\frac{11}{2}$ or $4x = 3$

$x = \frac{3}{4}$

The solution set is $\left\{-\frac{11}{2}, \frac{3}{4}\right\}$.

57. $|x + 5| = |x - 2|$

$x + 5 = x - 2$ or $x + 5 = -(x - 2)$

$5 = -2$ or $x + 5 = -x + 2$

False - yields no solution

$2x + 5 = 2$

$2x = -3$

$x = -\frac{3}{2}$

The solution set is $\left\{-\frac{3}{2}\right\}$.

59. $|2a + 4| = |3a - 1|$

$2a + 4 = 3a - 1$ or $2a + 4 = -(3a - 1)$

$-a + 4 = -1$ or $2a + 4 = -3a + 1$

$-a = -5$ or $5a + 4 = 1$

$a = 5$ or $5a = -3$

$a = -\frac{3}{5}$

The solution set is $\left\{5, -\frac{3}{5}\right\}$.

61. $|y - 3| = |3 - y|$

$y - 3 = 3 - y$ or $y - 3 = -(3 - y)$

$2y - 3 = 3$ or $y - 3 = -3 + y$

$2y = 6$ or $-3 = -3$

$y = 3$ True for all real values of y

All real numbers are solutions.

63. $|5 - p| = |p + 8|$

$5 - p = p + 8$ or $5 - p = -(p + 8)$

$5 - 2p = 8$ or $5 - p = -p - 8$

$-2p = 3$ or $5 = -8$

$p = -\frac{3}{2}$ False

The solution set is $\left\{-\frac{3}{2}\right\}$.

65. $\left|\frac{2x - 3}{6}\right| = \left|\frac{4 - 5x}{8}\right|$

$\frac{2x - 3}{6} = \frac{4 - 5x}{8}$ or $\frac{2x - 3}{6} = -\left(\frac{4 - 5x}{8}\right)$

$24\left(\frac{2x - 3}{6}\right) = 24\left(\frac{4 - 5x}{8}\right)$ or $\frac{2x - 3}{6} = \frac{-4 + 5x}{8}$

$8x - 12 = 12 - 15x$ or $24\left(\frac{2x - 3}{6}\right) = 24\left(\frac{-4 + 5x}{8}\right)$

$23x - 12 = 12$ or $8x - 12 = -12 + 15x$

$23x = 24$ or $-7x - 12 = -12$

$x = \frac{24}{23}$ or $-7x = 0$

$x = 0$

The solution set is $\left\{\frac{24}{23}, 0\right\}$.

67. $\left|\frac{1}{2}x - 5\right| = \left|\frac{1}{4}x + 3\right|$

$\frac{1}{2}x - 5 = \frac{1}{4}x + 3$ or $\frac{1}{2}x - 5 = -\left(\frac{1}{4}x + 3\right)$

$\frac{1}{4}x - 5 = 3$ or $\frac{1}{2}x - 5 = -\frac{1}{4}x - 3$

$\frac{1}{4}x = 8$ or $\frac{3}{4}x - 5 = -3$

$x = 32$ or $\frac{3}{4}x = 2$

$x = \frac{8}{3}$

The solution set is $\left\{32, \frac{8}{3}\right\}$.

69. $|x| < 3$

$-3 < x < 3$ Part (b)

The solution set is $\{x|-3 < x < 3\}$.

71. $|x| \geqslant 2$

$x \leqslant -2$ or $x \geqslant 2$ Part (c)

The solution set is $\{x|x \leqslant -2 \text{ or } x \geqslant 2\}$.

73. $|x - 3| < 1$

$-1 < x - 3 < 1$ Part (b)

$2 < x < 4$ Adding 3

The solution set is $\{x|2 < x < 4\}$.

75. $|x + 4| \leqslant 1$

$-1 \leqslant x + 4 \leqslant 1$ Part (b)

$-5 \leqslant x \leqslant -3$ Subtracting 4

The solution set is $\{x|-5 \leqslant x \leqslant -3\}$.

77. $|2x - 3| \leqslant 4$

$-4 \leqslant 2x - 3 \leqslant 4$ Part (b)

$-1 \leqslant 2x \leqslant 7$ Adding 3

$-\frac{1}{2} \leqslant x \leqslant \frac{7}{2}$ Dividing by 2

The solution set is $\left\{x\middle|-\frac{1}{2} \leqslant x \leqslant \frac{7}{2}\right\}$.

79. $|2y - 7| > 10$

$2y - 7 < -10$ or $2y - 7 > 10$ Part (c)

$2y < -3$ or $2y > 17$ Adding 7

$y < -\frac{3}{2}$ or $y > \frac{17}{2}$ Dividing by 2

The solution set is $\left\{y\middle|y < -\frac{3}{2} \text{ or } y > \frac{17}{2}\right\}$.

81. $|4x - 9| \geqslant 14$

$4x - 9 \leqslant -14$ or $4x - 9 \geqslant 14$ Part (c)

$4x \leqslant -5$ or $4x \geqslant 23$

$x \leqslant -\frac{5}{4}$ or $x \geqslant \frac{23}{4}$

The solution set is $\left\{x\middle|x \leqslant -\frac{5}{4} \text{ or } x \geqslant \frac{23}{4}\right\}$.

83. $|y - 3| < 12$

$-12 < y - 3 < 12$ Part (b)

$-9 < y < 15$ Adding 3

The solution set is $\{y|-9 < y < 15\}$.

85. $|2x + 3| \leqslant 4$

$-4 \leqslant 2x + 3 \leqslant 4$ Part (b)

$-7 \leqslant 2x \leqslant 1$ Subtracting 3

$-\frac{7}{2} \leqslant x \leqslant \frac{1}{2}$ Dividing by 2

The solution set is $\left\{x\middle|-\frac{7}{2} \leqslant x \leqslant \frac{1}{2}\right\}$.

87. $|4 - 3y| > 8$

$4 - 3y < -8$ or $4 - 3y > 8$ Part (c)

$-3y < -12$ or $-3y > 4$ Subtracting 4

$y > 4$ or $y < -\frac{4}{3}$ Dividing by -3

The solution set is $\left\{y\middle|y < -\frac{4}{3} \text{ or } y > 4\right\}$.

89. $|9 - 4x| \geqslant 14$

$9 - 4x \leqslant -14$ or $9 - 4x \geqslant 14$ Part (c)

$-4x \leqslant -23$ or $-4x \geqslant 5$ Subtracting 9

$x \geqslant \frac{23}{4}$ or $x \leqslant -\frac{5}{4}$ Dividing by -4

The solution set is $\left\{x\middle|x \leqslant -\frac{5}{4} \text{ or } x \geqslant \frac{23}{4}\right\}$.

91. $|3 - 4x| < 21$

$-21 < 3 - 4x < 21$ Part (b)

$-24 < -4x < 18$ Subtracting 3

$6 > x > -\frac{9}{2}$ Dividing by -4 and simplifying

The solution set is $\left\{x\middle|-\frac{9}{2} < x < 6\right\}$.

93. $\left|\frac{1}{2} + 3x\right| \geq 12$

$\frac{1}{2} + 3x \leq -12$ or $\frac{1}{2} + 3x \geq 12$ Part (c)

$3x \leq -\frac{25}{2}$ or $3x \geq \frac{23}{2}$ Subtracting $\frac{1}{2}$

$x \leq -\frac{25}{6}$ or $x \geq \frac{23}{6}$ Dividing by 3

The solution set is $\left\{x \middle| x \leq -\frac{25}{6} \text{ or } x \geq \frac{23}{6}\right\}$.

95. $\left|\frac{x - 7}{3}\right| < 4$

$-4 < \frac{x - 7}{3} < 4$ Part (b)

$-12 < x - 7 < 12$ Multiplying by 3

$-5 < x < 19$ Adding 7

The solution set is $\{x|-5 < x < 19\}$.

97. $\left|\frac{2 - 5x}{4}\right| \geq \frac{2}{3}$

$\frac{2 - 5x}{4} \leq -\frac{2}{3}$ or $\frac{2 - 5x}{4} \geq \frac{2}{3}$ Part (c)

$2 - 5x \leq -\frac{8}{3}$ or $2 - 5x \geq \frac{8}{3}$ Multiplying by 4

$-5x \leq -\frac{14}{3}$ or $-5x \geq \frac{2}{3}$ Subtracting 2

$x \geq \frac{14}{15}$ or $x \leq -\frac{2}{15}$ Dividing by -5

The solution set is $\left\{x \middle| x \leq -\frac{2}{15} \text{ or } x \geq \frac{14}{15}\right\}$.

99. $|m + 5| + 9 \leq 16$

$|m + 5| \leq 7$ Subtracting 9

$-7 \leq m + 5 \leq 7$

$-12 \leq m \leq 2$

The solution set is $\{m|-12 \leq m \leq 2\}$.

101. $7 - |3 - 2x| \geq 5$

$-|3 - 2x| \geq -2$ Subtracting 7

$|3 - 2x| \leq 2$ Multiplying by -1

$-2 \leq 3 - 2x \leq 2$ Part (b)

$-5 \leq -2x \leq -1$ Subtracting 3

$\frac{5}{2} \geq x \geq \frac{1}{2}$ Dividing by -2

The solution set is $\left\{x \middle| \frac{5}{2} \geq x \geq \frac{1}{2}\right\}$, or $\left\{x \middle| \frac{1}{2} \leq x \leq \frac{5}{2}\right\}$.

103. $\left|\frac{2x - 1}{0.0059}\right| \leq 1$

$-1 \leq \frac{2x - 1}{0.0059} \leq 1$

$-0.0059 \leq 2x - 1 \leq 0.0059$

$0.9941 \leq 2x \leq 1.0059$

$0.49705 \leq x \leq 0.50295$

The solution set is $\{x|0.49705 \leq x \leq 0.50295\}$.

105. $-43.5 + (-5.8) = -49.3$

(Add absolute values and make the result negative.)

107. $-43.5(-5.8) = 252.3$

(Multiply absolute values and make the result positive.)

109. From the definition of absolute value, $|2x - 5| = 2x - 5$ only when $2x - 5 \geq 0$. Solve $2x - 5 \geq 0$.

$2x - 5 \geq 0$

$2x \geq 5$

$x \geq \frac{5}{2}$

The solution set is $\left\{x \middle| x \geq \frac{5}{2}\right\}$.

111. $|x + 5| = x + 5$

From the definition of absolute value, $|x + 5| = x + 5$ only when $x + 5 \geq 0$, or $x \geq -5$. The solution set is $\{x|x \geq -5\}$.

113. $|7x - 2| = x + 4$

From the definition of absolute value, we know $x + 4 \geq 0$, or $x \geq -4$. So we have $x \geq -4$ and

$7x - 2 = x + 4$ or $7x - 2 = -(x + 4)$

$6x = 6$ or $7x - 2 = -x - 4$

$x = 1$ or $8x = -2$

$x = -\frac{1}{4}$

The solution set is $\left\{x \middle| x \geq -4 \text{ and } x = 1 \text{ or } x = -\frac{1}{4}\right\}$, or $\left\{1, -\frac{1}{4}\right\}$.

115. $|x - 6| \leq -8$

From the definition of absolute value we know that $|x - 6| \geq 0$. Thus $|x - 6| \leq -8$ is false for all x. The solution set is $\emptyset$.

117. $|x + 5| > x$

The inequality is true for all $x < 0$ (because absolute value must be nonnegative). The solution set in this case is $\{x|x < 0\}$. If $x = 0$, we have $|0 + 5| > 0$, which is true. The solution set in this case is $\{0\}$. If $x > 0$, we have the following:

$x + 5 < -x$ or $x + 5 > x$

$2x < -5$ or $5 > 0$

$x < -\frac{5}{2}$

Although $x > 0$ and $x < -\frac{5}{2}$ yields no solution, $x > 0$ and $5 > 0$ (true for all x) yield the solution set $\{x|x > 0\}$ in this case. The solution set for the inequality is $\{x|x < 0\} \cup \{0\} \cup \{x|x > 0\}$, or all real numbers.

119. Using part (b), we find that $-3 < x < 3$ is equivalent to $|x| < 3$.

121. Using part (c), we find that $x \leq -6$ or $x \geq 6$ is equivalent to $|x| \geq 6$.

123.

$x < -8$ or $x > 2$	
$x < -5$ or $x > 5$	Adding 3
$|x + 3| > 5$	Part (c)

Exercise Set 5.1

1.

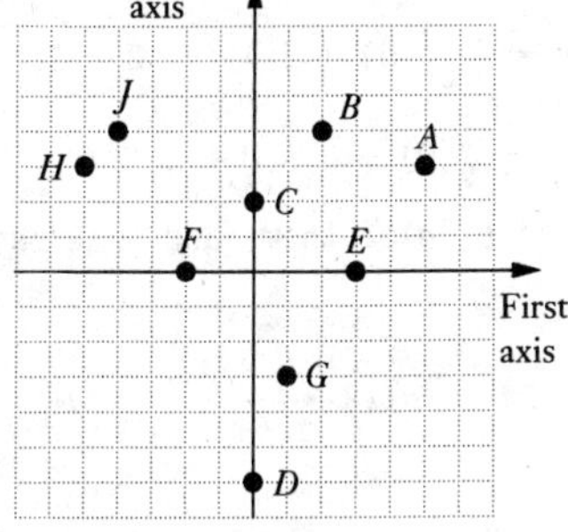

A(5,3) is 5 units right and 3 units up.
B(2,4) is 2 units right and 4 units up.
C(0,2) is 0 units left or right and 2 units up.
D(0,-6) is 0 units left or right and 6 units down.
E(3,0) is 3 units right and 0 units up or down.
F(-2,0) is 2 units left and 0 units up or down.
G(1,-3) is 1 unit right and 3 units down.
H(-5,3) is 5 units left and 3 units up.
J(-4,4) is 4 units left and 4 units up.

3.

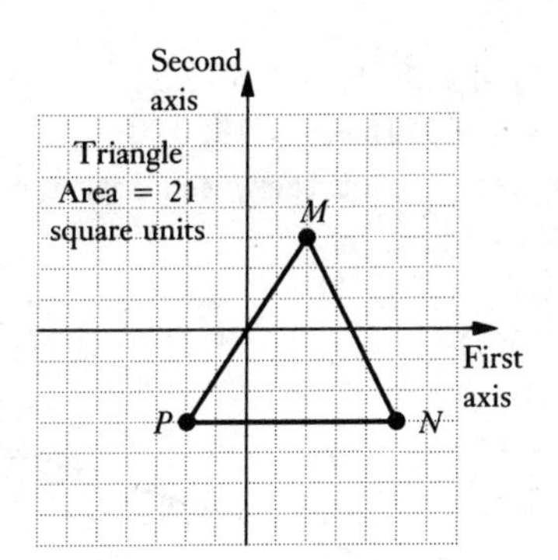

A triangle is formed. The area of a triangle is found by using the formula $A = \frac{1}{2}bh$. In this triangle the base and height are respectively 7 units and 6 units.

$A = \frac{1}{2}bh = \frac{1}{2} \cdot 7 \cdot 6 = \frac{42}{2} = 21$ square units

5. We substitute 10 for x and 3 for y.

$3x - 7y = 9$	
$3 \cdot 10 - 7 \cdot 3$	9
$30 - 21$	
9	

The equation becomes true: (10,3) is a solution.

7. We substitute 4 for x and -2 for y.

$4y - 5x = 7 + 3x$	
$4(-2) - 5(4)$	$7 + 3(4)$
$-8 - 20$	$7 + 12$
-28	19

The equation becomes false: (4,-2) is not a solution.

9. We substitute 2 for p and 3 for q.

$2p + q = 5$	
$2 \cdot 2 + 3$	5
$4 + 3$	
7	

The equation becomes false: (2,3) is not a solution.

11. Graph: $y = 5x$

We choose any number for x and then determine y. We find several ordered pairs in this manner, plot them, and draw the line.

When $x = 0$, $y = 5 \cdot 0 = 0$.

When $x = -1$, $y = 5(-1) = -5$.

When $x = 1$, $y = 5 \cdot 1 = 5$.

When $x = \frac{1}{2}$, $y = 5 \cdot \frac{1}{2} = \frac{5}{2}$, or $2\frac{1}{2}$.

x	y	(x,y)
0	0	(0,0)
-1	-5	(-1,-5)
1	5	(1,5)
$\frac{1}{2}$	$\frac{5}{2}$	$(\frac{1}{2},\frac{5}{2})$

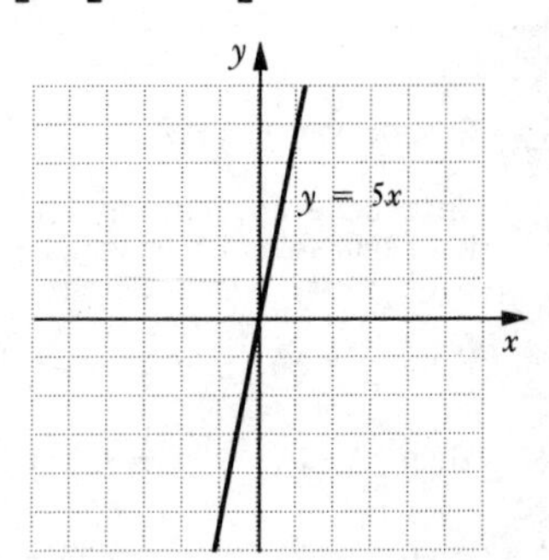

13. Graph: $y = -3x$

We choose any number for x and then determine y. We find several ordered pairs in this manner, plot them, and draw the line.

When $x = 0$, $y = -3 \cdot 0 = 0$.

When $x = -2$, $y = -3(-2) = 6$.

When $x = 1$, $y = -3 \cdot 1 = -3$.

When $x = 2$, $y = -3 \cdot 2 = -6$.

x	y	(x,y)
0	0	(0,0)
-2	6	(-2,6)
1	-3	(1,-3)
2	-6	(2,-6)

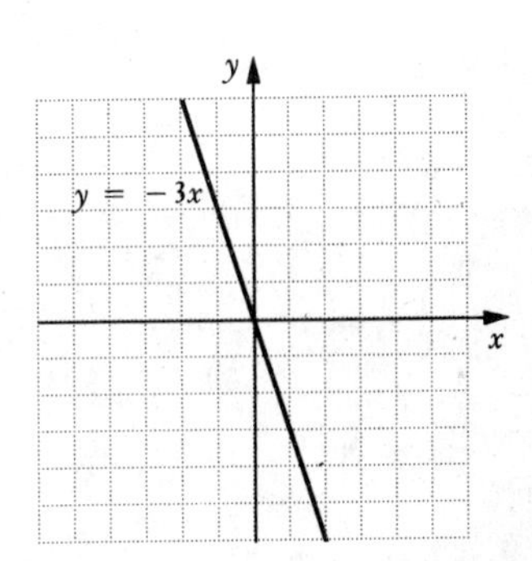

15. Graph: $y = x + 3$

We choose any number for x and then determine y. We find several ordered pairs in this manner, plot them, and draw the line.

When $x = -4$, $y = -4 + 3 = -1$.

When $x = -1$, $y = -1 + 3 = 2$.

When $x = 0$, $y = 0 + 3 = 3$.

When $x = 2$, $y = 2 + 3 = 5$.

x	y	(x,y)
-4	-1	(-4,-1)
-1	2	(-1,2)
0	3	(0,3)
2	5	(2,5)

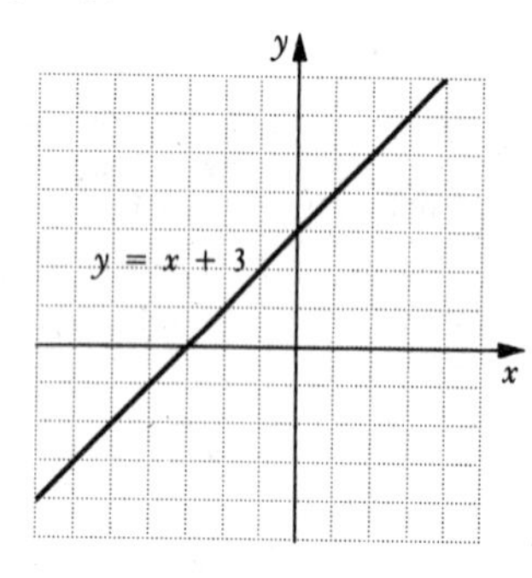

17. Graph: $y = \frac{1}{4}x + 2$

We choose any number for x and then determine y. We find several ordered pairs in this manner, plot them, and draw the line.

When $x = -4$, $y = \frac{1}{4}(-4) + 2 = -1 + 2 = 1$.

When $x = 0$, $y = \frac{1}{4} \cdot 0 + 2 = 0 + 2 = 2$.

When $x = 2$, $y = \frac{1}{4} \cdot 2 + 2 = \frac{1}{2} + 2 = 2\frac{1}{2}$.

When $x = 4$, $y = \frac{1}{4} \cdot 4 + 2 = 1 + 2 = 3$.

x	y	(x,y)
-4	1	(-4,1)
0	2	(0,2)
2	$2\frac{1}{2}$	$(2,2\frac{1}{2})$
4	3	(4,3)

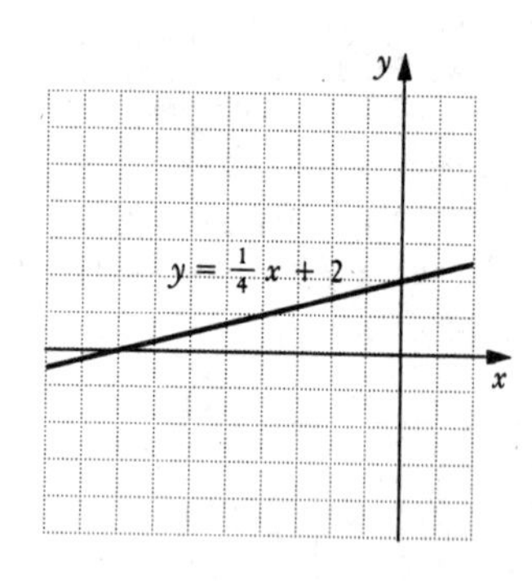

19. Graph: $y = -\frac{1}{5}x + 2$

We choose any number for x and then determine y. For example,

when $x = -5$, $y = -\frac{1}{5}(-5) + 2 = 1 + 2 = 3$;

when $x = 0$, $y = -\frac{1}{5}(0) + 2 = 0 + 2 = 2$.

We compute other pairs, plot them, and draw the line.

x	y	(x,y)
-5	3	(-5,3)
0	2	(0,2)
2	$\frac{8}{5}$	$\left(2,\frac{8}{5}\right)$
5	1	(5,1)

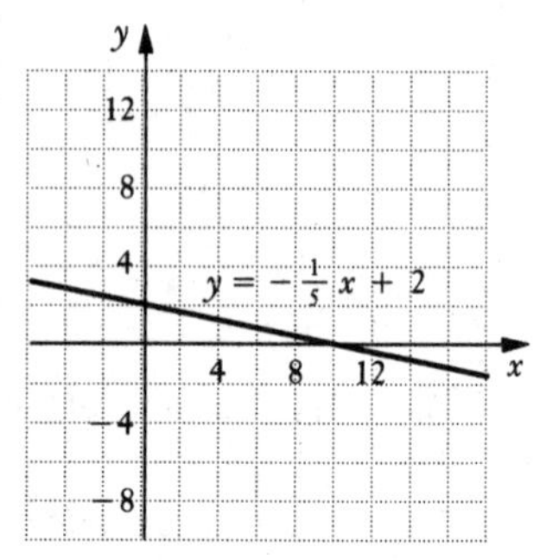

21. Graph: $y = 0.3x - 5$

We choose any number for x and then determine y. For example,

when $x = -1$, $y = 0.3(-1) - 5 = -5.3$;

when $x = 0$, $y = 0.3(0) - 5 = -5$.

We compute other pairs, plot them, and draw the line.

x	y	(x,y)
-1	-5.3	(-1,-5.3)
0	-5	(0,-5)
2	-4.4	(2,-4.4)
5	-3.5	(5,-3.5)

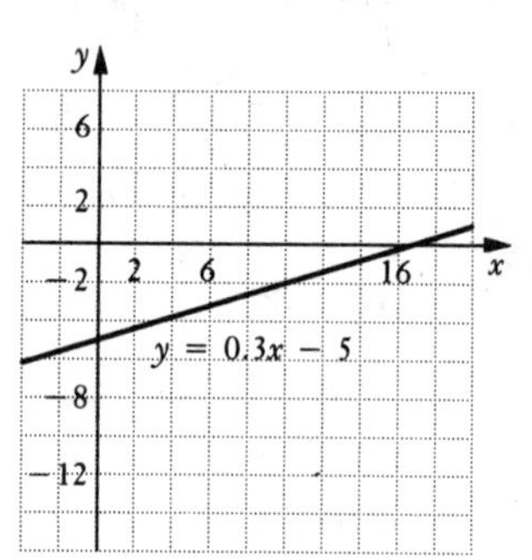

23. a) $C = \frac{5}{9}(F - 32)$

We let F = 0: $C = \frac{5}{9}(0 - 32) = \frac{5}{9}(-32) = -\frac{160}{9} = -17\frac{7}{9}°$

We let F = 32: $C = \frac{5}{9}(32 - 32) = \frac{5}{9} \cdot 0 = 0°$

We let F = 41: $C = \frac{5}{9}(41 - 32) = \frac{5}{9} \cdot 9 = 5°$

We let F = 212: $C = \frac{5}{9}(212 - 32) = \frac{5}{9} \cdot 180 = 100°$

b) Using the values computed in part (a), we make a table of values. (Note that this is a situation where alphabetical order of variables is not used.) Because the range of values of F is larger than the range of values of C, we scale the axes differently. We plot the points and draw the graph.

23. (continued)

F	C	(F, C)
0	$-17\frac{7}{9}$	$\left(0,-17\frac{7}{9}\right)$
32	0	(32,0)
41	5	(41,5)
212	100	(212,100)

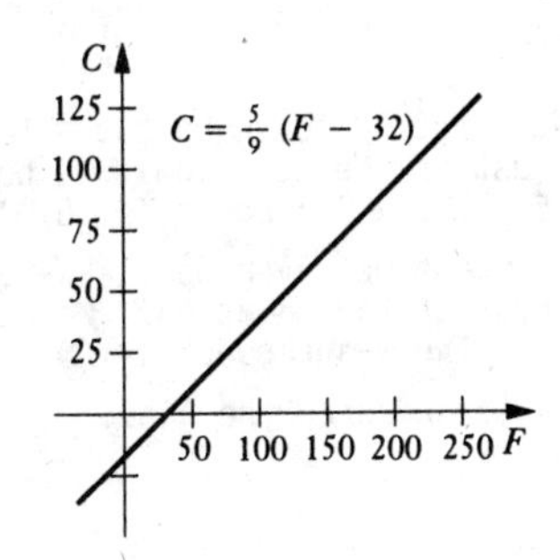

25. $-6x + 2x - 32 < 64$

$-4x - 32 < 64$ Collecting like terms

$-4x < 96$ Adding 32

$x > -24$ Dividing by -4

The solution set is $\{x|x > -24\}$.

27. $128 \div \left(-\frac{1}{2}\right) \div (-64) = 128\cdot(-2) \div (-64)$

$= -256 \div (-64)$

$= 4$

Exercise Set 5.2

1. $x - 3y = 6$

To find the x-intercept we cover up the y-term and look at the rest of the equation. We have $x = 6$. The x-intercept is (6,0).

To find the y-intercept we cover up the x-term and look at the rest of the equation. We have $-3y = 6$, or $y = -2$. The y-intercept is (0,-2).

We plot these points and draw the line.

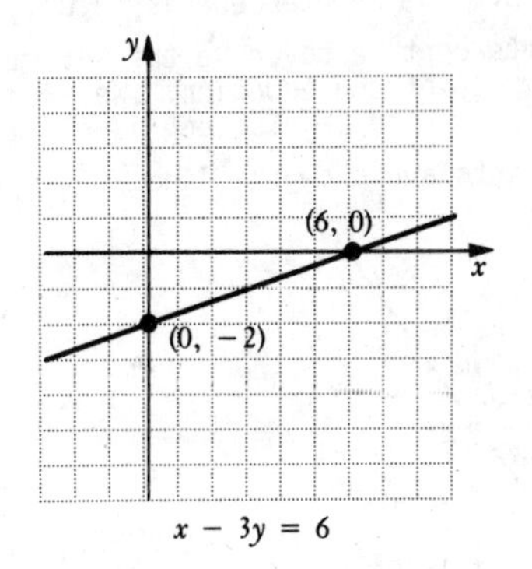

1. (continued)

We use a third point as a check. We choose $x = -3$ and solve for y.

$-3 - 3y = 6$

$-3y = 9$

$y = -3$

We plot (-3,-3) and note that it is on the line.

3. $x + 2y = 4$

To find the x-intercept we cover up the y-term and look at the rest of the equation. We have $x = 4$. The x-intercept is (4,0).

To find the y-intercept we cover up the x-term and look at the rest of the equation. We have $2y = 4$, or $y = 2$. The y-intercept is (0,2).

We plot these points and draw the line.

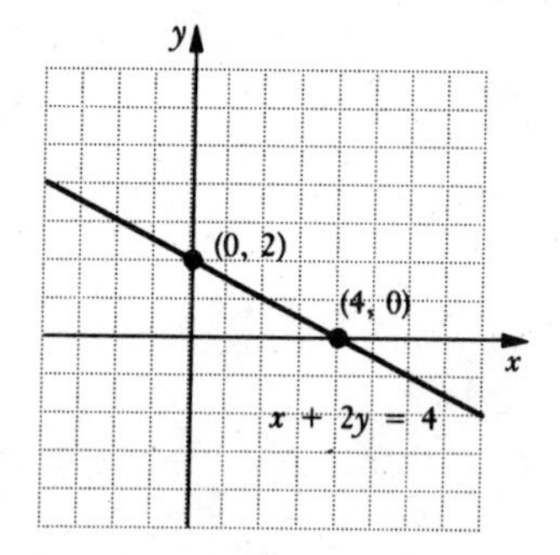

We use a third point as a check. We choose $x = -4$ and solve for y.

$-4 + 2y = 4$

$2y = 8$

$y = 4$

We plot (-4,4) and note that it is on the line.

5. $5x - 2y = 10$

To find the x-intercept we cover up the y-term and look at the rest of the equation. We have $5x = 10$, or $x = 2$. The x-intercept is (2,0).

To find the y-intercept we cover up the x-term and look at the rest of the equation. We have $-2y = 10$, or $y = -5$. The y-intercept is (0,-5).

We plot these points and draw the line.

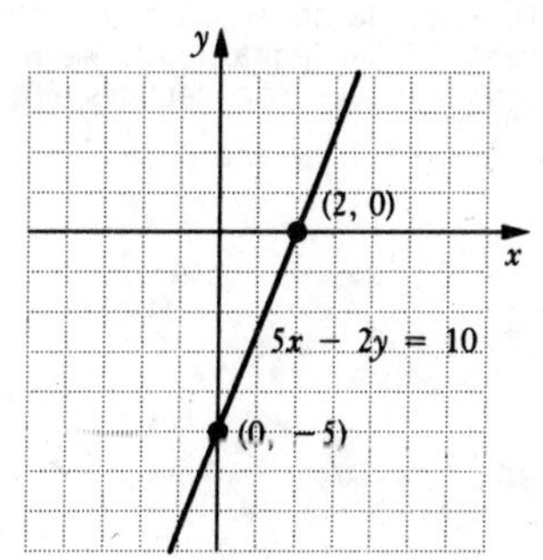

5. (continued)

We use a third point as a check. We choose $x = 4$ and solve for y.

$$5(4) - 2y = 10$$
$$20 - 2y = 10$$
$$-2y = -10$$
$$y = 5$$

We plot (4,5) and note that it is on the line.

7. $5y = -15 + 3x$

To find the x-intercept we let $y = 0$ and solve for x. We have $0 = -15 + 3x$, or $15 = 3x$, or $5 = x$. The x-intercept is (5,0).

To find the y-intercept we cover up the x-term and look at the rest of the equation. We have $5y = -15$, or $y = -3$. The y-intercept is (0,-3).

We plot these points and draw the line.

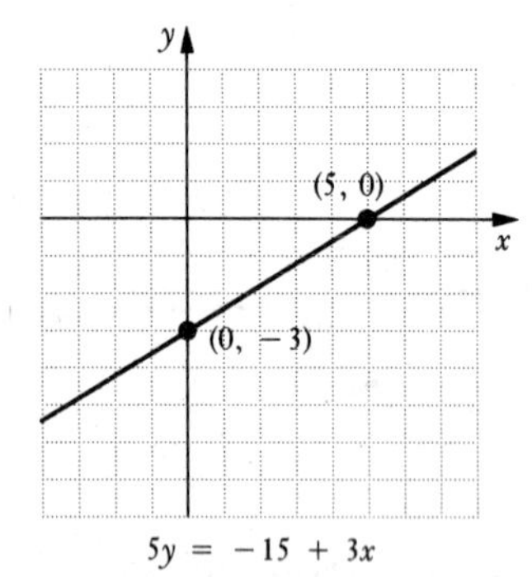

$5y = -15 + 3x$

We use a third point as a check. We choose $x = -5$ and solve for y.

$$5y = -15 + 3(-5)$$
$$5y = -15 - 15$$
$$5y = -30$$
$$y = -6$$

We plot (-5,-6) and note that it is on the line.

9. $5x - 10 = 5y$

To find the x-intercept we let $y = 0$ and solve for x. We have $5x - 10 = 0$, or $5x = 10$, or $x = 2$. The x-intercept is (2,0).

To find the y-intercept we cover up the x-term and look at the rest of the equation. We have $-10 = 5y$, or $-2 = y$. The y-intercept is (0,-2).

We plot these points and draw the line.

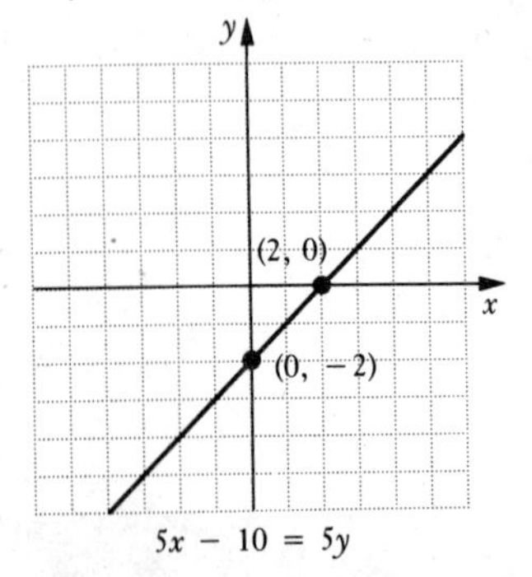

$5x - 10 = 5y$

9. (continued)

We use a third point as a check. We choose $x = 5$ and solve for y.

$$5(5) - 10 = 5y$$
$$25 - 10 = 5y$$
$$15 = 5y$$
$$3 = y$$

We plot (5,3) and note that it is on the line.

11. $4x + 5y = 20$

To find the x-intercept we cover up the y-term and look at the rest of the equation. We have $4x = 20$, or $x = 5$. The x-intercept is (5,0).

To find the y-intercept we cover up the x-term and look at the rest of the equation. We have $5y = 20$, or $y = 4$. The y-intercept is (0,4).

We plot these points and draw the line.

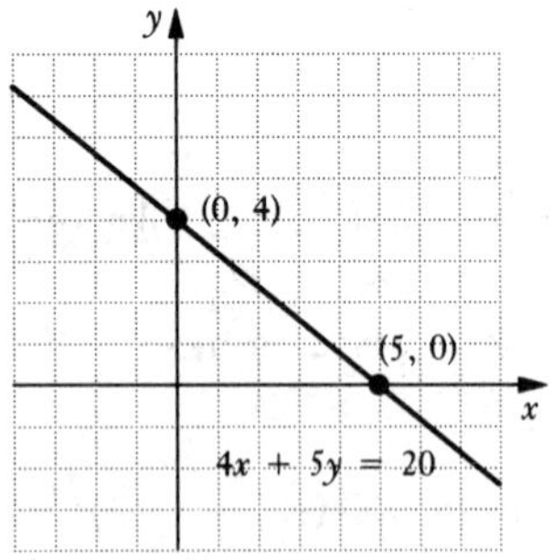

We use a third point as a check. We choose $x = 3$ and solve for y.

$$4(3) + 5y = 20$$
$$12 + 5y = 20$$
$$5y = 8$$
$$y = \frac{8}{5}$$

We plot $\left(3,\frac{8}{5}\right)$ and note that it is on the line.

13. $4x - 3y = 12$

To find the x-intercept we cover up the y-term and look at the rest of the equation. We have $4x = 12$, or $x = 3$. The x-intercept is (3,0).

To find the y-intercept we cover up the x-term and look at the rest of the equation. We have $-3y = 12$, or $y = -4$. The y-intercept is (0,-4).

We plot these points and draw the line.

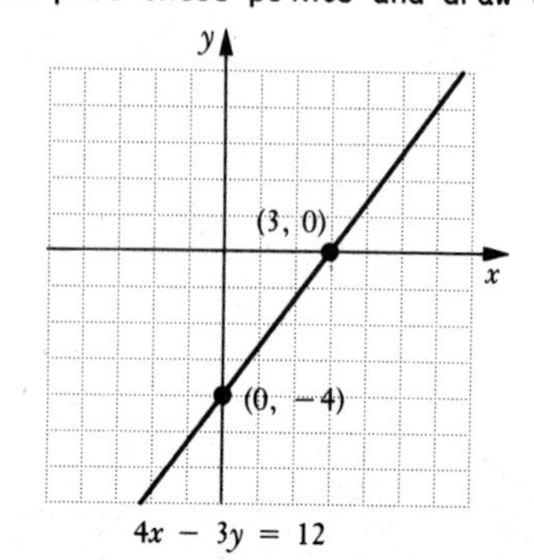

$4x - 3y = 12$

13. (continued)

We use a third point as a check. We choose x = 5 and solve for y.

$$4 \cdot 5 - 3y = 12$$
$$20 - 3y = 12$$
$$-3y = -8$$
$$y = \frac{8}{3}$$

We plot $\left(5, \frac{8}{3}\right)$ and note that it is on the line.

15. y - 3x = 0

To find the x-intercept we cover up the y-term and look at the rest of the equation. We have -3x = 0, or x = 0. The x-intercept is (0,0). This is also the y-intercept. To find another point we let x = 2 and solve for y.

$$y - 3 \cdot 2 = 0$$
$$y - 6 = 0$$
$$y = 6$$

We plot (0,0) and (2,6) and draw the line.

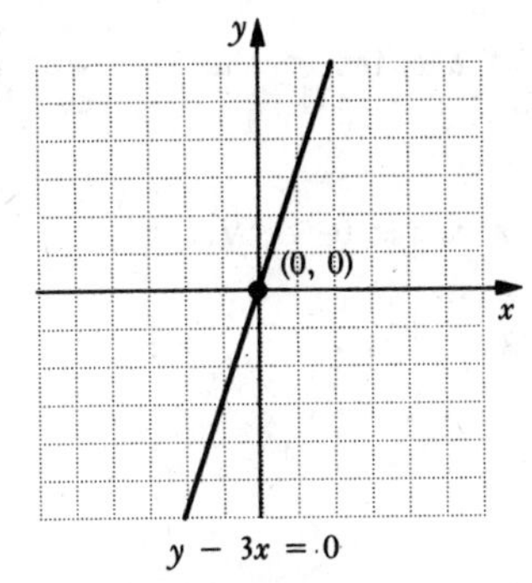

y − 3x = 0

We use a third point as a check. We choose x = -1 and solve for y.

$$y - 3(-1) = 0$$
$$y + 3 = 0$$
$$y = -3$$

We plot (-1,-3) and note that it is on the line.

17. 6x - 7 + 3y = 9x - 2y + 8

-3x + 5y = 15   Collecting like terms

To find the x-intercept we cover up the y-term and look at the rest of the equation. We have -3x = 15, or x = -5. The x-intercept is (-5,0).

To find the y-intercept we cover up the x-term and look at the rest of the equation. We have 5y = 15, or y = 3. The y-intercept is (0,3).

We plot these points and draw the line.

17. (continued)

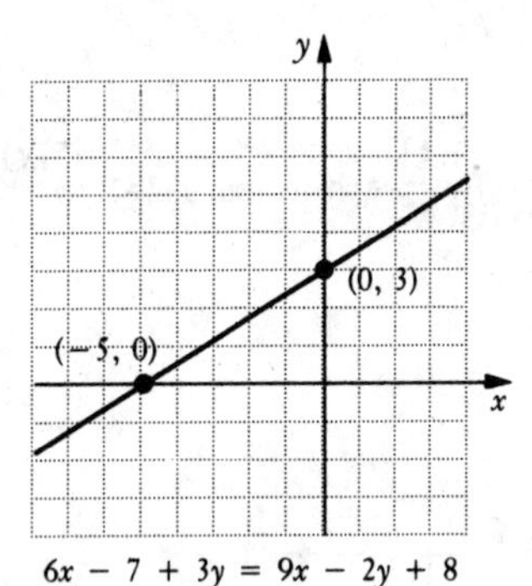

6x − 7 + 3y = 9x − 2y + 8

We use a third point as a check. We choose x = 5 and solve for y.

$$-3 \cdot 5 + 5y = 15$$
$$-15 + 5y = 15$$
$$5y = 30$$
$$y = 6$$

We plot (5,6) and note that it is on the line.

19. x = 4

Since y is missing any number for y will do. Thus all ordered pairs (4,y) are solutions. The graph is parallel to the y-axis.

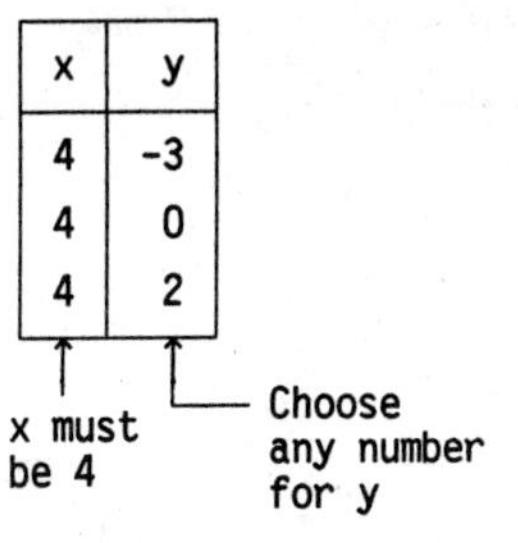

x	y
4	-3
4	0
4	2

x must be 4

Choose any number for y

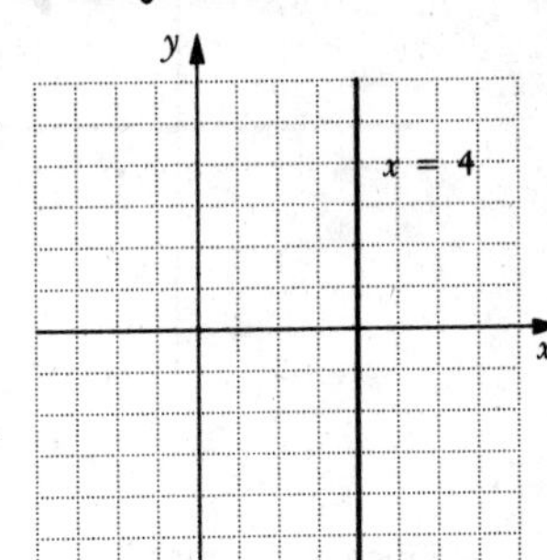

21. y = -2

Since x is missing any number for x will do. Thus all ordered pairs (x,-2) are solutions. The graph is parallel to the x-axis.

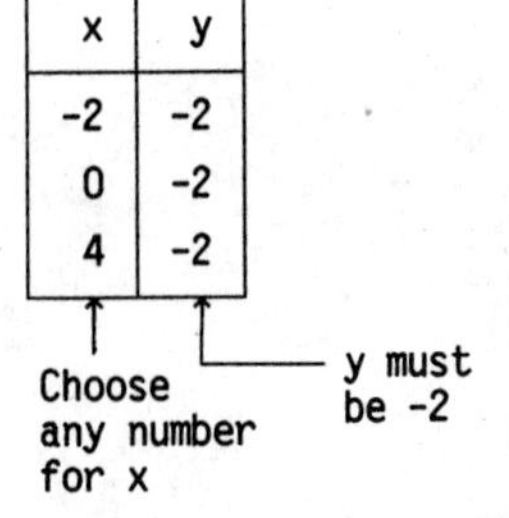

x	y
-2	-2
0	-2
4	-2

Choose any number for x

y must be -2

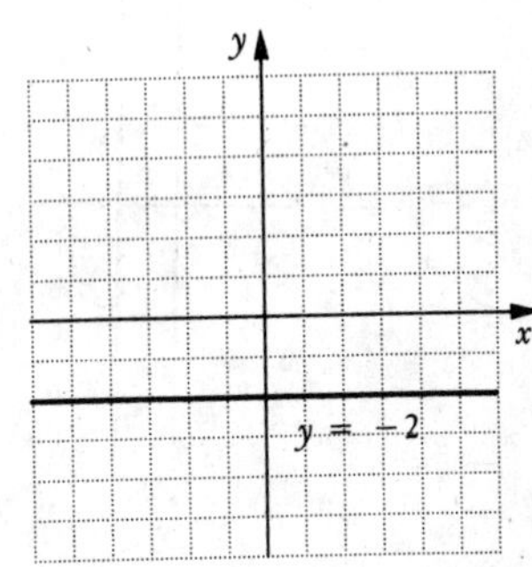

23. $3x + 15 = 0$

$$3x = -15$$
$$x = -5$$

Since y is missing all ordered pairs $(-5,y)$ are solutions. The graph is parallel to the y-axis.

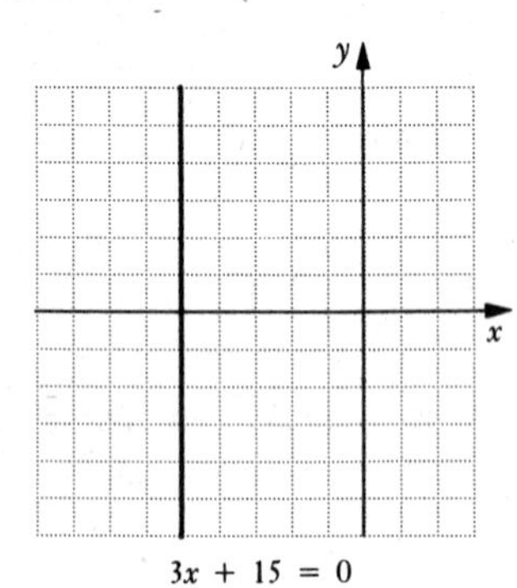

$3x + 15 = 0$

25. $y = 0$

Since x is missing all ordered pairs $(x,0)$ are solutions. The graph is the x-axis.

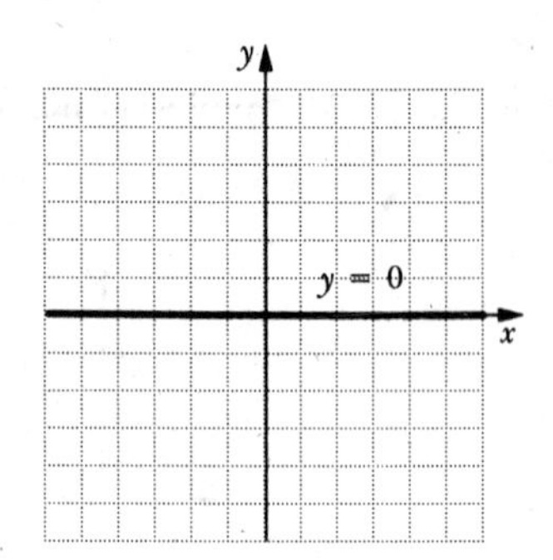

27. $x = \frac{3}{2}$

Since y is missing all ordered pairs $\left(\frac{3}{2},y\right)$ are solutions. The graph is parallel to the y-axis.

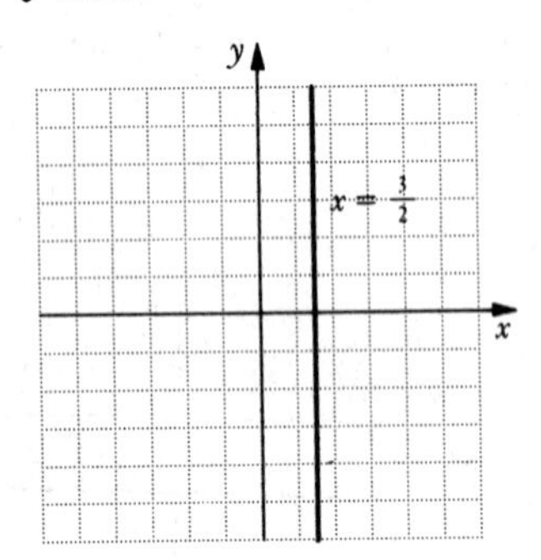

29. Familiarize. We will use the formula $I = Prt$.

Translate. We substitute \$17.60 for I, \$320 for P, and $\frac{1}{2}$ for t.

$$I = Prt$$
$$\$17.60 = \$320(r)\left(\frac{1}{2}\right)$$

Solve.

$$\$17.60 = \$320(r)\left(\frac{1}{2}\right)$$
$$\$17.60 = \$160r$$
$$0.11 = r \qquad \text{Dividing by \$160}$$

Check. $\$320(0.11)\left(\frac{1}{2}\right) = \$17.60$. The number checks.

State. The interest rate would have to be 0.11, or 11%.

31. All points on the y-axis are pairs of the form $(0,y)$. Thus any number for y will do and x must be 0. The equation is $x = 0$.

33. The x-coordinate must be -4, and the y-coordinate must be 5. The point is $(-4,5)$.

35. The x-coordinate of a point on the line must be 12, and any number for y will do. The equation is $x = 12$.

37. We substitute 4 for x and 0 for y.

$$y = mx + 3$$
$$0 = m(4) + 3$$
$$-3 = 4m$$
$$-\frac{3}{4} = m$$

Exercise Set 5.3

1. Let $(3,8) = (x_1,y_1)$ and $(9,-4) = (x_2,y_2)$.

$$\text{Slope} = \frac{y_2 - y_1}{x_2 - x_1} = \frac{-4 - 8}{9 - 3} = \frac{-12}{6} = -2$$

3. Let $(-8,-7) = (x_1,y_1)$ and $(-9,-12) = (x_2,y_2)$.

$$\text{Slope} = \frac{y_2 - y_1}{x_2 - x_1} = \frac{-12 - (-7)}{-9 - (-8)} = \frac{-5}{-1} = 5$$

5. Let $(-16.3,12.4) = (x_1,y_1)$ and $(-5.2,8.7) = (x_2,y_2)$.

$$\text{Slope} = \frac{y_2 - y_1}{x_2 - x_1} = \frac{8.7 - 12.4}{-5.2 - (-16.3)} = \frac{-3.7}{11.1} = -\frac{37}{111} = -\frac{1}{3}$$

7. Let $(3.2,-12.8) = (x_1,y_1)$ and $(3.2,2.4) = (x_2,y_2)$.

$$\text{Slope} = \frac{y_2 - y_1}{x_2 - x_1} = \frac{2.4 - (-12.8)}{3.2 - 3.2} = \frac{15.2}{0}$$

Since we cannot divide by 0, the slope is undefined.

9. $3x = 12 + y$

$3x - 12 = y$

$y = 3x - 12$ ($y = mx + b$)

The slope is 3.

11. $5x - 6 = 15$

$5x = 21$

$x = \frac{21}{5}$

When y is missing, the line is parallel to the y-axis. The line is vertical, and the slope is undefined.

13. $5y = 6$

$y = \frac{6}{5}$

When x is missing, the line is parallel to the x-axis. The line is horizontal and has a slope of 0.

15. $y - 6 = 14$

$y = 20$

When x is missing, the line is parallel to the x-axis. The line is horizontal and has a slope of 0.

17. $12 - 4x = 9 + x$

$3 = 5x$

$\frac{3}{5} = x$

When y is missing, the line is parallel to the y-axis. The line is vertical, and the slope is undefined.

19. $2y - 4 = 35 + x$

$2y = x + 39$

$y = \frac{1}{2}x + \frac{39}{2}$ ($y = mx + b$)

The slope is $\frac{1}{2}$.

21. $3y + x = 3y + 2$

$x = 2$

When y is missing, the line is parallel to the y-axis. The line is vertical, and the slope is undefined.

23. $3y - 2x = 5 + 9y - 2x$

$3y = 5 + 9y$

$0 = 5 + 6y$

$-5 = 6y$

$-\frac{5}{6} = y$

When x is missing, the line is parallel to the x-axis. The line is horizontal and has a slope of 0.

25. $2y - 7x = 10 - 3x$

$2y = 4x + 10$

$y = 2x + 5$ ($y = mx + b$)

The slope is 2.

27. $y = \underbrace{-8}x\underbrace{-9}$

The slope is -8. The y-intercept is (0,-9).

29. $y = 3.8x$

Think of this as $y = 3.8x + 0$.

The slope is 3.8; the y-intercept is (0,0).

31. $2x + 3y = 8$

$3y = -2x + 8$

$y = -\frac{2}{3}x + \frac{8}{3}$

The slope is $-\frac{2}{3}$; the y-intercept is $\left(0,\frac{8}{3}\right)$.

33. $-8x - 7y = 24$

$-7y = 8x + 24$

$y = -\frac{8}{7}x - \frac{24}{7}$

The slope is $-\frac{8}{7}$; the y-intercept is $\left(0,-\frac{24}{7}\right)$.

35. $9x = 3y + 6$

$9x - 6 = 3y$

$3x - 2 = y$

The slope is 3; the y-intercept is (0,-2).

37. $-6x = 4y + 3$

$-6x - 3 = 4y$

$-\frac{3}{2}x - \frac{3}{4} = y$

The slope is $-\frac{3}{2}$; the y-intercept is $\left(0,-\frac{3}{4}\right)$.

39. We use the slope-intercept equation and substitute 5 for m and 8 for b.

$y = mx + b$

$y = 5x + 8$

41. We use the slope-intercept equation and substitute 5.8 for m and -1 for b.

$y = mx + b$

$y = 5.8x - 1$

43. We use the slope-intercept equation and substitute $-\frac{7}{3}$ for m and -5 for b.

$y = mx + b$

$y = -\frac{7}{3}x - 5$

45. $y = \frac{5}{2}x + 1$

First we plot the y-intercept (0,1). Then we consider the slope $\frac{5}{2}$. Starting at the y-intercept and using the slope, we find another point by moving 5 units up and 2 units to the right. We get to a new point (2,6).

We can also think of the slope as $\frac{-5}{-2}$. We again start at the y-intercept (0,1). We move 5 units down and 2 units to the left. We get to another new point (-2,-4). We plot the points and draw the line.

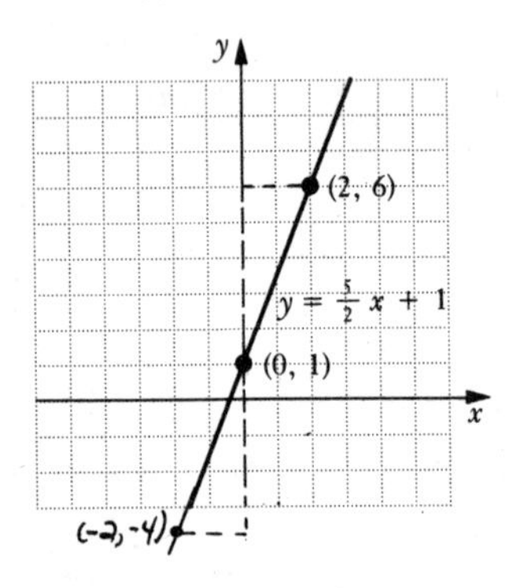

47. $y = -\frac{5}{2}x - 4$

First we plot the y-intercept (0,-4). We can think of the slope as $\frac{-5}{2}$. Starting at the y-intercept and using the slope, we find another point by moving 5 units down and 2 units to the right. We get to a new point (2,-9).

We can also think of the slope as $\frac{5}{-2}$. We again start at the y-intercept (0,-4). We move 5 units up and 2 units to the left. We get to another new point (-2,1). We plot the points and draw the line.

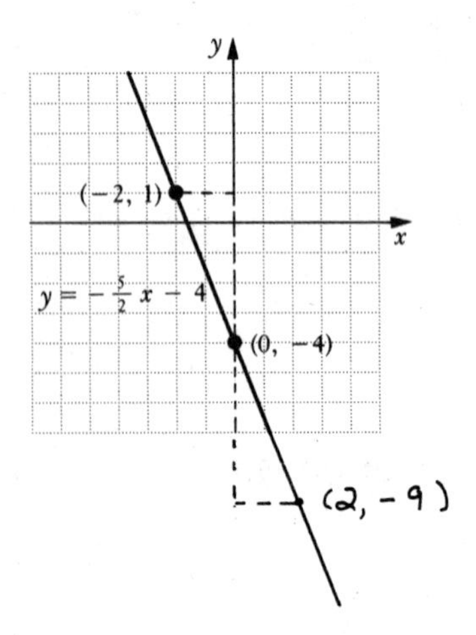

49. $y = 2x - 5$

First we plot the y-intercept (0,-5). We can think of the slope as $\frac{2}{1}$. Starting at the y-intercept and using the slope, we find another point by moving 2 units up and 1 unit to the right. We get to a new point (1,-3).

We can also think of the slope as $\frac{-2}{-1}$. We again start at the y-intercept (0,-5). We move 2 units down and 1 unit to the left. We get to another new point (-1,-7). We plot the points and draw the line.

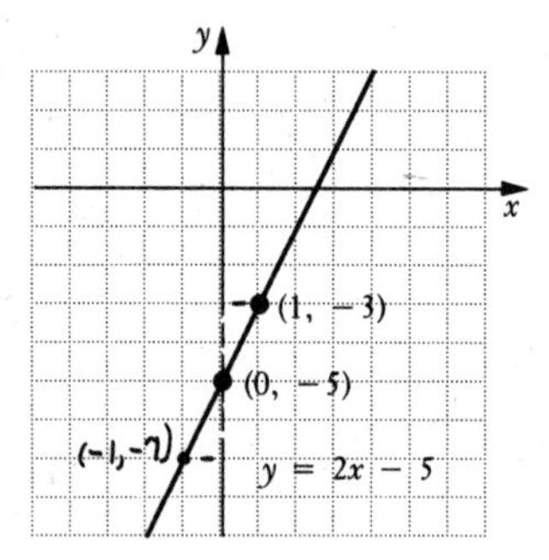

51. $y = \frac{1}{3}x + 6$

First we plot the y-intercept (0,6). Then we consider the slope $\frac{1}{3}$. Starting at the y-intercept and using the slope, we find another point by moving 1 unit up and 3 units to the right. We get to a new point (3,7).

We can also think of the slope as $\frac{-1}{-3}$. We again start at the y-intercept (0,6). We move 1 unit down and 3 units to the left. We get to another new point (-3,5). We plot the points and draw the line.

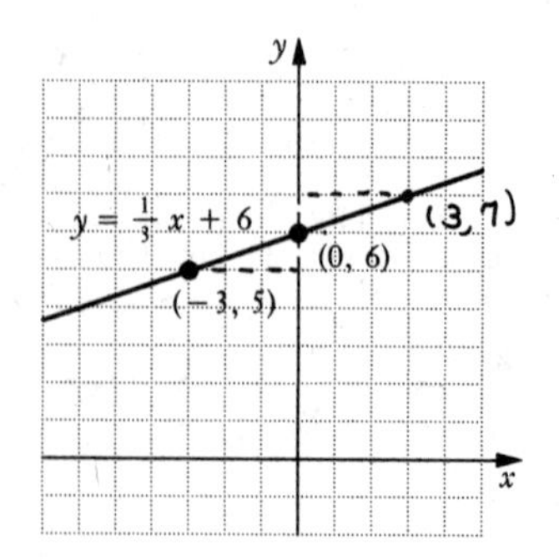

53. $y = -0.25x + 2$, or $y = -\frac{1}{4}x + 2$

First we plot the y-intercept (0,2). We can think of the slope as $\frac{-1}{4}$. Starting at the y-intercept and using the slope, we move 1 unit down and 4 units to the right. We get to a new point (4,1). We can also think of the slope as $\frac{1}{-4}$. We again start at the y-intercept (0,2). We move 1 unit up and 4 units to the left. We get to another new point (-4,3). We plot the points and draw the graph.

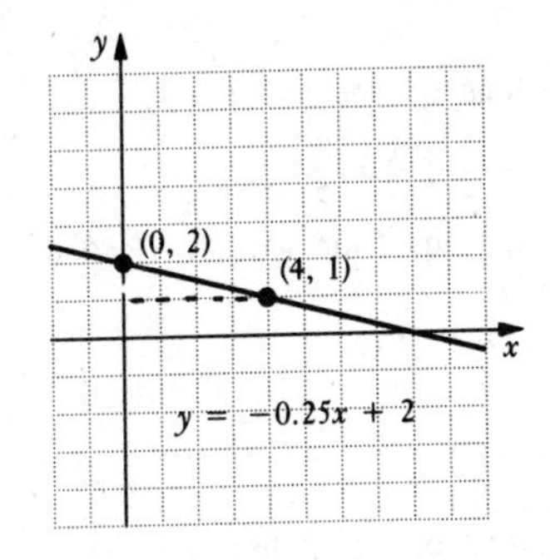

55. $y = -\frac{3}{4}x$, or $y = -\frac{3}{4}x + 0$

First we plot the y-intercept (0,0). We can think of the slope as $\frac{-3}{4}$. Starting at the y-intercept we move 3 units down and 4 units to the right. We get to a new point (4,-3).

We can also think of the slope as $\frac{3}{-4}$. We again start at the y-intercept (0,0). We move 3 units up and 4 units to the left. We get to another new point (-4,3). We plot the points and draw the graph.

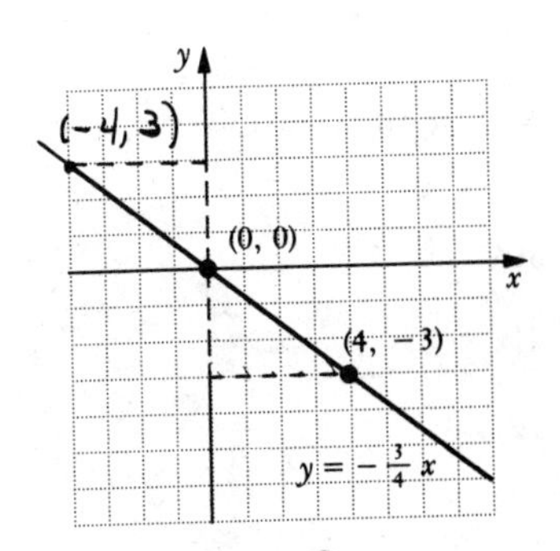

57. Grade $= \frac{\text{vertical change}}{\text{horizontal change}} = \frac{211.2 \text{ ft}}{5280 \text{ ft}} = 0.04$

The grade of the road is 4%.

59. The grade is 13%, or 0.13. Let v represent the height of the end of the treadmill.

$$\text{Grade} = \frac{\text{vertical change}}{\text{horizontal change}}$$

$0.13 = \frac{v}{6 \text{ ft}}$ Substituting

$0.78 \text{ ft} = v$ Multiplying by 6 ft

61. Familiarize. Let t represent the length of a side of the triangle. Then t - 5 represents the length of a side of the square.

Translate.

Perimeter of the square is the same as perimeter of the triangle.

$$4(t - 5) = 3t$$

Solve.

$$4(t - 5) = 3t$$
$$4t - 20 = 3t$$
$$t - 20 = 0$$
$$t = 20$$

Check. If 20 is the length of a side of the triangle, then the length of a side of the square is 20 - 5, or 15. The perimeter of the square is 4·15, or 60, and the perimeter of the triangle is 3·20, or 60. The numbers check.

State. The square and triangle have sides of length 15 and 20, respectively.

63. $|5x - 8| \geq 32$

$5x - 8 \leq -32$ or $5x - 8 \geq 32$

$5x \leq -24$ or $5x \geq 40$

$x \leq -\frac{24}{5}$ or $x \geq 8$

The solution set is $\left\{x \middle| x \leq -\frac{24}{5} \text{ or } x \geq 8\right\}$.

65. $\frac{1}{8}y = -x - \frac{7}{16}$

$$8\left(\frac{1}{8}y\right) = 8\left(-x - \frac{7}{16}\right)$$
$$y = -8x - \frac{7}{2}$$

The slope is -8, and the y-intercept is $\left(0, -\frac{7}{2}\right)$.

67. $x = -\frac{7}{3}y - \frac{2}{11}$

$$\frac{7}{3}y + x = -\frac{2}{11}$$
$$\frac{7}{3}y = -x - \frac{2}{11}$$
$$\frac{3}{7}\left(\frac{7}{3}y\right) = \frac{3}{7}\left(-x - \frac{2}{11}\right)$$
$$y = -\frac{3}{7}x - \frac{6}{77}$$

The slope is $-\frac{3}{7}$, and the y-intercept is $\left(0, -\frac{6}{77}\right)$.

69. a) $m = \frac{-6c - (-c)}{5b - b} = \frac{-6c + c}{4b} = \frac{-5c}{4b}$, or $-\frac{5c}{4b}$

b) $m = \frac{d + e - d}{b - b} = \frac{e}{0}$ Undefined

The slope is undefined.

c) $m = \frac{a + d - (-a - d)}{c + f - (c - f)} = \frac{a + d + a + d}{c + f - c + f} =$

$\frac{2a + 2d}{2f} = \frac{2(a + d)}{2f} = \frac{a + d}{f}$

Exercise Set 5.4

1. $y - y_1 = m(x - x_1)$ Point-slope equation

$y - 2 = 4(x - 3)$ Substituting 4 for m, 3 for x_1, and 2 for y_1

$y - 2 = 4x - 12$

$y = 4x - 10$

3. $y - y_1 = m(x - x_1)$ Point-slope equation

$y - 7 = -2(x - 4)$ Substituting -2 for m, 4 for x_1, and 7 for y_1

$y - 7 = -2x + 8$

$y = -2x + 15$

5. $y - y_1 = m(x - x_1)$ Point-slope equation

$y - (-4) = 3[x - (-2)]$ Substituting 3 for m, -2 for x_1, and -4 for y_1

$y + 4 = 3(x + 2)$

$y + 4 = 3x + 6$

$y = 3x + 2$

7. $y - y_1 = m(x - x_1)$ Point-slope equation

$y - 0 = -2(x - 8)$ Substituting -2 for m, 8 for x_1, and 0 for y_1

$y = -2x + 16$

9. $y - y_1 = m(x - x_1)$ Point-slope equation

$y - (-7) = 0(x - 0)$ Substituting 0 for m, 0 for x_1, and -7 for y_1

$y + 7 = 0$

$y = -7$

11. $y - y_1 = m(x - x_1)$

$y - (-2) = \frac{2}{3}(x - 1)$

$y + 2 = \frac{2}{3}x - \frac{2}{3}$

$y = \frac{2}{3}x - \frac{8}{3}$

13. First find the slope of the line:

$m = \frac{6 - 4}{5 - 1} = \frac{2}{4} = \frac{1}{2}$

Use the point-slope equation with $m = \frac{1}{2}$ and $(1,4) = (x_1,y_1)$. (We could let $(5,6) = (x_1,y_1)$ instead and obtain an equivalent equation.)

$y - 4 = \frac{1}{2}(x - 1)$

$y - 4 = \frac{1}{2}x - \frac{1}{2}$

$y = \frac{1}{2}x + \frac{7}{2}$

15. First find the slope of the line:

$m = \frac{2 - (-1)}{2 - (-1)} = \frac{2 + 1}{2 + 1} = \frac{3}{3} = 1$

Use the point-slope equation with m = 1 and $(-1,-1) = (x_1,y_1)$.

$y - (-1) = 1[(x - (-1)]$

$y + 1 = x + 1$

$y = x$

17. First find the slope of the line:

$m = \frac{5 - 0}{0 - (-2)} = \frac{5}{2}$

Use the point-slope equation with $m = \frac{5}{2}$ and $(-2,0) = (x_1,y_1)$.

$y - 0 = \frac{5}{2}[x - (-2)]$

$y = \frac{5}{2}(x + 2)$

$y = \frac{5}{2}x + 5$

19. First find the slope of the line:

$m = \frac{-6 - (-3)}{-4 - (-2)} = \frac{-6 + 3}{-4 + 2} = \frac{-3}{-2} = \frac{3}{2}$

Use the point-slope equation with $m = \frac{3}{2}$ and $(-2,-3) = (x_1,y_1)$.

$y - (-3) = \frac{3}{2}[x - (-2)]$

$y + 3 = \frac{3}{2}(x + 2)$

$y + 3 = \frac{3}{2}x + 3$

$y = \frac{3}{2}x$

21. First find the slope of the line:

$m = \frac{2 - 0}{5 - 0} = \frac{2}{5}$

Use the point-slope equation with $m = \frac{2}{5}$ and $(0,0) = (x_1,y_1)$.

$y - 0 = \frac{2}{5}(x - 0)$

$y = \frac{2}{5}x$

23. First find the slope of the line:

$$m = \frac{-\frac{1}{2} - 6}{\frac{1}{4} - \frac{3}{4}} = \frac{-\frac{13}{2}}{-\frac{1}{2}} = 13$$

Use the point-slope equation with $m = 13$ and $\left(\frac{3}{4},6\right) = (x_1,y_1)$.

$$y - 6 = 13\left(x - \frac{3}{4}\right)$$

$$y - 6 = 13x - \frac{39}{4}$$

$$y = 13x - \frac{15}{4}$$

25. We first solve for y and determine the slope of each line.

$x + 6 = y$

$y = x + 6$ Reversing the order

The slope of $y = x + 6$ is 1.

$y - x = -2$

$y = x - 2$

The slope of $y = x - 2$ is 1.

The slopes are the same; the lines are parallel.

27. We first solve for y and determine the slope of each line.

$y + 3 = 5x$

$y = 5x - 3$

The slope of $y = 5x - 3$ is 5.

$3x - y = -2$

$3x + 2 = y$

$y = 3x + 2$ Reversing the order

The slope of $y = 3x + 2$ is 3.

The slopes are not the same; the lines are not parallel.

29. We determine the slope of each line.

The slope of $y = 3x + 9$ is 3.

$2y = 6x - 2$

$y = 3x - 1$

The slope of $y = 3x - 1$ is 3.

The slopes are the same; the lines are parallel.

31. We determine the slope of each line.

The slope of $y = 4x - 5$ is 4.

$4y = 8 - x$

$4y = -x + 8$

$y = -\frac{1}{4}x + 2$

The slope of $4y = 8 - x$ is $-\frac{1}{4}$.

The product of their slopes is $4\left(-\frac{1}{4}\right)$, or -1; the lines are perpendicular.

33. We determine the slope of each line.

$x + 2y = 5$

$2y = -x + 5$

$y = -\frac{1}{2}x + \frac{5}{2}$

The slope of $x + 2y = 5$ is $-\frac{1}{2}$.

$2x + 4y = 8$

$4y = -2x + 8$

$y = -\frac{1}{2}x + 2.$

The slope of $2x + 4y = 8$ is $-\frac{1}{2}$.

The product of their slopes is $\left(-\frac{1}{2}\right)\left(-\frac{1}{2}\right)$, or $\frac{1}{4}$; the lines are not perpendicular. For the lines to be perpendicular, the product must be -1.

35. We determine the slope of each line.

$2x - 3y = 7$

$-3y = -2x + 7$

$y = \frac{2}{3}x - \frac{7}{3}$

The slope of $2x - 3y = 7$ is $\frac{2}{3}$.

$2y - 3x = 10$

$2y = 3x + 10$

$y = \frac{3}{2}x + 5$

The slope of $2y - 3x = 10$ is $\frac{3}{2}$.

The product of their slopes is $\frac{2}{3} \cdot \frac{3}{2} = 1$; the lines are not perpendicular. For the lines to be perpendicular, the product must be -1.

37. First solve the equation for y and determine the slope of the given line.

$x + 2y = 6$ Given line

$2y = -x + 6$

$y = -\frac{1}{2}x + 3$

The slope of the given line is $-\frac{1}{2}$.

The line through (3,7) must have slope $-\frac{1}{2}$. We find an equation of this new line using the point-slope equation.

$y - y_1 = m(x - x_1)$ Point-slope equation

$y - 7 = -\frac{1}{2}(x - 3)$ Substituting

$y - 7 = -\frac{1}{2}x + \frac{3}{2}$

$y = -\frac{1}{2}x + \frac{17}{2}$

39. First solve the equation for y and determine the slope of the given line.

$5x - 7y = 8$ Given line

$5x - 8 = 7y$

$\frac{5}{7}x - \frac{8}{7} = y$

$y = \frac{5}{7}x - \frac{8}{7}$

The slope of the given line is $\frac{5}{7}$.

The line through (2,-1) must have slope $\frac{5}{7}$. We find an equation of this new line using the point-slope equation.

$y - y_1 = m(x - x_1)$ Point-slope equation

$y - (-1) = \frac{5}{7}(x - 2)$ Substituting

$y + 1 = \frac{5}{7}x - \frac{10}{7}$

$y = \frac{5}{7}x - \frac{17}{7}$

41. First solve the equation for y and determine the slope of the given line.

$3x - 9y = 2$ Given line

$3x - 2 = 9y$

$\frac{1}{3}x - \frac{2}{9} = y$

The slope of the given line is $\frac{1}{3}$.

The line through (-6,2) must have slope $\frac{1}{3}$. We find an equation of this new line using the point-slope equation.

$y - y_1 = m(x - x_1)$ Point-slope equation

$y - 2 = \frac{1}{3}[x - (-6)]$ Substituting

$y - 2 = \frac{1}{3}(x + 6)$

$y - 2 = \frac{1}{3}x + 2$

$y = \frac{1}{3}x + 4$

43. First solve the equation for y and determine the slope of the given line.

$2x + y = -3$ Given line

$y = -2x - 3$

The slope of the given line is -2.

To find the slope of a perpendicular line, take the reciprocal of -2 and change the sign. The slope is $\frac{1}{2}$.

43. (continued)

We find the equation of the line with slope $\frac{1}{2}$ containing the point (2,5).

$y - y_1 = m(x - x_1)$ Point-slope equation

$y - 5 = \frac{1}{2}(x - 2)$ Substituting

$y - 5 = \frac{1}{2}x - 1$

$y = \frac{1}{2}x + 4$

45. First solve the equation for y and determine the slope of the given line.

$3x + 4y = 5$ Given line

$4y = -3x + 5$

$y = -\frac{3}{4}x + \frac{5}{4}$

The slope of the given line is $-\frac{3}{4}$.

To find the slope of the perpendicular line, take the reciprocal of $-\frac{3}{4}$ and change the sign. The slope is $\frac{4}{3}$.

We find the equation of the line with slope $\frac{4}{3}$ and containing the point (3,-2).

$y - y_1 = m(x - x_1)$ Point-slope equation

$y - (-2) = \frac{4}{3}(x - 3)$ Substituting

$y + 2 = \frac{4}{3}x - 4$

$y = \frac{4}{3}x - 6$

47. First solve the equation for y and determine the slope of the given line.

$2x + 5y = 7$ Given line

$5y = -2x + 7$

$y = -\frac{2}{5}x + \frac{7}{5}$

The slope of the given line is $-\frac{2}{5}$.

To find the slope of the perpendicular line, take the reciprocal of $-\frac{2}{5}$ and change the sign. The slope is $\frac{5}{2}$.

We find the equation of the line with slope $\frac{5}{2}$ and containing the point (0,9).

$y - y_1 = m(x - x_1)$ Point-slope equation

$y - 9 = \frac{5}{2}(x - 0)$ Substituting

$y - 9 = \frac{5}{2}x$

$y = \frac{5}{2}x + 9$

49. $2x + 3 \leq 5x - 4$

$-3x + 3 \leq -4$

$-3x \leq -7$

$x \geq \frac{7}{3}$ Dividing by -3

The solution set is $\left\{x \middle| x \geq \frac{7}{3}\right\}$.

51. $|2x + 3| = |x - 4|$

$2x + 3 = x - 4$ or $2x + 3 = -(x - 4)$

$x + 3 = -4$ or $2x + 3 = -x + 4$

$x = -7$ or $3x + 3 = 4$

$x = -7$ or $3x = 1$

$x = -7$ or $x = \frac{1}{3}$

The solution set is $\left\{-7, \frac{1}{3}\right\}$.

53. Find the slope of the line containing (3,-5) and (-2,7).

$m = \frac{7 - (-5)}{-2 - 3} = \frac{12}{-5} = -\frac{12}{5}$

To find the slope of the perpendicular line, take the reciprocal of $-\frac{12}{5}$ and change the sign. The slope is $\frac{5}{12}$.

We find the equation of the line with slope $\frac{5}{12}$ and containing the point (-1,3).

$y - y_1 = m(x - x_1)$

$y - 3 = \frac{5}{12}[x - (-1)]$

$y - 3 = \frac{5}{12}(x + 1)$

$y - 3 = \frac{5}{12}x + \frac{5}{12}$

$y = \frac{5}{12}x + \frac{41}{12}$

55. Find the slope of each line.

$x + 7y = 70$

$7y = -x + 70$

$y = -\frac{1}{7}x + 10$

The slope of $x + 7y = 70$ is $-\frac{1}{7}$.

$y + 3 = kx$

$y = kx - 3$

The slope of $y + 3 = kx$ is k.

In order for the graphs to be perpendicular, the product of the slopes must be -1.

$-\frac{1}{7} \cdot k = -1$

$k = 7$ Multiplying by -7

57. See the answer section in the text.

Exercise Set 5.5

1. a) In 1950 $t = 0$, so one data point is (0,72). In 1970 $t = 1970 - 1950$, or 20, so the other data point is (20,75).

b) We first find the slope of the line:

$m = \frac{75 - 72}{20 - 0} = \frac{3}{20}$, or 0.15

Then we substitute into the point-slope equation, using (0,72) for the point.

$E - 72 = 0.15(t - 0)$

$E - 72 = 0.15t$

$E = 0.15t + 72$

c) In 1996, $t = 1996 - 1950$, or 46. We find E when $t = 46$:

$E = 0.15(46) + 72 = 78.9$

In 2000, $t = 2000 - 1950$, or 50. We find E when $t = 50$:

$E = 0.15(50) + 72 = 79.5$

3. a) The data points (W,H) are (165,70) and (145,67).

b) We first find the slope of the line:

$m = \frac{70 - 67}{165 - 145} = \frac{3}{20}$, or 0.15

Then we substitute into the point-slope equation, using (165,70) for the point.

$H - 70 = 0.15(W - 165)$

$H - 70 = 0.15W - 24.75$

$H = 0.15W + 45.25$

c) We find H when $W = 130$:

$H = 0.15(130) + 45.25 = 64.75$ in.

5. a) In 1930, $t = 1930 - 1930$, or 0. One data point is (0,3.85). In 1950, $t = 1950 - 1930$, or 20. The other data point is (20,3.70).

b) We first find the slope of the line:

$m = \frac{3.70 - 3.85}{20 - 0} = \frac{-0.15}{20} = -0.0075$

Then we substitute into the point-slope equation, using (0,3.85) for the point.

$R - 3.85 = -0.0075(t - 0)$

$R = -0.0075t + 3.85$

c) In 1998, $t = 1998 - 1930$, or 68. We find R when $t = 68$:

$R = -0.0075(68) + 3.85 = 3.34$ min

In 2002, $t = 2002 - 1930$, or 72. We find R when $t = 72$:

$R = -0.0075(72) + 3.85 = 3.31$ min

5. (continued)

d) We find t when R = 3.3:

$$3.3 = -0.0075t + 3.85$$
$$-0.55 = -0.0075t$$
$$73\tfrac{1}{3} = t$$
$$1930 + 73\tfrac{1}{3} = 2003\tfrac{1}{3}$$

The record will be 3.3 minutes one-third of the way through 2003.

7. a) The data points (M,C) are (2,1.75) and (3,2.00).

b) We first find the slope of the line:

$$m = \frac{2.00 - 1.75}{3 - 2} = 0.25$$

Then we substitute into the point-slope equation, using (3,2.00) for the point.

$$C - 2.00 = 0.25(M - 3)$$
$$C - 2.00 = 0.25M - 0.75$$
$$C = 0.25M + 1.25$$

c) Find C when M = 7:

$$C = 0.25(7) + 1.25 = \$3.00$$

9. The data points (t,V) are (0,5200) and (2,4225). We find the slope of the line:

$$m = \frac{4225 - 5200}{2 - 0} = \frac{-975}{2} = -487.5$$

Then we find the equation of the line:

$$V - 5200 = -487.5(t - 0)$$
$$V = -487.5t + 5200$$

Finally we find V when t = 8:

$$V = -487.5(8) + 5200 = 1300$$

The value after 8 years is $1300.

11. The data points (T,L) are (18,100) and (20,100.00356). We find the slope of the line:

$$m = \frac{100.00356 - 100}{20 - 18} = \frac{0.00356}{2} = 0.00178$$

Then we find the equation of the line:

$$L - 100 = 0.00178(T - 18)$$
$$L - 100 = 0.00178T - 0.03204$$
$$L = 0.00178T + 99.96796$$

Find L when T = 40:

$$L = 0.00178(40) + 99.96796 = 100.03916$$

At 40° C, the length of the wire is 100.03916 cm.

Find L when T = 0:

$$L = 0.00178(0) + 99.96796 = 99.96796$$

At 0° C, the length of the wire is 99.96796 cm.

Exercise Set 5.6

1. We use alphabetical order of variables. We replace x by -4 and y by 2.

$2x + y < -5$	
$2(-4) + 2$	-5
$-8 + 2$	
-6	TRUE

Since $-6 < -5$ is true, (-4,2) is a solution.

3. We use alphabetical order of variables. We replace x by 8 and y by 14.

$2y - 3x > 5$	
$2\cdot14 - 3\cdot8$	5
$28 - 24$	
4	FALSE

Since $4 > 5$ is false, (8,14) is not a solution.

5. Graph: $y > 2x$

We first graph the line $y = 2x$. We draw the line dashed since the inequality symbol is $>$. To determine which half-plane to shade, test a point not on the line. We try (1,1) and substitute:

$y > 2x$	
1	$2\cdot1$
FALSE	2

Since $1 > 2$ is false, (1,1) is not a solution, nor are any points in the half-plane containing (1,1). The points in the opposite half-plane are solutions, so we shade that half-plane and obtain the graph.

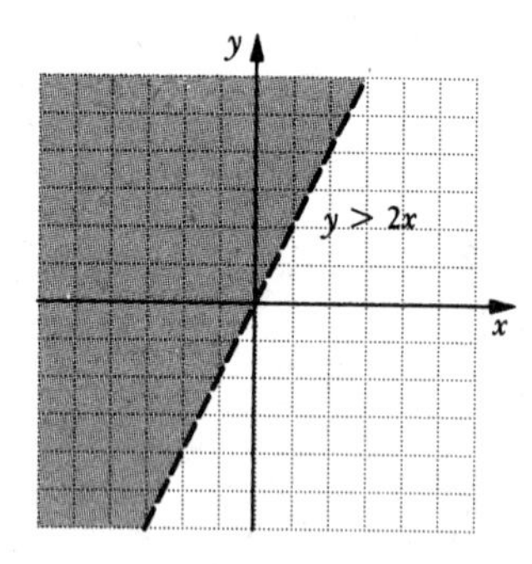

7. Graph: $y < x + 1$

First graph the line $y = x + 1$. Draw it dashed since the inequality symbol is <. Test the point (0,0) to determine if it is a solution.

$y < x + 1$	
0	0 + 1
TRUE	1

Since $0 < 1$ is true, we shade the half-plane containing (0,0) and obtain the graph.

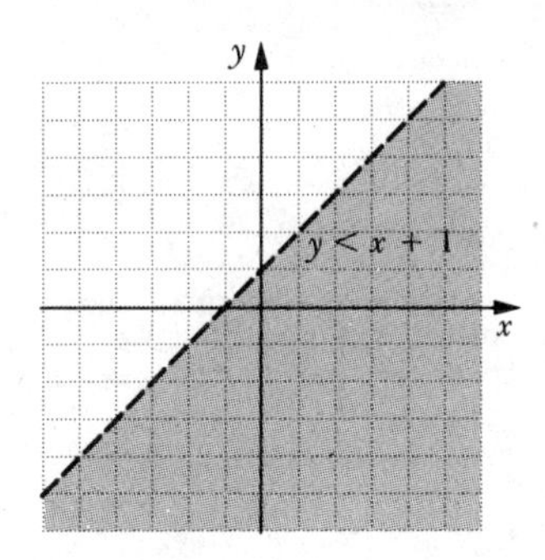

9. Graph: $y > x - 2$

We first graph $y = x - 2$. Draw a dashed line since the inequality symbol is >. Test the point (0,0) to determine if it is a solution.

$y > x - 2$	
0	0 - 2
TRUE	-2

Since $0 > -2$ is true, we shade the half-plane containing (0,0) and obtain the graph.

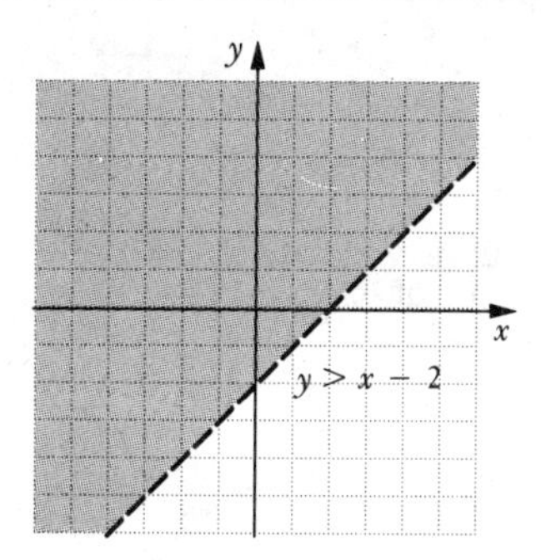

11. Graph: $x + y < 4$

First graph $x + y = 4$. Draw the line dashed since the inequality symbol is <. Test the point (0,0) to determine if it is a solution.

$x + y < 4$	
0 + 0	4
0	TRUE

Since $0 < 4$ is true, we shade the half-plane containing (0,0) and obtain the graph.

11. (continued)

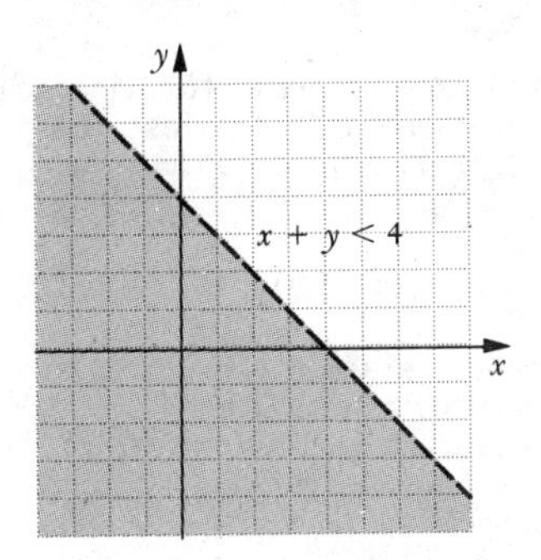

13. Graph: $3x + 4y \leq 12$

We first graph $3x + 4y = 12$. Draw the line solid since the inequality symbol is ≤. Test the point (0,0) to determine if it is a solution.

$3x + 4y \leq 12$	
$3 \cdot 0 + 4 \cdot 0$	12
0	TRUE

Since $0 \leq 12$ is true, we shade the half-plane containing (0,0) and obtain the graph.

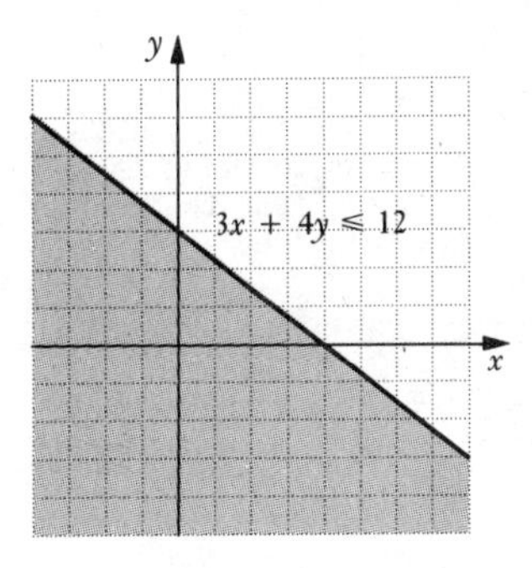

15. Graph: $2y - 3x > 6$

We first graph $2y - 3x = 6$. Draw the line dashed since the inequality symbol is >. Test the point (0,0) to determine if it is a solution.

$2y - 3x > 6$	
$2 \cdot 0 - 3 \cdot 0$	6
0	FALSE

Since $0 > 6$ is false, we shade the half-plane that does not contain (0,0) and obtain the graph.

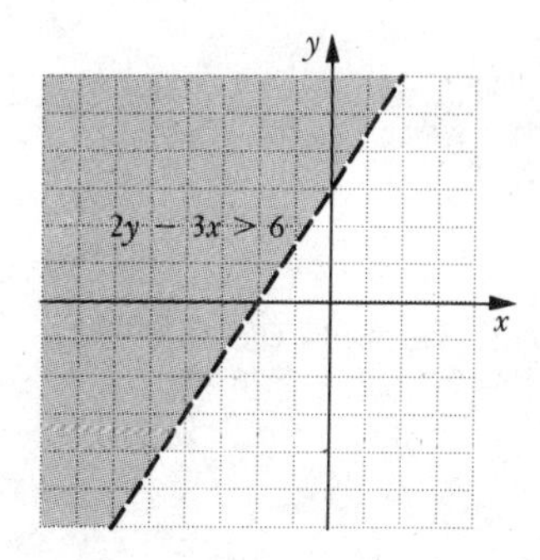

17. Graph: $3x - 2 \leqslant 5x + y$

$-2 \leqslant 2x + y$

We first graph $-2 = 2x + y$. Draw the line solid since the inequality symbol is $\leqslant$. Test the point (0,0) to determine if it is a solution.

$-2 \leqslant 2x + y$	
-2	$2 \cdot 0 + 0$
TRUE	0

Since $-2 \leqslant 0$ is true, we shade the half-plane containing (0,0) and obtain the graph.

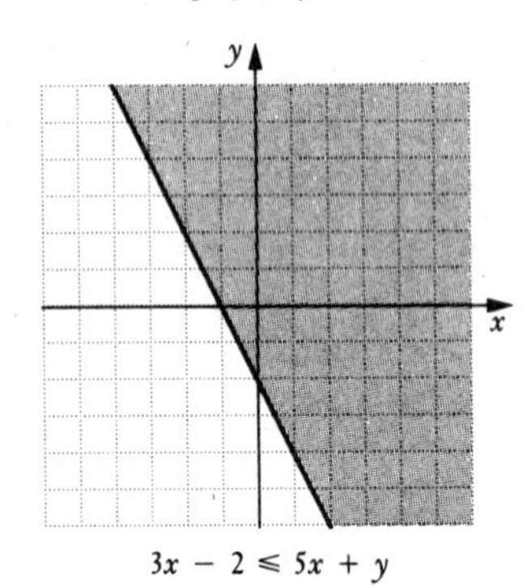

19. Graph: $x < -4$

We first graph $x = -4$. Draw the line dashed since the inequality symbol is $<$. Test the point (0,0) to determine if it is a solution.

$x < -4$		
0	-4	FALSE

Since $0 < -4$ is false, we shade the half-plane that does not contain (0,0) and obtain the graph.

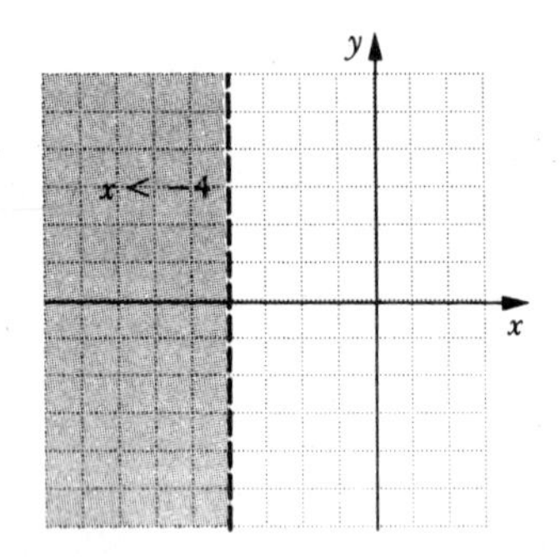

21. Graph: $y > -2$

We first graph $y = -2$. We draw the line dashed since the inequality symbol is $>$. Test the point (0,0) to determine if it is a solution.

$y > -2$		
0	-2	TRUE

Since $0 > -2$ is true, we shade the half-plane containing (0,0) and obtain the graph.

21. (continued)

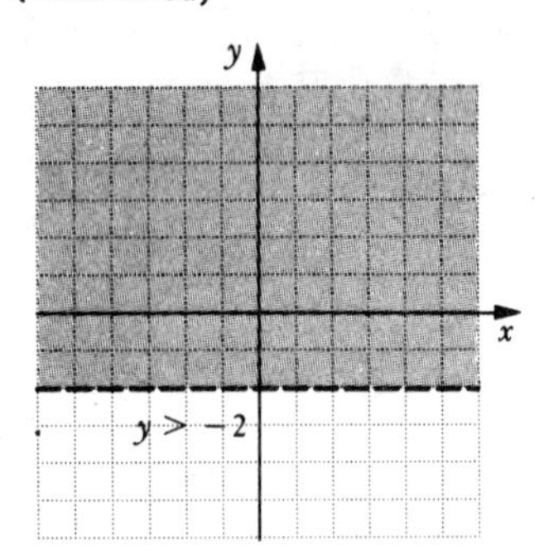

23. Graph: $2x + 3y \leqslant 6$

We first graph $2x + 3y = 6$. We draw the line solid since the inequality symbol is $\leqslant$. Test the point (0,0) to determine if it is a solution.

$2x + 3y \leqslant 6$	
$2 \cdot 0 + 3 \cdot 0$	6
0	TRUE

Since $0 \leqslant 6$ is true, we shade the half-plane containing (0,0) and obtain the graph.

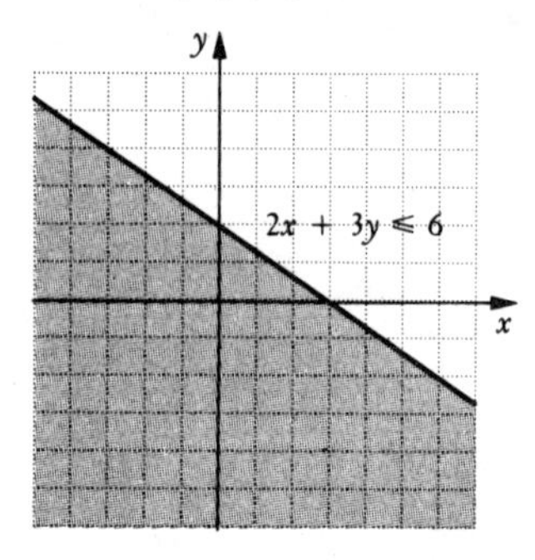

25. Familiarize. We make a drawing. We let x represent the length of the shorter piece of rope. Then $x + 5$ represents the length of the longer piece.

x | x + 5

← 78 ft →

Translate.

Shorter length	plus	longer length	is	78 ft.
↓	↓	↓	↓	↓
x	$+$	$x + 5$	$=$	78

Solve.

$$x + x + 5 = 78$$
$$2x + 5 = 78$$
$$2x = 73$$
$$x = 36.5$$

Check. If $x = 36.5$, then $x + 5 = 36.5 + 5$, or 41.5, and $36.5 + 41.5 = 78$. The numbers check.

State. The lengths of the pieces are 36.5 ft and 41.5 ft.

27. The length is at most 94 ft, so

$L \leq 94$ and

$2L \leq 188.$ (Multiplying by 2)

The width is at most 50 ft, so

$W \leq 50$ and

$2W \leq 100.$ (Multiplying by 2)

(Of course, L and W would also have to be positive, but we will consider only the upper bound on the dimensions here.)

Then we have

$2L + 2W \leq 188 + 100$, or

$2L + 2W \leq 288.$

(See Exercise 90, Exercise Set 2.4.)

To graph the inequality, we first graph $2L + 2W = 288$ using a solid line since the inequality symbol is $\leq$. (We will let W be the first coordinate and L the second. This is another case where alphabetical order of variables is not used.) Test the point (0,0) to determine if it is a solution.

$2L + 2W \leq 288$	
$2\cdot 0 + 2\cdot 0$	288
0	TRUE

Since $0 \leq 288$ is true, we shade the half-plane containing (0,0) and obtain the graph.

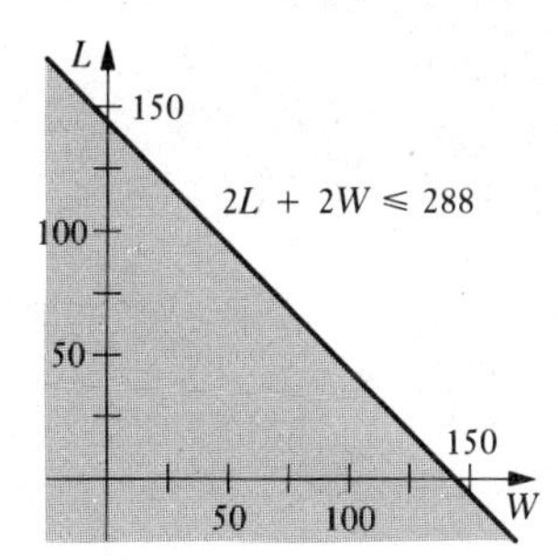

29. The total weight of c children is 35c kg, and the total weight of a adults is 75a kg. When the total weight is more than 1000 kg the elevator is overloaded, so we have

$35c + 75a > 1000.$

(Of course, c and a would also have to be nonnegative, but we will not deal with those constraints here.)

To graph the inequality, we first graph $35c + 75a = 1000$ using a dashed line since the inequality symbol is >. Test the point (0,0) to determine if it is a solution.

$35c + 75a > 1000$	
$35\cdot 0 + 75\cdot 0$	1000
0	FALSE

Since $0 > 1000$ is false, we shade the half-plane that does not contain (0,0) and obtain the graph.

29. (continued)

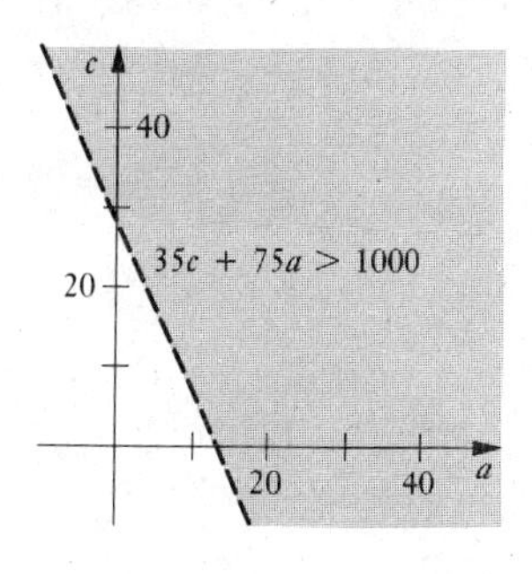

Exercise Set 6.1

1. We use alphabetical order for the variables. We replace x by 1 and y by 2.

$4x - y = 2$	
$4\cdot 1 - 2$	2
$4 - 2$	
2	TRUE

$10x - 3y = 4$	
$10\cdot 1 - 3\cdot 2$	4
$10 - 6$	
4	TRUE

The pair (1,2) makes both equations true, so it is a solution of the system.

3. We use alphabetical order for the variables. We replace x by 2 and y by 5.

$y = 3x - 1$	
5	$3\cdot 2 - 1$
	$6 - 1$
TRUE	5

$2x + y = 4$	
$2\cdot 2 + 5$	4
$4 + 5$	
9	FALSE

The pair (2,5) is not a solution of $2x + y = 4$. Therefore it is not a solution of the system of equations.

5. We replace x by 1 and y by 5.

$x + y = 6$	
$1 + 5$	6
6	TRUE

$y = 2x + 3$	
5	$2\cdot 1 + 3$
	$2 + 3$
TRUE	5

The pair (1,5) makes both equations true, so it is a solution of the system.

7. Graph both lines on the same set of axes.

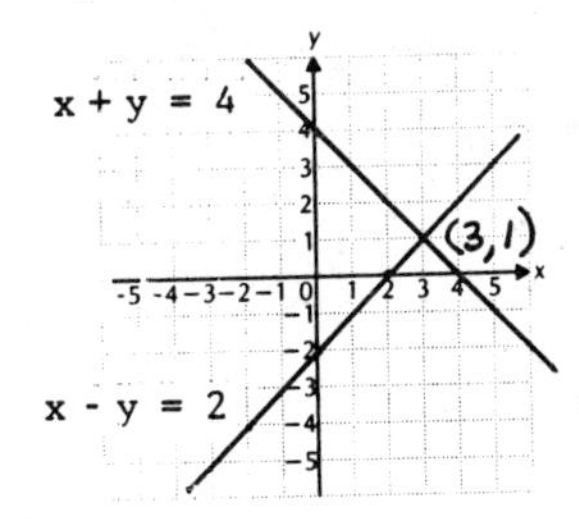

The solution (point of intersection) seems to be the point (3,1).

Check:

$x + y = 4$	
$3 + 1$	4
4	TRUE

$x - y = 2$	
$3 - 1$	2
2	TRUE

The solution is (3,1).

Since the system of equations has a solution it is consistent. Since there is exactly one solution, the system is independent.

9. Graph both lines on the same set of axes.

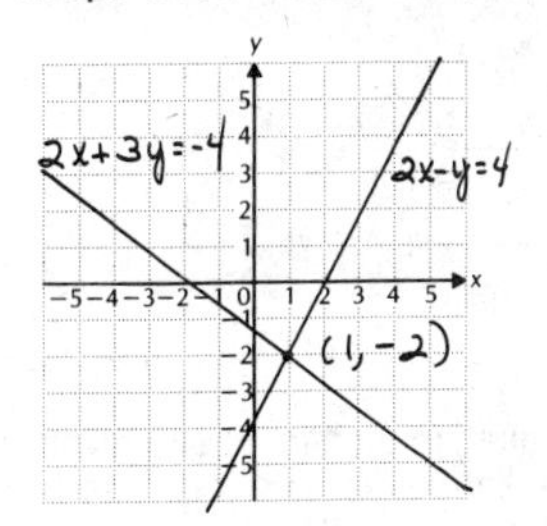

The solution (point of intersection) seems to be the point (1,-2).

Check:

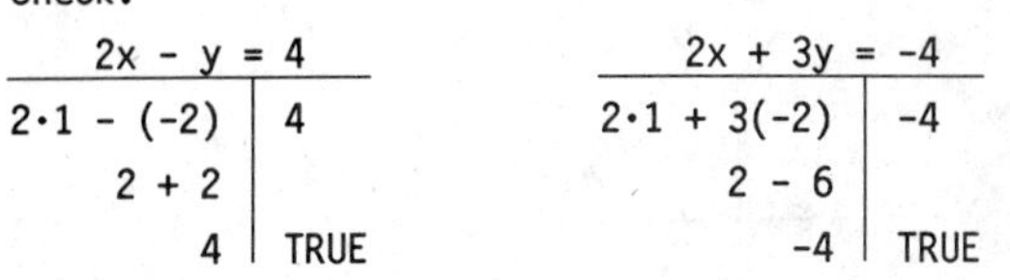

$2x - y = 4$	
$2\cdot 1 - (-2)$	4
$2 + 2$	
4	TRUE

$2x + 3y = -4$	
$2\cdot 1 + 3(-2)$	-4
$2 - 6$	
-4	TRUE

The solution is (1,-2).

Since the system of equations has a solution, it is consistent. Since there is exactly one solution, the system is independent.

11. Graph both lines on the same set of axes.

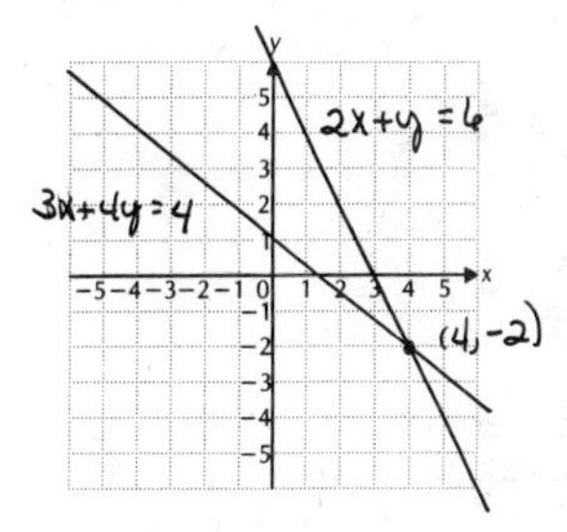

The solution (point of intersection) seems to be the point (4,-2).

Check:

$2x + y = 6$	
$2\cdot 4 + (-2)$	6
$8 - 2$	
6	TRUE

$3x + 4y = 4$	
$3\cdot 4 + 4(-2)$	4
$12 - 8$	
4	TRUE

The solution is (4,-2).

Since the system of equations has a solution, it is consistent. Since there is exactly one solution, the system is independent.

13. Graph both lines on the same set of axes.

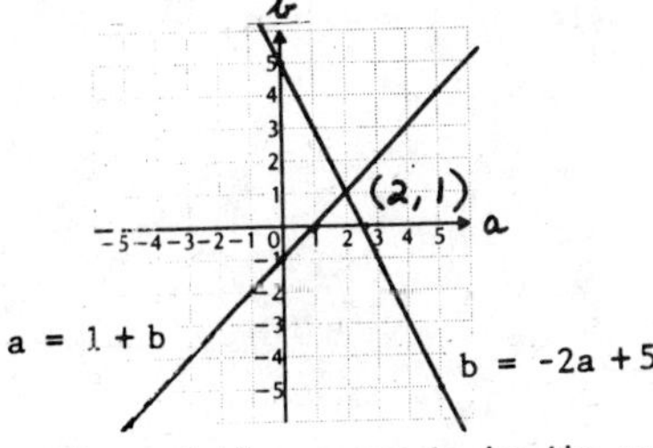

The solution seems to be the point (2,1).

13. (continued)

Check:

$a = 1 + b$	
2	1 + 1
TRUE	2

$b = -2a + 5$	
1	$-2\cdot 2 + 5$
	$-4 + 5$
TRUE	1

The solution is (2,1).

Since the system of equations has a solution, it is consistent. Since there is exactly one solution, the system is independent.

15. Graph both lines on the same set of axes.

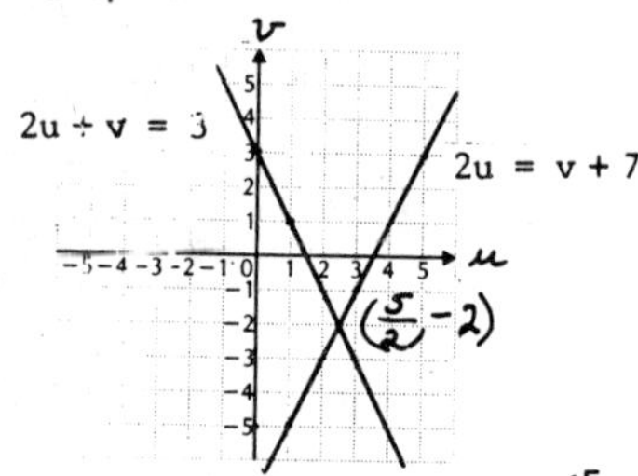

The solution seems to be $\left(\frac{5}{2},-2\right)$.

Check:

$2u + v = 3$	
$2\cdot\frac{5}{2} + (-2)$	3
$5 - 2$	
3	TRUE

$2u = v + 7$		
$2\cdot\frac{5}{2}$	$-2 + 7$	
5	5	TRUE

The solution is $\left(\frac{5}{2},-2\right)$.

Since the system of equations has a solution, it is consistent. Since there is exactly one solution, the system is independent.

17. Graph both lines on the same set of axes.

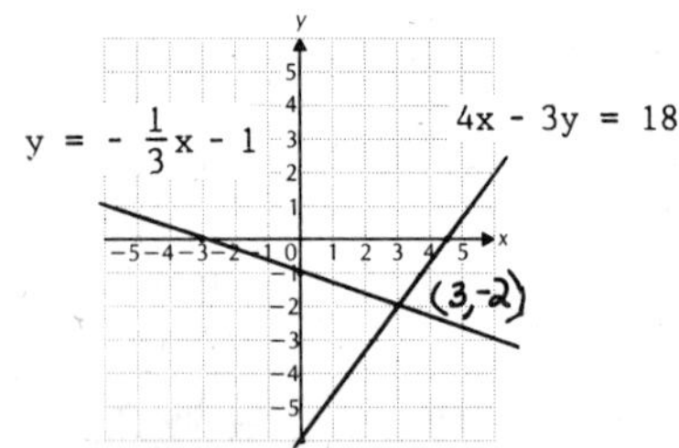

The ordered pair (3,-2) checks in both equations. It is the solution.

Since the system of equations has a solution, it is consistent. Since there is exactly one solution, the system is independent.

19. Graph both lines on the same set of axes.

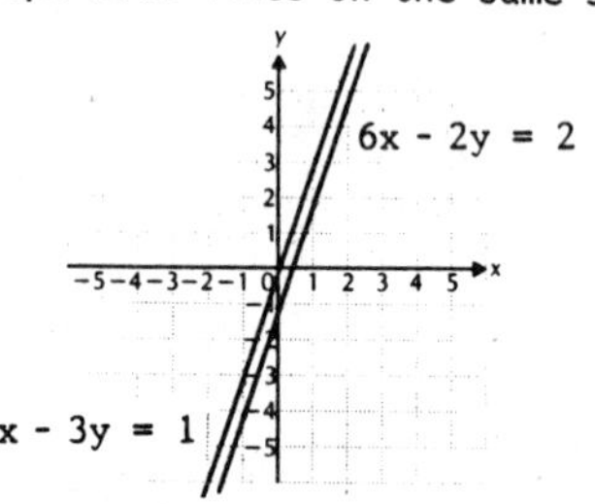

The lines are parallel. There is no solution.

Since the system of equations has no solution, it is inconsistent. Since there is no solution, the system is independent.

21. Graph both lines on the same set of axes.

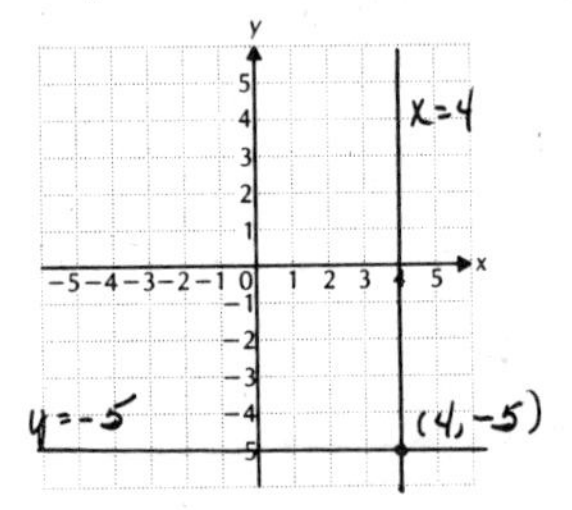

The ordered pair (4,-5) checks in both equations. It is the solution. Since the system of equations has a solution, it is consistent. Since there is exactly one solution, the system is independent.

23. Graph both lines on the same set of axes.

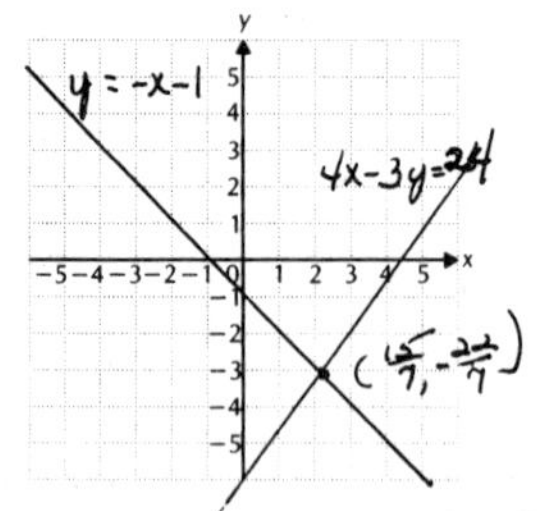

The ordered pair $\left(\frac{15}{7},-\frac{22}{7}\right)$ checks in both equations. It is the solution.

Since the system of equations has a solution, it is consistent. Since there is exactly one solution, the system is independent.

25. Graph both lines on the same set of axes.

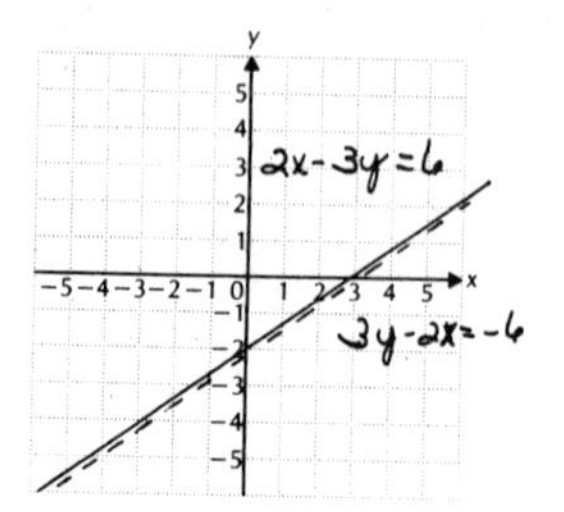

25. (continued)

The graphs are the same. Any solution of one of the equations is also a solution of the other. Each equation has an infinite number of solutions. Thus the system of equations has an infinite number of solutions. Since the system of equations has a solution, it is consistent. Since there are infinitely many solutions, the system is dependent.

27. Substitute 4 for x and -5 for y in the first equation:

$$A(4) - 6(-5) = 13$$
$$4A + 30 = 13$$
$$4A = -17$$
$$A = -\frac{17}{4}$$

Substitute 4 for x and -5 for y in the second equation:

$$4 - B(-5) = -8$$
$$4 + 5B = -8$$
$$5B = -12$$
$$B = -\frac{12}{5}$$

We have $A = -\frac{17}{4}$, $B = -\frac{12}{5}$.

29. There are many correct answers. One can be found by expressing the sum and difference of the two numbers:

$$x + y = 9$$
$$x - y = -3$$

31. There are many correct answers. One can be found by writing an equation in two variables and then writing a constant multiple of that equation:

$$x + y = 1$$
$$2x + 2y = 2$$

Exercise Set 6.2

1. $3x + 5y = 3$ (1)

$x = 8 - 4y$ (2)

We substitute $8 - 4y$ for x in the first equation and solve for y.

$3x + 5y = 3$ (1)

$3(8 - 4y) + 5y = 3$ Substituting

$24 - 12y + 5y = 3$

$24 - 7y = 3$

$-7y = -21$

$y = 3$

1. (continued)

Next we substitute 3 for y in either equation of the original system and solve for x.

$x = 8 - 4y$ (2)

$x = 8 - 4\cdot3$ Substituting

$x = 8 - 12$

$x = -4$

We check the ordered pair (-4,3).

$3x + 5y = 3$	
$3(-4) + 5\cdot3$	3
$-12 + 15$	
3	TRUE

$x = 8 - 4y$	
-4	$8 - 4\cdot3$
	$8 - 12$
TRUE	-4

Since (-4,3) checks, it is the solution.

3. $9x - 2y = 3$ (1)

$3x - 6 = y$ (2)

We substitute $3x - 6$ for y in the first equation and solve for x.

$9x - 2y = 3$ (1)

$9x - 2(3x - 6) = 3$ Substituting

$9x - 6x + 12 = 3$

$3x + 12 = 3$

$3x = -9$

$x = -3$

Next we substitute -3 for x in either equation of the original system and solve for y.

$3x - 6 = y$ (2)

$3(-3) - 6 = y$ Substituting

$-9 - 6 = y$

$-15 = y$

We check the ordered pair (-3,-15).

$9x - 2y = 3$	
$9(-3) - 2(-15)$	3
$-27 + 30$	
3	TRUE

$3x - 6 = y$	
$3(-3) - 6$	-15
$-9 - 6$	
-15	TRUE

Since (-3,-15) checks, it is the solution.

5. $5m + n = 8$ (1)

$3m - 4n = 14$ (2)

We solve the first equation for n.

$5m + n = 8$ (1)

$n = 8 - 5m$

We substitute $8 - 5m$ for n in the second equation and solve for m.

$3m - 4n = 14$ (2)

$3m - 4(8 - 5m) = 14$ Substituting

$3m - 32 + 20m = 14$

$23m - 32 = 14$

$23m = 46$

$m = 2$

5. (continued)

Now we substitute 2 for m in either equation of the original system and solve for n.

$$\begin{aligned} 5m + n &= 8 && (1)\\ 5\cdot 2 + n &= 8 && \text{Substituting}\\ 10 + n &= 8\\ n &= -2 \end{aligned}$$

We check the ordered pair (2,-2).

$$\begin{array}{c|c} \multicolumn{2}{c}{5m + n = 8}\\ \hline 5\cdot 2 + (-2) & 8\\ 10 - 2 & \\ 8 & \text{TRUE} \end{array} \qquad \begin{array}{c|c} \multicolumn{2}{c}{3m - 4n = 14}\\ \hline 3\cdot 2 - 4(-2) & 14\\ 6 + 8 & \\ 14 & \text{TRUE} \end{array}$$

Since (2,-2) checks, it is the solution.

7. $4x + 12y = 4$ (1)

$-5x + y = 11$ (2)

We solve the second equation for y.

$$\begin{aligned} -5x + y &= 11 && (2)\\ y &= 5x + 11 \end{aligned}$$

We substitute 5x + 11 for y in the first equation and solve for x.

$$\begin{aligned} 4x + 12y &= 4 && (1)\\ 4x + 12(5x + 11) &= 4 && \text{Substituting}\\ 4x + 60x + 132 &= 4\\ 64x + 132 &= 4\\ 64x &= -128\\ x &= -2 \end{aligned}$$

Now substitute -2 for x in either equation of the original system and solve for y.

$$\begin{aligned} -5x + y &= 11 && (2)\\ -5(-2) + y &= 11 && \text{Substituting}\\ 10 + y &= 11\\ y &= 1 \end{aligned}$$

We check the ordered pair (-2,1).

$$\begin{array}{c|c} \multicolumn{2}{c}{4x + 12y = 4}\\ \hline 4(-2) + 12\cdot 1 & 4\\ -8 + 12 & \\ 4 & \text{TRUE} \end{array} \qquad \begin{array}{c|c} \multicolumn{2}{c}{-5x + y = 11}\\ \hline -5(-2) + 1 & 11\\ 10 + 1 & \\ 11 & \text{TRUE} \end{array}$$

Since (-2,1) checks, it is the solution.

9.
$$\begin{aligned} x + 3y &= 7\\ \underline{-x + 4y} &\underline{= 7}\\ 0 + 7y &= 14 && \text{Adding}\\ 7y &= 14\\ y &= 2 \end{aligned}$$

Substitute 2 for y in one of the original equations and solve for x.

$$\begin{aligned} x + 3y &= 7\\ x + 3\cdot 2 &= 7 && \text{Substituting}\\ x + 6 &= 7\\ x &= 1 \end{aligned}$$

9. (continued)

Check: For (1,2)

$$\begin{array}{c|c} \multicolumn{2}{c}{x + 3y = 7}\\ \hline 1 + 3\cdot 2 & 7\\ 1 + 6 & \\ 7 & \text{TRUE} \end{array} \qquad \begin{array}{c|c} \multicolumn{2}{c}{-x + 4y = 7}\\ \hline -1 + 4\cdot 2 & 7\\ -1 + 8 & \\ 7 & \text{TRUE} \end{array}$$

Since (1,2) checks, it is the solution.

11.
$$\begin{aligned} 9x + 3y &= -3\\ \underline{2x - 3y} &\underline{= -8}\\ 11x + 0 &= -11 && \text{Adding}\\ 11x &= -11\\ x &= -1 \end{aligned}$$

Substitute -1 for x in one of the original equations and solve for y.

$$\begin{aligned} 9x + 3y &= -3\\ 9(-1) + 3y &= -3 && \text{Substituting}\\ -9 + 3y &= -3\\ 3y &= 6\\ y &= 2 \end{aligned}$$

We obtain (-1,2). This checks, so it is the solution.

13. $5x + 3y = 19$ (1)

$2x - 5y = 11$ (2)

We multiply twice to make two terms become additive inverses.

$$\begin{aligned} &\text{From (1):} & 25x + 15y &= 95 && \text{Multiplying by 5}\\ &\text{From (2):} & \underline{6x - 15y} &\underline{= 33} && \text{Multiplying by 3}\\ && 31x &= 128 && \text{Adding}\\ && x &= \frac{128}{31} \end{aligned}$$

Substitute $\frac{128}{31}$ for x in one of the original equations and solve for y.

$$\begin{aligned} 5x + 3y &= 19\\ 5\cdot\frac{128}{31} + 3y &= 19 && \text{Substituting}\\ \frac{640}{31} + 3y &= \frac{589}{31}\\ 3y &= -\frac{51}{31}\\ \frac{1}{3}\cdot 3y &= \frac{1}{3}\cdot\left(-\frac{51}{31}\right)\\ y &= -\frac{17}{31} \end{aligned}$$

We obtain $\left(\frac{128}{31}, -\frac{17}{31}\right)$. This checks, so it is the solution.

15. $5r - 3s = 24$ (1)

$3r + 5s = 28$ (2)

We multiply twice to make two terms become additive inverses.

From (1): $25r - 15s = 120$ Multiplying by 5

From (2): $9r + 15s = 84$ Multiplying by 3

$34r = 204$ Adding

$r = 6$

Substitute 6 for r in one of the original equations and solve for s.

$3r + 5s = 28$

$3\cdot 6 + 5s = 28$ Substituting

$18 + 5s = 28$

$5s = 10$

$s = 2$

We obtain (6,2). This checks, so it is the solution.

17. $0.3x - 0.2y = 4$

$0.2x + 0.3y = 1$

We first multiply each equation by 10 to clear decimals.

$3x - 2y = 40$ (1)

$2x + 3y = 10$ (2)

We use the multiplication principle with both equations of the resulting system.

From (1): $9x - 6y = 120$ Multiplying by 3

From (2): $4x + 6y = 20$ Multiplying by 2

$13x = 140$ Adding

$x = \frac{140}{13}$

Substitute $\frac{140}{13}$ for x in one of the equations in which the decimals were cleared and solve for y.

$2x + 3y = 10$

$2 \cdot \frac{140}{13} + 3y = 10$ Substituting

$\frac{280}{13} + 3y = \frac{130}{13}$

$3y = -\frac{150}{13}$

$y = -\frac{50}{13}$

We obtain $\left(\frac{140}{13}, -\frac{50}{13}\right)$. This checks, so it is the solution.

19. $\frac{1}{2}x + \frac{1}{3}y = 4$

$\frac{1}{4}x + \frac{1}{3}y = 3$

We first multiply each equation by the LCM of the denominators to clear fractions.

$3x + 2y = 24$ Multiplying by 6

$3x + 4y = 36$ Multiplying by 12

We multiply by -1 on both sides of the first equation and then add.

$-3x - 2y = -24$ Multiplying by -1

$3x + 4y = 36$

$2y = 12$ Adding

$y = 6$

Substitute 6 for y in one of the equations in which the fractions were cleared and solve for x.

$3x + 2y = 24$

$3x + 2\cdot 6 = 24$ Substituting

$3x + 12 = 24$

$3x = 12$

$x = 4$

We obtain (4,6). This checks, so it is the solution.

21. $\frac{2}{5}x + \frac{1}{2}y = 2$

$\frac{1}{2}x - \frac{1}{6}y = 3$

We first multiply each equation by the LCM of the denominators to clear fractions.

$4x + 5y = 20$ Multiplying by 10

$3x - y = 18$ Multiplying by 6

We multiply by 5 on both sides of the second equation and then add.

$4x + 5y = 20$

$15x - 5y = 90$ Multiplying by 5

$19x = 110$ Adding

$x = \frac{110}{19}$

Substitute $\frac{110}{19}$ for x in one of the equations in which the fractions were cleared and solve for y.

$3x - y = 18$

$3\left(\frac{110}{19}\right) - y = 18$ Substituting

$\frac{330}{19} - y = \frac{342}{19}$

$-y = \frac{12}{19}$

$y = -\frac{12}{19}$

We obtain $\left(\frac{110}{19}, -\frac{12}{19}\right)$. This checks, so it is the solution.

23. $2x + 3y = 1$

$4x + 6y = 2$

Multiply the first equation by -2 and then add.

$$\begin{aligned} -4x - 6y &= -2 \\ 4x + 6y &= 2 \\ \hline 0 &= 0 \quad \text{Adding} \end{aligned}$$

We have an equation that is true for all numbers x and y. The system is dependent and has an infinite number of solutions.

25. $2x - 4y = 5$

$2x - 4y = 6$

Multiply the first equation by -1 and then add.

$$\begin{aligned} -2x + 4y &= -5 \\ 2x - 4y &= 6 \\ \hline 0 &= 1 \end{aligned}$$

We have a false equation. The system has no solution.

27. $5x - 9y = 7$

$7y - 3x = -5$

We first write the second equation in the form $Ax + By = C$.

$5x - 9y = 7$

$-3x + 7y = -5$

We use the multiplication principle with both equations and then add.

$$\begin{aligned} 15x - 27y &= 21 \quad \text{Multiplying by 3} \\ -15x + 35y &= -25 \quad \text{Multiplying by 5} \\ \hline 8y &= -4 \quad \text{Adding} \\ y &= -\frac{1}{2} \end{aligned}$$

Substitute $-\frac{1}{2}$ for y in one of the original equations and solve for x.

$$\begin{aligned} 5x - 9y &= 7 \\ 5x - 9\left(-\frac{1}{2}\right) &= 7 \quad \text{Substituting} \\ 5x + \frac{9}{2} &= \frac{14}{2} \\ 5x &= \frac{5}{2} \\ x &= \frac{1}{2} \end{aligned}$$

We obtain $\left(\frac{1}{2}, -\frac{1}{2}\right)$. This checks, so it is the solution.

29. $3(a - b) = 15$

$4a = b + 1$

We first write each equation in the form $Ax + By = C$.

$3a - 3b = 15$

$4a - b = 1$

We multiply by -3 on both sides of the second equation and then add.

$$\begin{aligned} 3a - 3b &= 15 \\ -12a + 3b &= -3 \quad \text{Multiplying by -3} \\ \hline -9a &= 12 \\ a &= -\frac{12}{9} \\ a &= -\frac{4}{3} \end{aligned}$$

Substitute $-\frac{4}{3}$ for a in one of the original equations and solve for b.

$$\begin{aligned} 4a - b &= 1 \\ 4\left(-\frac{4}{3}\right) - b &= 1 \\ -\frac{16}{3} - b &= \frac{3}{3} \\ -b &= \frac{19}{3} \\ b &= -\frac{19}{3} \end{aligned}$$

We obtain $\left(-\frac{4}{3}, -\frac{19}{3}\right)$. This checks, so it is the solution.

31. $x - \frac{1}{10}y = 100$

$y - \frac{1}{10}x = -100$

We first write the second equation in the form $Ax + By = C$.

$x - \frac{1}{10}y = 100$

$-\frac{1}{10}x + y = -100$

Next we multiply each equation by 10 to clear fractions.

$10x - y = 1000$

$-x + 10y = -1000$

We multiply by 10 on both sides of the first equation and then add.

$$\begin{aligned} 100x - 10y &= 10{,}000 \quad \text{Multiplying by 10} \\ -x + 10y &= -1000 \\ \hline 99x &= 9000 \\ x &= \frac{9000}{99} \\ x &= \frac{1000}{11} \end{aligned}$$

31. (continued)

Substitute $\frac{1000}{11}$ for x in one of the equations in which the fractions were cleared and solve for y.

$$10x - y = 1000$$

$$10\left(\frac{1000}{11}\right) - y = 1000 \quad \text{Substituting}$$

$$\frac{10{,}000}{11} - y = \frac{11{,}000}{11}$$

$$-y = \frac{1000}{11}$$

$$y = -\frac{1000}{11}$$

We obtain $\left(\frac{1000}{11}, -\frac{1000}{11}\right)$. This checks, so it is the solution.

33. $0.05x + 0.25y = 22$

$0.15x + 0.05y = 24$

We first multiply each equation by 100 to clear decimals.

$5x + 25y = 2200$

$15x + 5y = 2400$

We multiply by -5 on both sides of the second equation and add.

$$\begin{aligned} 5x + 25y &= 2200 \\ -75x - 25y &= -12{,}000 \quad \text{Multiplying by -5} \\ \hline -70x &= -9800 \quad \text{Adding} \\ x &= \frac{-9800}{-70} \\ x &= 140 \end{aligned}$$

Substitute 140 for x in one of the equations in which the decimals were cleared and solve for y.

$$\begin{aligned} 5x + 25y &= 2200 \\ 5\cdot 140 + 25y &= 2200 \quad \text{Substituting} \\ 700 + 25y &= 2200 \\ 25y &= 1500 \\ y &= 60 \end{aligned}$$

We obtain (140,60). This checks, so it is the solution.

35. $y = 1.3x - 7$

The equation is in slope-intercept form, $y = mx + b$. The slope is 1.3.

37. $3.5x - 2.1y = 106.2$

$4.1x + 16.7y = -106.28$

Since this is a calculator exercise, you may choose not to clear the decimals. We will do so here, however.

$35x - 21y = 1062$ Multiplying by 10

$410x + 1670y = -10{,}628$ Multiplying by 100

Multiply twice to make two terms become additive inverses.

$$\begin{aligned} 58{,}450x - 35{,}070y &= 1{,}773{,}540 \quad \text{Multiplying by 1670} \\ 8610x + 35{,}070y &= -223{,}188 \quad \text{Multiplying by 21} \\ \hline 67{,}060x &= 1{,}550{,}352 \quad \text{Adding} \\ x &\approx 23.118879 \end{aligned}$$

37. (continued)

Substitute 23.118879 for x in one of the equations in which the decimals were cleared and solve for y.

$$\begin{aligned} 35x - 21y &= 1062 \\ 35(23.118879) - 21y &= 1062 \quad \text{Substituting} \\ 809.160765 - 21y &= 1062 \\ -21y &= 252.839235 \\ y &\approx -12.039964 \end{aligned}$$

The numbers check, so the solution is (23.118879,-12.039964).

39. $5x + 2y = a$

$x - y = b$

We multiply by 2 on both sides of the second equation and then add.

$$\begin{aligned} 5x + 2y &= a \\ 2x - 2y &= 2b \quad \text{Multiplying by 2} \\ \hline 7x &= a + 2b \quad \text{Adding} \\ x &= \frac{a + 2b}{7} \end{aligned}$$

Next we multiply by -5 on both sides of the second equation and then add.

$$\begin{aligned} 5x + 2y &= a \\ -5x + 5y &= -5b \quad \text{Multiplying by -5} \\ \hline 7y &= a - 5b \\ y &= \frac{a - 5b}{7} \end{aligned}$$

We obtain $\left(\frac{a + 2b}{7}, \frac{a - 5b}{7}\right)$. This checks, so it is the solution.

41. $(0,-3)$ and $\left(-\frac{3}{2}, 6\right)$ are two solutions of $px - qy = -1$.

Substitute 0 for x and -3 for y.

$$\begin{aligned} p\cdot 0 - q\cdot(-3) &= -1 \\ 3q &= -1 \\ q &= -\frac{1}{3} \end{aligned}$$

Substitute $-\frac{3}{2}$ for x and 6 for y.

$$\begin{aligned} p\cdot\left(-\frac{3}{2}\right) - q\cdot 6 &= -1 \\ -\frac{3}{2}p - 6q &= -1 \end{aligned}$$

Substitute $-\frac{1}{3}$ for q and solve for p.

$$\begin{aligned} -\frac{3}{2}x - 6\cdot\left(-\frac{1}{3}\right) &= -1 \\ -\frac{3}{2}p + 2 &= -1 \\ -\frac{3}{2}p &= -3 \\ -\frac{2}{3}\cdot\left(-\frac{3}{2}p\right) &= -\frac{2}{3}\cdot(-3) \\ p &= 2 \end{aligned}$$

Thus, $p = 2$ and $q = -\frac{1}{3}$.

Exercise Set 6.3

1. Familiarize. Let x = the first number and y = the second number.

Translate.

The sum of two numbers is -42.

Rewording: The first number plus the second number is -42.

$x + y = -42$

The first number minus the second number is 52.

$x - y = 52$

We have a system of equations:

$x + y = -42,$

$x - y = 52$

Solve. We solve the system of equations. We use the elimination method.

$x + y = -42$

$x - y = 52$

$2x = 10$ Adding

$x = 5$

Substitute 5 for x in one of the equations and solve for y.

$x + y = -42$

$5 + y = -42$

$y = -47$

Check. The sum of the numbers is 5 + (-47), or -42. The difference is 5 - (-47), or 52. The numbers check.

State. The numbers are 5 and -47.

3. Familiarize. Let x = the number of white sweatshirts sold and y = the number of yellow sweatshirts sold. Organize the information in a table.

Kind of sweatshirt	White	Yellow	Total	
Number sold	x	y	30	→ $x + y = 30$
Price	\$9.95	\$10.50		
Amount taken in	9.95x	10.50y	310.60	→ $9.95x + 10.50y = 310.60$

Translate. Using the "Number sold" and "Amount taken in" rows we have a system of equations:

$x + y = 30$

$9.95x + 10.50y = 310.60$

After clearing decimals, we have

$x + y = 30$ (1)

$995x + 1050y = 31,060$ (2)

3. (continued)

Solve. We solve the system of equations. We use elimination.

$-995x - 995y = -29,850$ Multiplying (1) by -995

$995x + 1050y = 31,060$

$55y = 1210$ Adding

$y = 22$

$x + 22 = 30$ Substituting 22 for y in (1)

$x = 8$

Check. The total number of sweatshirts sold was 8 + 22, or 30.

Money from white: \$ 9.95 × 8 = \$ 79.60

Money from yellow: \$10.50 × 22 = \$231.00

Total = \$310.60

The numbers check.

State. 8 white sweatshirts and 22 yellow sweatshirts were sold.

5. Familiarize. The basketball court is a rectangle with perimeter 288 ft. Let ℓ = length and w = width. Recall that for a rectangle with length ℓ and width w, the perimeter P is given by $P = 2\ell + 2w$.

Translate. The formula for perimeter gives us one equation:

$2\ell + 2w = 288$

The statement relating length and width gives us a second equation:

Length is 44 ft longer than width

$\ell = 44 + w$

We have a system of equations:

$2\ell + 2w = 288,$

$\ell = 44 + w$

Solve. We solve the system of equations. We use substitution.

$2(44 + w) + 2w = 288$ Substituting 44 + w for ℓ in (1)

$88 + 2w + 2w = 288$

$88 + 4w = 288$

$4w = 200$

$w = 50$

$\ell = 44 + 50$ Substituting 50 for w in (2)

$\ell = 94$

Check. The perimeter of a 94 ft by 50 ft rectangle is $2\cdot 94 + 2\cdot 50 = 188 + 100 = 288$. Also, 94 ft is 44 ft longer than 50 ft. The numbers check.

State. The length is 94 ft, and the width is 50 ft.

7. Familiarize. Let x = the measure of one angle and y = the measure of the other angle. Recall that two angles are supplementary if the sum of their measures is 180°.

Translate. The fact that the angles are supplementary gives us one equation.

Rewording: The sum of the measures is 180°.

$$x + y = 180$$

The second statement gives us another equation:

One angle is 3° less than twice the other.

$$x = 2y - 3$$

We have a system of equations:

$$x + y = 180,$$
$$x = 2y - 3$$

Solve. We solve the system of equations. We use substitution.

$(2y - 3) + y = 180$ Substituting $2y - 3$ for x in (1)

$3y - 3 = 180$

$3y = 183$

$y = 61$

$x + 61 = 180$ Substituting 61 for y in (1)

$x = 119$

Check. The sum of the angles is 61° + 119°, or 180°, so they are supplementary. Also, two times the 61° angle minus 3° is 119°, the other angle. The numbers check.

State. The measure of one angle is 61°, and the measure of the other is 119°.

9. Familiarize. List the information in a table.

Type of score	Field goal	Free throw	Total
Number scored	x	y	18
Points per score	2	1	
Points scored	$2x$	$1\cdot y$, or y	30

Translate. The "Number scored" row of the table gives us one equation:

$$x + y = 18$$

The "Points scored" row gives us a second equation:

$$2x + y = 30$$

We have a system of equations:

$$x + y = 18,$$
$$2x + y = 30$$

Solve. We solve the system of equations. We use the elimination method.

$-x - y = -18$ Multiplying (1) by -1

$2x + y = 30$

$x = 12$

$12 + y = 18$ Substituting 12 for x in (1)

$y = 6$

9. (continued)

Check. The total number of times the player scored is 12 + 6, or 18.

Points from field goals: $12 \times 2 = 24$

Points from free throws: $6 \times 1 = 6$

Total 30

The numbers check.

State. The player made 12 field goals and 6 free throws.

11. Familiarize. Let x = number of games won and y = number of games tied. The total points earned in x wins is $2x$; the total points earned in y ties is $1\cdot y$, or y.

Translate.

Points from wins plus points from ties is 60.

$$2x + y = 60$$

Number of wins is 9 more than number of ties.

$$x = 9 + y$$

We have a system of equations:

$$2x + y = 60,$$
$$x = 9 + y$$

Solve. We solve the system of equations. We use substitution.

$2(9 + y) + y = 60$ Substituting $9 + y$ for x in (1)

$18 + 2y + y = 60$

$18 + 3y = 60$

$3y = 42$

$y = 14$

$x = 9 + 14$ Substituting 14 for y in (2)

$x = 23$

Check. The number of wins, 23, is 9 more than the number of ties, 14.

Points from wins: $23 \times 2 = 46$

Points from ties: $14 \times 1 = 14$

Total 60

The numbers check.

State. The team had 23 wins and 14 ties.

13. Familiarize. Let x = number of 30-sec commercials and y = number of 60-sec commercials. The total time used by x 30-sec commercials is $30x$; the total time used by y 60-sec commercials is $60y$. Also note that 10 min = 10×60, or 600 sec.

Translate.

Total number of commercials is 12.

$$x + y = 12$$

Total commercial time is 10 min, or 600 sec.

$$30x + 60y = 600$$

We have a system of equations:

$$x + y = 12,$$
$$30x + 60y = 600$$

Solve. We solve the system of equations. We use the elimination method.

$$-30x - 30y = -360 \quad \text{Multiplying (1) by -30}$$
$$30x + 60y = 600$$
$$30y = 240 \quad \text{Adding}$$
$$y = 8$$

$$x + 8 = 12 \quad \text{Substituting 8 for y in (1)}$$
$$x = 4$$

Check. The total number of commercials is 4 + 8, or 12.

Time for 30-sec commercials: $30 \times 4 = 120$ sec

Time for 60-sec commercials: $60 \times 8 = 480$ sec

600 sec, or 10 min

The numbers check.

State. There were 4 30-sec commercials and 8 60-sec commercials.

15. Familiarize. Let x = the larger number and y = the smaller number.

Translate.

The difference of the numbers is 16.

$$x - y = 16$$

Three times the larger is nine times the smaller.

$$3x = 9y$$

We have a system of equations:

$$x - y = 16, \quad (1)$$
$$3x = 9y \quad (2)$$

Solve. Solve the system of equations. We use the substitution method.

$$x = 3y \quad \text{Solving (2) for x}$$
$$3y - y = 16 \quad \text{Substituting 3y for x in (1)}$$
$$2y = 16$$
$$y = 8$$

$$x - 8 = 16 \quad \text{Substituting 8 for y in (1)}$$
$$x = 24$$

Check. The difference of the numbers is 24 - 8, or 16. Also, $3 \cdot 24 = 72 = 9 \cdot 8$. The numbers check.

State. The larger number is 24 and the smaller is 8.

17. Familiarize. We organize the information in a table.

Let x = the number of pounds of soybean meal and y = the number of pounds of corn meal.

Type of meal	Pounds of meal	Percent of protein	Pounds protein in meal
Soybean	x	16%	0.16x
Corn	y	9%	0.09y
Mixture	350	12%	0.12 × 350 or 42

Translate. The "Pounds of meal" column gives us one equation: $x + y = 350$

The last column gives us a second equation: $0.16x + 0.09y = 42$

After clearing decimals, we have this system:

$$x + y = 350, \quad (1)$$
$$16x + 9y = 4200 \quad (2)$$

Solve. Solve the system of equations.

$$-9x - 9y = -3150 \quad \text{Multiplying (1) by -9}$$
$$16x + 9y = 4200$$
$$7x = 1050$$
$$x = 150$$

$$150 + y = 350 \quad \text{Substituting 150 for x in (1)}$$
$$y = 200$$

Check. The total number of pounds is 150 + 200, or 350. Also, 16% of 150 is 24, and 9% of 200 is 18. Their total is 42. The numbers check.

State. 150 lb of soybean meal and 200 lb of corn meal should be mixed.

19. Familiarize. We can organize the information in a table. Let x = the number of liters of the drink containing 15% orange juice and y = the number of liters of the drink containing 5% orange juice.

Type of canned juice drink	Amount of drink	Percent of orange juice	Amount of orange juice in drink
15% juice	x	15%	0.15x
5% juice	y	5%	0.05y
Mixture	10	10%	0.1 × 10 or 1

Translate. The "Amount of drink" column gives us one equation: $x + y = 10$

The last column gives us a second equation: $0.15x + 0.05y = 1$

After clearing decimals, we have this system:

$$x + y = 10, \quad (1)$$
$$15x + 5y = 100 \quad (2)$$

19. (continued)

Solve. Solve the system of equations

$-5x - 5y = -50$ Multiplying (1) by -5

$15x + 5y = 100$

$10x = 50$ Adding

$x = 5$

$5 + y = 10$ Substituting 5 for x in (1)

$y = 5$

Check. The total number of liters is 5 + 5, or 10. Also, 15% of 5 is 0.75, and 5% of 5 is 0.25. Their sum is 1. The numbers check.

State. 5 L of each drink should be used.

21. Familiarize. Let x = one investment and y = the other investment. We organize the information in a table.

	Principal	Rate	Time	Interest (I = Prt)
1st investment	x	14%	1 yr	0.14x
2nd investment	y	16%	1 yr	0.16y
Total	$8800			$1326

Translate. The first column gives us one equation: $x + y = 8800$

The last column gives us a second equation: $0.14x + 0.16y = 1326$

After clearing decimals, we have this system:

$x + y = 8800,$ (1)

$14x + 16y = 132,600$ (2)

Solve. We solve the system of equations.

$-14x - 14y = -123,200$ Multiplying (1) by -14

$14x + 16y = 132,600$

$2y = 9400$ Adding

$y = 4700$

$x + 4700 = 8800$ Substituting 4700 for y in (1)

$x = 4100$

Check. The sum of the investments is $4100 + $4700, or $8800. The amounts of interest earned are 14% of $4100, or $574, and 16% of $4700, or $752. The total interest earned is $574 + $752, or $1326. The values check.

State. $4100 is invested at 14% and $4700 is invested at 16%.

23. Familiarize. Let x = one investment and y = the other investment. Organize the information in a table.

	Principal	Rate	Time	Interest (I = Prt)
1st investment	x	12%	1 yr	0.12x
2nd investment	y	11%	1 yr	0.11y
Total	$1150			$133.75

Translate. The first column gives us one equation: $x + y = 1150$

The last column gives us a second equation: $0.12x + 0.11y = 133.75$

After clearing decimals, we have this system:

$x + y = 1150,$ (1)

$12x + 11y = 13,375$ (2)

Solve. We solve the system of equations.

$-11x - 11y = -12,650$ Multiplying (1) by -11

$12x + 11y = 13,375$

$x = 725$ Adding

$725 + y = 1150$ Substituting 725 for x in (1)

$y = 425$

Check. The sum of the investments is $725 + $425, or $1150. The amounts of interest earned are 12% of $725, or $87 and 11% of $425, or $46.75. The total interest earned is $87 + $46.75, or $133.75. The values check.

State. $725 was invested at 12% and $425 was invested at 11%.

25. Familiarize. Let x = the cost of one hot dog and y = the cost of one hamburger.

Translate. The first statement gives us one equation.

Three hot-dogs and five hamburgers cost $18.50.

$3x + 5y = 18.50$

The second statement gives us another equation.

Five hot-dogs and three hamburgers cost $16.70.

$5x + 3y = 16.70$

After clearing decimals, we have this system:

$30x + 50y = 185,$ (1)

$50x + 30y = 167$ (2)

Solve. We solve the system of equations.

$900x + 1500y = 5550$ Multiplying (1) by 30

$-2500x - 1500y = -8350$ Multiplying (2) by -50

$-1600x = -2800$ Adding

$x = 1.75$

25. (continued)

$30(1.75) + 50y = 185$ Substituting 1.75 for x in (1)

$52.5 + 50y = 185$

$50y = 132.5$

$y = 2.65$

Check. If one hot dog costs \$1.75 and one hamburger costs \$2.65, then 3 hot dogs and 5 hamburgers cost 3(\$1.75) + 5(\$2.65) = \$5.25 + \$13.25 = \$18.50. Also, 5 hot dogs and 3 hamburgers cost 5(\$1.75) + 3(\$2.65) = \$8.75 + \$7.95, or \$16.70. The numbers check.

State. One hot dog costs \$1.75 and one hamburger costs \$2.65.

27. Familiarize. Let x = Carlos' age now and y = Maria's age now. Four years ago, Carlos' age was x - 4 and Maria's age was y - 4.

Translate.

Carlos' age now	is	8	more than	Maria's age now.
x	$=$	8	$+$	y

Four years ago

Maria's age	was	$\frac{2}{3}$	of	Carlos' age.
$y - 4$	$=$	$\frac{2}{3}$	$\cdot$	$(x - 4)$

We have a system of equations:

$x = 8 + y,$ (1)

$y - 4 = \frac{2}{3}(x - 4)$ (2)

Solve. We use the substitution method.

$y - 4 = \frac{2}{3}(8 + y - 4)$ Substituting 8 + y for x in (2)

$y - 4 = \frac{2}{3}(4 + y)$

$3(y - 4) = 2(4 + y)$ Clearing the fraction

$3y - 12 = 8 + 2y$

$y - 12 = 8$

$y = 20$

$x = 8 + 20$ Substituting 20 for y in (1)

$x = 28$

Check. Carlos, who is 28, is 8 years older than his sister Maria, who is 20. Four years ago, Carlos was 24 and Maria 16, and 16 is $\frac{2}{3}$ of 24. The numbers check.

State. Now Carlos is 28 years old and Maria is 20 years old.

29. Familiarize. The amount of change is \$20 - \$9.25, or \$10.75. Let q = the number of quarters and f = the number of fifty-cent pieces. Organize the information in a table.

Kind of coin	Quarter	Fifty-cent piece	Total	
Number	q	f	30	→ $q + f = 30$
Amount of change	0.25q	0.50f	\$10.75	→ $0.25q + 0.50f = 10.75$

Translate. The rows of the table give us two equations:

$q + f = 30,$

$0.25q + 0.50f = 10.75$

After clearing decimals, we have

$q + f = 30$ (1)

$25q + 50f = 1075.$ (2)

Solve. Solve the system of equations.

$-25q - 25f = -750$ Multiplying (1) by -25

$25q + 50f = 1075$

$25f = 325$ Adding

$f = 13$

$q + 13 = 30$ Substituting 13 for f in (1)

$q = 17$

Check. The total number of coins is 17 + 13, or 30. The value of 17 quarters is \$4.25, and the value of 13 fifty-cent pieces is \$6.50, so the total value of the coins is \$4.25 + \$6.50, or \$10.75. The numbers check.

State. There are 17 quarters and 13 fifty-cent pieces.

31. Familiarize. We first make a drawing.

Slow train
d kilometers 75 km/h (t + 2) hours

Fast train
d kilometers 125 km/h t hours

From the drawing we see that the distances are the same. Now complete the chart.

$d = r \cdot t$

	Distance	Rate	Time	
Slow train	d	75	t + 2	→ $d = 75(t + 2)$
Fast train	d	125	t	→ $d = 125t$

Translate. Using d = rt in each row of the table, we get a system of equations:

$d = 75(t + 2),$

$d = 125t$

31. (continued)

Solve. We solve the system of equations.

$125t = 75(t + 2)$ Using substitution

$125t = 75t + 150$

$50t = 150$

$t = 3$

The time for the fast train should be 3 hr, and the time for the slow train 3 + 2, or 5 hr.

Then $d = 125t = 125 \cdot 3 = 375$

Check. At 125 km/h, in 3 hr the fast train will travel $125 \cdot 3 = 375$ km. At 75 km/h, in 5 hr the slow train will travel $75 \cdot 5 = 375$ km. The numbers check.

State. The trains will meet 375 km from the station.

33. Familiarize. We first make a drawing.

Chicago 110 km/h t hours → ← t hours 90 km/h Indianapolis

350 km

The sum of the distances is 350 km. The times are the same. We organize the information in a table.

$d = r \cdot t$

Motorcycle	Distance	Rate	Time	
From Chicago	d	110	t	→ $d = 110t$
From Indpls.	350 - d	90	t	→ $350 - d = 90t$

We let t = the time, d = the distance from Chicago, and 350 - d = the distance from Indianapolis.

Translate. Using $d = rt$ in each row of the table, we get a system of equations:

$d = 110t,$ (1)

$350 - d = 90t$ (2)

Solve. We use the substitution method.

$350 - 110t = 90t$ Substituting 110t for d in (2)

$350 = 200t$

$\frac{350}{200} = t$

$\frac{7}{4} = t$

Check. The motorcycle from Chicago will travel $110 \cdot \frac{7}{4}$, or 192.5 km. The motorcycle from Indianapolis will travel $90 \cdot \frac{7}{4}$, or 157.5 km. The sum of the two distances is 192.5 + 157.5, or 350 km. The value checks.

State. In $1\frac{3}{4}$ hours the motorcycles will meet.

35. Familiarize. We first make a drawing.

Downstream, 6 mph current →

d mi, r + 6, 3 hr

Upstream, 6 mph current ←

d mi, r - 6, 5 hr

Let d = the distance and r = the speed of the boat in still water. Then when the boat travels downstream its speed is r + 6, and its speed upstream is r - 6. From the drawing we see that the distances are the same. Organize the information in a table.

$d = r \cdot t$

	Distance	Rate	Time	
Downstream	d	r + 6	3	→ $d = (r + 6)3$
Upstream	d	r - 6	5	→ $d = (r - 6)5$

Translate. Using $d = rt$ in each row of the table, we get a system of equations:

$d = 3r + 18,$

$d = 5r - 30$

Solve. Solve the system of equations.

$3r + 18 = 5r - 30$ Using substitution

$18 = 2r - 30$

$48 = 2r$

$24 = r$

Check. When r = 24, r + 6 = 30, and the distance traveled in 3 hr is $30 \cdot 3 = 90$ mi. Also, r - 6 = 18, and the distance traveled in 5 hr is $18 \cdot 5 = 90$ mi. The value checks.

State. The speed of the boat in still water is 24 mph.

37. Familiarize. We first make a drawing.

Stronger head wind, w mph →

2900 mi, r - w, 5 hr

Weaker head wind, w/2 mph ←

2900 mi, $r - \frac{w}{2}$, 4 hr 50 min

Let w = the speed of the stronger head wind and r = the plane's air speed in still air. Then w/2 = the speed of the weaker head wind. We will express 4 hr 50 min as $4\frac{5}{6}$ hr, or $\frac{29}{6}$ hr. Organize the information in a table.

$d = r \cdot t$

	Distance	Rate	Time
Stonger head wind	2900	r - w	5
Weaker head wind	2900	$r - \frac{w}{2}$	$\frac{29}{6}$

37. (continued)

Translate. Using $d = rt$ in each row of the table, we get a system of equations:

$$2900 = 5r - 5w,$$
$$2900 = \frac{29}{6}r - \frac{29}{12}w$$

After clearing fractions and rearranging we have

$$5r - 5w = 2900, \quad (1)$$
$$58r - 29w = 34,800. \quad (2)$$

Solve. We use the elimination method.

$145r - 145w = 84,100$ Multiplying (1) by 29
$-290r + 145w = -174,000$ Multiplying (2) by -5
$-145r = -89,900$ Adding
$r = 620$

$5(620) - 5w = 2900$ Substituting 620 for r in (1)
$3100 - 5w = 2900$
$-5w = -200$
$w = 40$

Check. When Gary flies 2900 mi in 5 hr his speed is 2900/5, or 580 mph. This is 620 - 40. When he flies 2900 mi in $\frac{29}{6}$ hr his speed is $\frac{2900}{\frac{29}{6}}$, or 600 mph. This is $620 - \frac{40}{2}$. The values check.

State. The head wind is 40 mph. The plane's speed in still air is 620 mph.

39. Familiarize. We first make a drawing.

d km, 420 km/h, t hr   1000 - d km, 330 km/h, t hr

|———— 1000 km ————|

Let d = the distance traveled at 420 km/h and t = the time traveled. Then 1000 - d = the distance traveled at 330 km/h. We organize the information in a table.

$$d = r \cdot t$$

	Distance	Rate	Time
Faster airplane	d	420	t
Slower airplane	1000 - d	330	t

Translate. Using $d = rt$ in each row of the table, we get a system of equations:

$$d = 420t,$$
$$1000 - d = 330t$$

Solve. We use substitution.

$1000 - 420t = 330t$ Substituting
$1000 = 750t$
$\frac{4}{3} = t$

39. (continued)

Check. If $t = \frac{4}{3}$, then $420 \cdot \frac{4}{3} = 560$, the distance traveled by the faster airplane. Also, $330 \cdot \frac{4}{3} = 440$, the distance traveled by the slower plane. The sum of the distances is 560 + 440, or 1000 km. The values check.

State. The airplanes will meet after $\frac{4}{3}$ hr, or $1\frac{1}{3}$ hr.

41. Familiarize. We first make a drawing. Let ℓ = the length and w = the width of the original piece of posterboard. Then w - 6 = the width after cutting off 6 in.

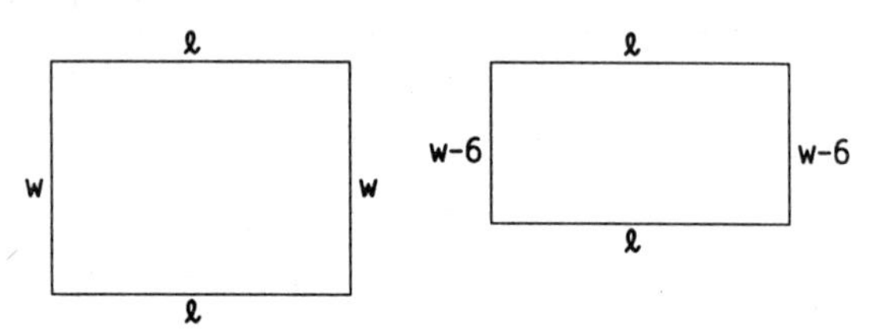

Before cutting   After cutting

Translate. The first statement gives us one equation.

The perimeter of the original piece of posterboard is 156 in.

$$2\ell + 2w = 156$$

Rewording the second statement, we get another equation.

The length is 4 times the width after cutting off 6 in.

$$\ell = 4 \cdot (w - 6)$$

We have a system of equations:

$$2\ell + 2w = 156, \quad (1)$$
$$\ell = 4(w - 6) \quad (2)$$

Solve. We solve the system of equations.

$2 \cdot 4(w - 6) + 2w = 156$ Substituting $4(w - 6)$ for ℓ in (1)
$8w - 48 + 2w = 156$
$10w - 48 = 156$
$10w = 204$
$w = \frac{204}{10}$, or $\frac{102}{5}$

$\ell = 4\left(\frac{102}{5} - 6\right)$ Substituting $\frac{102}{5}$ for w in (2)
$\ell = 4\left(\frac{102}{5} - \frac{30}{5}\right)$
$\ell = 4\left(\frac{72}{5}\right)$
$\ell = \frac{288}{5}$

41. (continued)

Check. The perimeter of a rectangle with width $\frac{102}{5}$ in. and length $\frac{288}{5}$ in. is $2\left(\frac{288}{5}\right) + 2\left(\frac{102}{5}\right) = \frac{576}{5} + \frac{204}{5} = \frac{780}{5} = 156$ in. If 6 in. is cut off the width, the new width is $\frac{102}{5} - 6 = \frac{102}{5} - \frac{30}{5} = \frac{72}{5}$. The length, $\frac{288}{5}$, is $4\left(\frac{72}{5}\right)$. The numbers check.

State. The original piece of posterboard has width $\frac{102}{5}$ in. or $20\frac{2}{5}$ in. and length $\frac{288}{5}$ in., or $57\frac{3}{5}$ in.

43. Familiarize. Let x and y represent the number of members who ordered one and two books, respectively. Note that the y members ordered 2y books.

Translate.

The number of books sold was 880.

$$x + 2y = 880$$

The amount taken in was $9840.

$$12x + 20y = 9840$$

We have a system of equations:

$$x + 2y = 880, \quad (1)$$
$$12x + 20y = 9840 \quad (2)$$

Solve. We use the elimination method.

$$-12x - 24y = -10,560 \quad \text{Multiplying (1) by -12}$$
$$12x + 20y = 9840$$
$$-4y = -720 \quad \text{Adding}$$
$$y = 180$$

Check. If 180 members each buy 2 books, they buy a total of 360 books. Then 880 - 360, or 520 members, each buy one book. The amount taken in from 520 single book orders and 180 orders of 2 books is 520 ($12) + 180($20), or $9840. The result checks.

State. 180 members ordered two books.

45. Familiarize. Let x = the amount of the original solution remaining after some is drained and replaced with pure antifreeze. Let y = the amount of pure antifreeze added. In a table we organize the information regarding the solution after some of the original solution is drained and replaced with pure antifreeze.

	Original solution	Pure antifreeze	Mixture
Amount of solution	x	y	16 L
Percent of antifreeze	30%	100%	50%
Amount of antifreeze in solution	0.3x	1·y, or y	0.5(16), or 8 L

45. (continued)

Translate. The first row of the table gives us one equation.

$$x + y = 16$$

The third row of the table gives us another equation.

$$0.3x + y = 8$$

After clearing decimals, we have this system:

$$x + y = 16, \quad (1)$$
$$3x + 10y = 80 \quad (2)$$

Solve. We use the elimination method.

$$-3x - 3y = -48 \quad \text{Multiplying (1) by -3}$$
$$3x + 10y = 80$$
$$7y = 32 \quad \text{Adding}$$
$$y = \frac{32}{7}$$

Check. If $\frac{32}{7}$ L of the 30% solution are drained, then $16 - \frac{32}{7}$, or $\frac{80}{7}$ L, of 30% solution remain. The amount of alcohol in the new solution is $0.3\left(\frac{80}{7}\right) + 1 \cdot \frac{32}{7}$, or 8 L. The result checks.

State. $\frac{32}{7}$ L, or $4\frac{4}{7}$ L, of the 30% mixture should be drained and replaced with pure antifreeze.

47. Familiarize. Let x and y represent the number of miles of city driving and highway driving, respectively.

Translate. We write two equations.

The car was driven 465 mi.

$$x + y = 465$$

The gallons used in city driving plus the gallons used in highway driving are 23 gal.

$$\frac{x}{18} + \frac{y}{24} = 23$$

After clearing fractions, we have this system of equations:

$$x + y = 465, \quad (1)$$
$$4x + 3y = 1656 \quad (2)$$

Solve. We use the elimination method.

$$-3x - 3y = -1395 \quad \text{Multiplying (1) by -3}$$
$$4x + 3y = 1656$$
$$x = 261 \quad \text{Adding}$$

$$261 + y = 465 \quad \text{Substituting 261 for x in (1)}$$
$$y = 204$$

Check. The total numbers of miles driven is 261 + 204, or 465. In city driving, 261/18, or 14.5 gal of gasoline were used. In highway driving, 204/24, or 8.5 gal of gasoline were used. The total amount of gasoline used was 14.5 + 8.5, or 23 gal. The numbers check.

State. 261 mi were driven in the city, and 204 mi were driven on the highway.

49. Familiarize. Let x = the amount of salt in the original solution, and let y = the amount of freshwater to be added. We organize the information in a table.

	Original solution	Freshwater	Mixture
Amount of solution	2000 gal	y	2000 + y
Percent of salt	7.5%	0%	7%
Amount of salt in solution	x	0·y, or 0	x

Translate. The "Original solution" column of the table gives us one equation.

$0.075(2000) = x$, or $150 = x$

The "Mixture" column gives us another equation.

$0.07(2000 + y) = x$, or $140 + 0.07y = x$

After clearing decimals, we have this system of equations:

$$150 = x,$$
$$14{,}000 + 7y = 100x$$

Solve. We use substitution.

$$14{,}000 + 7y = 100(150)$$
$$14{,}000 + 7y = 15{,}000$$
$$7y = 1000$$
$$y = \frac{1000}{7}$$

Check. The amount of salt in the tank is $0.075(2000)$, or 150 gal. If the amount of solution is $2000 + \frac{1000}{7}$, or $\frac{15{,}000}{7}$ gal, then $\dfrac{150}{\frac{15{,}000}{7}} = 0.07$. We have a 7% solution, so the result checks.

State. The amount of freshwater that should be added is $\frac{1000}{7} = 142\frac{6}{7}$, or about 143 gal, to the nearest gallon.

Exercise Set 6.4

1. Substitute (1,-2,3) into the three equations, using alphabetical order.

$$\begin{array}{c|c} x + y + z & 2 \\ \hline 1 + (-2) + 3 & 2 \\ 2 & \end{array}$$

$$\begin{array}{c|c} x - 2y - z & 2 \\ \hline 1 - 2(-2) - 3 & 2 \\ 1 + 4 - 3 & \\ 2 & \end{array}$$

$$\begin{array}{c|c} 3x + 2y + z & 2 \\ \hline 3\cdot 1 + 2(-2) + 3 & 2 \\ 3 - 4 + 3 & \\ 2 & \end{array}$$

The triple (1,-2,3) makes all three equations true, so it is a solution.

3.
$$x + y + z = 6, \quad (1)$$
$$2x - y + 3z = 9, \quad (2)$$
$$-x + 2y + 2z = 9 \quad (3)$$

Add equations (1) and (2) to eliminate y:

$$x + y + z = 6 \quad (1)$$
$$\underline{2x - y + 3z = 9} \quad (2)$$
$$3x + 4z = 15 \quad (3) \text{ Adding}$$

Use a different pair of equations and eliminate y:

$$4x - 2y + 6z = 18 \quad \text{Multiplying (2) by 2}$$
$$\underline{-x + 2y + 2z = 9} \quad (3)$$
$$3x + 8z = 27 \quad (5)$$

Now solve the system of equations (4) and (5).

$$3x + 4z = 15 \quad (4)$$
$$3x + 8z = 27 \quad (5)$$

$$-3x - 4z = -15 \quad \text{Multiplying (4) by -1}$$
$$\underline{3x + 8z = 27}$$
$$4z = 12$$
$$z = 3$$

$$3x + 4\cdot 3 = 15 \quad \text{Substituting 3 for z in (4)}$$
$$3x + 12 = 15$$
$$3x = 3$$
$$x = 1$$

$$1 + y + 3 = 6 \quad \text{Substituting 1 for x and 3 for z in (1)}$$
$$y + 4 = 6$$
$$y = 2$$

We obtain (1,2,3). This checks, so it is the solution.

5.
$$2x - y - 3z = -1, \quad (1)$$
$$2x - y + z = -9, \quad (2)$$
$$x + 2y - 4z = 17 \quad (3)$$

We start by eliminating z from two different pairs of equations.

$$2x - y - 3z = -1 \quad (1)$$
$$\underline{6x - 3y + 3z = -27} \quad \text{Multiplying (2) by 3}$$
$$8x - 4y = -28 \quad (4) \text{ Adding}$$

$$8x - 4y + 4z = -36 \quad \text{Multiplying (2) by 4}$$
$$\underline{x + 2y - 4z = 17}$$
$$9x - 2y = -19 \quad (5) \text{ Adding}$$

Now solve the system of equations (4) and (5).

$$8x - 4y = -28 \quad (4)$$
$$9x - 2y = -19 \quad (5)$$

$$8x - 4y = -28 \quad (4)$$
$$\underline{-18x + 4y = 38} \quad \text{Multiplying (5) by -2}$$
$$-10x = 10 \quad \text{Adding}$$
$$x = -1$$

5. (continued)

$8(-1) - 4y = -28$ Substituting -1 for x in (4)

$-8 - 4y = -28$

$-4y = -20$

$y = 5$

$2(-1) - 5 + z = -9$

$-2 - 5 + z = -9$

$-7 + z = -9$

$z = -2$

We obtain (-1,5,-2). This checks, so it is the solution.

7. $2x - 3y + z = 5,$ (1)

$x + 3y + 8z = 22,$ (2)

$3x - y + 2z = 12$ (3)

We start by eliminating y from two different pairs of equations.

$2x - 3y + z = 5$ (1)

$x + 3y + 8z = 22$ (2)

$3x + 9z = 27$ (4) Adding

$x + 3y + 8z = 22$ (2)

$9x - 3y + 6z = 36$ Multiplying (3) by 3

$10x + 14z = 58$ (5) Adding

Solve the system of equations (4) and (5).

$3x + 9z = 27$ (4)

$10x + 14z = 58$ (5)

$30x + 90z = 270$ Multiplying (4) by 10

$-30x - 42z = -174$ Multiplying (5) by -3

$48z = 96$ Adding

$z = 2$

$3x + 9\cdot 2 = 27$ Substituting 2 for z in (4)

$3x + 18 = 27$

$3x = 9$

$x = 3$

$2\cdot 3 - 3y + 2 = 5$ Substituting 3 for x and 2 for z in (1)

$-3y + 8 = 5$

$-3y = -3$

$y = 1$

We obtain (3,1,2). This checks, so it is the solution.

9. $3a - 2b + 7c = 13,$ (1)

$a + 8b - 6c = -47,$ (2)

$7a - 9b - 9c = -3$ (3)

We start by eliminating a from two different pairs of equations.

$3a - 2b + 7c = 13$ (1)

$-3a - 24b + 18c = 141$ Multiplying (2) by -3

$-26b + 25c = 154$ (4) Adding

$-7a - 56b + 42c = 329$ Multiplying (2) by -7

$7a - 9b - 9c = -3$ (3)

$-65b + 33c = 326$ (5) Adding

Now solve the system of equations (4) and (5).

$-26b + 25c = 154$ (4)

$-65b + 33c = 326$ (5)

$-130b + 125c = 770$ Multiplying (4) by 5

$130b - 66c = -652$ Multiplying (5) by -2

$59c = 118$

$c = 2$

$-26b + 25\cdot 2 = 154$ Substituting 2 for c in (4)

$-26b + 50 = 154$

$-26b = 104$

$b = -4$

$a + 8(-4) - 6(2) = -47$ Substituting -4 for b and 2 for c in (2)

$a - 32 - 12 = -47$

$a - 44 = -47$

$a = -3$

We obtain (-3,-4,2). This checks, so it is the solution.

11. $2x + 3y + z = 17,$ (1)

$x - 3y + 2z = -8,$ (2)

$5x - 2y + 3z = 5$ (3)

We start by eliminating y from two different pairs of equations.

$2x + 3y + z = 17$ (1)

$x - 3y + 2z = -8$ (2)

$3x + 3z = 9$ (4) Adding

$4x + 6y + 2z = 34$ Multiplying (1) by 2

$15x - 6y + 9z = 15$ Multiplying (3) by 3

$19x + 11z = 49$ (5) Adding

Now solve the system of equations (4) and (5).

$3x + 3z = 9$ (4)

$19x + 11z = 49$ (5)

$33x + 33z = 99$ Multiplying (4) by 11

$-57x - 33z = -147$ Multiplying (5) by -3

$-24x = -48$

$x = 2$

11. (continued)

$3\cdot 2 + 3z = 9$ Substituting 2 for x in (4)

$6 + 3z = 9$

$3z = 3$

$z = 1$

$2\cdot 2 + 3y + 1 = 17$ Substituting 2 for x and 1 for z in (1)

$3y + 5 = 17$

$3y = 12$

$y = 4$

We obtain (2,4,1). This checks, so it is the solution.

13. $2x + y + z = -2,$ (1)

$2x - y + 3z = 6,$ (2)

$3x - 5y + 4z = 7$ (3)

We start by eliminating y from two different pairs of equations.

$2x + y + z = -2$ (1)

$2x - y + 3z = 6$ (2)

$4x + 4z = 4$ (4) Adding

$10x + 5y + 5z = -10$ Multiplying (1) by 5

$3x - 5y + 4z = 7$ (3)

$13x + 9z = -3$ (5) Adding

Now solve the system of equations (4) and (5).

$4x + 4z = 4$ (4)

$13x + 9z = -3$ (5)

$36x + 36z = 36$ Multiplying (4) by 9

$-52x - 36z = 12$ Multiplying (5) by -4

$-16x = 48$ Adding

$x = -3$

$4(-3) + 4z = 4$ Substituting -3 for x in (4)

$-12 + 4z = 4$

$4z = 16$

$z = 4$

$2(-3) + y + 4 = -2$ Substituting -3 for x and 4 for z in (1)

$y - 2 = -2$

$y = 0$

We obtain (-3,0,4). This checks, so it is the solution.

15. $x - y + z = 4,$ (1)

$5x + 2y - 3z = 2,$ (2)

$3x - 7y + 4z = 8$ (3)

We start by eliminating z from two different pairs of equations.

$3x - 3y + 3z = 12$ Multiplying (1) by 3

$5x + 2y - 3z = 2$ (2)

$8x - y = 14$ (4) Adding

$-4x + 4y - 4z = -16$ Multiplying (1) by -4

$3x - 7y + 4z = 8$ (3)

$-x - 3y = -8$ (5) Adding

Now solve the system of equations (4) and (5).

$8x - y = 14$ (4)

$-x - 3y = -8$ (5)

$8x - y = 14$ (4)

$-8x - 24y = -64$ Multiplying (5) by 8

$-25y = -50$

$y = 2$

$8x - 2 = 14$ Substituting 2 for y in (4)

$8x = 16$

$x = 2$

$2 - 2 + z = 4$ Substituting 2 for x and 2 for y in (1)

$z = 4$

We obtain (2,2,4). This checks, so it is the solution.

17. $4x - y - z = 4,$ (1)

$2x + y + z = -1,$ (2)

$6x - 3y - 2z = 3$ (3)

We start by eliminating y from two different pairs of equations.

$4x - y - z = 4$ (1)

$2x + y + z = -1$ (2)

$6x = 3$ (4) Adding

At this point we can either continue by eliminating y from a second pair of equations or we can solve (4) for x and substitute that value in a different pair of the original equations to obtain a system of two equations in two variables. We take the second option.

$6x = 3$ (4)

$x = \frac{1}{2}$

Substitute $\frac{1}{2}$ for x in (1):

$4\left(\frac{1}{2}\right) - y - z = 4$

$2 - y - z = 4$

$-y - z = 2$ (5)

17. (continued)

Substitute $\frac{1}{2}$ for x in (3):

$$6\left(\frac{1}{2}\right) - 3y - 2z = 3$$
$$3 - 3y - 2z = 3$$
$$-3y - 2z = 0 \quad (6)$$

Solve the system of equations (5) and (6).

$$2y + 2z = -4 \quad \text{Multiplying (5) by -2}$$
$$-3y - 2z = 0 \quad (6)$$
$$-y = -4$$
$$y = 4$$

$$-4 - z = 2 \quad \text{Substituting 4 for y in (5)}$$
$$-z = 6$$
$$z = -6$$

We obtain $\left(\frac{1}{2},4,-6\right)$. This checks, so it is the solution.

19. $2r + 3s + 12t = 4$, (1)
$4r - 6s + 6t = 1$, (2)
$r + s + t = 1$ (3)

We start by eliminating s from two different pairs of equations.

$$4r + 6s + 24t = 8 \quad \text{Multiplying (1) by 2}$$
$$4r - 6s + 6t = 1 \quad (2)$$
$$8r + 30t = 9 \quad (4) \quad \text{Adding}$$

$$4r - 6s + 6t = 1 \quad (2)$$
$$6r + 6s + 6t = 6 \quad \text{Multiplying (3) by 6}$$
$$10r + 12t = 7 \quad (5) \quad \text{Adding}$$

Solve the system of equations (4) and (5).

$$40r + 150t = 45 \quad \text{Multiplying (4) by 5}$$
$$-40r - 48t = -28 \quad \text{Multiplying (5) by -4}$$
$$102t = 17$$
$$t = \frac{17}{102}$$
$$t = \frac{1}{6}$$

$$8r + 30\left(\frac{1}{6}\right) = 9 \quad \text{Substituting } \frac{1}{6} \text{ for t in (4)}$$
$$8r + 5 = 9$$
$$8r = 4$$
$$r = \frac{1}{2}$$

$$\frac{1}{2} + s + \frac{1}{6} = 1 \quad \text{Substituting } \frac{1}{2} \text{ for r and } \frac{1}{6} \text{ for t in (3)}$$
$$s + \frac{2}{3} = 1$$
$$s = \frac{1}{3}$$

We obtain $\left(\frac{1}{2},\frac{1}{3},\frac{1}{6}\right)$. This checks, so it is the solution.

21. $4a + 9b = 8$, (1)
$8a + 6c = -1$, (2)
$6b + 6c = -1$ (3)

We will use the elimination method. Note that there is no c in equation (1). We will use equations (2) and (3) to obtain another equation with no c terms.

$$8a + 6c = -1 \quad (2)$$
$$-6b - 6c = 1 \quad \text{Multiplying (3) by -1}$$
$$8a - 6b = 0 \quad (4) \quad \text{Adding}$$

Now solve the system of equations (1) and (4).

$$-8a - 18b = -16 \quad \text{Multiplying (1) by -2}$$
$$8a - 6b = 0$$
$$-24b = -16$$
$$b = \frac{2}{3}$$

$$8a - 6\left(\frac{2}{3}\right) = 0 \quad \text{Substituting } \frac{2}{3} \text{ for b in (4)}$$
$$8a - 4 = 0$$
$$8a = 4$$
$$a = \frac{1}{2}$$

$$8\left(\frac{1}{2}\right) + 6c = -1 \quad \text{Substituting } \frac{1}{2} \text{ for a in (2)}$$
$$4 + 6c = -1$$
$$6c = -5$$
$$c = -\frac{5}{6}$$

We obtain $\left(\frac{1}{2},\frac{2}{3},-\frac{5}{6}\right)$. This checks, so it is the solution.

23. $x + y + z = 57$, (1)
$-2x + y = 3$, (2)
$x - z = 6$ (3)

We will use the substitution method. Solve equations (2) and (3) for y and z, respectively. Then substitute in equation (1) to solve for x.

$$-2x + y = 3 \quad \text{Solving (2) for y}$$
$$y = 2x + 3$$
$$x - z = 6 \quad \text{Solving (3) for z}$$
$$-z = -x + 6$$
$$z = x - 6$$

$$x + (2x + 3) + (x - 6) = 57 \quad \text{Substituting in (1)}$$
$$4x - 3 = 57$$
$$4x = 60$$
$$x = 15$$

To find y, substitute 15 for x in $y = 2x + 3$:

$$y = 2\cdot 15 + 3 = 33$$

To find z, substitute 15 for x in $z = x - 6$:

$$z = 15 - 6 = 9$$

We obtain (15,33,9). This checks, so it is the solution.

25. $a - 3c = 6$, (1)
$b + 2c = 2$, (2)
$7a - 3b - 5c = 14$ (3)

We will use the elimination method. Note that there is no b in equation (1). We will use equations (2) and (3) to obtain another equation with no b term.

$3b + 6c = 6$ Multiplying (2) by 3
$7a - 3b - 5c = 14$ (3)
$7a + c = 20$ (4)

Now solve the system of equations (1) and (4).

$a - 3c = 6$ (1)
$21a + 3c = 60$ Multiplying (4) by 3
$22a = 66$
$a = 3$

$3 - 3c = 6$ Substituting 3 for a in (1)
$c = -1$

$b + 2(-1) = 2$ Substituting -1 for c in (2)
$b = 4$

We obtain (3,4,-1). This checks, so it is the solution.

27. $F = \frac{1}{2}t(c - d)$
$2F = t(c - d)$
$2F = tc - td$
$2F + td = tc$
$\frac{2F + td}{t} = c$, or
$\frac{2F}{t} + d = c$

29. $w + x + y + z = 2$, (1)
$w + 2x + 2y + 4z = 1$, (2)
$w - x + y + z = 6$, (3)
$w - 3x - y + z = 2$, (4)

Start by eliminating w from three different pairs of equations.

$w + x + y + z = 2$ (1)
$-w - 2x - 2y - 4z = -1$ Multiplying (2) by -1
$-x - y - 3z = 1$ (5) Adding

$w + x + y + z = 2$ (1)
$-w + x - y - z = -6$ Multiplying (3) by -1
$2x = -4$ (6) Adding

$w + x + y + z = 2$ (1)
$-w + 3x + y - z = -2$ Multiplying (4) by -1
$4x + 2y = 0$ (7) Adding

We can solve (6) for x:
$2x = -4$
$x = -2$

29. (continued)

Substitute -2 for x in (7):
$4(-2) + 2y = 0$
$-8 + 2y = 0$
$2y = 8$
$y = 4$

Substitute -2 for x and 4 for y in (5):
$-(-2) - 4 - 3z = 1$
$-2 - 3z = 1$
$-3z = 3$
$z = -1$

Substitute -2 for x, 4 for y, and -1 for z in (1):
$w - 2 + 4 - 1 = 2$
$w + 1 = 2$
$w = 1$

We obtain (1,-2,4,-1). This checks, so it is the solution.

Exercise Set 6.5

1. Familiarize. Let x = the first number, y = the second number, and z = the third number.

Translate.

The sum of the three numbers is 5.
$x + y + z = 5$

The first number minus the second plus the third is 1.
$x - y + z = 1$

The first number minus the third is 3 more than the second.
$x - z = y + 3$

We now have a system of equations.

$x + y + z = 5$, or $x + y + z = 5$,
$x - y + z = 1$, $x - y + z = 1$,
$x - z = y + 3$ $x - y - z = 3$

Solve. Solving the system we get (4,2,-1).

Check. The sum of the numbers is 5. The first minus the second plus the third is $4 - 2 + (-1)$, or 1. The first minus the third is 5, which is three more than the second. The numbers check.

State. The numbers are 4, 2, and -1.

3. Familiarize. We first make a drawing.

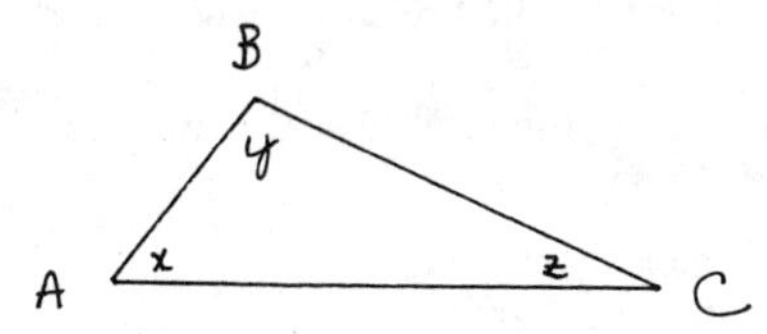

We let x, y, and z represent the measures of angles A, B, and C, respectively. The measures of the angles of a triangle add up to 180°.

Translate.

The sum of the measures is 180°.

$x + y + z = 180$

The measure of angle B	is	2° more than three times the measure of angle A.
y	$=$	$3x + 2$

The measure of angle C	is	8° more than the measure of angle A.
z	$=$	$x + 8$

We now have a system of equations.

$x + y + z = 180,$

$y = 3x + 2,$

$z = x + 8$

Solve. Solving the system we get (34,104,42).

Check. The sum of the numbers is 180, so that checks. Three times the measure of angle A is 3·34, or 102°, and 2° added to 102° is 104°, the measure of angle B. The measure of angle C, 42°, is 8° more than 34°, the measure of angle A. These values check.

State. Angles A, B, and C measure 34°, 104°, and 42°, respectively.

5. Familiarize. Let x, y, and z represent the amount spent on newspaper, television and radio ads, respectively, in billions of dollars.

Translate.

The total expenditure was $84.8 billion.

$x + y + z = 84.8$

The total amount spent on television and radio ads	was	$2.6 billion	more than	the amount spent on newspaper ads.
$y + z$	$=$	2.6	$+$	x

The amount spent on newspaper ads	was	$5.1 billion	more than	the amount spent on television ads.
x	$=$	5.1	$+$	y

We have a system of equations.

$x + y + z = 84.8,$

$y + z = 2.6 + x,$

$x = 5.1 + y$

Solve. Solving the system we get (41.1,36,7.7).

5. (continued)

Check. The sum of the numbers is 84.8, so that checks. Also, 36 + 7.7 = 43.7 which is 2.6 + 41.1, and 41.1 = 5.1 + 36. These values check.

State. $41.1 billion was spent on newspaper ads, $36 billion was spent on television ads, and $7.7 billion was spent on radio ads.

7. Familiarize. We first make a drawing.

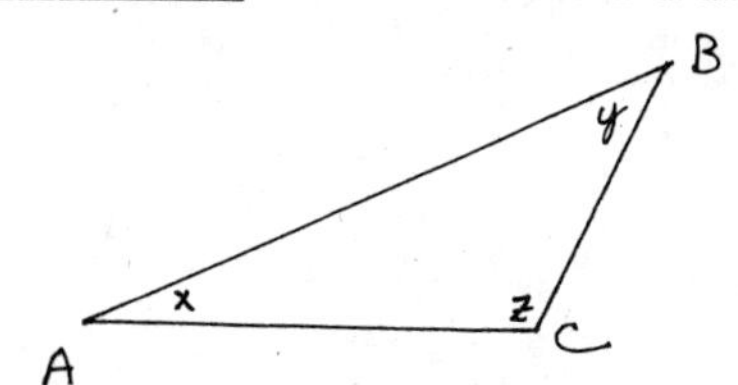

We let x, y, and z represent the measures of angles A, B, and C, respectively. The measures of the angles of a triangle add up to 180°.

Translate.

The sum of the measures is 180°.

$x + y + z = 180$

The measure of angle B	is twice	the measure of angle A.
y	$= 2\cdot$	x

The measure of angle C	is	80° more than the measure of angle A.
z	$=$	$x + 80$

We now have a system of equations.

$x + y + z = 180,$

$y = 2x,$

$z = x + 80$

Solve. Solving the system we get (25,50,105).

Check. The sum of the numbers is 180, so that checks. The measure of angle B, 50°, is twice 25°, the measure of angle A. The measure of angle C, 105°, is 80° more than 25°, the measure of angle A. The values check.

State. Angles A, B, and C measure 25°, 50°, and 105°, respectively.

9. Familiarize. Let x, y, and z represent the amount of cholesterol in one egg, one cupcake, and one slice of pizza, respectively, in mg.

Translate.

The cholesterol in 1 egg and 1 cupcake and 1 slice of pizza is 302 mg.

$$x + y + z = 302$$

The cholesterol in 2 cupcakes and 3 slices of pizza is 65 mg.

$$2y + 3z = 65$$

The cholesterol in 2 eggs and 1 cupcake is 567 mg.

$$2x + y = 567$$

We have a system of equations.

$$x + y + z = 302,$$
$$2y + 3z = 65,$$
$$2x + y = 567$$

Solve. Solving the system we get (274,19,9).

Check. The cholesterol in 1 egg, 1 cupcake, and 1 slice of pizza would be 274 + 19 + 9, or 302 mg. The cholesterol in 2 cupcakes and 3 slices of pizza would be $2\cdot19 + 3\cdot9$, or 65 mg. The cholesterol in 2 eggs and 1 cupcake would be $2\cdot274 + 19$, or 567 mg. These values check.

State. One egg contains 274 mg of cholesterol, 1 cupcake contains 19 mg, and 1 slice of pizza contains 9 mg.

11. Familiarize. Let s = the number of servings of steak, p = the number of baked potatoes, and a = the number of servings of asparagus. Then s servings of steak contain 300s calories, 20s g of protein, and no vitamin C. In p baked potatoes there are 100p calories, 5p g of protein, and 20p mg of vitamin C. And a servings of asparagus contain 50a calories, 5a g of protein, and 40a mg of vitamin C. The patient requires 800 calories, 55 g of protein, and 220 mg of vitamin C.

Translate. Write equations for the total number of calories, the total amount of protein, and the total amount of vitamin C.

$$300s + 100p + 50a = 800 \quad \text{(calories)}$$
$$20s + 5p + 5a = 55 \quad \text{(protein)}$$
$$20p + 40a = 220 \quad \text{(vitamin C)}$$

We now have a system of equations.

Solve. Solving the system we get s = 1, p = 3, and a = 4.

Check. One serving of steak provides 300 calories, 20 g of protein, and no vitamin C. Three baked potatoes provide 300 calories, 15 g of protein, and 60 mg of vitamin C. Four servings of asparagus provide 200 calories, 20 g of protein, and 160 mg of vitamin C. Together, then, they provide 800 calories, 55 g of protein, and 220 mg of vitamin C. The values check.

State. One serving of steak, 3 baked potatoes, and 4 servings of asparagus are required.

13. Familiarize. Let x, y, and z represent the number of fraternal twin births for Orientals, blacks, and whites in the U.S., respectively, out of every 15,400 births.

Translate. Out of every 15,400 births, we have the following statistics:

The total number of fraternal twin births is 739.

$$x + y + z = 739$$

The number of fraternal twin births for Orientals is 185 more than the number for blacks.

$$x = 185 + y$$

The number of fraternal twin births for Orientals is 231 more than the number for whites.

$$x = 231 + z$$

We have a system of equations.

$$x + y + z = 739,$$
$$x = 185 + y,$$
$$x = 231 + z$$

Solve. Solving the system we get (385,200,154).

Check. The total of the numbers is 739. Also 385 is 185 more than 200, and it is 231 more than 154.

State. Out of every 15,400 births, there are 385 births of fraternal twins for Orientals, 200 for blacks, and 154 for whites.

15. Familiarize. It helps to organize the information in a table.

Pumps Working	A	B	C	A + B	A + C	A, B, & C
Gallons Per hour	x	y	z	2200	2400	3700

We let x, y, and z represent the number of gallons per hour which can be pumped by pumps A, B, and C, respectively.

Translate. From the table, we obtain three equations.

$$x + y + z = 3700 \quad \text{(All three pumps working)}$$
$$x + y = 2200 \quad \text{(A and B working)}$$
$$x + z = 2400 \quad \text{(A and C working)}$$

Solve. Solving the system we get (900,1300,1500).

Check. The sum of the gallons per hour pumped when all three are pumping is 900 + 1300 + 1500, or 3700. The sum of the gallons per hour pumped when only pump A and pump B are pumping is 900 + 1300, or 2200. The sum of the gallons per hour pumped when only pump A and pump C are pumping is 900 + 1500, or 2400. The numbers check.

State. The pumping capacities of pumps A, B, and C are respectively 900, 1300, and 1500 gallons per hour.

17. Familiarize. Let x, y, and z represent the amount invested at 8%, 6%, and 9%, respectively. The interest earned is 0.08x, 0.06y, and 0.09z.

Translate.

The total invested was $80,000.

$x + y + z \quad = \quad 80{,}000$

The total interest was $6300.

$0.08x + 0.06y + 0.09z \quad = \quad 6300$

The interest at 8% was 4 times the interest at 6%.

$0.08x \quad = 4 \quad \cdot \quad 0.06y$

We have a system of equations.

$x + y + z = 80{,}000,$

$0.08x + 0.06y + 0.09z = 6300,$

$0.08x = 4(0.06y)$

Solve. Solving the system we get (45,000, 15,000, 20,000).

Check. The numbers add up to 80,000. Also, $0.08(45{,}000) + 0.06(15{,}000) + 0.09(20{,}000) = 6300$. In addition, $0.08(45{,}000) = 3600$ which is $4[0.6(15{,}000)]$. The values check.

State. $45,000 was invested at 8%, $15,000 at 6%, and $20,000 at 9%.

19. Familiarize. Let x, y, and z represent the number of par-3, par-4, and par-5 holes, respectively. Then a par golfer shoots 3x on the par-3 holes, 4x on the par-4 holes, and 5x on the par-5 holes.

Translate.

The total number of holes is 18.

$x + y + z \quad = 18$

A par golfer's score is 70.

$3x + 4y + 5z \quad = 70$

The number of par-4 holes is 2 times the number of par-5 holes.

$y \quad = 2 \quad \cdot \quad z$

We have a system of equations.

$x + y + z = 18,$

$3x + 4y + 5x = 70,$

$y = 2z$

Solve. Solving the system we get (6,8,4).

Check. The numbers add up to 18. A par golfer would shoot $3 \cdot 6 + 4 \cdot 8 + 5 \cdot 4$, or 70. The number of par-4 holes, 8, is twice the number of par-5 holes, 4. The numbers check.

State. There are 6 par-3 holes, 8 par-4 holes, and 4 par-5 holes.

21. Familiarize. We first make a drawing with additional labels.

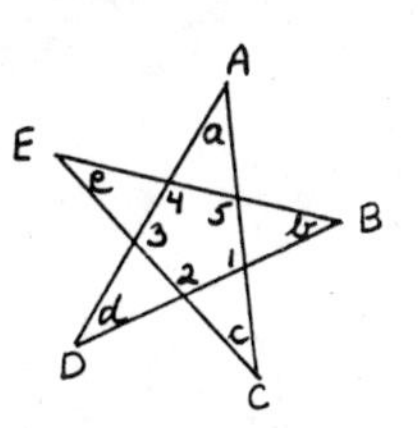

We let a, b, c, d, and e represent the angle measures at the tips of the star. We also label the interior angles of the pentagon 1, 2, 3, 4, and 5. We must recall the following geometric fact:

The sum of the measures of the interior angles of a polygon of n sides is given by $(n - 2)180°$.

Using this fact we know:

1. The sum of the angle measures of a triangle is $(3 - 2)180°$, or $180°$.
2. The sum of the angle measures of a pentagon is $(5 - 2)180°$, or $3(180°)$.

Translate. Using fact (1) listed above we obtain a system of 5 equations.

$a + 1 + d = 180,$

$b + 2 + e = 180,$

$c + 3 + a = 180,$

$d + 4 + b = 180,$

$e + 5 + c = 180$

Solve. Adding we obtain

$2a + 2b + 2c + 2d + 2e + 1 + 2 + 3 + 4 + 5 = 5(180)$

$2(a + b + c + d + e) + (1 + 2 + 3 + 4 + 5) = 5(180)$

Using fact (2) listed above we substitute 3(180) for (1 + 2 + 3 + 4 + 5) and solve for (a + b + c + d + e).

$2(a + b + c + d + e) + 3(180) = 5(180)$

$2(a + b + c + d + e) = 2(180)$

$a + b + c + d + e = 180$

Check. We should repeat the above calculations.

State. The sum of the angle measures at the tips of the star is $180°$.

23. Familiarize. Let T, G, and H represent the number of tickets Tom, Gary, and Hal begin with, respectively. After Hal gives tickets to Tom and Gary, each has the following number of tickets:

Tom: $T + T$, or $2T$,

Gary: $G + G$, or $2G$,

Hal: $H - T - G$.

After Tom gives tickets to Gary and Hal, each has the following number of tickets:

Gary: $2G + 2G$, or $4G$,

Hal: $(H - T - G) + (H - T - G)$, or $2(H - T - G)$

Tom: $2T - 2G - (H - T - G)$, or $3T - H - G$

23. (continued)

After Gary gives tickets to Hal and Tom, each has the following number of tickets:

Hal: $2(H - T - G) + 2(H - T - G)$, or
$4(H - T - G)$,

Tom: $(3T - H - G) + (3T - H - G)$, or
$2(3T - H - G)$,

Gary: $4G - 2(H - T - G) - (3T - H - G)$, or
$7G - H - T$.

Translate. Since Hal, Tom, and Gary each finish with 40 tickets, we write the following system of equations:

$4(H - T - G) = 40$,
$2(3T - H - G) = 40$,
$7G - H - T = 40$

Solve. Solving the system we find that $T = 35$.

Check. The check is left to the student.

State. Tom started with 35 tickets.

Exercise Set 6.6

1. Graph: $y < x$, (1)
$y > -x + 3$ (2)

We graph the lines $y = x$ and $y = -x + 3$, using dashed lines. We indicate the region for each inequality by the arrows at the ends of the lines. Note where the regions overlap and shade the region of solutions.

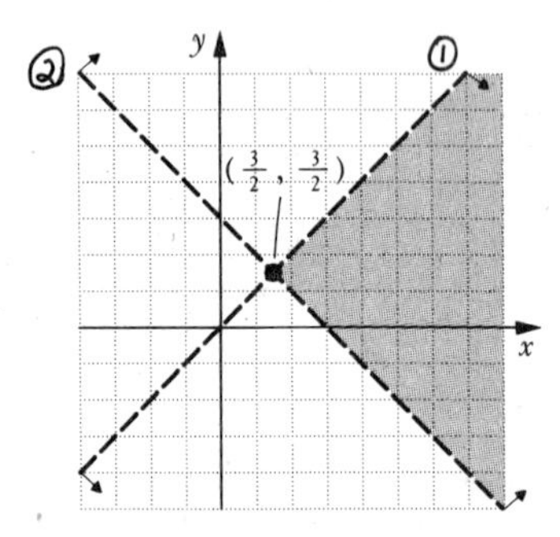

To find the vertex we solve the system of related equations:

$y = x$,
$y = -x + 3$

Solving, we obtain the vertex $\left(\frac{3}{2},\frac{3}{2}\right)$.

3. Graph: $y \geqslant x$, (1)
$y \leqslant -x + 4$ (2)

We graph the lines $y = x$ and $y = -x + 4$, using solid lines. We indicate the region for each inequality by the arrows at the ends of lines. Note where the regions overlap, and shade the region of solutions.

3. (continued)

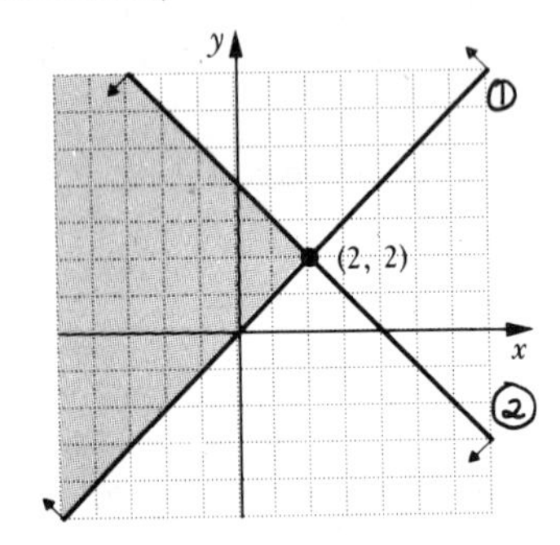

To find the vertex we solve the system of related equations:

$y = x$,
$y = -x + 4$

Solving, we obtain the vertex (2,2).

5. Graph: $y \geqslant -2$, (1)
$x \geqslant 1$ (2)

We graph the lines $y = -2$ and $x = 1$, using solid lines. We indicate the region for each inequality by arrows. Shade the region where they overlap.

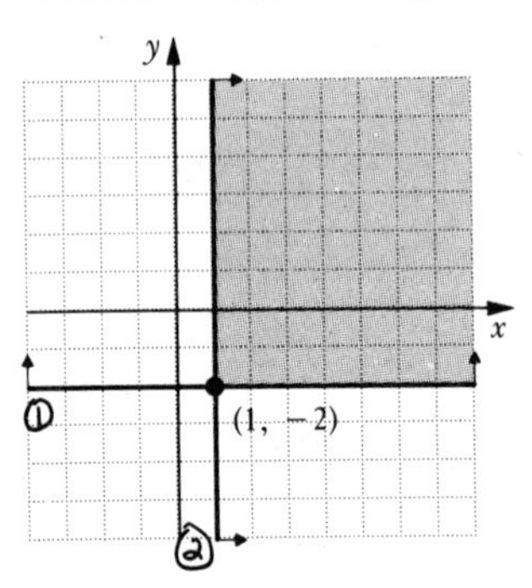

To find the vertex, we solve the system of related equations:

$y = -2$,
$x = 1$

Solving, we obtain the vertex (1,-2).

7. Graph: $x \leqslant 3$, (1)

$y \geqslant -3x + 2$ (2)

Graph the lines $x = 3$ and $y = -3x + 2$, using solid lines. Indicate the region for each inequality by arrows, and shade the region where they overlap.

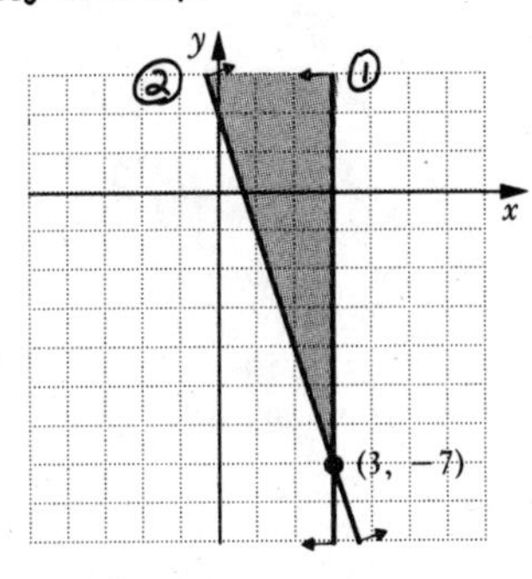

To find the vertex we solve the system of related equations:

$x = 3$,

$y = -3x + 2$

Solving, we obtain the vertex (3,-7).

9. Graph: $y \geqslant -2$, (1)

$y \geqslant x + 3$ (2)

Graph the lines $y = -2$ and $y = x + 3$, using solid lines. Indicate the region for each inequality by arrows, and shade the region where they overlap.

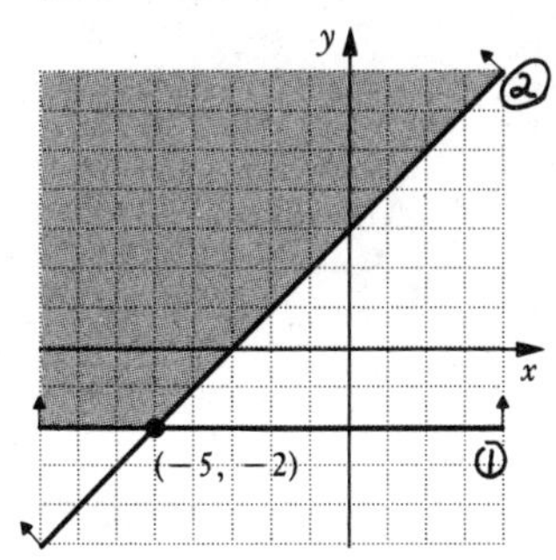

To find the vertex we solve the system of related equations:

$y = -2$,

$y = x + 3$

The vertex is (-5,-2).

11. Graph: $x + y \leqslant 1$, (1)

$x - y \leqslant 2$ (2)

Graph the lines $x + y = 1$ and $x - y = 2$, using solid lines. Indicate the region for each inequality by arrows, and shade the region where they overlap.

11. (continued)

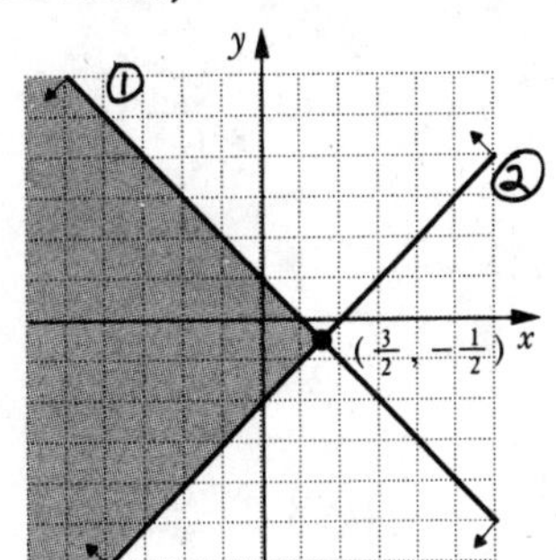

To find the vertex we solve the system of related equations:

$x + y = 1$,

$x - y = 2$

The vertex is $\left(\frac{3}{2}, -\frac{1}{2}\right)$.

13. Graph: $y - 2x \geqslant 1$, (1)

$y - 2x \leqslant 3$ (2)

Graph the lines $y - 2x = 1$ and $y - 2x = 3$, using solid lines. Indicate the region for each inequality by arrows, and shade the region where they overlap.

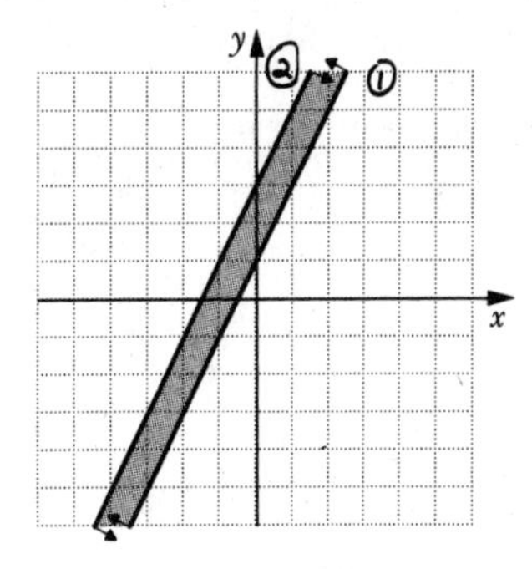

We can see from the graph that the lines are parallel. Hence there are no vertices.

15. Graph: $y \leqslant 2x + 1$, (1)

$y \geqslant -2x + 1$, (2)

$x \leqslant 2$ (3)

Shade the intersection of the graphs of $y \leqslant 2x + 1$, $y \geqslant -2x + 1$, and $x \leqslant 2$.

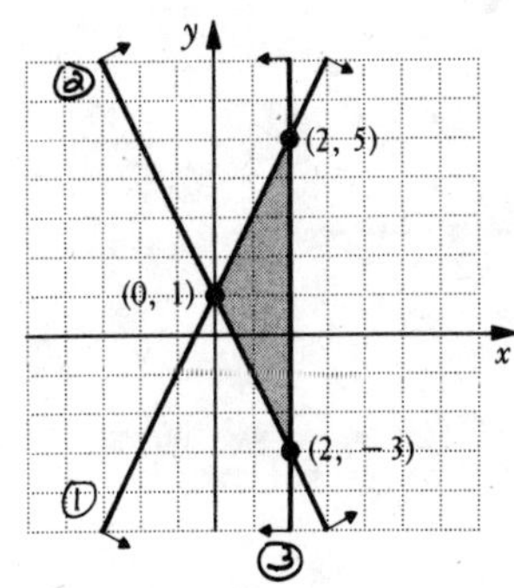

15. (continued)

To find the vertices we solve three different systems of equations. From (1) and (2) we obtain the vertex (0,1). From (1) and (3) we obtain the vertex (2,5). From (2) and (3) we obtain the vertex (2,-3).

17. Graph: $x + 2y \leqslant 12$, (1)

$2x + y \leqslant 12$ (2)

$x \geqslant 0$, (3)

$y \geqslant 0$ (4)

Shade the interesection of the graphs of the four inequalities above.

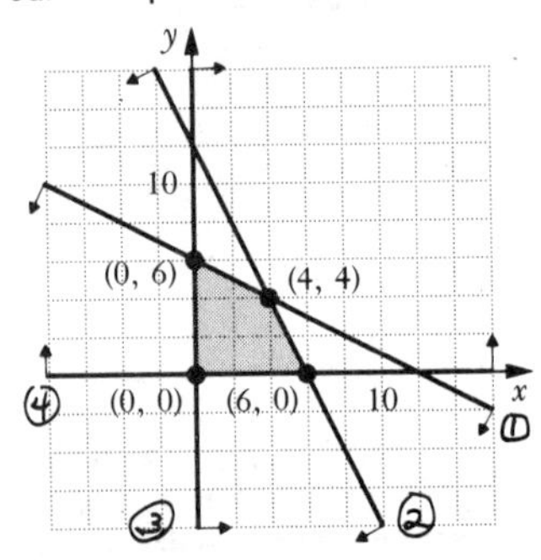

To find the vertices we solve four different systems of equations, as follows:

System of equations	Vertex
From (1) and (2)	(4,4)
From (1) and (3)	(0,6)
From (2) and (4)	(6,0)
From (3) and (4)	(0,0)

19. Graph: $8x + 5y \leqslant 40$, (1)

$x + 2y \leqslant 8$, (2)

$x \geqslant 0$, (3)

$y \geqslant 0$ (4)

Shade the intersection of the graphs of the four inequalities above.

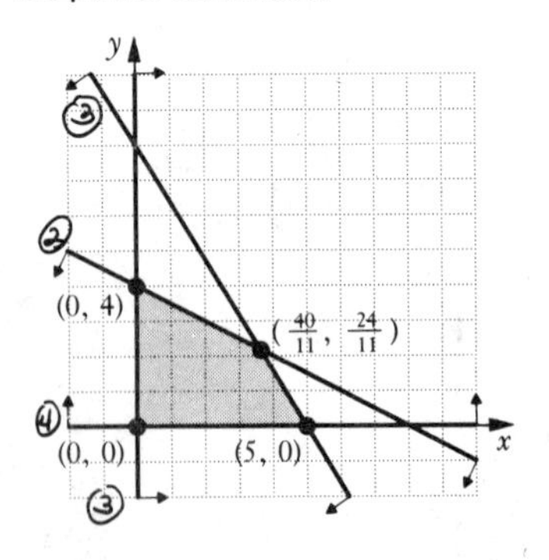

To find the vertices we solve four different systems of equations as follows:

19. (continued)

System of equations	Vertex
From (1) and (2)	$\left(\frac{40}{11}, \frac{24}{11}\right)$
From (1) and (4)	(5,0)
From (2) and (3)	(0,4)
From (3) and (4)	(0,0)

21. $\dfrac{4^3 + 5\cdot 6 - 7\cdot 8}{|3 - 5|^5 + 4(9 - 2)} = \dfrac{64 + 5\cdot 6 - 7\cdot 8}{|-2|^5 + 4\cdot 7}$

$= \dfrac{64 + 30 - 56}{32 + 28}$

$= \dfrac{38}{60} = \dfrac{19}{30}$

23. $5(3x - 4) = -2(x + 5)$

$15x - 20 = -2x - 10$

$17x - 20 = -10$

$17x = 10$

$x = \dfrac{10}{17}$

25. Graph: $x + y \geqslant 5$, (1)

$x + y \leqslant -3$ (2)

Graph the lines $x + y = 5$ and $x + y = -3$, using solid lines, and indicate the region for each inequality by arrows. The regions do not overlap (the solution set is Ø), so we do not shade any portion of the graph.

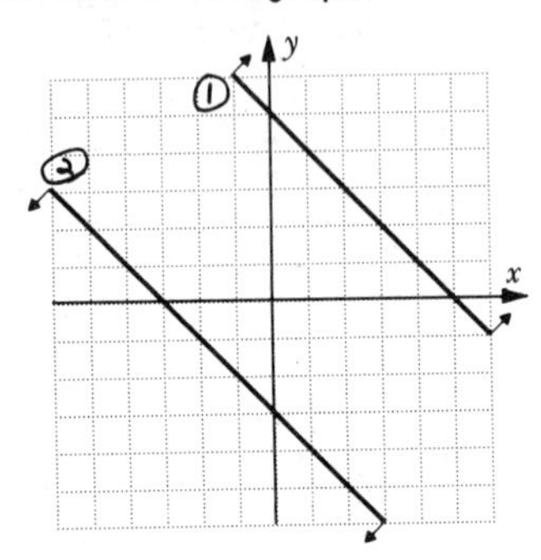

27. Graph: $x - 2y \leqslant 0$, (1)

$-2x + y \leqslant 2$, (2)

$x \leqslant 2$, (3)

$y \leqslant 2$, (4)

$x + y \leqslant 4$ (5)

Graph the five inequalities above, and shade the region where they overlap.

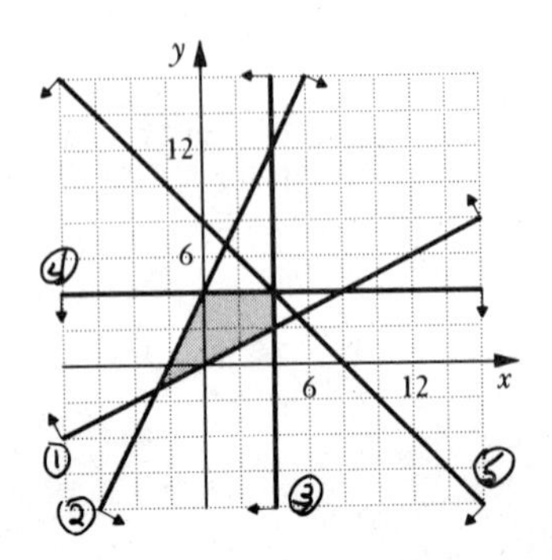

Exercise Set 7.1

1. The segment consists of the endpoints G and H and all points between them.

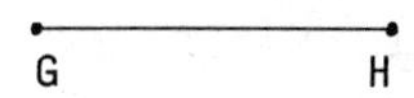

It can be named $\overline{GH}$ or $\overline{HG}$.

3. The ray with endpoint Q extends forever in the direction of point D.

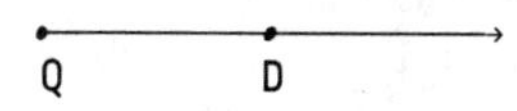

In naming a ray, the endpoint is always given first. This ray is named $\overrightarrow{QD}$.

5. 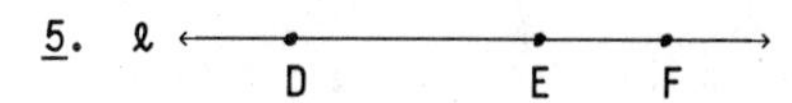

The line can be named with the small letter ℓ, or it can be named by any two points on it. This line can be named

ℓ, $\overleftrightarrow{DE}$, $\overleftrightarrow{ED}$, $\overleftrightarrow{EF}$, $\overleftrightarrow{FD}$, $\overleftrightarrow{EF}$, or $\overleftrightarrow{FE}$

7. The angle can be named in five different ways: angle GHI, angle IHG, $\angle$ GHI, $\angle$ IHG, or $\angle$ H.

9. Place the Δ of the protractor at the vertex of the angle, and line up one of the sides at 0°. We choose the horizontal side. Since 0° is on the inside scale, we check where the other side of the angle crosses the inside scale. It crosses at 10°. Thus, the measure of the angle is 10°.

11. Place the Δ of the protractor at the vertex of the angle, point B. Line up one of the sides at 0°. We choose the side that contains point A. Since 0° is on the outside scale, we check where the other side crosses the outside scale. It crosses at 180°. Thus, the measure of the angle is 180°.

13. Place the Δ of the protractor at the vertex of the angle, and line up one of the sides at 0°. We choose the horizontal side. Since 0° is on the inside scale, we check where the other side crosses the inside scale. It crosses at 130°. Thus, the measure of the angle is 130°.

15. The measure of the angle in Exercise 9 is 10°. Since its measure is greater than 0° and less than 90°, it is an acute angle.

17. The measure of the angle in Exercise 11 is 180°. It is a straight angle.

19. The measure of the angle in Exercise 13 is 130°. Since its measure is greater than 90° and less than 180°, it is an obtuse angle.

21. Using a protractor, we find that the lines do not intersect to form a right angle. They are not perpendicular.

23. Using a protractor, we find that the lines intersect to form a right angle. They are perpendicular.

25. All the sides are of different lengths. The triangle is a scalene triangle.

One angle is an obtuse angle. The triangle is an obtuse triangle.

27. All the sides are of different lengths. The triangle is a scalene triangle.

One angle is a right angle. The triangle is a right triangle.

29. All sides are the same length. The triangle is an equilateral triangle.

All three angles are acute. The triangle is an acute triangle.

31. All the sides are of different lengths. The triangle is a scalene triangle.

One angle is an obtuse angle. The triangle is an obtuse triangle.

33. The polygon has 4 sides. It is a quadrilateral.

35. The polygon has 5 sides. It is a pentagon.

37. The polygon has 3 sides. It is a triangle.

39. The polygon has 5 sides. It is a pentagon.

41. The polygon has 6 sides. It is a hexagon.

43. If a polygon has n sides, the sum of its angle measures is $(n - 2)\cdot 180°$. A decagon has 10 sides. Substituting 10 for n in the formula, we get

$$\begin{aligned}(n - 2)\cdot 180° &= (10 - 2)\cdot 180° \\ &= 8\cdot 180° \\ &= 1440°\end{aligned}$$

45. If a polygon has n sides, the sum of its angle measures is $(n - 2)\cdot 180°$. To find the sum of the angle measures for a 14-sided polygon, substitute 14 for n in the formula.

$$\begin{aligned}(n - 2)\cdot 180° &= (14 - 2)\cdot 180° \\ &= 12\cdot 180° \\ &= 2160°\end{aligned}$$

47. $$\begin{aligned}m(\angle A) + m(\angle B) + m(\angle C) &= 180° \\ 42° + 92° + x &= 180° \\ 134° + x &= 180° \\ x &= 180° - 134° \\ x &= 46°\end{aligned}$$

49.
$$\begin{array}{r} 1.75 \\ 12\overline{)21.00} \\ \underline{12} \\ 90 \\ \underline{84} \\ 60 \\ \underline{60} \\ 0 \end{array}$$

51. To divide by 100, move the decimal point 2 places to the left.

23.4 .23.4

23.4 ÷ 100 = 0.234

53.
$$\begin{array}{rl} 3.14 & \text{(2 decimal places)} \\ \underline{4.41} & \text{(2 decimal places)} \\ 314 & \\ 12560 & \\ \underline{125600} & \\ 13.8474 & \text{(4 decimal places)} \end{array}$$

Round to the nearest hundredth:

13.84[7]4 Thousandths digit is 5 or higher. Round up.

13.85

55. $48 \cdot \frac{1}{12} = \frac{48 \cdot 1}{12} = \frac{48}{12} = 4$

Exercise Set 7.2

1. Perimeter = 4 mm + 6 mm + 7 mm
= (4 + 6 + 7) mm
= 17 mm

3. Perimeter = 3.5 cm + 3.5 cm + 4.25 cm + 0.5 cm + 3.5 cm
= (3.5 + 3.5 + 4.25 + 0.5 + 3.5) cm
= 15.25 cm

5. P = 4·s Perimeter of a square
P = 4·3.25 m
= 13 m

7. P = 2·(ℓ + w) Perimeter of a rectangle
P = 2·(5 ft + 10 ft)
P = 2·(15 ft)
P = 30 ft

9. P = 2·(ℓ + w) Perimeter of a rectangle
P = 2·(34.67 cm + 4.9 cm)
P = 2·(39.57 cm)
P = 79.14 cm

11. P = 4·s Perimeter of a square
P = 4·22 ft
P = 88 ft

13. P = 4·s Perimeter of a square
P = 4·45.5 mm
P = 182 mm

15. Familiarize. First we will find the perimeter of the field. Then we will multiply to find the cost of the fence wire. We make a drawing.

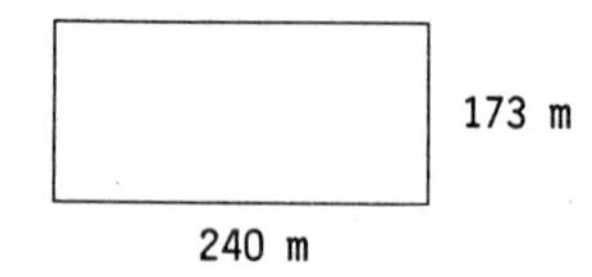

Translate. The perimeter of the field is given by
P = 2·(ℓ + w) = 2·(240 m + 173 m).

Solve. We calculate the perimeter.
P = 2·(240 m + 173 m) = 2·(413 m) = 826 m

Then we multiply to find the cost of the fence wire.
Cost = \$1.45/m × Perimeter =
\$1.45/m × 826 m = \$1197.70

Check. Repeat the calculations.

State. The perimeter of the field is 826 m. The fence wire will cost \$1197.70.

17. Familiarize. We make a drawing.

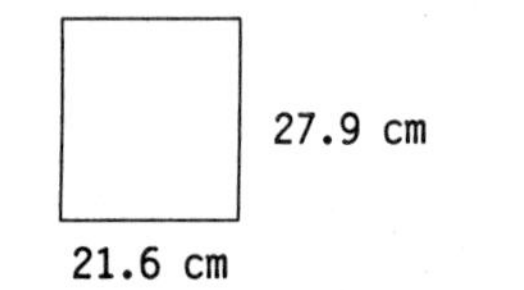

Translate. The perimeter of the sheet of paper is given by
P = 2·(ℓ + w) = 2(27.9 cm + 21.6 cm)

Solve. We do the calculation.
2(27.9 cm + 21.6 cm) = 2·(49.5 cm) = 99 cm

Check. We repeat the calculation.

State. The perimeter is 99 cm.

19. Familiarize. We first find the perimeter of the garden. We make a drawing.

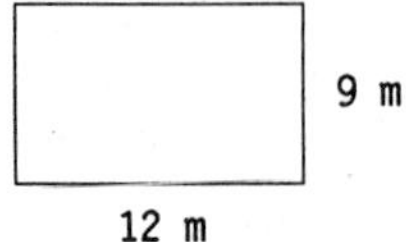

Translate. The perimeter is given by
P = 2·(ℓ + w) = 2·(12 m + 9 m).

Solve. We do the calculation.
P = 2·(12 m + 9 m) = 2·(21 m) = 42 m

a) We divide to find n, the number of posts needed.

19. (continued)

$n = 42 \div 3 = 14$

Thus, 14 posts are needed.

b) We multiply to find the cost of the posts.

Cost = $2.40 × 14 = $33.60

The posts will cost $33.60.

c) We subtract to find the length F of the fence.

$F = 42\text{ m} - 3\text{ m} = (42 - 3)\text{ m} = 39\text{ m}$

The fence will be 39 m long.

d) We multiply to find the cost of the fence.

Cost = $0.85/m × 39 m = $33.15

The fence will cost $33.15.

e) We add to find the total cost of the materials.

Total cost = Cost of posts +
Cost of fence + Cost of gate =
$33.60 + $33.15 + $9.95 = $76.70

Check. Repeat the calculations.

State. See the results stated in (a) - (e) above.

21. 56.1%

a) Drop the percent symbol. 56.1

b) Move the decimal point two places to the left. 0.56.1

56.1% = 0.561

23. $31^2 = 31 \times 31 = 961$

Exercise Set 7.3

1. $A = l \cdot w$ Area of a rectangular region

$A = (5\text{ km}) \cdot (3\text{ km})$

$= 5 \cdot 3 \cdot \text{km} \cdot \text{km}$

$= 15\text{ km}^2$

3. $A = l \cdot w$ Area of a rectangular region

$A = (2\text{ cm}) \cdot (0.7\text{ cm})$

$= 2 \cdot 0.7 \cdot \text{cm} \cdot \text{cm}$

$= 1.4\text{ cm}^2$

5. $A = s \cdot s$ Area of a square

$A = (2.5\text{ mm}) \cdot (2.5\text{ mm})$

$= 2.5 \cdot 2.5 \cdot \text{mm} \cdot \text{mm}$

$= 6.25\text{ mm}^2$

7. $A = s \cdot s$ Area of a square

$A = (90\text{ ft}) \cdot (90\text{ ft})$

$= 90 \cdot 90 \cdot \text{ft} \cdot \text{ft}$

$= 8100\text{ ft}^2$

9. $A = l \cdot w$

$A = (10\text{ ft}) \cdot (5\text{ ft})$

$= 10 \cdot 5 \cdot \text{ft} \cdot \text{ft}$

$= 50\text{ ft}^2$

11. $A = l \cdot w$

$A = (34.67\text{ cm}) \cdot (4.9\text{ cm})$

$= 34.67 \cdot 4.9 \cdot \text{cm} \cdot \text{cm}$

$= 169.883\text{ cm}^2$

13. $A = s \cdot s$

$A = (22\text{ ft}) \cdot (22\text{ ft})$

$= 22 \cdot 22 \cdot \text{ft} \cdot \text{ft}$

$= 484\text{ ft}^2$

15. $A = s \cdot s$

$A = (56.9\text{ km}) \cdot (56.9\text{ km})$

$= 56.9 \cdot 56.9 \cdot \text{km} \cdot \text{km}$

$= 3237.61\text{ km}^2$

17. Familiarize. We draw a picture.

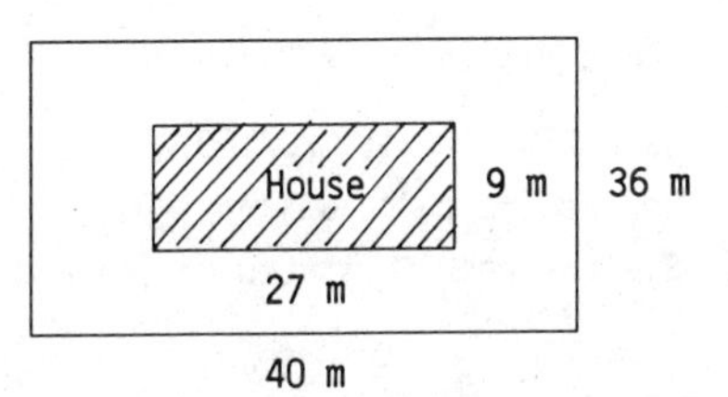

Translate. We let A = the area left over.

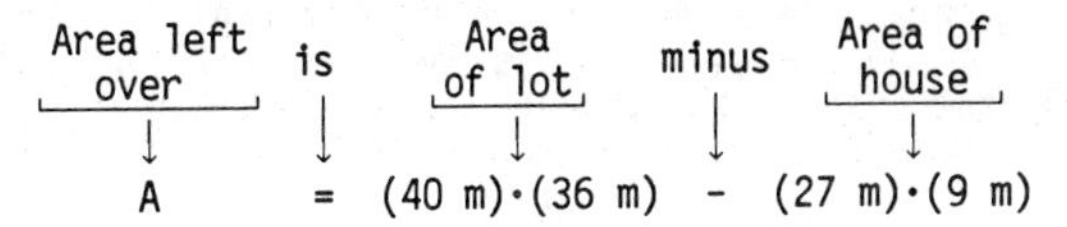

Solve. The area of the lot is

$(40\text{ m}) \cdot (36\text{ m}) = 40 \cdot 36 \cdot \text{m} \cdot \text{m} = 1440\text{ m}^2$.

The area of the house is

$(27\text{ m}) \cdot (9\text{ m}) = 27 \cdot 9 \cdot \text{m} \cdot \text{m} = 243\text{ m}^2$.

The area left over is

$A = 1440\text{ m}^2 - 243\text{ m}^2 = 1197\text{ m}^2$.

Check. Repeat the calculations.

State. The area left over is 1197 m^2.

19. Familiarize. We use the drawing in the text.

Translate. We let A = the area of the sidewalk.

Area of sidewalk is Total area minus Area of building

$A = (113.4\text{ m}) \times (75.4\text{ m}) - (110\text{ m}) \times (72\text{ m})$

Solve. The total area is

$(113.4\text{ m}) \times (75.4\text{ m}) = 113.4 \times 75.4 \times \text{m} \times \text{m} =$

$= 8550.36\text{ m}^2$.

19. (continued)

The area of the building is

$(110\ m) \times (72\ m) = 110 \times 72 \times m \times m = 7920\ m^2$.

The area of the sidewalk is

$A = 8550.36\ m^2 - 7920\ m^2 = 630.36\ m^2$.

Check. Repeat the calculations.

State. The area of the sidewalk is $630.36\ m^2$.

21. Familiarize. First we will find the area of the room. Then we will find the cost of the carpeting.

Translate. To find how many square meters of carpeting will be needed, find the area of the room.

$$A = \ell \cdot w = (5.5\ m) \times (4.5\ m)$$

Solve. a) We find the area.

$A = (5.5\ m) \times (4.5\ m) = 5.5 \times 4.5 \times m \times m$

$= 24.75\ m^2$

Thus, $24.75\ m^2$ of carpeting will be needed.

b) We multiply to find the cost of the carpet.

$$\text{Cost} = \$8.40/m^2 \times 24.75\ m^2 = \$207.90$$

The carpet will cost \$207.90.

Check. Repeat the calculations.

State. See the results stated in (a) and (b) above.

23. Think of this figure as a large rectangular region containing a smaller rectangular region. The area of the shaded region is the area left over when the area of the smaller region is taken away from the area of the larger region.

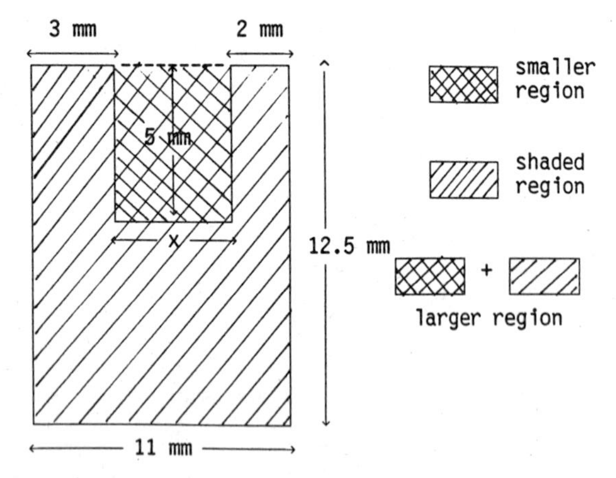

The larger rectangle measures 12.5 mm by 11 mm. One dimension of the smaller rectangle is 5 mm. Let x represent the other dimension. To find x we subtract:

$$x = 11\ m - 3\ m - 2\ m = (11 - 3 - 2)\ mm = 6\ mm$$

23. (continued)

(Area of shaded region) is (Area of larger region) minus (Area of smaller region)

$$A = (12.5\ mm \times 11\ mm) - (6\ mm \times 5\ mm)$$

The area of the larger region is

$(12.5\ mm) \times (11\ mm) = 12.5 \times 11 \times mm \times mm = 137.5\ mm^2$.

The area of the smaller region is

$(6\ mm) \times (5\ mm) = 6 \times 5 \times mm \times mm = 30\ mm^2$.

The area of the shaded region is

$A = 137.5\ mm^2 - 30\ mm^2 = 107.5\ mm^2$.

25. 0.452 — 0.45.2 (Move decimal point 2 places to the right) — 45.2% (Write a % symbol)

$0.452 = 45.2\%$

27. We multiply by 1 to get 100 in the denominator.

$$\frac{11}{20} = \frac{11}{20} \cdot \frac{5}{5} = \frac{55}{100} = 55\%$$

Exercise Set 7.4

1. $A = b \cdot h$ — Area of a parallelogram

$A = 8\ cm \cdot 4\ cm$ — Substituting 8 cm for b and 4 cm for h

$= 32\ cm^2$

3. $A = \frac{1}{2} \cdot b \cdot h$ — Area of a triangle

$A = \frac{1}{2} \cdot 12\ m \cdot 6\ m$ — Substituting 12 m for b and 6 m for h

$= \frac{12 \cdot 6}{2} \cdot m^2$

$= 36\ m^2$

5. $A = \frac{1}{2} \cdot h \cdot (a + b)$ — Area of a trapezoid

$A = \frac{1}{2} \cdot 6\ ft \cdot (5 + 12)\ ft$ — Substituting 6 ft for h, 5 ft for a, and 12 ft for b

$= \frac{6 \cdot 17}{2}\ ft^2 = 51\ ft^2$

7. $A = b \cdot h$ — Area of a parallelogram

$A = 8\ m \cdot 8\ m$

$= 64\ m^2$

9. $A = \frac{1}{2}\cdot h\cdot(a + b)$ Area of a trapezoid

$A = \frac{1}{2}\cdot 7\text{ mm}\cdot(4.5\text{ mm} + 8.5\text{ mm})$ Substituting 7 mm for h, 4.5 mm for a, and 8.5 mm for b

$= \frac{7\cdot 13}{2}\cdot\text{mm}^2$

$= \frac{91}{2}\text{ mm}^2$

$= 45.5\text{ mm}^2$

11. $A = b\cdot h$ Area of a parallelogram

$A = 2.3\text{ cm}\cdot 3.5\text{ cm}$ Substituting 2.3 cm for b and 3.5 cm for h

$= 8.05\text{ cm}^2$

13. $A = \frac{1}{2}\cdot h\cdot(a + b)$ Area of a trapezoid

$A = \frac{1}{2}\cdot 18\text{ cm}\cdot(9 + 24)\text{ cm}$ Substituting 18 cm for h, 9 cm for a, and 24 cm for b

$= \frac{18\cdot 33}{2}\text{ cm}^2$

$= 297\text{ cm}^2$

15. $A = \frac{1}{2}\cdot b\cdot h$ Area of a triangle

$A = \frac{1}{2}\cdot 4\text{ m}\cdot 3.5\text{ m}$ Substituting 4 m for b and 3.5 m for h

$= \frac{4\cdot 3.5}{2}\text{ m}^2$

$= 7\text{ m}^2$

17. $A = b\cdot h$ Area of a parallelogram

$A = 12\frac{1}{4}\text{ ft}\cdot 4\frac{1}{2}\text{ ft}$ Substituting $12\frac{1}{4}$ ft for b and $4\frac{1}{2}$ ft for h

$= \frac{49}{4}\cdot\frac{9}{2}\cdot\text{ft}$

$= \frac{441}{8}\text{ ft}^2$

$= 55\frac{1}{8}\text{ ft}^2$

19. Familiarize. We look for the kinds of figures whose areas we can calculate using area formulas that we already know.

Translate. The shaded region consists of a square region with a triangular region removed from it. The sides of the square are 30 cm, and the triangle has base 30 cm and height 15 cm. We find the area of the square using the formula $A = s\cdot s$ and the area of the triangle using $A = \frac{1}{2}\cdot b\cdot h$. Then we subtract.

Solve.

Area of the square:

$A = 30\text{ cm}\cdot 30\text{ cm} = 900\text{ cm}^2$

19. (continued)

Area of the triangle:

$A = \frac{1}{2}\cdot 30\text{ cm}\cdot 15\text{ cm} = 225\text{ cm}^2$

Area of the shaded region:

$A = 900\text{ cm}^2 - 225\text{ cm}^2 = 675\text{ cm}^2$

Check. We repeat the calculations.

State. The area of the shaded region is 675 cm^2.

21. Familiarize. We look for the kinds of figures whose areas we can calculate using area formulas that we already know.

Translate. The shaded region consists of 8 triangles, each with base 52 in. and height 52 in. We will find the area of one triangle using the formula $A = \frac{1}{2}\cdot b\cdot h$. Then we will multiply by 8.

Solve.

$A = \frac{1}{2}\cdot 52\text{ in.}\cdot 52\text{ in.} = 1352\text{ in}^2$

Then we multiply by 8:

$8\cdot 1352\text{ in}^2 = 10{,}816\text{ in}^2$

Check. We repeat the calculations.

State. The area of the shaded region is 10,816 in^2.

23. Familiarize. We make a drawing, shading the area left over after the pool is constructed.

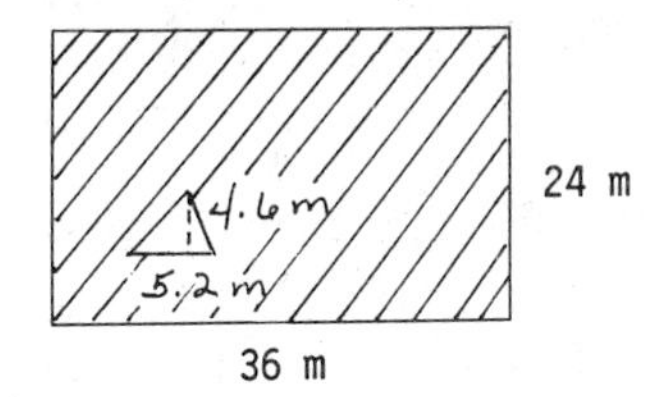

Translate. The shaded region consists of a rectangular region with a triangular region removed from it. The rectangular region has dimensions 36 m by 24 m, and the triangular region has base 5.2 m and height 4.6 m. We wil find the area of the rectangular region using the formula $A = b\cdot h$ and the area of the triangular region using $A = \frac{1}{2}\cdot b\cdot h$. Then we will subtract to find the area of the shaded region.

Solve.

Area of rectangle:

$A = 36\text{ m}\cdot 24\text{ m} = 864\text{ m}^2$

Area of triangle:

$A = \frac{1}{2}\cdot(5.2\text{ m})\cdot(4.6\text{ m}) = 11.96\text{ m}^2$

Area of shaded region:

$A = 864\text{ m}^2 - 11.96\text{ m}^2 = 852.04\text{ m}^2$

Check. We repeat the calculations.

State. The area left over is 852.04 m^2.

25. $9.25\% = \frac{9.25}{100}$ Definition of percent

$= \frac{9.25}{100} \cdot \frac{100}{100}$ Multiplying by 1 to get rid of the decimal point in the numerator

$= \frac{925}{10,000}$

$= \frac{37}{400} \cdot \frac{25}{25}$ Simplifying

$= \frac{37}{400}$

27.
```
     1.3 7 5
  8)1 1.0 0 0
     8
     3 0
     2 4
       6 0
       5 6
         4 0
         4 0
           0
```

$\frac{11}{8} = 1.375$

a) Move the decimal point 2 places to the right. 1.37.5

b) Add a percent symbol. 137.5%

$\frac{11}{8} = 137.5\%$

Exercise Set 7.5

1. $d = 2 \cdot r$

$d = 2 \cdot 7 \text{ cm} = 14 \text{ cm}$

3. $d = 2 \cdot r$

$d = 2 \cdot \frac{3}{4} \text{ in.} = \frac{6}{4} \text{ in.} = \frac{3}{2} \text{ in.}$, or $1\frac{1}{2}$ in.

5. $r = \frac{d}{2}$

$r = \frac{32 \text{ ft}}{2} = 16 \text{ ft}$

7. $r = \frac{d}{2}$

$r = \frac{1.4 \text{ cm}}{2} = 0.7 \text{ cm}$

9. $C = 2 \cdot \pi \cdot r$

$C \approx 2 \cdot \frac{22}{7} \cdot 7 \text{ cm} = \frac{2 \cdot 22 \cdot 7}{7} \text{ cm} = 44 \text{ cm}$

11. $C = 2 \cdot \pi \cdot r$

$C \approx 2 \cdot \frac{22}{7} \cdot \frac{3}{4} \text{ in.} = \frac{2 \cdot 22 \cdot 3}{7 \cdot 4} \text{ in.} = \frac{132}{28} \text{ in.}$

$= \frac{33}{7}$ in., or $4\frac{5}{7}$ in.

13. $C = \pi \cdot d$

$C \approx 3.14 \cdot 32 \text{ ft} = 100.48 \text{ ft}$

15. $C = \pi \cdot d$

$C \approx 3.14 \cdot 1.4 \text{ cm} = 4.396 \text{ cm}$

17. $A = \pi \cdot r \cdot r$

$A \approx \frac{22}{7} \cdot 7 \text{ cm} \cdot 7 \text{ cm} = \frac{22}{7} \cdot 49 \text{ cm}^2 = 154 \text{ cm}^2$

19. $A = \pi \cdot r \cdot r$

$A \approx \frac{22}{7} \cdot \frac{3}{4} \text{ in.} \cdot \frac{3}{4} \text{ in.} = \frac{22 \cdot 3 \cdot 3}{7 \cdot 4 \cdot 4} \text{ in}^2$

$= \frac{99}{56} \text{ in}^2$, or $1\frac{43}{56} \text{ in}^2$

21. $A = \pi \cdot r \cdot r$

$A \approx 3.14 \cdot 16 \text{ ft} \cdot 16 \text{ ft}$ $\left[r = \frac{d}{2};\ r = \frac{32 \text{ ft}}{2} = 16 \text{ ft}\right]$

$A = 3.14 \cdot 256 \text{ ft}^2$

$A = 803.84 \text{ ft}^2$

23. $A = \pi \cdot r \cdot r$

$A \approx 3.14 \cdot 0.7 \text{ cm} \cdot 0.7 \text{ cm}$ $\left[r = \frac{d}{2};\ r = \frac{1.4 \text{ cm}}{2} = 0.7 \text{ cm}\right]$

$A = 3.14 \cdot 0.49 \text{ cm}^2 = 1.5386 \text{ cm}^2$

25. $r = \frac{d}{2}$

$r = \frac{6 \text{ cm}}{2} = 3 \text{ cm}$

The radius is 3 cm.

$C = \pi \cdot d$

$C \approx 3.14 \cdot 6 \text{ cm} = 18.84 \text{ cm}$

The circumference is about 18.84 cm.

$A = \pi \cdot r \cdot r$

$A \approx 3.14 \cdot 3 \text{ cm} \cdot 3 \text{ cm} = 28.26 \text{ cm}^2$

The area is about 28.26 cm^2.

27. $A = \pi \cdot r \cdot r$

$A \approx 3.14 \cdot 220 \text{ km} \cdot 220 \text{ km} = 151,976 \text{ km}^2$

The broadcast area is about 151,976 km^2.

29. $C = \pi \cdot d$

$C \approx 3.14 \cdot 1.1 \text{ m} = 3.454 \text{ m}$

The circumference of the elm tree is about 3.454 m.

31. $C = \pi \cdot d$

$7.85 \text{ cm} \approx 3.14 \cdot d$ Substituting 7.85 cm for C and 3.14 for π

$\frac{7.85 \text{ cm}}{3.14} = d$ Dividing on both sides by 3.14

$2.5 \text{ cm} = d$

The diameter is about 2.5 cm.

31. (continued)

$r = \frac{d}{2}$

$r = \frac{2.5\text{ cm}}{2} = 1.25\text{ cm}$

The radius is about 1.25 cm.

$A = \pi \cdot r \cdot r$

$A \approx 3.14 \cdot 1.25\text{ cm} \cdot 1.25\text{ cm} = 4.90625\text{ cm}^2$

The area is about 4.90625 cm^2.

33. 

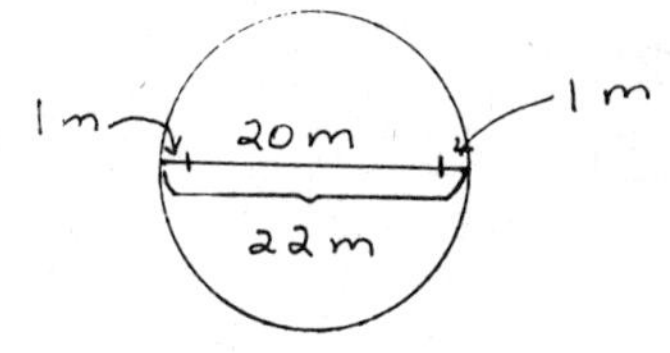

Find the area of the larger circle (pool plus wall). Its diameter is 1 m + 20 m + 1 m, or 22 m. Thus, its radius is $\frac{22}{2}$ m or 11 m.

$A = \pi \cdot r \cdot r$

$A \approx 3.14 \cdot 11\text{ m} \cdot 11\text{ m} = 379.94\text{ m}^2$

Find the area of the pool. Its diameter is 20 m. Thus, its radius is $\frac{20}{2}$ m, or 10 m.

$A = \pi \cdot r \cdot r$

$A \approx 3.14 \cdot 10\text{ m} \cdot 10\text{ m} = 314\text{ m}^2$

We subtract to find the area of the walk:

$A = 379.94\text{ m}^2 - 314\text{ m}^2$

$A = 65.94\text{ m}^2$

The area of the walk is 65.94 m^2.

35. The perimeter consists of the circumferences of three semicircles, each with diameter 8 ft, and one side of a square of length 8 ft. We first find the circumference of one semicircle. This is one-half the circumference of a circle with diameter 8 ft:

$\frac{1}{2} \cdot \pi \cdot d \approx \frac{1}{2} \cdot 3.14 \cdot 8\text{ ft} = 12.56\text{ ft}$

Then we multiply by 3:

$3 \cdot (12.56\text{ ft}) = 37.68\text{ ft}$

Finally we add the circumferences of the semicircles and the length of the side of the square:

$37.68\text{ ft} + 8\text{ ft} = 45.68\text{ ft}$

The perimeter is 45.68 ft.

37. The perimeter consists of three-fourths of the circumference of a circle with radius 4 yd and two sides of a square with sides of length 4 yd. We first find three-fourths of the circumference of the circle:

$\frac{3}{4} \cdot 2 \cdot \pi \cdot r \approx 0.75 \cdot 2 \cdot 3.14 \cdot 4\text{ yd} = 18.84\text{ yd}$

Then we add this length to the lengths of two sides of the square:

$18.84\text{ yd} + 4\text{ yd} + 4\text{ yd} = 26.84\text{ yd}$

The perimeter is 26.84 yd.

39. The perimeter consists of three-fourths of the perimeter of a square with side of length 10 yd and the circumference of a semicircle with diameter 10 yd. First we find three-fourths of the perimeter of the square:

$\frac{3}{4} \cdot 4 \cdot s = \frac{3}{4} \cdot 4 \cdot 10\text{ yd} = 30\text{ yd}$

Then we find one-half of the circumference of a circle with diameter 10 yd:

$\frac{1}{2} \cdot \pi \cdot d \approx \frac{1}{2} \cdot 3.14 \cdot 10\text{ yd} = 15.7\text{ yd}$

Then we add:

$30\text{ yd} + 15.7\text{ yd} = 45.7\text{ yd}$

The perimeter is 45.7 yd.

41. The shaded region consists of a circle, with radius 8 m, with two circles, each with diameter 8 m, removed. First we find the area of the large circle:

$A = \pi \cdot r \cdot r \approx 3.14 \cdot 8\text{ m} \cdot 8\text{ m} = 200.96\text{ m}^2$

Then find the area of one of the small circles:

The radius is $\frac{8\text{ m}}{2} = 4\text{ m}$.

$A = \pi \cdot r \cdot r \approx 3.14 \cdot 4\text{ m} \cdot 4\text{ m} = 50.24\text{ m}^2$

We multiply this area by 2 to find the area of the two small circles:

$2 \cdot 50.24\text{ m}^2 = 100.48\text{ m}^2$

Finally we subtract to find the area of the shaded region:

$200.96\text{ m}^2 - 100.48\text{ m}^2 = 100.48\text{ m}^2$

The area of the shaded region is 100.48 m^2.

43. The shaded region consists of one-half of a circle with diameter 2.8 cm and a triangle with base 2.8 cm and height 2.8 cm. First we find the area of the semicircle. The radius is $\frac{2.8\text{ cm}}{2} = 1.4\text{ cm}$.

$\frac{1}{2} \cdot \pi \cdot r \cdot r \approx \frac{1}{2} \cdot 3.14 \cdot 1.4\text{ cm} \cdot 1.4\text{ cm} = 3.0772\text{ cm}^2$

Then we find the area of the triangle.

$\frac{1}{2} \cdot b \cdot h = \frac{1}{2} \cdot 2.8\text{ cm} \cdot 2.8\text{ cm} = 3.92\text{ cm}^2$

Finally we add to find the area of the shaded region.

$3.0772\text{ cm}^2 + 3.92\text{ cm}^2 = 6.9972\text{ cm}^2$

The area of the shaded region is 6.9972 cm^2.

45. The shaded region consists of a rectangle, with dimensions 10.2 cm by 12.8 cm, with the area of two semicircles, each with diameter 10.2 cm, removed. This is equivalent to removing one circle with diameter 10.2 cm from the rectangle. First we find the area of the rectangle.

$\ell \cdot w = (10.2 \text{ cm}) \cdot (12.8 \text{ cm}) = 130.56 \text{ cm}^2$

Then we find the area of the circle. The radius is $\frac{10.2 \text{ cm}}{2} = 5.1$ cm.

$\pi \cdot r \cdot r \approx 3.14 \cdot 5.1 \text{ cm} \cdot 5.1 \text{ cm} = 81.6714 \text{ cm}^2$

Finally we subtract to find the area of the shaded region.

$130.56 \text{ cm}^2 - 81.6714 \text{ cm}^2 = 48.8886 \text{ cm}^2$

47. 0.875

a) Move the decimal point 2 places to the right. 0.87.5

b) Add a percent symbol. 87.5%

0.875 = 87.5%

49. $0.\overline{6}$

a) Move the decimal point 2 places to the right. $0.66.\overline{6}$

b) Add a percent symbol. $66.\overline{6}\%$

$0.\overline{6} = 66.\overline{6}\%$

51. Find 3927 ÷ 1250 using a calculator.

$\frac{3927}{1250} = 3.1416 \approx 3.142$ Rounding

53. The height of the stack of tennis balls is three times the diameter of one ball, or $3 \cdot d$.

The circumference of one ball is given by $\pi \cdot d$.

The circumference of one ball is greater than the height of the stack of balls, because $\pi > 3$.

Exercise Set 7.6

1. $V = \ell \cdot w \cdot h$

$V = 12 \text{ cm} \cdot 8 \text{ cm} \cdot 8 \text{ cm}$

$V = 12 \cdot 64 \text{ cm}^3$

$V = 768 \text{ cm}^3$

$SA = 2\ell w + 2\ell h + 2wh$

$= 2 \cdot 12 \text{ cm} \cdot 8 \text{ cm} + 2 \cdot 12 \text{ cm} \cdot 8 \text{ cm} + 2 \cdot 8 \text{ cm} \cdot 8 \text{ cm}$

$= 192 \text{ cm}^2 + 192 \text{ cm}^2 + 128 \text{ cm}^2$

$= 512 \text{ cm}^2$

3. $V = \ell \cdot w \cdot h$

$V = 7.5 \text{ cm} \cdot 2 \text{ cm} \cdot 3 \text{ cm}$

$V = 7.5 \cdot 6 \text{ cm}^3$

$V = 45 \text{ cm}^3$

$SA = 2\ell w + 2\ell h + 2wh$

$= 2 \cdot 7.5 \text{ cm} \cdot 2 \text{ cm} + 2 \cdot 7.5 \text{ cm} \cdot 3 \text{ cm} + 2 \cdot 2 \text{ cm} \cdot 3 \text{ cm}$

$= 30 \text{ cm}^2 + 45 \text{ cm}^2 + 12 \text{ cm}^2$

$= 87 \text{ cm}^2$

5. $V = \ell \cdot w \cdot h$

$V = 10 \text{ m} \cdot 5 \text{ m} \cdot 1.5 \text{ m}$

$V = 10 \cdot 7.5 \text{ m}^3$

$V = 75 \text{ m}^3$

$SA = 2\ell w + 2\ell h + 2wh$

$= 2 \cdot 10 \text{ m} \cdot 5 \text{ m} + 2 \cdot 10 \text{ m} \cdot 1.5 \text{ m} + 2 \cdot 5 \text{ m} \cdot 1.5 \text{ m}$

$= 100 \text{ m}^2 + 30 \text{ m}^2 + 15 \text{ m}^2$

$= 145 \text{ m}^2$

7. $V = \ell \cdot w \cdot h$

$V = 6\frac{1}{2} \text{ yd} \cdot 5\frac{1}{2} \text{ yd} \cdot 10 \text{ yd}$

$V = \frac{13}{2} \cdot \frac{11}{2} \cdot 10 \text{ yd}^3$

$V = \frac{715}{2} \text{ yd}^3$

$V = 357\frac{1}{2} \text{ yd}^3$

$SA = 2\ell w + 2\ell h + 2wh$

$= 2 \cdot 6\frac{1}{2} \text{ yd} \cdot 5\frac{1}{2} \text{ yd} + 2 \cdot 6\frac{1}{2} \text{ yd} \cdot 10 \text{ yd} + 2 \cdot 5\frac{1}{2} \text{ yd} \cdot 10 \text{ yd}$

$= \frac{2}{1} \cdot \frac{13}{2} \cdot \frac{11}{2} \text{ yd}^2 + \frac{2}{1} \cdot \frac{13}{2} \cdot \frac{10}{1} \text{ yd}^2 + \frac{2}{1} \cdot \frac{11}{2} \cdot \frac{10}{1} \text{ yd}^2$

$= \frac{143}{2} \text{ yd}^2 + 130 \text{ yd}^2 + 110 \text{ yd}^2$

$= 71\frac{1}{2} \text{ yd}^2 + 130 \text{ yd}^2 + 110 \text{ yd}^2$

$= 311\frac{1}{2} \text{ yd}^2$

9. $V = Bh = \pi \cdot r^2 \cdot h$

$\approx 3.14 \times 8 \text{ in.} \times 8 \text{ in.} \times 4 \text{ in.}$

$= 803.84 \text{ in}^3$

11. $V = Bh = \pi \cdot r^2 \cdot h$

$\approx 3.14 \times 5 \text{ cm} \times 5 \text{ cm} \times 4.5 \text{ cm}$

$= 353.25 \text{ cm}^3$

13. $V = Bh = \pi \cdot r^2 \cdot h$

$\approx \frac{22}{7} \times 210 \text{ yd} \times 210 \text{ yd} \times 300 \text{ yd}$

$= 41{,}580{,}000 \text{ yd}^3$

15. $V = Bh = \pi \cdot r^2 \cdot h$

$\approx \frac{22}{7} \times \frac{1}{2} \text{ ft} \times \frac{1}{2} \text{ ft} \times 3\frac{1}{2} \text{ ft}$

$= \frac{22}{7} \times \frac{1}{2} \text{ ft} \times \frac{1}{2} \text{ ft} \times \frac{7}{2} \text{ ft}$

$= \frac{11}{4} \text{ ft}^3$, or $2\frac{3}{4} \text{ ft}^3$

17. $V = \frac{4}{3} \cdot \pi \cdot r^3$

$\approx \frac{4}{3} \times 3.14 \times (100 \text{ in.})^3$

$= 4{,}186{,}666.67 \text{ in}^3$

19. $V = \frac{4}{3} \cdot \pi \cdot r^3$

$\approx \frac{4}{3} \times 3.14 \times (3.1 \text{ m})^3$

$\approx 124.72 \text{ m}^3$

21. $V = \frac{4}{3} \cdot \pi \cdot r^3$

$\approx \frac{4}{3} \times \frac{22}{7} \times (7 \text{ km})^3$

$\approx \frac{4}{3} \times \frac{22}{7} \times 343 \text{ km}^3$

$= 1437\frac{1}{3} \text{ km}^3$

23. $V = \frac{1}{3}\pi \cdot r^2 \cdot h$

$\approx \frac{1}{3} \times 3.14 \times 33 \text{ ft} \times 33 \text{ ft} \times 100 \text{ ft}$

$= 113{,}982 \text{ ft}^3$

25. $V = \frac{1}{3}\pi \cdot r^2 \cdot h$

$\approx \frac{1}{3} \times \frac{22}{7} \times 10.5 \text{ in.} \times 10.5 \text{ in.} \times 14 \text{ in.}$

$= 1617 \text{ in}^3$

27. $V = \frac{1}{3}\pi \cdot r^2 \cdot h$

$\approx \frac{1}{3} \times \frac{22}{7} \times 1.4 \text{ cm} \times 1.4 \text{ cm} \times 12 \text{ cm}$

$= 24.64 \text{ cm}^3$

29. We must find the radius of the base in order to use the formula for the volume of a circular cylinder.

$r = \frac{d}{2} = \frac{14 \text{ m}}{2} = 7 \text{ m}$

$V = Bh = \pi \cdot r^2 \cdot h$

$\approx \frac{22}{7} \times 7 \text{ m} \times 7 \text{ m} \times 220 \text{ m}$

$= 33{,}880 \text{ m}^3$

31. We must find the radius of the silo in order to use the formula for the volume of a circular cylinder.

$r = \frac{d}{2} = \frac{6 \text{ m}}{2} = 3 \text{ m}$

$V = Bh = \pi \cdot r^2 \cdot h$

$\approx 3.14 \times 3 \text{ m} \times 3 \text{ m} \times 13 \text{ m}$

$= 367.38 \text{ m}^3$

33. We must find the radius of the tank in order to use the formula for the volume of a sphere.

$r = \frac{d}{2} = \frac{6 \text{ m}}{2} = 3 \text{ m}$

$V = \frac{4}{3} \cdot \pi \cdot r^3$

$\approx \frac{4}{3} \times 3.14 \times (3 \text{ m})^3$

$\approx 113.0 \text{ m}^3$ Rounding to the nearest tenth of a cubic meter.

35. We must find the radius of the earth in order to use the formula for the volume of a sphere.

$r = \frac{d}{2} = \frac{6400 \text{ km}}{2} = 3200 \text{ km}$

$V = \frac{4}{3} \cdot \pi \cdot r^3$

$\approx \frac{4}{3} \cdot 3.14 \times (3200 \text{ km})^3$

$\approx 137{,}188{,}693{,}333.33 \text{ km}^3$

37. $5 \times 22\frac{1}{2} = \frac{5}{1} \times \frac{45}{2} = \frac{5 \times 45}{2} = \frac{225}{2} = 112\frac{1}{2}$

39. $10^3 = 10 \cdot 10 \cdot 10 = 1000$

41. First find the volume of one one-dollar bill in cubic inches:

$V = \ell \cdot w \cdot h$

$V = 6.0625 \text{ in.} \times 2.3125 \text{ in.} \times 0.0041 \text{ in.}$

$V = 0.05748 \text{ in}^3$ Rounding

Then multiply to find the volume of one million one-dollar bills in cubic inches:

$1{,}000{,}000 \times 0.05748 \text{ in}^3 = 57{,}480 \text{ in}^3$

The volume of one million one-dollar bills is about $57{,}480 \text{ in}^3$.

Next we convert to cubic feet:

$57{,}480 \text{ in}^3 = 57{,}480 \times \text{in.} \times \text{in.} \times \text{in.}$

$= 57{,}480 \times \text{in.} \times \text{in.} \times \text{in.} \times \frac{1 \text{ ft}}{12 \text{ in.}} \times \frac{1 \text{ ft}}{12 \text{ in.}} \times \frac{1 \text{ ft}}{12 \text{ in.}}$

$= \frac{57{,}480}{12 \times 12 \times 12} \times \frac{\text{in.}}{\text{in.}} \times \frac{\text{in.}}{\text{in.}} \times \frac{\text{in.}}{\text{in.}} \times \text{ft} \times \text{ft} \times \text{ft}$

$= 33.3 \text{ ft}^3$

Exercise Set 7.7

1. Two angles are complementary if the sum of their measures is 90°.

$90° - 11° = 79°$

The measure of the complement is 79°.

3. Two angles are complementary if the sum of their measures is 90°.

$90° - 67° = 23°$

The measure of the complement is 23°.

5. Two angles are supplementary if the sum of their measures is 180°.

$$180° - 3° = 177°$$

The measure of the supplement is 177°.

7. Two angles are supplementary if the sum of their measures is 180°.

$$180° - 139° = 41°$$

The measure of the supplement is 41°.

9. The segments have different lengths. They are not congruent.

11. $m\angle G = m\angle R$, so $\angle G \cong \angle R$.

13. Since $\angle 2$ and $\angle 5$ are vertical angles, $m\angle 2 = 67°$. Likewise, $\angle 1$ and $\angle 4$ are vertical angles, so $m\angle 4 = 80°$.

$m\angle 1 + m\angle 2 + m\angle 3 = 180°$

$80° + 67° + m\angle 3 = 180°$ Substituting

$147° + m\angle 3 = 180°$

$m\angle 3 = 180° - 147°$

$m\angle 3 = 33°$

Since $\angle 3$ and $\angle 6$ are vertical angles, $m\angle 6 = 33°$.

15. a) The pairs of corresponding angles are

$\angle 1$ and $\angle 3$

$\angle 2$ and $\angle 4$

$\angle 8$ and $\angle 6$

$\angle 7$ and $\angle 5$.

b) The interior angles are $\angle 2$, $\angle 3$, $\angle 6$, and $\angle 7$.

c) The pairs of alternate interior angles are

$\angle 2$ and $\angle 6$

$\angle 3$ and $\angle 7$.

17. $\angle 4$ and $\angle 6$ are vertical angles, so $m\angle 6 = 125°$.

$\angle 4$ and $\angle 2$ are corresponding angles. By Property 1, $m\angle 2 = 125°$.

$\angle 6$ and $\angle 8$ are corresponding angles. By Property 1, $m\angle 8 = 125°$.

$\angle 2$ and $\angle 3$ are interior angles on the same side of the transversal. Using Property 4 and $m\angle 2 = 125°$, $m\angle 3 = 55°$.

$\angle 6$ and $\angle 7$ are interior angles on the same side of the transversal. Using Property 4 and $m\angle 6 = 125°$, $m\angle 7 = 55°$.

$\angle 3$ and $\angle 5$ are vertical angles so $m\angle 5 = 55°$.

$\angle 7$ and $\angle 1$ are vertical angles so $m\angle 1 = 55°$.

19. Considering the transversal $\overleftrightarrow{BC}$, $\angle ABE$ and $\angle DCE$ are alternate interior angles. By Property 2, $\angle ABE \cong \angle DCE$. Then $m\angle ABE = m\angle DCE = 95°$.

Considering the transversal $\overleftrightarrow{AD}$, $\angle BAE$ and $\angle CDE$ are alternate interior angles. By Property 2, $\angle BAE \cong \angle CDE$. We cannot determine the measure of these angles.

$\angle AEB$ and $\angle DEC$ are vertical angles, so $\angle AEB \cong \angle DEC$. We cannot determine the measure of these angles.

21. Considering the transversal $\overleftrightarrow{CE}$, $\angle AEC$ and $\angle DCE$ are alternate interior angles. By Property 2, $\angle AEC \cong \angle DEC$. Then $m\angle AEC = m\angle DEC = 50°$.

Considering the transversal $\overleftrightarrow{DE}$, $\angle BED$ and $\angle EDC$ are alternate interior angles. By Property 2, $\angle BED \cong \angle EDC$. Then $m\angle BED = m\angle EDC = 41°$.

Exercise Set 7.8

1. The notation tells us the way in which the vertices of the two triangles are matched.

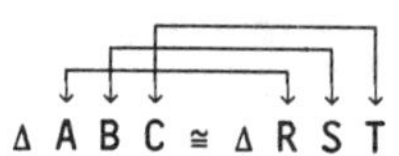

$\triangle ABC \cong \triangle RST$ means

$\angle A \cong \angle R$	and	$\overline{AB} \cong \overline{RS}$
$\angle B \cong \angle S$		$\overline{AC} \cong \overline{RT}$
$\angle C \cong \angle T$		$\overline{BC} \cong \overline{ST}$

3. The notation tells us the way in which the vertices of the two triangles are matched.

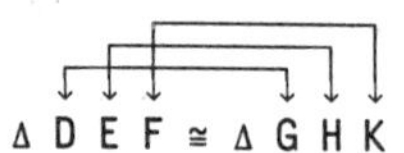

$\triangle DEF \cong \triangle GHK$ means

$\angle D \cong \angle G$	and	$\overline{DE} \cong \overline{GH}$
$\angle E \cong \angle H$		$\overline{DF} \cong \overline{GK}$
$\angle F \cong \angle K$		$\overline{EF} \cong \overline{HK}$

5. The notation tells us the way in which the vertices of the two triangles are matched.

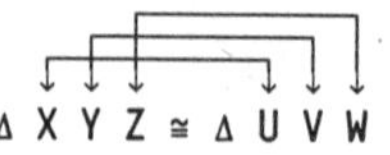

$\triangle XYZ \cong \triangle UVW$ means

$\angle X \cong \angle U$	and	$\overline{XY} \cong \overline{UV}$
$\angle Y \cong \angle V$		$\overline{XZ} \cong \overline{UW}$
$\angle Z \cong \angle W$		$\overline{YZ} \cong \overline{VW}$

7.

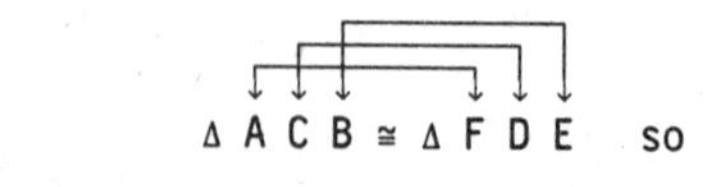

$\angle A \cong \angle F$	and	$\overline{AC} \cong \overline{FD}$
$\angle C \cong \angle D$		$\overline{AB} \cong \overline{FE}$
$\angle B \cong \angle E$		$\overline{CB} \cong \overline{DE}$

9. $\triangle MNO \cong \triangle QPS$ so

$\angle M \cong \angle Q$ and $\overline{MN} \cong \overline{QP}$

$\angle N \cong \angle P$ $\overline{MO} \cong \overline{QS}$

$\angle O \cong \angle S$ $\overline{NO} \cong \overline{PS}$

11. We cannot determine from the information given that two sides of one triangle and the included angle are congruent to two sides and the included angle of the other triangle. Therefore, we cannot use the SAS Property.

13. Two sides of one triangle and the included angle are congruent to two sides and the included angle of the other triangle. They are congruent by the SAS Property.

15. Two sides of one triangle and the included angle are congruent to two sides and the included angle of the other triangle. They are congruent by the SAS Property.

17. We cannot determine from the information given that three sides of one triangle are congruent to three sides of the other triangle. Therefore, we cannot use the SSS Property.

19. Three sides of one triangle are congruent to three sides of the other triangle. They are congruent by the SSS Property.

21. Three sides of one triangle are congruent to three sides of the other triangle. They are congruent by the SSS Property.

23. Two angles and the included side of one triangle are congruent to two angles and the included side of the other triangle. They are congruent by the ASA Property.

25. Two angles and the included side of one triangle are congruent to two angles and the included side of the other triangle. They are congruent by the ASA Property.

27. The vertical angles are congruent so two angles and the included side of one triangle are congruent to two angles and the included side of the other triangle. They are congruent by the ASA Property.

29. Two angles and the included side of one triangle are congruent to two angles and the included side of the other triangle. They are congruent by the ASA Property.

31. Two sides of one triangle and the included angle are congruent to two sides and the included angle of the other triangle. They are congruent by the SAS Property.

33. Three sides of one triangle are congruent to three sides of the other triangle. In addition, two sides of one triangle and the included angle are congruent to two sides and the included angle of the other triangle. Therefore, we can use either the SSS Property or the SAS Property to show that they are congruent.

35. Since R is the midpoint of $\overline{PT}$, $\overline{PR} \cong \overline{TR}$.

Since R is the midpoint of $\overline{QS}$, $\overline{RQ} \cong \overline{RS}$.

$\angle PRQ$ and $\angle TRS$ are vertical angles, so

$\angle PRQ \cong \angle TRS$.

Two sides and the included angle of $\triangle PRQ$ are congruent to two sides and the included angle of $\triangle TRS$, so $\triangle PRQ \cong \triangle TRS$ by the SAS Property.

37. Since $\overline{GL} \perp \overline{KM}$, $m\angle GLK = m\angle GLM = 90°$. Then $\angle GLK \cong \angle GLM$.

Since L is the midpoint of $\overline{KM}$, $\overline{KL} \cong \overline{LM}$.

$GL \cong GL$.

Two sides and the included angle of $\triangle KLG$ are congruent to two sides and the included angle of $\triangle MLG$, so $\triangle KLG \cong \triangle MLG$ by the SAS Property.

39. The information given tells us that

$\overline{AE} \cong \overline{CB}$ and $\overline{AB} \cong \overline{CD}$.

Since B is the midpoint of $\overline{ED}$, $\overline{EB} \cong \overline{BD}$.

Three sides of $\triangle AEB$ are congruent to three sides of $\triangle CDB$, so $\triangle AEB \cong \triangle CDB$ by the SSS Property.

41. The information given tells us that $\overline{HK} \cong \overline{KJ}$ and $\overline{GK} \cong \overline{LK}$.

Since $\overline{GK} \perp \overline{LJ}$, $m\angle HKL = m\angle GKJ = 90°$.

Then $\angle HKL \cong \angle GKJ$.

Two sides and the included angle of $\triangle LKH$ are congruent to two sides and the included angle of $\triangle GKJ$, so $\triangle LKH \cong \triangle GKJ$ by the SAS Property. This means that the remaining corresponding parts of the two triangles are congruent. That is, $\angle HLK \cong \angle JGK$, $\angle LHK \cong \angle GJK$, and $\overline{LH} \cong \overline{GJ}$.

43. Two angles and the included side of $\triangle PED$ are congruent to two angles and the included side of $\triangle PFG$, so $\triangle PED \cong \triangle PFG$ by the ASA Property. Then corresponding parts of the two triangles are congruent, so $\overline{EP} \cong \overline{FP}$. Therefore, P is the midpoint of $\overline{EF}$.

45. $\angle A$ and $\angle C$ are opposite angles, so $m\angle A = 70°$ by Property 2.

$\angle C$ and $\angle B$ are consecutive angles, so they are supplementary by Property 4. Then

$m\angle B = 180° - m\angle C$

$m\angle B = 180° - 70°$

$m\angle B = 110°$

$\angle B$ and $\angle D$ are opposite angles, so $m\angle D = 110°$ by Property 2.

47. $\angle M$ and $\angle K$ are opposite angles, so $m\angle M = 71°$ by Property 2.

$\angle K$ and $\angle L$ are consecutive angles, so they are supplementary by Property 4. Then

$$m\angle L = 180° - m\angle K$$
$$m\angle L = 180° - 71°$$
$$m\angle L = 109°$$

$\angle J$ and $\angle L$ are opposite angles, so $m\angle J = 109°$ by Property 2.

49. $\overline{ON}$ and $\overline{TU}$ are opposite sides of the parallelogram. So are $\overline{OT}$ and $\overline{NU}$. The opposite sides of a parallelogram are congruent (Property 3), so $TU = 9$ and $NU = 15$.

51. $\overline{JM}$ and $\overline{KL}$ are opposite sides of the parallelogram. So are $\overline{JK}$ and $\overline{ML}$. The opposite sides of a parallelogram are congruent (Property 3). Then $KL = 3\frac{1}{2}$, and $JK + LM = 22 - 3\frac{1}{2} - 3\frac{1}{2} = 15$. Thus, $JK = LM = \frac{1}{2} \cdot 15 = 7\frac{1}{2}$.

53. The diagonals of a parallelogram bisect each other (Property 5). Then

$$AC = 2 \cdot AB = 2 \cdot 14 = 28$$
$$ED = 2 \cdot BD = 2 \cdot 19 = 38$$

Exercise Set 7.9

1. Vertex R is matched with vertex A, vertex S is matched with vertex B, and vertex T is matched with vertex C. Then

$\overline{RS} \longleftrightarrow \overline{AB}$ and $\angle R \longleftrightarrow \angle A$
$\overline{ST} \longleftrightarrow \overline{BC}$ $\quad \angle S \longleftrightarrow \angle B$
$\overline{TR} \longleftrightarrow \overline{CA}$ $\quad \angle T \longleftrightarrow \angle C$

3. Vertex C is matched with vertex W, vertex B is matched with vertex J, and vertex S is matched with vertex Z. Then

$\overline{CB} \longleftrightarrow \overline{WJ}$ and $\angle C \longleftrightarrow \angle W$
$\overline{BS} \longleftrightarrow \overline{JZ}$ $\quad \angle B \longleftrightarrow \angle J$
$\overline{SC} \longleftrightarrow \overline{ZW}$ $\quad \angle S \longleftrightarrow \angle Z$

5. The notation tells us the way in which the vertices are matched.

$\triangle ABC \sim \triangle RST$

$\triangle ABC \sim \triangle RST$ means

$\angle A \cong \angle R$
$\angle B \cong \angle S$ and $\frac{AB}{RS} = \frac{AC}{RT} = \frac{BC}{ST}$
$\angle C \cong \angle T$

7. The notation tells us the way in which the vertices are matched.

$\triangle MES \sim \triangle CLF$

$\triangle MES \sim \triangle CLF$ means

$\angle M \cong \angle C$
$\angle E \cong \angle L$ and $\frac{ME}{CL} = \frac{MS}{CF} = \frac{ES}{LF}$
$\angle S \cong \angle F$

9. If we match P with N, S with D, and Q with M, the corresponding angles will be congruent. That is, $\triangle PSQ \sim \triangle NDM$. Then

$$\frac{PS}{ND} = \frac{PQ}{NM} = \frac{SQ}{DM}$$

11. If we match T with G, A with F, and W with C, the corresponding angles will be congruent. That is, $\triangle TAW \sim \triangle GFC$. Then

$$\frac{TA}{GF} = \frac{TW}{GC} = \frac{AW}{FC}$$

13. Since $\triangle ABC \sim \triangle PQR$, the corresponding sides are proportional. Then

$\frac{3}{6} = \frac{4}{PR}$ and $\frac{3}{6} = \frac{5}{QR}$

$3(PR) = 6 \cdot 4$ $\quad 3(QR) = 6 \cdot 5$
$3(PR) = 24$ $\quad 3(QR) = 30$
$PR = 8$ $\quad QR = 10$

15. Recall that if a transversal intersects two parallel lines, then the alternate interior angles are congruent. Thus,

$\angle A \cong \angle B$ and $\angle D \cong \angle C$.

Since $\angle AED$ and $\angle CEB$ are vertical angles, they are congruent. Thus,

$\angle AED \cong \angle CEB$.

Then $\triangle AED \sim \triangle CEB$, and the lengths of the corresponding sides are proportional.

$$\frac{AD}{CB} = \frac{ED}{EC}$$
$$\frac{7}{28} = \frac{6}{EC}$$
$$7 \cdot EC = 168$$
$$EC = 24$$

17. $17\frac{3}{4} \times 5\frac{1}{2} = \frac{71}{4} \times \frac{11}{2} = \frac{71 \times 11}{4 \times 2} = \frac{781}{8} = 97\frac{5}{8}$

19. $6^4 = 6 \cdot 6 \cdot 6 \cdot 6 = 1296$

Exercise Set 8.1

1. $-11x^4 - x^3 + x^2 + 3x - 9$

Term	$-11x^4$	$-x^3$	x^2	$3x$	-9
Degree	4	3	2	1	0
Degree of polynomial	4				
Leading term	$-11x^4$				
Leading coefficient	-11				

3. $y^3 + 2y^7 + x^2y^4 - 8$

Term	y^3	$2y^7$	x^2y^4	-8
Degree	3	7	6	0
Degree of polynomial	7			
Leading term	$2y^7$			
Leading coefficient	2			

5. $a^5 + 4a^2b^4 + 6ab + 4a - 3$

Term	a^5	$4a^2b^4$	$6ab$	$4a$	-3
Degree	5	6	2	1	0
Degree of polynomial	6				
Leading term	$4a^2b^4$				
Leading coefficient	4				

7. $-4y^3 - 6y^2 + 7y + 23$

9. $-xy^3 + x^2y^2 + x^3y + 1$

11. $-5a^5y^5 - 4a^2y^3 + 3ay$

13. $12 + 4x - 5x^2 + 3x^4$

15. $3xy^3 + x^2y^2 - 9x^3y + 2x^4$

17. $-7ab + 4ax - 7ax^2 + 4x^6$

19. $P = 4x^2 - 3x + 2$

When $x = 4$, $P = 4\cdot4^2 - 3\cdot4 + 2$
$= 64 - 12 + 2$
$= 54$

When $x = 0$, $P = 4\cdot0^2 - 3\cdot0 + 2$
$= 0 - 0 + 2$
$= 2$

21. $P = 8y^3 - 12y - 5$

When $y = -2$, $P = 8(-2)^3 - 12(-2) - 5$
$= -64 + 24 - 5$
$= -45$

When $y = \frac{1}{3}$, $P = 8\left(\frac{1}{3}\right)^3 - 12\cdot\frac{1}{3} - 5$
$= 8\cdot\frac{1}{27} - 4 - 5$
$= \frac{8}{27} - 9$
$= \frac{8}{27} - \frac{243}{27}$
$= -\frac{235}{27}$, or $-8\frac{19}{27}$

23. To find the number of games played when there are 8 teams in the league, we substitute 8 for n in the polynomial and calculate:

$N = \frac{1}{2}n^2 - \frac{1}{2}n$
$= \frac{1}{2}(8)^2 - \frac{1}{2}(8)$
$= \frac{1}{2}\cdot64 - 4$
$= 32 - 4$
$= 28$

There will be 28 games.

For 20 teams, we substitute 20 for n and calculate:

$N = \frac{1}{2}n^2 - \frac{1}{2}n$
$= \frac{1}{2}(20)^2 - \frac{1}{2}(20)$
$= \frac{1}{2}\cdot400 - 10$
$= 200 - 10$
$= 190$

There will be 190 games.

25. We substitute 4.7 for h, 1.2 for r, and 3.14 for π:

$2\pi rh + 2\pi r^2 = 2(3.14)(1.2)(4.7) + 2(3.14)(1.2)^2$
$= 2(3.14)(1.2)(4.7) + 2(3.14)(1.44)$
$= 35.4192 + 9.0432$
$= 44.4624$

The surface area is about 44.5 in^2.

27. Substitute 200 for x:

$R = 280x - 0.4x^2$
$= 280(200) - 0.4(200)^2$
$= 56{,}000 - 0.4(40{,}000)$
$= 56{,}000 - 16{,}000$
$= 40{,}000$

The total revenue is \$40,000.

29. Substitute 200 for x:

$C = 7000 + 0.6x^2$
$= 7000 + 0.6(200)^2$
$= 7000 + 0.6(40{,}000)$
$= 7000 + 24{,}000$
$= 31{,}000$

The total cost is \$31,000.

31. We subtract:

$P = R - C$
$= 280x - 0.4x^2 - (7000 + 0.6x^2)$
$= 280x - 0.4x^2 + (-7000 - 0.6x^2)$ Adding the opposite
$= -x^2 + 280x - 7000$

The total profit is given by $P = -x^2 + 280x - 7000$.

33. $6x^2 - 7x^2 + 3x^2 = (6 - 7 + 3)x^2 = 2x^2$

35. $5x - 4y - 2x + 5y$
$= (5 - 2)x + (-4 + 5)y$
$= 3x + y$

37. $5a + 7 - 4 + 2a - 6a + 3$
$= (5 + 2 - 6)a + (7 - 4 + 3)$
$= a + 6$

39. $3a^2b + 4b^2 - 9a^2b - 6b^2$
$= (3 - 9)a^2b + (4 - 6)b^2$
$= -6a^2b - 2b^2$

41. $8x^2 - 3xy + 12y^2 + x^2 - y^2 + 5xy + 4y^2$
$= (8 + 1)x^2 + (-3 + 5)xy + (12 - 1 + 4)y^2$
$= 9x^2 + 2xy + 15y^2$

43. $4x^2y - 3y + 2xy^2 - 5x^2y + 7y + 7xy^2$
$= (4 - 5)x^2y + (-3 + 7)y + (2 + 7)xy^2$
$= -x^2y + 4y + 9xy^2$

45. $(3x^2 + 5y^2 + 6) + (2x^2 - 3y^2 - 1)$
$= (3 + 2)x^2 + (5 - 3)y^2 + (6 - 1)$
$= 5x^2 + 2y^2 + 5$

47. $(2a + 3b - c) + (4a - 2b + 2c)$
$= (2 + 4)a + (3 - 2)b + (-1 + 2)c$
$= 6a + b + c$

49. $(a^2 - 3b^2 + 4c^2) + (-5a^2 + 2b^2 - c^2)$
$= (1 - 5)a^2 + (-3 + 2)b^2 + (4 - 1)c^2$
$= -4a^2 - b^2 + 3c^2$

51. $(x^2 + 2x - 3xy - 7) + (-3x^2 - x + 2xy + 6)$
$= (1 - 3)x^2 + (2 - 1)x + (-3 + 2)xy + (-7 + 6)$
$= -2x^2 + x - xy - 1$

53. $(7x^2y - 3xy^2 + 4xy) + (-2x^2y - xy^2 + xy)$
$= (7 - 2)x^2y + (-3 - 1)xy^2 + (4 + 1)xy$
$= 5x^2y - 4xy^2 + 5xy$

55. $(2r^2 + 12r - 11) + (6r^2 - 2r + 4) + (r^2 - r - 2)$
$= (2 + 6 + 1)r^2 + (12 - 2 - 1)r + (-11 + 4 - 2)$
$= 9r^2 + 9r - 9$

57. $\left(\frac{2}{3}xy + \frac{5}{6}xy^2 + 5.1x^2y\right) + \left(-\frac{4}{5}xy + \frac{3}{4}xy^2 - 3.4x^2y\right)$
$= \left(\frac{2}{3} - \frac{4}{5}\right)xy + \left(\frac{5}{6} + \frac{3}{4}\right)xy^2 + (5.1 - 3.4)x^2y$
$= \left(\frac{10}{15} - \frac{12}{15}\right)xy + \left(\frac{10}{12} + \frac{9}{12}\right)xy^2 + 1.7x^2y$
$= -\frac{2}{15}xy + \frac{19}{12}xy^2 + 1.7x^2y$

59. $5x^3 - 7x^2 + 3x - 6$

a) $-(5x^3 - 7x^2 + 3x - 6)$ Writing an inverse sign in front

b) $-5x^3 + 7x^2 - 3x + 6$ Writing the opposite of each term

61. $-12y^5 + 4ay^4 - 7by^2$

a) $-(-12y^5 + 4ay^4 - 7by^2)$

b) $12y^5 - 4ay^4 + 7by^2$

63. $(8x - 4) - (-5x + 2)$
$= (8x - 4) + (5x - 2)$ Adding the opposite
$= 13x - 6$

65. $(-3x^2 + 2x + 9) - (x^2 + 5x - 4)$
$= (-3x^2 + 2x + 9) + (-x^2 - 5x + 4)$ Adding the opposite
$= -4x^2 - 3x + 13$

67. $(5a - 2b + c) - (3a + 2b - 2c)$
$= (5a - 2b + c) + (-3a - 2b + 2c)$
$= 2a - 4b + 3c$

69. $(3x^2 - 2x - x^3) - (5x^2 - 8x - x^3)$
$= (3x^2 - 2x - x^3) + (-5x^2 + 8x + x^3)$
$= -2x^2 + 6x$

71. $(5a^2 + 4ab - 3b^2) - (9a^2 - 4ab + 2b^2)$
$= (5a^2 + 4ab - 3b^2) + (-9a^2 + 4ab - 2b^2)$
$= -4a^2 + 8ab - 5b^2$

73. $P - Q = 2 - 3y - (6y - 7)$
$= 2 - 3y + (-6y + 7)$
$= -9y + 9$

75. $(6ab - 4a^2b + 6ab^2) - (3ab^2 - 10ab - 12a^2b)$
$= (6ab - 4a^2b + 6ab^2) + (-3ab^2 + 10ab + 12a^2b)$
$= 8a^2b + 16ab + 3ab^2$

77. $(0.09y^4 - 0.052y^3 + 0.93) - (0.03y^4 - 0.084y^3 + 0.94y^2)$
$= (0.09y^4 - 0.052y^3 + 0.93) + (-0.03y^4 + 0.084y^3 - 0.94y^2)$
$= 0.06y^4 + 0.032y^3 - 0.94y^2 + 0.93$

79. $\left(\frac{5}{8}x^4 - \frac{1}{4}x^2 - \frac{1}{2}\right) - \left(-\frac{3}{8}x^4 + \frac{3}{4}x^2 + \frac{1}{2}\right)$
$= \left(\frac{5}{8}x^4 - \frac{1}{4}x^2 - \frac{1}{2}\right) + \left(\frac{3}{8}x^4 - \frac{3}{4}x^2 - \frac{1}{2}\right)$
$= x^4 - x^2 - 1$

81. $|4 - 2x| < 18$

$-18 < 4 - 2x < 18$
$-22 < -2x < 14$ Subtracting 4
$11 > x > -7$ Dividing by -2

The solution set is $\{x|11 > x > -7\}$, or $\{x|-7 < x < 11\}$.

83. The area of the base is $x \cdot x$, or x^2.
The area of each side is $x \cdot (x - 2)$.
The total area of all four sides is $4x(x - 2)$.

The surface area of this box can be expressed as a polynomial.

$x^2 + 4x(x - 2) = x^2 + 4x^2 - 8x = 5x^2 - 8x$

85. Writing in columns is helpful.

Add:

$$\begin{array}{r} 47x^{4a} + 3x^{3a} + 22x^{2a} + x^a + 1 \\ 37x^{3a} + 8x^{2a} \quad\quad + 3 \\ \hline 47x^{4a} + 40x^{3a} + 30x^{2a} + x^a + 4 \end{array}$$

87. Writing in columns is helpful.

Subtract:

$$\begin{array}{l} 2x^{5b} + 4x^{4b} + 3x^{3b} \quad\quad\quad\quad + 8 \\ \underline{-(x^{5b} \quad\quad + 2x^{3b} + 6x^{2b} + 9x^b + 8)} \end{array}$$

We rewrite as an addition problem.

Add:

$$\begin{array}{l} 2x^{5b} + 4x^{4b} + 3x^{3b} \quad\quad\quad\quad + 8 \\ \underline{-x^{5b} \quad\quad - 2x^{3b} \quad 6x^{2b} - 9x^b - 8} \\ x^{5b} + 4x^{4b} + x^{3b} - 6x^{2b} - 9x^b \end{array}$$

Exercise Set 8.2

1. $2y^2(5y) = (2 \cdot 5)(y^2 \cdot y) = 10y^3$

3. $5x(-4x^2y) = 5(-4)(x \cdot x^2)y = -20x^3y$

5. $2x^3y^2(-5x^2y^4) = 2(-5)(x^3 \cdot x^2)(y^2 \cdot y^4) = -10x^5y^6$

7. $2x(3 - x)$
$= 2x \cdot 3 - 2x \cdot x$
$= 6x - 2x^2$

9. $3ab(a + b)$
$= 3ab \cdot a + 3ab \cdot b$
$= 3a^2b + 3ab^2$

11. $5cd(3c^2d - 5cd^2)$
$= 5cd \cdot 3c^2d - 5cd \cdot 5cd^2$
$= 15c^3d^2 - 25c^2d^3$

13. $(5x + 2)(3x - 1)$
$= 15x^2 - 5x + 6x - 2$ FOIL
$= 15x^2 + x - 2$

15. $(s + 3t)(s - 3t)$
$= s^2 - (3t)^2$ $(A + B)(A - B) = A^2 - B^2$
$= s^2 - 9t^2$

17. $(x - y)(x - y)$
$= x^2 - 2xy + y^2$ $(A - B)^2 = A^2 - 2AB + B^2$

19. $(y + 8x)(2y - 7x)$
$= 2y^2 - 7xy + 16xy - 56x^2$ FOIL
$= 2y^2 + 9xy - 56x^2$

21. $(a^2 - 2b^2)(a^2 - 3b^2)$
$= a^4 - 3a^2b^2 - 2a^2b^2 + 6b^4$ FOIL
$= a^4 - 5a^2b^2 + 6b^4$

23. $(x - 4)(x^2 + 4x + 16)$
$= (x - 4)(x^2) + (x - 4)(4x) + (x - 4)(16)$ Distributive law
$= x(x^2) - 4(x^2) + x(4x) - 4(4x) + x(16) - 4(16)$ Distributive law
$= x^3 - 4x^2 + 4x^2 - 16x + 16x - 64$ Multiplying monomials
$= x^3 - 64$ Collecting like terms

25. $(x + y)(x^2 - xy + y^2)$
$= (x + y)x^2 + (x + y)(-xy) + (x + y)(y^2)$
$= x(x^2) + y(x^2) + x(-xy) + y(-xy) + x(y^2) + y(y^2)$
$= x^3 + x^2y - x^2y - xy^2 + xy^2 + y^3$
$= x^3 + y^3$

27.
$$\begin{array}{rl}
a^2 + a - 1 & \\
a^2 + 4a - 5 & \\
\hline
-5a^2 - 5a + 5 & \text{Multiplying by } -5 \\
4a^3 + 4a^2 - 4a & \text{Multiplying by } 4a \\
a^4 + a^3 - a^2 & \text{Multiplying by } a^2 \\
\hline
a^4 + 5a^3 - 2a^2 - 9a + 5 & \text{Adding}
\end{array}$$

29.
$$\begin{array}{rl}
4a^2b - 2ab + 3b^2 & \\
ab - 2b + a & \\
\hline
4a^3b - 2a^2b + 3ab^2 & (1) \\
-6b^3 \quad + 4ab^2 - 8a^2b^2 & (2) \\
3ab^3 \quad - 2a^2b^2 + 4a^3b^2 & (3) \\
\hline
3ab^3 - 6b^3 + 4a^3b - 2a^2b + 7ab^2 - 10a^2b^2 + 4a^3b^2 & (4)
\end{array}$$

(1) Multiplying by a
(2) Multiplying by -2b
(3) Multiplying by ab
(4) Adding

31. $\left(x + \frac{1}{4}\right)\left(x + \frac{1}{4}\right)$
$= x^2 + \frac{1}{4}x + \frac{1}{4}x + \frac{1}{16}$ FOIL
$= x^2 + \frac{1}{2}x + \frac{1}{16}$

33. $(1.3x - 4y)(2.5x + 7y)$
$= 3.25x^2 + 9.1xy - 10xy - 28y^2$ FOIL
$= 3.25x^2 - 0.9xy - 28y^2$

35. $(a + 2)(a + 3)$
$= a^2 + 3a + 2a + 6$ FOIL
$= a^2 + 5a + 6$

37. $(y + 3)(y - 2)$
$= y^2 - 2y + 3y - 6$ FOIL
$= y^2 + y - 6$

39. $\left(2a + \frac{1}{3}\right)^2$
$= (2a)^2 + 2(2a)\left(\frac{1}{3}\right) + \left(\frac{1}{3}\right)^2$ $(A + B)^2 = A^2 + 2AB + B^2$
$= 4a^2 + \frac{4}{3}a + \frac{1}{9}$

41. $(x - 2y)^2$
$= x^2 - 2(x)(2y) + (2y)^2$ $(A - B)^2 = A^2 - 2AB + B^2$
$= x^2 - 4xy + 4y^2$

43. $\left(b - \frac{1}{3}\right)\left(b - \frac{1}{2}\right)$
$= b^2 - \frac{1}{2}b - \frac{1}{3}b + \frac{1}{6}$ FOIL
$= b^2 - \frac{3}{6}b - \frac{2}{6}b + \frac{1}{6}$
$= b^2 - \frac{5}{6}b + \frac{1}{6}$

45. $(2x + 9)(x + 2)$
$= 2x^2 + 4x + 9x + 18$ FOIL
$= 2x^2 + 13x + 18$

47. $(20a - 0.16b)^2$
$= (20a)^2 - 2(20a)(0.16b) + (0.16b)^2$
$(A - B)^2 = A^2 - 2AB + B^2$
$= 400a^2 - 6.4ab + 0.0256b^2$

49. $(2x - 3y)(2x + y)$
$= 4x^2 + 2xy - 6xy - 3y^2$ FOIL
$= 4x^2 - 4xy - 3y^2$

51. $(x + 3)^2$
$= x^2 + 2 \cdot x \cdot 3 + 3^2$ $(A + B)^2 = A^2 + 2AB + B^2$
$= x^2 + 6x + 9$

53. $(2x^2 - 3y^2)^2$
$= (2x^2)^2 - 2(2x^2)(3y^2) + (3y^2)^2$
$(A - B)^2 = A^2 - 2AB + B^2$
$= 4x^4 - 12x^2y^2 + 9y^4$

55. $(a^2b^2 + 1)^2$
$= (a^2b^2)^2 + 2(a^2b^2) \cdot 1 + 1^2$
$(A + B)^2 = A^2 + 2AB + B^2$
$= a^4b^4 + 2a^2b^2 + 1$

57. $(0.1a^2 - 5b)^2$
$= (0.1a^2)^2 - 2(0.1a^2)(5b) + (5b)^2$
$(A - B)^2 = A^2 - 2AB + B^2$
$= 0.01a^4 - a^2b + 25b^2$

59. $(c + 2)(c - 2) = c^2 - 2^2$
$(A + B)(A - B) = A^2 - B^2$
$= c^2 - 4$

61. $(2a + 1)(2a - 1) = (2a)^2 - 1^2$
$(A + B)(A - B) = A^2 - B^2$
$= 4a^2 - 1$

63. $(3m - 2n)(3m + 2n) = (3m)^2 - (2n)^2$
$(A + B)(A - B) = A^2 - B^2$
$= 9m^2 - 4n^2$

65. $(x^2 + yz)(x^2 - yz) = (x^2)^2 - (yz)^2$
$(A + B)(A - B) = A^2 - B^2$
$= x^4 - y^2z^2$

67. $(-mn + m^2)(mn + m^2) = (m^2 - mn)(m^2 + mn)$
$= (m^2)^2 - (mn)^2$ $(A + B)(A - B) = A^2 - B^2$
$= m^4 - m^2n^2$

69. $\left(\frac{1}{2}p - \frac{2}{3}q\right)\left(\frac{1}{2}p + \frac{2}{3}q\right)$

$= \left(\frac{1}{2}p\right)^2 - \left(\frac{2}{3}q\right)^2 = \frac{1}{4}p^2 - \frac{4}{9}q^2$

71. $(x + 1)(x - 1)(x^2 + 1) = (x^2 - 1^2)(x^2 + 1)$
$= (x^2 - 1)(x^2 + 1)$
$= (x^2)^2 - 1^2$
$= x^4 - 1$

73. $(a - b)(a + b)(a^2 - b^2) = (a^2 - b^2)(a^2 - b^2)$
$= (a^2 - b^2)^2$
$= (a^2)^2 - 2(a^2)(b^2) + (b^2)^2$
$= a^4 - 2a^2b^2 + b^4$

75. $(a + b + 1)(a + b - 1)$
$= [(a + b) + 1][(a + b) - 1]$
$= (a + b)^2 - 1^2$
$= a^2 + 2ab + b^2 - 1$

77. $(2x + 3y + 4)(2x + 3y - 4)$
$= [(2x + 3y) + 4][(2x + 3y) - 4]$
$= (2x + 3y)^2 - 4^2$
$= 4x^2 + 12xy + 9y^2 - 16$

79. $A = P(1 + i)^2$
$A = P(1 + 2i + i^2)$
$A = P + 2Pi + Pi^2$

81. Familiarize. Let a, b, and c represent the daily production of machines A, B, and C, respectively.

Translate. Rewording, we have:

Production of A, B, and C is 222 per day.
$a + b + c = 222$

Production of A and B alone is 159.
$a + b = 159$

Production of B and C alone is 147.
$b + c = 147$

We have a system of equations.
$a + b + c = 222,$
$a + b = 159,$
$b + c = 147$

Solve. Solving the system we get (75,84,63).

Check. The daily production of the three machines together is 75 + 84 + 63, or 222. The daily production of A and B alone is 75 + 84, or 159. The daily production of B and C alone is 84 + 63, or 147. The numbers check.

State. The daily production of suitcases by machines A, B, and C is 75, 84, and 63, respectively.

83. $(6y)^2\left(-\frac{1}{3}x^2y^3\right)^3$
$= (36y^2)\left(-\frac{1}{27}x^6y^9\right)$
$= -\frac{36}{27}x^6y^{11}$
$= -\frac{4}{3}x^6y^{11}$

85. $(-r^6s^2)^3\left(-\frac{r^2}{6}\right)^2(9s^4)$
$= (-r^{18}s^6)\left(\frac{r^4}{36}\right)(9s^4)$
$= -\frac{9}{36}r^{22}s^{10}$
$= -\frac{1}{4}r^{22}s^{10}$

87. $(a^xb^{2y})\left(\frac{1}{2}a^{3x}b\right)^2$
$= (a^xb^{2y})\left(\frac{1}{4}a^{6x}b^2\right)$
$= \frac{1}{4}a^{7x}b^{2y+2}$

89. $y^3z^n(y^{3n}z^3 - 4yz^{2n})$
$= y^3z^n(y^{3n}z^3) - y^3z^n(4yz^{2n})$
$= y^{3+3n}z^{n+3} - 4y^4z^{3n}$

91. $(y - 1)^6(y + 1)^6 = [(y - 1)(y + 1)]^6$
$= (y^2 - 1)^6 = [(y^2 - 1)^2]^3 = (y^4 - 2y^2 + 1)^3$
$= [(y^4 - 2y^2) + 1]^2(y^4 - 2y^2 + 1)$
$= (y^8 - 4y^6 + 4y^4 + 2y^4 - 4y^2 + 1)(y^4 - 2y^2 + 1)$
$= (y^8 - 4y^6 + 6y^4 - 4y^2 + 1)(y^4 - 2y^2 + 1)$
$= y^{12} - 4y^{10} + 6y^8 - 4y^6 + y^4 - 2y^{10} + 8y^8 - 12y^6 + 8y^4 - 2y^2 + y^8 - 4y^6 + 6y^4 - 4y^2 + 1)$
$= y^{12} - 6y^{10} + 15y^8 - 20y^6 + 15y^4 - 6y^2 + 1$

93. $\left(3x^5 - \frac{5}{11}\right)^2$
$= (3x^5)^2 - 2(3x^5)\left(\frac{5}{11}\right) + \left(\frac{5}{11}\right)^2$
$= 9x^{10} - \frac{30}{11}x^5 + \frac{25}{121}$

95. $\left(x - \frac{1}{7}\right)\left(x^2 + \frac{1}{7}x + \frac{1}{49}\right)$
$= x^3 + \frac{1}{7}x^2 + \frac{1}{49}x - \frac{1}{7}x^2 - \frac{1}{49}x - \frac{1}{343}$
$= x^3 - \frac{1}{343}$

97. $(x^2 - 7x + 12)(x^2 + 7x + 12)$
$= [(x^2 + 12) - 7x][(x^2 + 12) + 7x]$
$= (x^2 + 12)^2 - (7x)^2$
$= x^4 + 24x^2 + 144 - 49x^2$
$= x^4 - 25x^2 + 144$

99. $[a - (b - 1)][(b - 1)^2 + a(b - 1) + a^2]$
$= (a - b + 1)(b^2 - 2b + 1 + ab - a + a^2)$
$= ab^2 - 2ab + a + a^2b - a^2 + a^3 - b^3 + 2b^2 -$
$b - ab^2 + ab - a^2b + b^2 - 2b + 1 + ab - a + a^2$
$= a^3 - b^3 + 3b^2 - 3b + 1$

101. $(x^{a-b})^{a+b} = x^{(a-b)(a+b)} = x^{a^2-b^2}$

Exercise Set 8.3

1. $4a^2 + 2a$
$= 2a\cdot 2a + 2a\cdot 1$
$= 2a(2a + 1)$

3. $y^3 + 9y^2$
$= y\cdot y^2 + 9\cdot y^2$
$= y^2(y + 9)$

5. $6x^2 - 3x^4$
$= 3x^2\cdot 2 - 3x^2\cdot x^2$
$= 3x^2(2 - x^2)$

7. $4x^2y - 12xy^2$
$= 4xy\cdot x - 4xy\cdot 3y$
$= 4xy(x - 3y)$

9. $3y^2 - 3y - 9$
$= 3\cdot y^2 - 3\cdot y - 3\cdot 3$
$= 3(y^2 - y - 3)$

11. $4ab - 6ac + 12ad$
$= 2a\cdot 2b - 2a\cdot 3c + 2a\cdot 6d$
$= 2a(2b - 3c + 6d)$

13. $10a^4 + 15a^2 - 25a - 30$
$= 5\cdot 2a^4 + 5\cdot 3a^2 - 5\cdot 5a - 5\cdot 6$
$= 5(2a^4 + 3a^2 - 5a - 6)$

15. $-3x + 12 = -3(x - 4)$

17. $-6y - 72 = -6(y + 12)$

19. $-2x^2 + 4x - 12 = -2(x^2 - 2x + 6)$

21. $-3y^2 + 24y = -3y(y - 8)$

23. $-3y^3 + 12y^2 - 15y = -3y(y^2 - 4y + 5)$

25. $-x^2 + 3x - 7 = -1(x^2 - 3x + 7)$

27. $-a^4 + 2a^3 - 13a = -a(a^3 - 2a^2 + 13)$

29. $a(b - 2) + c(b - 2)$
$= (a + c)(b - 2)$

31. $(x - 2)(x + 5) + (x - 2)(x + 8)$
$= (x - 2)[(x + 5) + (x + 8)]$
$= (x - 2)(2x + 13)$

33. $a^2(x - y) + a^2(x - y)$
$= 2a^2(x - y)$

35. $ac + ad + bc + bd$
$= a(c + d) + b(c + d)$
$= (a + b)(c + d)$

37. $b^3 - b^2 + 2b - 2$
$= b^2(b - 1) + 2(b - 1)$
$= (b^2 + 2)(b - 1)$

39. $y^3 - 8y^2 + y - 8$
$= y^2(y - 8) + 1(y - 8)$
$= (y^2 + 1)(y - 8)$

41. $24x^3 - 36x^2 + 72x - 108$
$= 12(2x^3 - 3x^2 + 6x - 9)$
$= 12[x^2(2x - 3) + 3(2x - 3)]$
$= 12(x^2 + 3)(2x - 3)$

43. $x^6 + x^5 - x^3 + x^2$
$= x^2\cdot x^4 + x^2\cdot x^3 - x^2\cdot x + x^2\cdot 1$
$= x^2(x^4 + x^3 - x + 1)$

45. $2y^4 + 6y^2 + 5y^2 + 15$
$= 2y^2(y^2 + 3) + 5(y^2 + 3)$
$= (2y^2 + 5)(y^2 + 3)$

47. Familiarize. Let x, y, and z represent the amounts invested at $5\frac{1}{2}\%$, 7%, and 8%, respectively. Then the interest is $5\frac{1}{2}\%x$, 7%y, and 8%z, or 0.055x, 0.07y, and 0.08z.

Translate.

The total interest is \$244.
$0.055x + 0.07y + 0.08z = 244$

The total investment is \$3500.
$x + y + z = 3500$

The amount invested at 8% is \$1100 more than the amount invested at 7%.
$z = 1100 + y$

We have a system of equations:
$0.055x + 0.07y + 0.08z = 244,$
$x + y + z = 3500,$
$z = 1100 + y$

Solve. Solving the system we get (1200, 600, 1700).

47. (continued)

Check. The sum of the numbers is 3500. Also $0.055(\$1200) + 0.07(\$600) + 0.08(\$1700) =$ $\$66 + \$42 + \$136 = \$244$, and \$1700 is \$1100 more than \$600. The numbers check.

State. \$1200 is invested at $5\frac{1}{2}\%$, \$600 is invested at 7%, and \$1700 is invested at 8%.

49. $4y^{4a} + 12y^{2a} + 10y^{2a} + 30$

$= 2(2y^{4a} + 6y^{2a} + 5y^{2a} + 15)$

$= 2[2y^{2a}(y^{2a} + 3) + 5(y^{2a} + 3)]$

$= 2(2y^{2a} + 5)(y^{2a} + 3)$

51. $R = 280x - 0.4x^2$

$R = 0.4x(700 - x)$

53. $D = \frac{1}{2}(8)^2 - \frac{3}{2}(8) = 32 - 12 = 20$

Exercise Set 8.4

1. $x^2 + 9x + 20$

We look for two numbers whose product is 20 and whose sum is 9. Since both 20 and 9 are positive, we need only consider positive factors.

Pairs of Factors	Sums of Factors
1, 20	21
2, 10	12
4, 5	9

The numbers we need are 4 and 5. The factorization is $(x + 4)(x + 5)$.

3. $t^2 - 8t + 15$

Since the constant term is positive and the coefficient of the middle term is negative, we look for a factorization of 15 in which both factors are negative. Their sum must be -8.

Pairs of Factors	Sums of Factors
-1, -15	-16
-3, -5	-8

The numbers we need are -3 and -5. The factorization is $(t - 3)(t - 5)$.

5. $x^2 - 27 - 6x = x^2 - 6x - 27$

Since the constant term is negative, we look for a factorization of -27 in which one factor is positive and one factor is negative. Their sum must be -6, so the negative factor must have the larger absolute value. Thus we consider only pairs of factors in which the negative factor has the larger absolute value.

Pairs of Factors	Sums of Factors
-27, 1	-26
-9, 3	-6

The numbers we need are -9 and 3. The factorization is $(x - 9)(x + 3)$.

7. $2y^2 - 16y + 32$

$= 2(y^2 - 8y + 16)$ Removing the common factor

We now factor $y^2 - 8y + 16$. We look for two numbers whose product is 16 and whose sum is -8. Since the constant term is positive and the coefficient of the middle term is negative, we look for a factorization of 16 in which both factors are negative.

Pairs of Factors	Sums of Factors
-1, -16	-17
-2, -8	-10
-4, -4	-8

The numbers we want are -4 and -4.

$y^2 - 8y + 16 = (y - 4)(y - 4)$

We must not forget to include the common factor 2.

$2y^2 - 16y + 32 = 2(y - 4)(y - 4)$.

9. $p^2 + 3p - 54$

Since the constant term is negative, we look for a factorization of -54 in which one factor is positive and one factor is negative. We consider only pairs of factors in which the positive factor has the larger absolute value, since the sum of the factors, 3, is positive.

Pairs of Factors	Sums of Factors
54, -1	53
27, -2	25
18, -3	15
9, -6	3

The numbers we need are 9 and -6. The factorization is $(p + 9)(p - 6)$.

11. $14x + x^2 + 45 = x^2 + 14x + 45$

Since the constant term and the middle term are both positive, we look for a factorization of 45 in which both factors are positive. Their sum must be 14.

Pairs of Factors	Sums of Factors
45, 1	46
15, 3	18
9, 5	14

The numbers we need are 9 and 5. The factorization is $(x + 9)(x + 5)$.

13. $y^2 + 2y - 63$

Since the constant term is negative, we look for a factorization of -63 in which one factor is positive and one factor is negative. We consider only pairs of factors in which the positive factor has the larger absolute value, since the sum of the factors, 2, is positive.

Pairs of Factors	Sums of Factors
63, -1	62
21, -3	18
9, -7	2

The numbers we need are 9 and -7. The factorization is $(y + 9)(y - 7)$.

15. $t^2 - 11t + 28$

Since the constant term is positive and the coefficient of the middle term is negative, we look for a factorization of 28 in which both factors are negative. Their sum must be -11.

Pairs of Factors	Sums of Factors
-28, -1	-29
-14, -2	-16
-7, -4	-11

The numbers we need are -7 and -4. The factorization is $(t - 7)(t - 4)$.

17. $3x + x^2 - 10 = x^2 + 3x - 10$

Since the constant term is negative, we look for a factorization of -10 in which one factor is positive and one factor is negative. We consider only pairs of factors in which the positive factor has the larger absolute value, since the sum of the factors, 3, is positive.

Pairs of Factors	Sums of Factors
10, -1	9
5, -2	3

The numbers we need are 5 and -2. The factorization is $(x + 5)(x - 2)$.

19. $x^2 + 5x + 6$

We look for two numbers whose product is 6 and whose sum is 5. Since 6 and 5 are both positive, we need consider only positive factors.

Pairs of Factors	Sums of Factors
1, 6	7
2, 3	5

The numbers we need are 2 and 3. The factorization is $(x + 2)(x + 3)$.

21. $56 + x - x^2 = -x^2 + x + 56 = -(x^2 - x - 56)$

We now factor $x^2 - x - 56$. Since the constant term is negative, we look for a factorization of -56 in which one factor is positive and one factor is negative. We consider only pairs of factors in which the negative factor has the larger absolute value, since the sum of the factors, -1, is negative.

Pairs of Factors	Sums of Factors
-56, 1	-55
-28, 2	-26
-14, 4	-10
-8, 7	-1

The numbers we need are -8 and 7. Thus, $x^2 - x - 56 = (x - 8)(x + 7)$. We must not forget to include the factor that was factored out earlier:

$56 + x - x^2 = -(x - 8)(x + 7)$, or

$(-x + 8)(x + 7)$, or $(8 - x)(7 + x)$

23. $32y + 4y^2 - y^3$

There is a common factor, y. We also factor out -1 in order to make the leading coefficient positive.

$32y + 4y^2 - y^3 = -y(-32 - 4y + y^2)$

$= -y(y^2 - 4y - 32)$

Now we factor $y^2 - 4y - 32$. Since the constant term is negative, we look for a factorization of -32 in which one factor is positive and one factor is negative. We consider only pairs of factors in which the negative factor has the larger absolute value, since the sum of the factors, -4, is negative.

Pairs of Factors	Sums of Factors
-32, 1	-31
-16, 2	-14
-8, 4	-4

The numbers we need are -8 and 4. Thus, $y^2 - 4y - 32 = (y - 8)(y + 4)$. We must not forget to include the common factor:

$32y + 4y^2 - y^3 = -y(y - 8)(y + 4)$, or

$y(-y + 8)(y + 4)$, or $y(8 - y)(4 + y)$

25. $x^4 + 11x^2 - 80$

First make a substitution. We let $u = x^2$, so $u^2 = x^4$. Then we consider $u^2 + 11u - 80$. We look for pairs of factors of -80, one positive and one negative, such that the positive factor has the larger absolute value and the sum of the factors is 11.

Pairs of Factors	Sums of Factors
80, -1	79
40, -2	38
20, -4	16
16, -5	11
10, -8	2

The numbers we need are 16 and -5. Then $u^2 + 11u - 80 = (u + 16)(u - 5)$. Replacing u by x^2 we obtain the factorization of the original trinomial: $(x^2 + 16)(x^2 - 5)$

27. $x^2 - 3x + 7$

There are no factors of 7 whose sum is -3. This trinomial is not factorable into binomials with integer coefficients.

29. $x^2 + 12xy + 27y^2$

We look for numbers p and q such that $x^2 + 12xy + 27y^2 = (x + py)(x + qy)$. Our thinking is much the same as if we were factoring $x^2 + 12x + 27$. We look for factors of 27 whose sum is 12. Those factors are 9 and 3. Then

$$x^2 + 12xy + 27y^2 = (x + 9y)(x + 3y).$$

31. $x^2 - 14x + 49$

Since the constant term is positive and the coefficient of the middle term is negative, we look for a factorization of 49 in which both factors are negative. Their sum must be -14.

Pairs of Factors	Sums of Factors
-49, -1	-50
-7, -7	-14

The numbers we need are -7 and -7. The factorization is $(x - 7)(x - 7)$.

33. $x^4 + 50x^2 + 49$

Substitute u for x^2 (and hence u^2 for x^4). Consider $u^2 + 50u + 49$. We look for a pair of positive factors of 49 whose sum is 50.

Pairs of Factors	Sums of Factors
7, 7	14
1, 49	50

The numbers we need are 1 and 49. Then $u^2 + 50 + 49 = (u + 1)(u + 49)$. Replacing u by x^2 we have

$$x^4 + 50x^2 + 49 = (x^2 + 1)(x^2 + 49).$$

35. $x^6 - x^3 - 42$

Substitute u for x^3 (and hence u^2 for x^6). Consider $u^2 - u - 42$. We look for a pair of factors of -42, one positive and one negative, such that the negative factor has the larger absolute value and the sum of the factors is -1.

Pairs of Factors	Sums of Factors
-42, 1	-41
-21, 2	-19
-14, 3	-11
-7, 6	-1

The numbers we need are -7 and 6. Then $u^2 - u - 42 = (u - 7)(u + 6)$. Replacing u by x^3 we obtain the factorization of the original trinomial: $(x^3 - 7)(x^3 + 6)$

37. $3x^2 - 16x - 12$

We will use the FOIL method.

a) There is no common factor (other than 1).

b) Factor the first term, $3x^2$. The factors are 3x and x. We have this possibility:

$$(3x \quad)(x \quad)$$

c) Factor the last term, -12. The factors are 12, -1 and -1, 12 and 6, -2 and -6, 2, and 4, -3 and -3, 4.

d) We look for factors in b) and c) such that the sum of their products is the middle term -16x. Since the constant term is negative, we know that one sign of the constant terms of the factors will be positive and the other will be negative. Try some possibilities and check by multiplying.

$$(3x - 3)(x + 4) = 3x^2 + 9x - 12$$

This gives a middle term with a positive coefficient. But, more significantly, the expression 3x - 3 has a common factor of 3. But we know there is no common factor (other than 1) so we can eliminate (3x - 3) and all other factors containing a common factor, such as (3x + 6), (3x - 12), and so on.

We try another possibility:

$$(3x + 2)(x - 6) = 3x^2 - 16x - 12$$

The factorization is $(3x + 2)(x - 6)$.

39. $6x^3 - 15x - x^2 = 6x^3 - x^2 - 15x$

We will use the grouping method.

a) Look for a common factor. We factor out x:

$$x(6x^2 - x - 15)$$

b) Factor the trinomial $6x^2 - x - 15$. Multiply the leading coefficient, 6, and the constant, -15.

$$6(-15) = -90$$

c) Try to factor -90 so the sum of the factors is -1. We need only consider pairs of factors in which the negative term has the larger absolute value, since their sum is negative.

Pairs of Factors	Sums of Factors
-90, 1	-89
-45, 2	-43
-30, 3	-27
-16, 5	-11
-10, 9	-1

d) We split the middle term, -x, as follows:

$$-x = -10x + 9x$$

e) Factor by grouping:

$$\begin{aligned} 6x^2 - x - 15 &= 6x^2 - 10x + 9x - 15 \\ &= 2x(3x - 5) + 3(3x - 5) \\ &= (3x - 5)(2x + 3) \end{aligned}$$

We must include the common factor to get a factorization of the original trinomial:

$$6x^3 - 15x - x^2 = x(3x - 5)(2x + 3)$$

41. $3a^2 - 10a + 8$

We will use the FOIL method.

a) There is no common factor (other than 1).

b) Factor the first term, $3a^2$. The factors are 3a and a. We have this possibility: $(3a \quad)(a \quad)$

c) Factor the last term, 8. The factors are 8, 1 and -8, -1 and 4, 2 and -4, -2.

d) Look for factors in b) and c) such that the sum of the products is the middle term, -10a. Try some possibilities and check by multiplying. Trial and error leads us to the correct factorization, $(3a - 4)(a - 2)$.

43. $35y^2 + 34y + 8$

We will use the grouping method.

a) There is no common factor (other than 1).

b) Multiply the leading coefficient, 35, and the constant, 8: $35(8) = 280$

c) Try to factor 280 so the sum of the factors is 34. We need only consider pairs of positive factors since 280 and 34 are both positive.

Pairs of Factors	Sums of Factors
280, 1	281
140, 2	142
70, 4	74
56, 5	61
40, 7	47
28, 10	38
20, 14	34

d) Split 34y as follows:

$$34y = 20y + 14y$$

e) Factor by grouping:

$$35y^2 + 34y + 8 = 35y^2 + 20y + 14y + 8$$
$$= 5y(7y + 4) + 2(7y + 4)$$
$$= (7y + 4)(5y + 2)$$

45. $4t + 10t^2 - 6 = 10t^2 + 4t - 6$

We will use the FOIL method.

a) Factor out the common factor, 2:

$$2(5t^2 + 2t - 3)$$

b) Now we factor the trinomial $5t^2 + 2t - 3$.

Factor the first term, $5t^2$. The factors are 5t and t. We have this possibility: (5t)(t)

c) Factor the last term, -3. The factors are 1, -3 and -1, 3.

d) Look for factors in b) and c) such that the sum of the products is the middle term, 2t. Trial and error leads us to the correct factorization:
$5t^2 + 2t - 3 = (5t - 3)(t + 1)$

We must include the common factor to get a factorization of the original trinomial:

$$4t + 10t^2 - 6 = 2(5t - 3)(t + 1)$$

47. $8x^2 - 16 - 28x = 8x^2 - 28x - 16$

We will use the grouping method.

a) Factor out the common factor, 4:

$$4(2x^2 - 7x - 4)$$

b) Now we factor the trinomial $2x^2 - 7x - 4$. Multiply the leading coefficient, 2, and the constant, -4: $2(-4) = -8$

c) Factor -8 so the sum of the factors is -7. We need only consider pairs of factors in which the negative factor has the larger absolute value, since their sum is negative.

47. (continued)

Pairs of Factors	Sums of Factors
-4, 2	-2
-8, 1	-7

d) Split -7x as follows:

$$-7x = -8x + x$$

e) Factor by grouping:

$$2x^2 - 7x - 4 = 2x^2 - 8x + x - 4$$
$$= 2x(x - 4) + (x - 4)$$
$$= (x - 4)(2x + 1)$$

We must include the common factor to get a factorization of the original trinomial:

$$8x^2 - 16 - 28x = 4(x - 4)(2x + 1)$$

49. $12x^3 - 31x^2 + 20x$

We will use the FOIL method.

a) Factor out the common factor, x:

$$x(12x^2 - 31x + 20)$$

b) We now factor the trinomial $12x^2 - 31x + 20$. Factor the first term, $12x^2$. The factors are 12x, x and 6x, 2x and 4x, 3x. We have these possibilities: (12x)(x), (6x)(2x), (4x)(3x)

c) Factor the last term, 20. The factors are 20, 1 and -20, -1 and 10, 2 and -10, -2 and 5, 4 and -5, -4.

d) Look for factors in b) and c) such that the sum of the products is the middle term, -31x. Trial and error leads us to the correct factorization:
$12x^2 - 31x + 20 = (4x - 5)(3x - 4)$

We must include the common factor to get a factorization of the original trinomial:

$$12x^3 - 31x^2 + 20x = x(4x - 5)(3x - 4)$$

51. $14x^4 - 19x^3 - 3x^2$

We will use the grouping method.

a) Factor out the common factor, x^2:

$$x^2(14x^2 - 19x - 3)$$

b) Now we factor the trinomial $14x^2 - 19x - 3$. Multiply the leading coefficient, 14, and the constant, -3: $14(-3) = -42$

c) Factor -42 so the sum of the factors is -19. We need only consider pairs of factors in which the negative factor has the larger absolute value, since the sum is negative.

Pairs of Factors	Sums of Factors
-42, 1	-41
-21, 2	-19
-14, 3	-11
-7, 6	-1

d) Split -19x as follows:

$$-19x = -21x + 2x$$

51\. (continued)

e) Factor by grouping:

$$14x^2 - 19x - 3 = 14x^2 - 21x + 2x - 3$$
$$= 7x(2x - 3) + 2x - 3$$
$$= (2x - 3)(7x + 1)$$

We must include the common factor to get a factorization of the original trinomial:

$$14x^4 - 19x^3 - 3x^2 = x^2(2x - 3)(7x + 1)$$

53\. $3a^2 - a - 4$

We will use the FOIL method.

a) There is no common factor (other than 1).

b) Factor the first term, $3a^2$. The factors are 3a and a. We have this possibility: (3a)(a)

c) Factor the last term, -4. The factors are 4, -1 and -4, 1 and 2, -2.

d) Look for factors in b) and c) such that the sum of the products is the middle term, -a. Trial and error leads us to the correct factorization: $(3a - 4)(a + 1)$

55\. $9x^2 + 15x + 4$

We will use the grouping method.

a) There is no common factor (other than 1).

b) Multiply the leading coefficient and the constant: $9(4) = 36$

c) Factor 36 so the sum of the factors is 15. We need only consider pairs of positive factors since 36 and 15 are both positive.

Pairs of Factors	Sums of Factors
36, 1	37
18, 2	20
12, 3	15
9, 4	13
6, 6	12

d) Split 15x as follows:

$$15x = 12x + 3x$$

e) Factor by grouping:

$$9x^2 + 15x + 4 = 9x^2 + 12x + 3x + 4$$
$$= 3x(3x + 4) + 3x + 4$$
$$= (3x + 4)(3x + 1)$$

57\. $3 + 35z - 12z^2 = -12z^2 + 35z + 3$

We will use the FOIL method.

a) Factor out -1 so the leading coefficient is positive: $-(12z^2 - 35z - 3)$

b) Now we factor the trinomial $12z^2 - 35z - 3$. Factor the first term, $12z^2$. The factors are 12z, z and 6z, 2z and 4z, 3z. We have these possibilities: (12z)(z), (6z)(2z), (4z)(3z)

c) Factor the last term, -3. The factors are 3, -1 and -3, 1.

57\. (continued)

d) Look for factors in b) and c) such that the sum of the products is the middle term, -35z. Trial and error leads us to the correct factorization: $(12z + 1)(z - 3)$

We must include the common factor to get a factorization of the original trinomial:

$3 + 35z - 12z^2 = -(12z + 1)(z - 3)$, or $(12z + 1)(-z + 3)$, or $(1 + 12z)(3 - z)$

59\. $-4t^2 - 4t + 15$

We will use the grouping method.

a) Factor out -1 so the leading coefficient is positive: $-(4t^2 + 4t - 15)$

b) Now we factor the trinomial $4t^2 + 4t - 15$. Multiply the leading coefficient and the constant: $4(-15) = -60$

c) Factor -60 so the sum of the factors is 4. The desired factorization is 10(-6).

d) Split 4t as follows:

$$4t = 10t - 6t$$

e) Factor by grouping:

$$4t^2 + 4t - 15 = 4t^2 + 10t - 6t - 15$$
$$= 2t(2t + 5) - 3(2t + 5)$$
$$= (2t + 5)(2t - 3)$$

We must include the common factor to get a factorization of the original trinomial:

$-4t^2 - 4t + 15 = -(2t + 5)(2t - 3)$, or $(2t + 5)(-2t + 3)$

61\. $3x^3 - 5x^2 - 2x$

We will use the FOIL method.

a) Factor out the common factor, x: $x(3x^2 - 5x - 2)$

b) Now we factor the trinomial $3x^2 - 5x - 2$. Factor the first term, $3x^2$. The factors are 3x and x. We have this possibility: (3x)(x)

c) Factor the last term, -2. The factors are 2, -1 and -2, 1.

d) Look for factors in b) and c) such that the sum of the products is the middle term, -5x. Trial and error leads us to the correct factorization: $(3x + 1)(x - 2)$

We must include the common factor to get a factorization of the original trinomial:

$$3x^3 - 5x^2 - 2x = x(3x + 1)(x - 2)$$

63. $24x^2 - 2 - 47x = 24x^2 - 47x - 2$

We will use the grouping method.

a) There is no common factor (other than 1).

b) Multiply the leading coefficient and the constant: $24(-2) = -48$

c) Factor -48 so the sum of the factors is -47. The desired factorization is $-48\cdot 1$.

d) Split $-47x$ as follows:

$$-47x = -48x + x$$

e) Factor by grouping:

$$\begin{aligned} 24x^2 - 47x - 2 &= 24x^2 - 48x + x - 2 \\ &= 24x(x - 2) + (x - 2) \\ &= (x - 2)(24x + 1) \end{aligned}$$

65. $21x^2 + 37x + 12$

We will use the FOIL method.

a) There is no common factor (other than 1).

b) Factor the first term $21x^2$. The factors are $21x$, x and $7x$, $3x$. We have these possibilities: $(21x \quad)(x \quad)$ and $(7x \quad)(3x \quad)$.

c) Factor the last term, 12. The factors are 12, 1 and -12, -1 and 6, 2 and -6, -2 and 4, 3 and -4, -3.

d) Look for factors in b) and c) such that the sum of the products is the middle term, $37x$. Trial and error leads us to the correct factorization: $(7x + 3)(3x + 4)$

67. $40x^4 + 16x^2 - 12$

We will use the grouping method.

a) Factor out the common factor, 4.

$$4(10x^4 + 4x^2 - 3)$$

Now we will factor the trinomial $10x^4 + 4x^2 - 3$. Substitute u for x^2 (and u^2 for x^4), and factor $10u^2 + 4u - 3$.

b) Multiply the leading coefficient and the constant: $10(-3) = -30$

c) Factor -30 so the sum of the factors is 4. This cannot be done. The trinomial $10u^2 + 4u - 3$ cannot be factored into binomials with integer coefficients. We have

$$40x^4 + 16x^2 - 12 = 4(10x^4 + 4x^2 - 3)$$

69. $12a^2 - 17ab + 6b^2$

We will use the FOIL method. (Our thinking is much the same as if we were factoring $12a^2 - 17a + 6$.)

a) There are no common factors (other than 1).

b) Factor the first term, $12a^2$. The factors are $12a$, a and $6a$, $2a$ and $4a$, $3a$. We have these possibilities: $(12a \quad)(a \quad)$ and $(6a \quad)(2a \quad)$ and $(4a \quad)(3a \quad)$.

69. (continued)

c) Factor the last term, $6b^2$. The factors are $6b$, b and $-6b$, $-b$ and $3b$, $2b$ and $-3b$, $-2b$.

d) Look for factors in b) and c) such that the sum of the products is the middle term, $-17ab$. Trial and error leads us to the correct factorization: $(4a - 3b)(3a - 2b)$

71. $2x^2 + xy - 6y^2$

We will use the grouping method.

a) There is no common factor (other than 1).

b) Multiply the coefficients of the first and last terms: $2(-6) = -12$

c) Factor -12 so the sum of the factors is 1. The desired factorization is $4(-3)$.

d) Split xy as follows:

$$xy = 4xy - 3xy$$

e) Factor by grouping:

$$\begin{aligned} 2x^2 + xy - 6y^2 &= 2x^2 + 4xy - 3xy - 6y^2 \\ &= 2x(x + 2y) - 3y(x + 2y) \\ &= (x + 2y)(2x - 3y) \end{aligned}$$

73. $6x^2 - 29xy + 28y^2$

We will use the FOIL method.

a) There is no common factor (other than 1).

b) Factor the first term, $6x^2$. The factors are $6x$, x and $3x$, $2x$. We have these possibilities: $(6x \quad)(x \quad)$ and $(3x \quad)(2x \quad)$.

c) Factor the last term, $28y^2$. The factors are $28y$, y and $-28y$, $-y$ and $14y$, $2y$ and $-14y$, $-2y$ and $7y$, $4y$ and $-7y$, $-4y$.

d) Look for factors in b) and c) such that the sum of the products is the middle term, $-29xy$. Trial and error leads us to the correct factorization: $(3x - 4y)(2x - 7y)$

75. $9x^2 - 30xy + 25y^2$

We will use the grouping method.

a) There are no common factors (other than 1).

b) Multiply the coefficients of the first and last terms: $9(25) = 225$

c) Factor 225 so the sum of the factors is -30. The desired factorization is $-15(-15)$.

d) Split $-30xy$ as follows:

$$-30xy = -15xy - 15xy$$

e) Factor by grouping:

$$\begin{aligned} 9x^2 - 30xy + 25y^2 &= 9x^2 - 15xy - 15xy + 25y^2 \\ &= 3x(3x - 5y) - 5y(3x - 5y) \\ &= (3x - 5y)(3x - 5y) \end{aligned}$$

77\. $3x^6 + x^3 - 2$

We will use the FOIL method.

a) There is no common factor (other than 1). Substitute u for x^3 (and u^2 for x^6). We factor $3u^2 + u - 2$.

b) Factor the first term, $3u^2$. The factors are 3u and u. We have this possibility: (3u   )(u   )

c) Factor the last term, -2. The factors are 2, -1 and -2, 1.

d) Look for factors in b) and c) such that the sum of the products is the middle term, u. Trial and error leads us to the correct factorization: $(3u - 2)(u + 1)$

Replacing u by x^3 we have

$$3x^6 + x^3 - 2 = (3x^3 - 2)(x^3 + 1)$$

79\. a) $h = -16(0)^2 + 80(0) + 224 = 224$ ft

$h = -16(1)^2 + 80(1) + 224 = 288$ ft

$h = -16(3)^2 + 80(3) + 224 = 320$ ft

$h = -16(4)^2 + 80(4) + 224 = 288$ ft

$h = -16(6)^2 + 80(6) + 224 = 128$ ft

b) $h = -16t^2 + 80t + 224$

We will use the grouping method.

A) Factor out -16 so the leading coefficient is positive: $-16(t^2 - 5t - 14)$

B) Factor the trinomial $t^2 - 5t - 14$. Multiply the leading coefficient and the constant: $1(-14) = -14$

C) Factor -14 so the sum of the factors is -5. The desired factorization is $-7 \cdot 2$.

D) Split -5t as follows:

$$-5t = -7t + 2t$$

E) Factor by grouping:

$$\begin{aligned} t^2 - 5t - 14 &= t^2 - 7t + 2t - 14 \\ &= t(t - 7) + 2(t - 7) \\ &= (t - 7)(t + 2) \end{aligned}$$

We must include the common factor to get a factorization of the original trinomial.

$$h = -16(t - 7)(t + 2)$$

81\. $p^2q^2 + 7pq + 12$

The factorization will be of the form (pq   )(pq   ). We look for factors of 12 whose sum is 7. The factors we need are 4 and 3. The factorization is $(pq + 4)(pq + 3)$.

83\. $x^2 - \frac{4}{25} + \frac{3}{5}x = x^2 + \frac{3}{5}x - \frac{4}{25}$

We look for factors of $-\frac{4}{25}$ whose sum is $\frac{3}{5}$.

The factors are $\frac{4}{5}$ and $-\frac{1}{5}$. The factorization is $\left(x + \frac{4}{5}\right)\left(x - \frac{1}{5}\right)$.

85\. $y^2 + 0.4y - 0.05$

We look for factors of -0.05 whose sum is 0.4. The factors are -0.1 and 0.5. The factorization is $(y - 0.1)(y + 0.5)$.

87\. $7a^2b^2 + 6 + 13ab = 7a^2b^2 + 13ab + 6$

We will use the grouping method. There are no common factors (other than 1). Multiply the leading coefficient and the constant: $7(6) = 42$. Factor 42 so the sum of the factors is 13. The desired factorization is $6 \cdot 7$. Split the middle term and factor by grouping.

$$\begin{aligned} 7a^2b^2 + 13ab + 6 &= 7a^2b^2 + 6ab + 7ab + 6 \\ &= ab(7ab + 6) + 7ab + 6 \\ &= (7ab + 6)(ab + 1) \end{aligned}$$

89\. $3x^2 + 12x - 495$

Factor out the common factor, 3.

$$3(x^2 + 4x - 165)$$

Now factor $x^2 + 4x - 165$. Find factors of -165 whose sum is 4. The factors are -11 and 15. Then $x^2 + 4x - 165 = (x - 11)(x + 15)$, and $3x^2 + 12x - 495 = 3(x - 11)(x + 15)$.

91\. $216x + 78x^2 + 6x^3 = 6x^3 + 78x^2 + 216x$

Factor out the common factor, 6x.

$$6x(x^2 + 13x + 36)$$

Now factor $x^2 + 13x + 36$. Look for factors of 36 whose sum is 13. The factors are 9 and 4. Then $x^2 + 13x + 36 = (x + 9)(x + 4)$, and $6x^3 + 78x^2 + 216x = 6x(x + 9)(x + 4)$.

93\. $x^2 + ax + bx + ab$

$= x(x + a) + b(x + a)$ Factoring by grouping

$= (x + a)(x + b)$

95\. $bdx^2 + adx + bcx + ac$

$= dx(bx + a) + c(bx + a)$ Factoring by grouping

$= (bx + a)(dx + c)$

97\. $(x + 3)^2 - 2(x + 3) - 35$

Substitute u for $x + 3$ (and u^2 for $(x + 3)^2$). Then factor $u^2 - 2u - 35$. Look for factors of -35 whose sum is -2. The factors are -7 and 5. Then $u^2 - 2u - 35 = (u - 7)(u + 5)$, and $(x + 3)^2 - 2(x + 3) - 35 = [(x + 3) - 7][(x + 3) + 5]$, or $(x - 4)(x + 8)$.

99\. All such m are the sums of the factors of 75.

Pairs of Factors	Sums of Factors
75, 1	76
-75, -1	-76
25, 3	28
-25, -3	-28
15, 5	20
-15, -5	-20

m can be 76, -76, 28, -28, 20, or -20.

Exercise Set 8.5

1. $y^2 - 6y + 9 = (y - 3)^2$ Find the square terms and write their square roots with a minus sign between them.

3. $x^2 + 14x + 49 = (x + 7)^2$ Find the square terms and write their square roots with a plus sign between them.

5. $x^2 + 1 + 2x = x^2 + 2x + 1$ Changing order
$= (x + 1)^2$ Factoring the trinomial square

7. $y^2 + 36 - 12y = y^2 - 12y + 36$ Changing order
$= (y - 6)^2$ Factoring the trinomial square

9. $-18y^2 + y^3 + 81y = y^3 - 18y^2 + 81y$ Changing order
$= y(y^2 - 18y + 81)$ Removing the common factor
$= y(y - 9)^2$ Factoring the trinomial square

11. $12a^2 + 36a + 27 = 3(4a^2 + 12a + 9)$ Removing the common factor
$= 3(2a + 3)^2$ Factoring the trinomial square

13. $2x^2 - 40x + 200 = 2(x^2 - 20x + 100)$
$= 2(x - 10)^2$

15. $1 - 8d + 16d^2 = (1 - 4d)^2$, or $(4d - 1)^2$ Find the square terms and write their square roots with a minus sign between them.

17. $y^4 - 8y^2 + 16 = (y^2 - 4)^2$ Find the square terms and write their square roots with a minus sign between them.
$= [(y + 2)(y - 2)]^2$ Factoring the difference of squares
$= (y + 2)^2(y - 2)^2$

19. $0.25x^2 + 0.30x + 0.09 = (0.5x + 0.3)^2$ Find the square terms and write their square roots with a plus sign between them.

21. $p^2 - 2pq + q^2 = (p - q)^2$

23. $a^2 + 4ab + 4b^2 = (a + 2b)^2$

25. $25a^2 - 30ab + 9b^2 = (5a - 3b)^2$

27. $y^6 + 26y^3 + 169 = (y^3 + 13)^2$ Find the square terms and write their square roots with a plus sign between them.

29. $16x^{10} - 8x^5 + 1 = (4x^5 - 1)^2$ $[16x^{10} = (4x^5)^2]$

31. $x^4 + 2y^2y^2 + y^4 = (x^2 + y^2)^2$

33. $x^2 - 16 = x^2 - 4^2 = (x + 4)(x - 4)$

35. $p^2 - 49 = p^2 - 7^2 = (p + 7)(p - 7)$

37. $p^2q^2 - 25 = (pq)^2 - 5^2 = (pq + 5)(pq - 5)$

39. $6x^2 - 6y^2 = 6(x^2 - y^2)$ Removing the common factor
$= 6(x + y)(x - y)$ Factoring the difference of squares

41. $4xy^4 - 4xz^4 = 4x(y^4 - z^4)$ Removing the common factor
$= 4x[(y^2)^2 - (z^2)^2]$
$= 4x(y^2 + z^2)(y^2 - z^2)$ Factoring the difference of squares
$= 4x(y^2 + z^2)(y + z)(y - z)$ Factoring $y^2 - z^2$

43. $4a^3 - 49a = a(4a^2 - 49)$
$= a[(2a)^2 - 7^2]$
$= a(2a + 7)(2a - 7)$

45. $3x^8 - 3y^8 = 3(x^8 - y^8)$
$= 3[(x^4)^2 - (y^4)^2]$
$= 3(x^4 + y^4)(x^4 - y^4)$
$= 3(x^4 + y^4)[(x^2)^2 - (y^2)^2]$
$= 3(x^4 + y^4)(x^2 + y^2)(x^2 - y^2)$
$= 3(x^4 + y^4)(x^2 + y^2)(x + y)(x - y)$

47. $9a^4 - 25a^2b^4 = a^2(9a^2 - 25b^4)$
$= a^2[(3a)^2 - (5b^2)^2]$
$= a^2(3a + 5b^2)(3a - 5b^2)$

49. $\frac{1}{25} - x^2 = \left(\frac{1}{5}\right)^2 - x^2$
$= \left(\frac{1}{5} + x\right)\left(\frac{1}{5} - x\right)$

51. $0.04x^2 - 0.09y^2 = (0.2x)^2 - (0.3y)^2$
$= (0.2x + 0.3y)(0.2x - 0.3y)$

53. $m^3 - 7m^2 - 4m + 28 = m^2(m - 7) - 4(m - 7)$ Factoring by grouping
$= (m - 7)(m^2 - 4)$
$= (m - 7)(m + 2)(m - 2)$ Factoring the difference of squares

55. $a^3 - ab^2 - 2a^2 + 2b^2 = a(a^2 - b^2) - 2(a^2 - b^2)$ Factoring by grouping
$= (a^2 - b^2)(a - 2)$
$= (a + b)(a - b)(a - 2)$ Factoring the difference of squares

57. $(a + b)^2 - 100 = (a + b)^2 - 10^2$
$= (a + b + 10)(a + b - 10)$

59. $a^2 + 2ab + b^2 - 9 = (a^2 + 2ab + b^2) - 9$ Grouping as a difference of squares
$= (a + b)^2 - 3^2$
$= (a + b + 3)(a + b - 3)$

61. $r^2 - 2r + 1 - 4s^2 = (r^2 - 2r + 1) - 4s^2$ Grouping as a difference of squares
$= (r - 1)^2 - (2s)^2$
$= (r - 1 + 2s)(r - 1 - 2s)$

63. $2m^2 + 4mn + 2n^2 - 50b^2$
$= 2(m^2 + 2mn + n^2 - 25b^2)$ Removing the common factor
$= 2[(m^2 + 2mn + n^2) - 25b^2]$ Grouping as a difference of squares
$= 2[(m + n)^2 - (5b)^2]$
$= 2(m + n + 5b)(m + n - 5b)$

65. $9 - (a^2 + 2ab + b^2) = 9 - (a + b)^2$
$= [3 + (a + b)][3 - (a + b)]$
$= (3 + a + b)(3 - a - b)$

67. $\frac{1}{4}p^2 - \frac{2}{5}p + \frac{4}{25} = \left(\frac{1}{2}p - \frac{2}{5}\right)^2$ Find the square terms and write their square roots with a minus sign between them.

69. $x^{4a} - y^{2b} = (x^{2a})^2 - (y^b)^2 = (x^{2a} + y^b)(x^{2a} - y^b)$

71. $9x^{2n} - 6x^n + 1 = (3x^n)^2 - 6x^n + 1$
$= (3x^n - 1)^2$

73. $(3a + 4)^2 - 49b^2 = (3a + 4)^2 - (7b)^2$
$= (3a + 4 + 7b)(3a + 4 - 7b)$

Exercise Set 8.6

1. $x^3 + 8 = x^3 + 2^3$
$= (x + 2)(x^2 - 2x + 4)$
$A^3 + B^3 = (A + B)(A^2 - AB + B^2)$

3. $y^3 - 64 = y^3 - 4^3$
$= (y - 4)(y^2 + 4y + 16)$
$A^3 - B^3 = (A - B)(A^2 + AB + B^2)$

5. $w^3 + 1 = w^3 + 1^3$
$= (w + 1)(w^2 - w + 1)$
$A^3 + B^3 = (A + B)(A^2 - AB + B^2)$

7. $8a^3 + 1 = (2a)^3 + 1^3$
$= (2a + 1)(4a^2 - 2a + 1)$
$A^3 + B^3 = (A + B)(A^2 - AB + B^2)$

9. $y^3 - 8 = y^3 - 2^3$
$= (y - 2)(y^2 + 2y + 4)$
$A^3 - B^3 = (A - B)(A^2 + AB + B^2)$

11. $8 - 27b^3 = 2^3 - (3b)^3$
$= (2 - 3b)(4 + 6b + 9b^2)$

13. $64y^3 + 1 = (4y)^3 + 1^3$
$= (4y + 1)(16y^2 - 4y + 1)$

15. $8x^3 + 27 = (2x)^3 + 3^3$
$= (2x + 3)(4x^2 - 6x + 9)$

17. $a^3 - b^3 = (a - b)(a^2 + ab + b^2)$

19. $a^3 + \frac{1}{8} = a^3 + \left(\frac{1}{2}\right)^3$
$= \left(a + \frac{1}{2}\right)\left(a^2 - \frac{1}{2}a + \frac{1}{4}\right)$

21. $2y^3 - 128 = 2(y^3 - 64)$
$= 2(y^3 - 4^3)$
$= 2(y - 4)(y^2 + 4y + 16)$

23. $24a^3 + 3 = 3(8a^3 + 1)$
$= 3[(2a)^3 + 1^3]$
$= 3(2a + 1)(4a^2 - 2a + 1)$

25. $rs^3 + 64r = r(s^3 + 64)$
$= r(s^3 + 4^3)$
$= r(s + 4)(s^2 - 4s + 16)$

27. $5x^3 - 40z^3 = 5(x^3 - 8z^3)$
$= 5[x^3 - (2z)^3]$
$= 5(x - 2z)(x^2 + 2xz + 4z^2)$

29. $x^3 + 0.001 = x^3 + (0.1)^3$
$= (x + 0.1)(x^2 - 0.1x + 0.01)$

31. $64x^6 - 8t^6 = 8(8x^6 - t^6)$
$= 8[(2x^2)^3 - (t^2)^3]$
$= 8(2x^2 - t^2)(4x^4 + 2x^2t^2 + t^4)$

33. $2y^4 - 128y = 2y(y^3 - 64)$
$= 2y(y^3 - 4^3)$
$= 2y(y - 4)(y^2 + 4y + 16)$

35. $z^6 - 1 = (z^3)^2 - 1^2$ Writing as a difference of squares
$= (z^3 + 1)(z^3 - 1)$ Factoring a difference of squares
$= (z + 1)(z^2 - z + 1)(z - 1)(z^2 + z + 1)$ Factoring a sum and a difference of cubes

37. $t^6 + 64y^6 = (t^2)^3 + (4y^2)^3$
$= (t^2 + 4y^2)(t^4 - 4t^2y^2 + 16y^4)$

39. Graph: $5x = 10 - 2y$
$2y + 5x = 10$ Rewriting

To find the x-intercept, we cover up the y-term and consider the rest of the equation. We have $5x = 10$, or $x = 2$. The x-intercept is (2,0). To find the y-intercept, we cover up the x-term and consider the rest of the equation. We have $2y = 10$, or $y = 5$. The y-intercept is (0,5). We plot these points and draw the line.

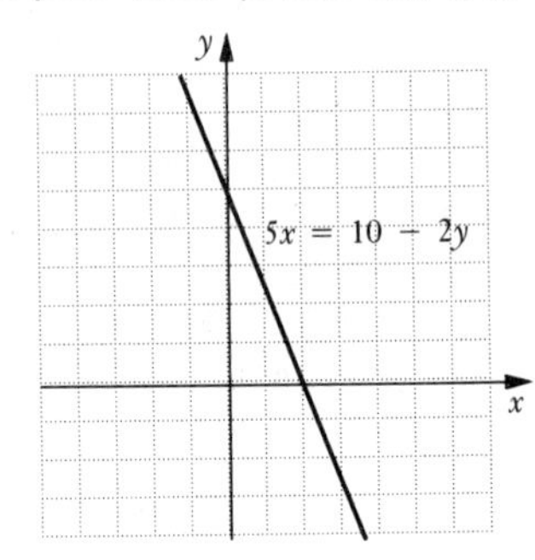

We find a third point as a check. We let x = 4 and solve for y:

$2y + 5\cdot 4 = 10$
$2y + 20 = 10$
$2y = -10$
$y = -5$

We plot the point (4,-5) and note that it is on the line.

41. $|5x - 6| \leqslant 39$
$-39 \leqslant 5x - 6 \leqslant 39$
$-33 \leqslant 5x \leqslant 45$
$-\frac{33}{5} \leqslant x \leqslant 9$

The solution set is $\left\{x \middle| -\frac{33}{5} \leqslant x \leqslant 9\right\}$.

43. $(a + b)^3 = (-2 + 3)^3 = 1^3 = 1$
$a^3 + b^3 = (-2)^3 + (3)^3 = -8 + 27 = 19$
$(a + b)(a^2 - ab + b^2)$
$= (-2 + 3)[(-2)^2 - (-2)(3) + (3)^2]$
$= 1(4 + 6 + 9)$
$= 19$
$(a + b)(a^2 + ab + b^2)$
$= (-2 + 3)[(-2)^2 + (-2)(3) + (3)^2]$
$= 1(4 - 6 + 9)$
$= 7$
$(a + b)(a - b)(a - b) = (-2 + 3)(-2 - 3)(-2 - 3)$
$= 1(-5)(-5)$
$= 25$

45. $x^{6a} + y^{3b} = (x^{2a})^3 + (y^b)^3$
$= (x^{2a} + y^b)(x^{4a} - x^{2a}y^b + y^{2b})$

47. $3x^{3a} + 24y^{3b} = 3(x^{3a} + 8y^{3b})$
$= 3[(x^a)^3 + (2y^b)^3]$
$= 3(x^a + 2y^b)(x^{2a} - 2x^ay^b + 4y^{2b})$

49. $\frac{1}{24}x^3y^3 + \frac{1}{3}z^3 = \frac{1}{3}\left(\frac{1}{8}x^3y^3 + z^3\right)$
$= \frac{1}{3}\left[\left(\frac{1}{2}xy\right)^3 + z^3\right]$
$= \frac{1}{3}\left(\frac{1}{2}xy + z\right)\left(\frac{1}{4}x^2y^2 - \frac{1}{2}xyz + z^2\right)$

51. $7x^3 - \frac{7}{8} = 7\left(x^3 - \frac{1}{8}\right)$
$= 7\left[x^3 - \left(\frac{1}{2}\right)^3\right]$
$= 7\left(x - \frac{1}{2}\right)\left(x^2 + \frac{1}{2}x + \frac{1}{4}\right)$

53. $(x + y)^3 - x^3$
$= [(x + y) - x][(x + y)^2 + x(x + y) + x^2]$
$= (x + y - x)(x^2 + 2xy + y^2 + x^2 + xy + x^2)$
$= y(3x^2 + 3xy + y^2)$

55. $(a + 2)^3 - (a - 2)^3$
$= [(a + 2) - (a - 2)][(a + 2)^2 + (a + 2)(a - 2) + (a - 2)^2]$
$= (a + 2 - a + 2)(a^2 + 4a + 4 + a^2 - 4 + a^2 - 4a + 4)$
$= 4(3a^2 + 4)$

Exercise Set 8.7

1. $x^2 - 144$
$= x^2 - 12^2$ Difference of squares
$= (x + 12)(x - 12)$

3. $2x^2 + 11x + 12$
$= (2x + 3)(x + 4)$ FOIL or grouping method

5. $3x^4 - 12$
$= 3(x^4 - 4)$ Difference of squares
$= 3(x^2 + 2)(x^2 - 2)$

7. $a^2 + 25 + 10a$
$= a^2 + 10a + 25$ Trinomial square
$= (a + 5)^2$

9. $2x^2 - 10x - 132$
$= 2(x^2 - 5x - 66)$
$= 2(x - 11)(x + 6)$ Trial and error

11. $9x^2 - 25y^2$
$= (3x)^2 - (5y)^2$ Difference of squares
$= (3x + 5y)(3x - 5y)$

13. $m^6 - 1$
$= (m^3)^2 - 1^2$ Difference of squares
$= (m^3 + 1)(m^3 - 1)$ Sum and difference of cubes
$= (m + 1)(m^2 - m + 1)(m - 1)(m^2 + m + 1)$

15. $x^2 + 6x - y^2 + 9$
$= x^2 + 6x + 9 - y^2$
$= (x + 3)^2 - y^2$ Difference of squares
$= [(x + 3) + y][(x + 3) - y]$
$= (x + y + 3)(x - y + 3)$

17. $250x^3 - 128y^3$
$= 2(125x^3 - 64y^3)$
$= 2[(5x)^3 - (4y)^3]$ Difference of cubes
$= 2(5x - 4y)(25x^2 + 20xy + 16y^2)$

19. $8m^3 + m^6 - 20$
$= (m^3)^2 + 8m^3 - 20$
$= (m^3 - 2)(m^3 + 10)$ Trial and error

21. $ac + cd - ab - bd$
$= c(a + d) - b(a + d)$ Factoring by grouping
$= (c - b)(a + d)$

23. $4c^2 - 4cd + d^2$ Trinomial square
$= (2c - d)^2$

25. $-7x^2 + 2x^3 + 4x - 14$
$= 2x^3 - 7x^2 + 4x - 14$
$= x^2(2x - 7) + 2(2x - 7)$ Factoring by grouping
$= (x^2 + 2)(2x - 7)$

27. $2x^3 + 6x^2 - 8x - 24$
$= 2(x^3 + 3x^2 - 4x - 12)$
$= 2[x^2(x + 3) - 4(x + 3)]$ Factoring by grouping
$= 2(x^2 - 4)(x + 3)$ Difference of squares
$= 2(x + 2)(x - 2)(x + 3)$

29. $16x^3 + 54y^3$
$= 2(8x^3 + 27y^3)$
$= 2[(2x)^3 + (3y)^3]$ Sum of cubes
$= 2(2x + 3y)(4x^2 - 6xy + 9y^2)$

31. $36y^2 - 35 + 12y$
$= 36y^2 + 12y - 35$
$= (6y - 5)(6y + 7)$ FOIL or grouping method

33. $a^8 - b^8$ Difference of squares
$= (a^4 + b^4)(a^4 - b^4)$ Difference of squares
$= (a^4 + b^4)(a^2 + b^2)(a^2 - b^2)$ Difference of squares
$= (a^4 + b^4)(a^2 + b^2)(a + b)(a - b)$

35. $a^3b - 16ab^3$
$= ab(a^2 - 16b^2)$ Difference of squares
$= ab(a + 4b)(a - 4b)$

37. $a(b - 2) + c(b - 2)$
$= (b - 2)(a + c)$ Removing a common factor

39. $5x^3 - 5x^2y - 5xy^2 + 5y^3$
$= 5(x^3 - x^2y - xy^2 + y^3)$
$= 5[x^2(x - y) - y^2(x - y)]$ Factoring by grouping
$= 5(x^2 - y^2)(x - y)$
$= 5(x + y)(x - y)(x - y)$ Factoring the difference of squares
$= 5(x + y)(x - y)^2$

41. $42ab + 27a^2b^2 + 8$
$= 27a^2b^2 + 42ab + 8$
$= (9ab + 2)(3ab + 4)$ FOIL or grouping method

43. $8y^4 - 125y$
$= y(8y^3 - 125)$
$= y[(2y)^3 - 5^3]$ Difference of cubes
$= y(2y - 5)(4y^2 + 10y + 25)$

45. $2x - 3y + 4z = 10,$ (1)
$4x + 6y - 4z = -5,$ (2)
$-8x - 9y + 8z = -2$ (3)

We will eliminate z.

$2x - 3y + 4z = 10$ (1)
$\underline{4x + 6y - 4z = -5}$ (2)
$6x + 3y = 5$ (4) Adding

$8x + 12y - 8z = -10$ Multiplying (2) by 2
$\underline{-8x - 9y + 8z = -2}$ (3)
$3y = -12$ (5) Adding

We can solve equation (5) for y.

$3y = -12$
$y = -4$

Substitute -4 for y in equation (4) and solve for x.

$6x + 3(-4) = 5$
$6x - 12 = 5$
$6x = 17$
$x = \frac{17}{6}$

Finally, substitute $\frac{17}{6}$ for x and -4 for y in one of the original equations and solve for z. We choose equation (1).

45. (continued)

$$2\left(\frac{17}{6}\right) - 3(-4) + 4z = 10$$

$$\frac{17}{3} + 12 + 4z = 10$$

$$4z = -\frac{23}{3}$$

$$z = -\frac{23}{12}$$

The triple $\left(\frac{17}{6}, -4, -\frac{23}{12}\right)$ checks. It is the solution.

47. $30y^4 - 97xy^2 + 60x^2$

$= (5y^2 - 12x)(6y^2 - 5x)$ FOIL or grouping method

49. $5x^3 - \frac{5}{27}$

$= 5\left(x^3 - \frac{1}{27}\right)$

$= 5\left[x^3 - \left(\frac{1}{3}\right)^3\right]$ Difference of cubes

$= 5\left(x - \frac{1}{3}\right)\left(x^2 + \frac{1}{3}x + \frac{1}{9}\right)$

51. $(x - p)^2 - p^2$ Difference of squares

$= (x - p + p)(x - p - p)$

$= x(x - 2p)$

53. $(y - 1)^4 - (y - 1)^2$

$= (y - 1)^2[(y - 1)^2 - 1]$ Removing the common factor

$= (y - 1)^2[(y - 1) + 1][(y - 1) - 1]$

Factoring the difference of squares

$= (y - 1)^2(y)(y - 2)$, or $y(y - 1)^2(y - 2)$

55. $x^6 - 2x^5 + x^4 - x^2 + 2x - 1$

$= x^4(x^2 - 2x + 1) - 1(x^2 - 2x + 1)$ Factoring by grouping

$= (x^2 - 2x + 1)(x^4 - 1)$ Trinomial square, difference of squares

$= (x - 1)^2(x^2 + 1)(x^2 - 1)$ Difference of squares

$= (x - 1)^2(x^2 + 1)(x + 1)(x - 1)$

$= (x - 1)^3(x^2 + 1)(x + 1)$

57. $4x^2 + 4xy + y^2 - r^2 + 6rs - 9s^2$

$= (4x^2 + 4xy + y^2) - 1(r^2 - 6rs + 9s^2)$ Grouping

$= (2x + y)^2 - (r - 3s)^2$ Difference of squares

$= [(2x + y) + (r - 3s)][(2x + y) - (r - 3s)]$

$= (2x + y + r - 3s)(2x + y - r + 3s)$

59. $c^{2w+1} + 2c^{w+1} + c$

$= c^{2w}\cdot c + 2c^{w}\cdot c + c$

$= c(c^{2w} + 2c^{w} + 1)$

$= c[(c^{w})^2 + 2(c^{w}) + 1]$ Trinomial square

$= c(c^{w} + 1)^2$

61. $y^9 - y$

$= y(y^8 - 1)$ Difference of squares

$= y(y^4 + 1)(y^4 - 1)$ Difference of squares

$= y(y^4 + 1)(y^2 + 1)(y^2 - 1)$ Difference of squares

$= y(y^4 + 1)(y^2 + 1)(y + 1)(y - 1)$

63. $3(x + 1)^2 + 9(x + 1) - 12$

$= 3[(x + 1)^2 + 3(x + 1) - 4]$

$= 3[(x + 1) + 4][(x + 1) - 1]$ Factor $u^2 + 3u - 4$ where $u = x + 1$

$= 3(x + 5)(x)$, or

$3x(x + 5)$

Exercise Set 8.8

1. $x^2 + 3x = 28$

$x^2 + 3x - 28 = 0$ Getting 0 on one side

$(x + 7)(x - 4) = 0$ Factoring

$x + 7 = 0$ or $x - 4 = 0$ Principle of zero products.

$x = -7$ or $x = 4$

The solutions are -7 and 4.

3. $y^2 + 16 = 8y$

$y^2 - 8y + 16 = 0$ Getting 0 on one side

$(y - 4)(y - 4) = 0$ Factoring

$y - 4 = 0$ or $y - 4 = 0$ Principle of zero products

$y = 4$ or $y = 4$ We have a repeated root.

The solution is 4.

5. $x^2 - 12x + 36 = 0$

$(x - 6)(x - 6) = 0$ Factoring

$x - 6 = 0$ or $x - 6 = 0$ Principle of zero products

$x = 6$ or $x = 6$ We have a repeated root.

The solution is 6.

7. $9x + x^2 + 20 = 0$

$x^2 + 9x + 20 = 0$ Changing order

$(x + 5)(x + 4) = 0$ Factoring

$x + 5 = 0$ or $x + 4 = 0$ Principle of zero products

$x = -5$ or $x = -4$

The solutions are -5 and -4.

9. $x^2 + 8x = 0$

$x(x + 8) = 0$ Factoring

$x = 0$ or $x + 8 = 0$ Principle of zero products

$x = 0$ or $x = -8$

The solutions are 0 and -8.

11. $x^2 - 9 = 0$
$(x + 3)(x - 3) = 0$
$x + 3 = 0$ or $x - 3 = 0$
$x = -3$ or $x = 3$
The solutions are -3 and 3.

13. $z^2 = 36$
$z^2 - 36 = 0$
$(z + 6)(z - 6) = 0$
$z + 6 = 0$ or $z - 6 = 0$
$z = -6$ or $z = 6$
The solutions are -6 and 6.

15. $y^2 + 2y = 63$
$y^2 + 2y - 63 = 0$
$(y + 9)(y - 7) = 0$
$y + 9 = 0$ or $y - 7 = 0$
$y = -9$ or $y = 7$
The solutions are -9 and 7.

17. $32 + 4x - x^2 = 0$
$0 = x^2 - 4x - 32$
$0 = (x - 8)(x + 4)$
$x - 8 = 0$ or $x + 4 = 0$
$x = 8$ or $x = -4$
The solutions are 8 and -4.

19. $3b^2 + 8b + 4 = 0$
$(3b + 2)(b + 2) = 0$
$3b + 2 = 0$ or $b + 2 = 0$
$3b = -2$ or $b = -2$
$b = -\frac{2}{3}$ or $b = -2$
The solutions are $-\frac{2}{3}$ and -2.

21. $8y^2 - 10y + 3 = 0$
$(4y - 3)(2y - 1) = 0$
$4y - 3 = 0$ or $2y - 1 = 0$
$4y = 3$ or $2y = 1$
$y = \frac{3}{4}$ or $y = \frac{1}{2}$
The solutions are $\frac{3}{4}$ and $\frac{1}{2}$.

23. $6z - z^2 = 0$
$0 = z^2 - 6z$
$0 = z(z - 6)$
$z = 0$ or $z - 6 = 0$
$z = 0$ or $z = 6$
The solutions are 0 and 6.

25. $12z^2 + z = 6$
$12z^2 + z - 6 = 0$
$(4z + 3)(3z - 2) = 0$
$4z + 3 = 0$ or $3z - 2 = 0$
$4z = -3$ or $3z = 2$
$z = -\frac{3}{4}$ or $z = \frac{2}{3}$
The solutions are $-\frac{3}{4}$ and $\frac{2}{3}$.

27. $5x^2 - 20 = 0$
$5(x^2 - 4) = 0$
$5(x + 2)(x - 2) = 0$
$x + 2 = 0$ or $x - 2 = 0$
$x = -2$ or $x = 2$
The solutions are -2 and 2.

29. $21r^2 + r - 10 = 0$
$(3r - 2)(7r + 5) = 0$
$3r - 2 = 0$ or $7r + 5 = 0$
$3r = 2$ or $7r = -5$
$r = \frac{2}{3}$ or $r = -\frac{5}{7}$
The solutions are $\frac{2}{3}$ and $-\frac{5}{7}$.

31. $15y^2 = 3y$
$15y^2 - 3y = 0$
$3y(5y - 1) = 0$
$3y = 0$ or $5y - 1 = 0$
$y = 0$ or $5y = 1$
$y = 0$ or $y = \frac{1}{5}$
The solutions are 0 and $\frac{1}{5}$.

33. $14 = x(x - 5)$
$14 = x^2 - 5x$
$0 = x^2 - 5x - 14$ Getting 0 on one side
$0 = (x - 7)(x + 2)$
$x - 7 = 0$ or $x + 2 = 0$
$x = 7$ or $x = -2$
The solutions are 7 and -2.

35. $2x^3 - 2x^2 = 12x$
$2x^3 - 2x^2 - 12x = 0$
$2x(x^2 - x - 6) = 0$
$2x(x - 3)(x + 2) = 0$
$2x = 0$ or $x - 3 = 0$ or $x + 2 = 0$
$x = 0$ or $x = 3$ or $x = -2$
The solutions are 0, 3, and -2.

37.
$$2x^3 = 128x$$
$$2x^3 - 128x = 0$$
$$2x(x^2 - 64) = 0$$
$$2x(x + 8)(x - 8) = 0$$
$$x = 0 \quad \text{or} \quad x + 8 = 0 \quad \text{or} \quad x - 8 = 0$$
$$x = 0 \quad \text{or} \quad x = -8 \quad \text{or} \quad x = 8$$
The solutions are 0, -8, and 8.

39. Familiarize. Let x represent the number.

Translate.

4 times	the square of a number	is	45	more than	8 times	the number.
4 $\cdot$	x^2	=	45	+	8 $\cdot$	x

Solve. We solve the equation:
$$4x^2 = 45 + 8x$$
$$4x^2 - 8x - 45 = 0$$
$$(2x - 9)(2x + 5) = 0$$
$$2x - 9 = 0 \quad \text{or} \quad 2x + 5 = 0$$
$$2x = 9 \quad \text{or} \quad 2x = -5$$
$$x = \frac{9}{2} \quad \text{or} \quad x = -\frac{5}{2}$$

Check. For $\frac{9}{2}$: Four times the square of $\frac{9}{2}$, or $4 \cdot \frac{81}{4}$, or 81, is 45 more than 8 times $\frac{9}{2}$, or $8 \cdot \frac{9}{2}$, or 36.

For $-\frac{5}{2}$: Four times the square of $-\frac{5}{2}$, or $4 \cdot \frac{25}{4}$, or 25, is 45 more than 8 times $-\frac{5}{2}$, or $8\left(-\frac{5}{2}\right)$, or -20.

Both numbers check.

State. The number is $\frac{9}{2}$ or $-\frac{5}{2}$.

41. Familiarize. We let w represent the width and w + 4 represent the length. We make a drawing and label it.

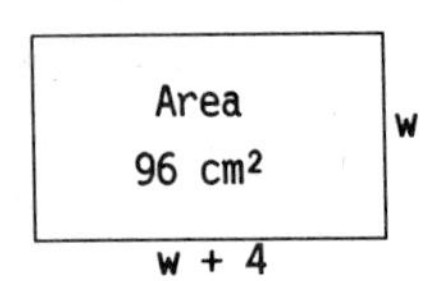

Recall that the formula for the area of a rectangle is Area = length × width.

Translate.

The area	is	96 cm².
$w(w + 4)$	=	96

41. (continued)

Solve. We solve the equation.
$$w(w + 4) = 96$$
$$w^2 + 4w = 96$$
$$w^2 + 4w - 96 = 0$$
$$(w + 12)(w - 8) = 0$$
$$w + 12 = 0 \quad \text{or} \quad w - 8 = 0$$
$$w = -12 \quad \text{or} \quad w = 8$$

Check. The number -12 is not a solution, because width cannot be negative. If the width is 8 cm, then the length is 8 + 4, or 12 cm, and the area is $8 \cdot 12$, or 96 cm². This value checks.

State. The length is 12 cm, and the width is 8 cm.

43. Familiarize. Let x represent the first positive odd integer. Then x + 2 represents the next positive odd integer.

Translate.

Square of first integer	plus	Square of second integer	is	202.
x^2	+	$(x + 2)^2$	=	202

Solve. We solve the equation:
$$x^2 + (x + 2)^2 = 202$$
$$x^2 + x^2 + 4x + 4 = 202$$
$$2x^2 + 4x - 198 = 0$$
$$2(x^2 + 2x - 99) = 0$$
$$2(x + 11)(x - 9) = 0$$
$$x + 11 = 0 \quad \text{or} \quad x - 9 = 0$$
$$x = -11 \quad \text{or} \quad x = 9$$

Check. We only check 9 since the problem asks for consecutive positive odd integers. If x = 9, then x + 2 = 11, and 9 and 11 are consecutive positive odd integers. The sum of the squares of 9 and 11 is 81 + 121, or 202. The numbers check.

State. The integers are 9 and 11.

45. Familiarize. We make a drawing and label it. We let x represent the length of a side of the original square.

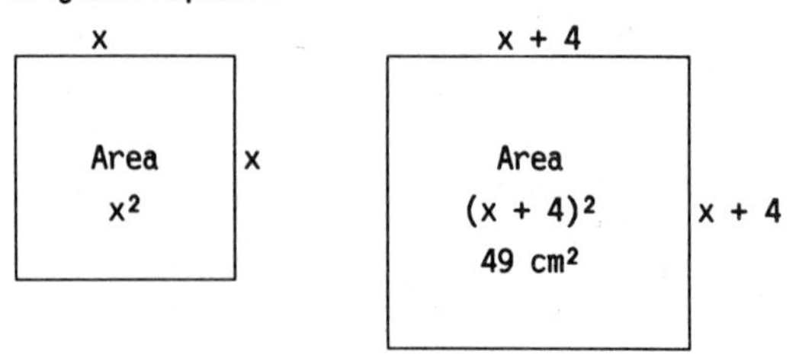

Translate.

Area of new square	is	49 cm².
$(x + 4)^2$	=	49

45. (continued)

Solve. We solve the equation:

$$(x + 4)^2 = 49$$
$$x^2 + 8x + 16 = 49$$
$$x^2 + 8x - 33 = 0$$
$$(x + 11)(x - 3) = 0$$
$$x + 11 = 0 \quad \text{or} \quad x - 3 = 0$$
$$x = -11 \quad \text{or} \quad x = 3$$

Check. We only check 3 since the length of a side cannot be negative. If we increase the length by 4, the new length is 3 + 4, or 7 cm. The new area is 7·7, or 49 cm^2. We have a solution.

State. The length of a side of the original square is 3 cm.

47. Familiarize. Let h represent the height of the triangle. Then h + 9 represents the base. Recall that the formula for the area of a triangle is $A = \frac{1}{2} \times \text{base} \times \text{height}$.

Translate.

Area is 56 cm^2.

$$\frac{1}{2}(h + 9)h = 56$$

Solve. We solve the equation:

$$\frac{1}{2}(h + 9)h = 56$$
$$(h + 9)h = 112 \quad \text{Multiplying by 2}$$
$$h^2 + 9h = 112$$
$$h^2 + 9h - 112 = 0$$
$$(h + 16)(h - 7) = 0$$
$$h + 16 = 0 \quad \text{or} \quad h - 7 = 0$$
$$h = -16 \quad \text{or} \quad h = 7$$

Check. We only check 7, since height cannot be negative. If the height is 7 cm, the base is 7 + 9, or 16 cm and the area is $\frac{1}{2} \cdot 16 \cdot 7$, or 56 cm^2. We have a solution.

State. The height is 7 cm, and the base is 16 cm.

49. Familiarize. Let x represent the length of a side of the square. Then the perimeter is x + x + x + x, or 4x, and the area is x·x, or x^2.

Translate.

Perimeter is 4 more than the area.

$$4x = 4 + x^2$$

Solve. We solve the equation:

$$4x = 4 + x^2$$
$$0 = x^2 - 4x + 4$$
$$0 = (x - 2)(x - 2)$$
$$x - 2 = 0 \quad \text{or} \quad x - 2 = 0$$
$$x = 2 \quad \text{or} \quad x = 2$$

49. (continued)

Check. If the length of a side is 2, the perimeter is 4·2 or 8, the area is 2·2 or 4, and 8 is four more than 4. The value checks.

State. The length of a side is 2.

51. Familiarize. Let x represent the first integer, x + 1 the second, and x + 2 the third.

Translate.

First · Third − Second = 1 + 10·Third

$$x \cdot (x + 2) - (x + 1) = 1 + 10(x + 2)$$

Solve. We solve the equation:

$$x(x + 2) - (x + 1) = 1 + 10(x + 2)$$
$$x^2 + 2x - x - 1 = 1 + 10x + 20$$
$$x^2 - 9x - 22 = 0$$
$$(x - 11)(x + 2) = 0$$
$$x - 11 = 0 \quad \text{or} \quad x + 2 = 0$$
$$x = 11 \quad \text{or} \quad x = -2$$

Check. If x = 11, the consecutive integers are 11, 12, and 13.

First · Third − Second	1 + 10·Third
11·13 − 12	1 + 10·13
143 − 12	1 + 130
131	131

If x = −2, the consecutive integers are −2, −1, and 0.

First · Third − Second	1 + 10·Third
−2·0 − (−1)	1 + 10·0
0 + 1	1 + 0
1	1

Both sets of integers check.

State. The three consecutive integers can be 11, 12, and 13 or −2, −1, and 0.

53. Familiarize. Let x represent the other leg of the triangle. Then x + 1 represents the hypotenuse.

Translate. We use the Pythagorean theorem.

$$a^2 + b^2 = c^2$$
$$9^2 + x^2 = (x + 1)^2$$

Solve. We solve the equation:

$$9^2 + x^2 = (x + 1)^2$$
$$81 + x^2 = x^2 + 2x + 1$$
$$80 = 2x$$
$$40 = x$$

Check. When x = 40, then x + 1 = 41, and $9^2 + 40^2 = 1681 = 41^2$. The numbers check.

State. The other sides have lengths of 40 m and 41 m.

55. Familiarize. We will use the function $h(t) = -16t^2 + 80t + 224$.

Translate.

Height	is	0 ft.
$-16t^2 + 80t + 224$	=	0

Solve. We solve the equation:

$$-16(t^2 - 5t - 14) = 0$$
$$-16(t - 7)(t + 2) = 0$$
$$t - 7 = 0 \quad \text{or} \quad t + 2 = 0$$
$$t = 7 \quad \text{or} \quad t = -2$$

Check. The number -2 is not a solution, since time cannot be negative. When $t = 7$, $h(7) = -16\cdot 7^2 + 80\cdot 7 + 224 = 0$. We have a solution.

State. The object reaches the ground after 7 sec.

57.
$$2x - 14y + 10z = 100, \quad (1)$$
$$5y - 8z = 80, \quad (2)$$
$$4z = 64 \quad (3)$$

Solve equation (3) for z:

$$4z = 64$$
$$z = 16$$

Substitute 16 for z in equation (2) and solve for y:

$$5y - 8(16) = 80$$
$$5y - 128 = 80$$
$$5y = 208$$
$$y = \frac{208}{5}$$

Substitute $\frac{208}{5}$ for y and 16 for z in equation (1) and solve for x:

$$2x - 14\left(\frac{208}{5}\right) + 10(16) = 100$$
$$2x - \frac{2912}{5} + 160 = 100$$
$$2x - \frac{2112}{5} = 100$$
$$2x = \frac{2612}{5}$$
$$x = \frac{1306}{5}$$

The solution is $\left(\frac{1306}{5}, \frac{208}{5}, 16\right)$.

59.
$$x(x + 8) = 16(x - 1)$$
$$x^2 + 8x = 16x - 16$$
$$x^2 - 8x + 16 = 0$$
$$(x - 4)(x - 4) = 0$$
$$x - 4 = 0 \quad \text{or} \quad x - 4 = 0$$
$$x = 4 \quad \text{or} \quad x = 4$$

The solution is 4.

61.
$$(a - 5)^2 = 36$$
$$a^2 - 10a + 25 = 36$$
$$a^2 - 10a - 11 = 0$$
$$(a - 11)(a + 1) = 0$$
$$a - 11 = 0 \quad \text{or} \quad a + 1 = 0$$
$$a = 11 \quad \text{or} \quad a = -1$$

The solutions are 11 and -1.

63.
$$(3x^2 - 7x - 20)(x - 5) = 0$$
$$(3x + 5)(x - 4)(x - 5) = 0$$
$$3x + 5 = 0 \quad \text{or} \quad x - 4 = 0 \quad \text{or} \quad x - 5 = 0$$
$$3x = -5 \quad \text{or} \quad x = 4 \quad \text{or} \quad x = 5$$
$$x = -\frac{5}{3} \quad \text{or} \quad x = 4 \quad \text{or} \quad x = 5$$

The solutions are $-\frac{5}{3}$, 4, and 5.

65.
$$2x^3 + 6x^2 = 8x + 24$$
$$2x^3 + 6x^2 - 8x - 24 = 0$$
$$2(x^3 + 3x^2 - 4x - 12) = 0$$
$$2[x^2(x + 3) - 4(x + 3)] = 0$$
$$2(x + 3)(x^2 - 4) = 0$$
$$2(x + 3)(x + 2)(x - 2) = 0$$
$$x + 3 = 0 \quad \text{or} \quad x + 2 = 0 \quad \text{or} \quad x - 2 = 0$$
$$x = -3 \quad \text{or} \quad x = -2 \quad \text{or} \quad x = 2$$

The solutions are -3, -2, and 2.

67. Familiarize. Let x represent one of the numbers. Then $17 - x$ represents the other number.

Translate.

The sum of the squares of the numbers	is	205.
$x^2 + (17 - x)^2$	=	205

Solve. We solve the equation.

$$x^2 + (17 - x)^2 = 205$$
$$x^2 + 289 - 34x + x^2 = 205$$
$$2x^2 - 34x + 289 = 205$$
$$2x^2 - 34x + 84 = 0$$
$$2(x^2 - 17x + 42) = 0$$
$$2(x - 3)(x - 14) = 0$$
$$x - 3 = 0 \quad \text{or} \quad x - 14 = 0$$
$$x = 3 \quad \text{or} \quad x = 14$$

Check. If one number is 3, then the other is $17 - 3$, or 14. If one number is 14, then the other is $17 - 14$, or 3. In either case, we have the numbers 3 and 14. Their sum is 17. Also, $3^2 + 14^2 = 9 + 196 = 205$. The numbers check.

State. The numbers are 3 and 14.

69. Familiarize. Let x represent the length of the hypotenuse. Then $x - 3$ and $x - 6$ represent the lengths of the legs of the triangle. We will first find the lengths of the legs (that is, the base and height of the triangle). Then we will find the area of the triangle using the formula Area $= \frac{1}{2} \times$ base $\times$ height.

Translate. We use the Pythagorean equation.

$$a^2 + b^2 = c^2$$
$$(x - 3)^2 + (x - 6)^2 = x^2$$

Solve. We solve the equation.

$$(x - 3)^2 + (x - 6)^2 = x^2$$
$$x^2 - 6x + 9 + x^2 - 12x + 36 = x^2$$
$$2x^2 - 18x + 45 = x^2$$
$$x^2 - 18x + 45 = 0$$
$$(x - 15)(x - 3) = 0$$

$$x - 15 = 0 \quad \text{or} \quad x - 3 = 0$$
$$x = 15 \quad \text{or} \quad x = 3$$

Check. If $x = 3$, then $x - 3 = 0$ and $x - 6 = -3$. Thus, 3 cannot be a solution since the lengths of the legs must be positive. If $x = 15$, $x - 3 = 12$ and $x - 6 = 9$, and $12^2 + 9^2 = 225 = 15^2$. These numbers check. We find the area of the triangle:

$$A = \frac{1}{2}bh = \frac{1}{2}(12)(9) = 54$$

State. The area of the triangle is 54 cm^2.

Exercise Set 8.9

1. $\dfrac{30x^8 - 15x^6 + 40x^4}{5x^4}$

$$= \frac{30x^8}{5x^4} - \frac{15x^6}{5x^4} + \frac{40x^4}{5x^4}$$
$$= 6x^4 - 3x^2 + 8$$

3. $\dfrac{9x^3y^4 - 18x^2y^3 + 27xy^2}{9xy}$

$$= \frac{9x^3y^4}{9xy} - \frac{18x^2y^3}{9xy} + \frac{27xy^2}{9xy}$$
$$= x^2y^3 - 2xy^2 + 3y$$

5.
$$\begin{array}{r} x + 7 \\ x + 3 \overline{) x^2 + 10x + 21} \\ \underline{x^2 + 3x} \\ 7x + 21 \\ \underline{7x + 21} \\ 0 \end{array}$$
$(x^2 + 10x) - (x^2 + 3x) = 7x$

The answer is $x + 7$.

7.
$$\begin{array}{r} a - 12 \\ a + 4 \overline{) a^2 - 8a - 16} \\ \underline{a^2 + 4a} \\ -12a - 16 \\ \underline{-12a - 48} \\ 32 \end{array}$$
$(a^2 - 8a) - (a^2 + 4a) = -12a$

$(-12a - 16) - (-12a - 48) = 32$

The answer is $a - 12$, R 32, or $a - 12 + \frac{32}{a + 4}$.

9.
$$\begin{array}{r} x + 2 \\ x + 5 \overline{) x^2 + 7x + 14} \\ \underline{x^2 + 5x} \\ 2x + 14 \\ \underline{2x + 10} \\ 4 \end{array}$$
$(x^2 + 7x) - (x^2 + 5x) = 2x$

$(2x + 14) - (2x + 10) = 4$

The answer is $x + 2$, R 4, or $x + 2 + \frac{4}{x + 5}$.

11.
$$\begin{array}{r} 2y^2 - y + 2 \\ 2y + 4 \overline{) 4y^3 + 6y^2 + 0y + 14} \\ \underline{4y^3 + 8y^2} \\ -2y^2 + 0y \\ \underline{-2y^2 - 4y} \\ 4y + 14 \\ \underline{4y + 8} \\ 6 \end{array}$$

The answer is $2y^2 - y + 2$, R 6, or $2y^2 - y + 2 + \frac{6}{2y + 4}$.

13.
$$\begin{array}{r} 2y^2 + 2y - 1 \\ 5y - 2 \overline{) 10y^3 + 6y^2 - 9y + 10} \\ \underline{10y^3 - 4y^2} \\ 10y^2 - 9y \\ \underline{10y^2 - 4y} \\ -5y + 10 \\ \underline{-5y + 2} \\ 8 \end{array}$$

The answer is $2y^2 + 2y - 1$, R 8, or $2y^2 + 2y - 1 + \frac{8}{5y - 2}$.

15.
$$\begin{array}{r} 2x^2 - x - 9 \\ x^2 + 2 \overline{) 2x^4 - x^3 - 5x^2 + x - 6} \\ \underline{2x^4 + 4x^2} \\ -x^3 - 9x^2 + x \\ \underline{-x^3 - 2x} \\ -9x^2 + 3x - 6 \\ \underline{-9x^2 - 18} \\ 3x + 12 \end{array}$$

The answer is $2x^2 - x - 9$, R $(3x + 12)$, or $2x^2 - x - 9 + \frac{3x + 12}{x^2 + 2}$.

17. $$\begin{array}{rl} & 2x^3 + 5x^2 + 17x + 51 \\ x^2 - 3x\,\big) & 2x^5 - x^4 + 2x^3 + 0x^2 - x \\ & \underline{2x^5 - 6x^4} \\ & 5x^4 + 2x^3 \\ & \underline{5x^4 - 15x^3} \\ & 17x^3 + 0x^2 \\ & \underline{17x^3 - 51x^2} \\ & 51x^2 - x \\ & \underline{51x^2 - 153x} \\ & 152x \end{array}$$

The answer is $2x^3 + 5x^2 + 17x + 51$, R $152x$, or $2x^3 + 5x^2 + 17x + 51 + \frac{152x}{x^2 - 3x}$.

19. $$\begin{array}{rl} & x^2 - x + 1 \\ x - 1\,\big) & x^3 - 2x^2 + 2x - 5 \\ & \underline{x^3 - x^2} \\ & -x^2 + 2x \\ & \underline{-x^2 + x} \\ & x - 5 \\ & \underline{x - 1} \\ & -4 \end{array}$$

The answer is $x^2 - x + 1$, R -4, or $x^2 - x + 1 + \frac{-4}{x - 1}$.

21. $$\begin{array}{rl} & a + 7 \\ a + 4\,\big) & a^2 + 11a - 19 \\ & \underline{a^2 + 4a} \\ & 7a - 19 \\ & \underline{7a + 28} \\ & -47 \end{array}$$

The answer is $a + 7$, R -47, or $a + 7 + \frac{-47}{a + 4}$.

23. $$\begin{array}{rl} & x^2 - 5x - 23 \\ x - 2\,\big) & x^3 - 7x^2 - 13x + 3 \\ & \underline{x^3 - 2x^2} \\ & -5x^2 - 13x \\ & \underline{-5x^2 + 10x} \\ & -23x + 3 \\ & \underline{-23x + 46} \\ & -43 \end{array}$$

The answer is $x^2 - 5x - 23$, R -43, or $x^2 - 5x - 23 + \frac{-43}{x - 2}$.

25. $$\begin{array}{rl} & 3x^2 - 2x + 2 \\ x + 3\,\big) & 3x^3 + 7x^2 - 4x + 3 \\ & \underline{3x^3 + 9x^2} \\ & -2x^2 - 4x \\ & \underline{-2x^2 - 6x} \\ & 2x + 3 \\ & \underline{2x + 6} \\ & -3 \end{array}$$

The answer is $3x^2 - 2x + 2$, R -3, or $3x^2 - 2x + 2 + \frac{-3}{x + 3}$.

27. $$\begin{array}{rl} & y^2 + 2y + 1 \\ y - 2\,\big) & y^3 + 0y^2 - 3y + 10 \\ & \underline{y^3 - 2y^2} \\ & 2y^2 - 3y \\ & \underline{2y^2 - 4y} \\ & y + 10 \\ & \underline{y - 2} \\ & 12 \end{array}$$

The answer is $y^2 + 2y + 1$, R 12, or $y^2 + 2y + 1 + \frac{12}{y - 2}$.

29. $$\begin{array}{rl} & 3x^3 + 9x^2 + 2x + 6 \\ x - 3\,\big) & 3x^4 + 0x^3 - 25x^2 + 0x - 18 \\ & \underline{3x^4 - 9x^3} \\ & 9x^3 - 25x^2 \\ & \underline{9x^3 - 27x^2} \\ & 2x^2 + 0x \\ & \underline{2x^2 - 6x} \\ & 6x - 18 \\ & \underline{6x - 18} \\ & 0 \end{array}$$

The answer is $3x^3 + 9x^2 + 2x + 6$.

31. $$\begin{array}{rl} & x^2 + 3x + 9 \\ x - 3\,\big) & x^3 + 0x^2 + 0x - 27 \\ & \underline{x^3 - 3x^2} \\ & 3x^2 + 0x \\ & \underline{3x^2 - 9x} \\ & 9x - 27 \\ & \underline{9x - 27} \\ & 0 \end{array}$$

The answer is $x^2 + 3x + 9$.

33. $$\begin{array}{rl} & y^3 + 2y^2 + 4y + 8 \\ y - 2\,\big) & y^4 + 0y^3 + 0y^2 + 0y - 16 \\ & \underline{y^4 - 2y^3} \\ & 2y^3 + 0y^2 \\ & \underline{2y^3 - 4y^2} \\ & 4y^2 + 0y \\ & \underline{4y^2 - 8y} \\ & 8y - 16 \\ & \underline{8y - 16} \\ & 0 \end{array}$$

The answer is $y^3 + 2y^2 + 4y + 8$.

35. Graph: $2x - 3y = 6$

To find the x-intercept we cover up the y-term and consider the rest of the equation. We have $2x = 6$, or $x = 3$. The x-intercept is (3,0). To find the y-intercept we cover up the x-term and consider the rest of the equation. We have $-3y = 6$, or $y = -2$. The y-intercept is (0,-2). We plot these points and draw the graph.

35. (continued)

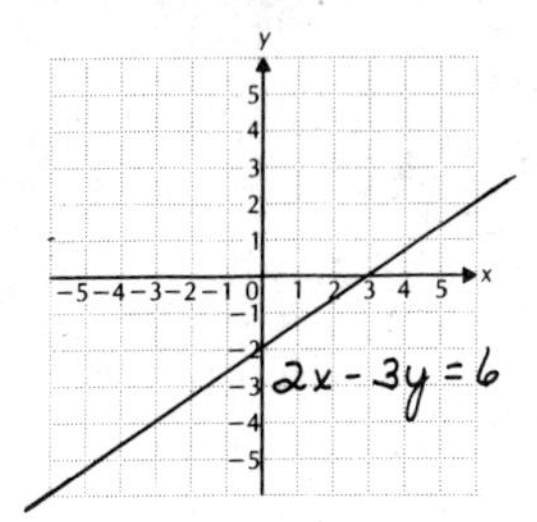

We find a third point as a check. We let $x = 6$ and solve for y.

$$2 \cdot 6 - 3y = 6$$
$$12 - 3y = 6$$
$$-3y = -6$$
$$y = 2$$

We plot the point (6,2) and see that is on the line.

37. $$35t^2 + 18t = 8$$
$$35t^2 + 18t - 8 = 0$$
$$(7t - 2)(5t + 4) = 0$$

$$7t - 2 = 0 \quad \text{or} \quad 5t + 4 = 0$$
$$7t = 2 \quad \text{or} \quad 5t = -4$$
$$t = \frac{2}{7} \quad \text{or} \quad t = -\frac{4}{5}$$

The solutions are $\frac{2}{7}$ and $-\frac{4}{5}$.

39. $x^2 + x + 1 \overline{)2x^6 + 0x^5 + 5x^4 - x^3 + 0x^2 + 0x + 1}$, quotient $2x^4 - 2x^3 + 5x^2 - 4x - 1$

$$\underline{2x^6 + 2x^5 + 2x^4}$$
$$-2x^5 + 3x^4 - x^3$$
$$\underline{-2x^5 - 2x^4 - 2x^3}$$
$$5x^4 + x^3 + 0x^2$$
$$\underline{5x^4 + 5x^3 + 5x^2}$$
$$-4x^3 - 5x^2 + 0x$$
$$\underline{-4x^3 - 4x^2 - 4x}$$
$$-x^2 + 4x + 1$$
$$\underline{-x^2 - x - 1}$$
$$5x + 2$$

The answer is $2x^4 - 2x^3 + 5x^2 - 4x - 1$, $R(5x + 2)$, or $2x^4 - 2x^3 + 5x^2 - 4x - 1 + \frac{5x + 2}{x^2 + x + 1}$.

41. $x^2 - xy + y^2 \overline{)x^4 - x^3y + x^2y^2 + 2x^2y - 2xy^2 + 2y^3}$, quotient $x^2 + 2y$

$$\underline{x^4 - x^3y + x^2y^2}$$
$$0 + 2x^2y - 2xy^2 + 2y^3$$
$$\underline{2x^2y - 2xy^2 + 2y^3}$$
$$0$$

The answer is $x^2 + 2y$.

43. $a + b \overline{)a^7 \qquad\qquad + b^7}$, quotient $a^6 - a^5b + a^4b^2 - a^3b^3 + a^2b^4 - ab^5 + b^6$

$$\underline{a^7 + a^6b}$$
$$-a^6b$$
$$\underline{-a^6b - a^5b^2}$$
$$a^5b^2$$
$$\underline{a^5b^2 + a^4b^3}$$
$$-a^4b^3$$
$$\underline{-a^4b^3 - a^3b^4}$$
$$a^3b^4$$
$$\underline{a^3b^4 + a^2b^5}$$
$$-a^2b^5$$
$$\underline{-a^2b^5 - ab^6}$$
$$ab^6 + b^7$$
$$\underline{ab^6 + b^7}$$
$$0$$

The answer is $a^6 - a^5b + a^4b^2 - a^3b^3 + a^2b^4 - ab^5 + b^6$.

Exercise Set 9.1

1. $\dfrac{3a^2 - 16}{5a - 20}$

The meaningful replacements are all real numbers for which the denominator is not 0. We set the denominator equal to 0 and solve:

$5a - 20 = 0$

$5a = 20$

$a = 4$

The meaningful replacements are all real numbers except 4.

3. $\dfrac{y^2 - y - 2}{y^2 - 15y + 54}$

To find the meaningful replacements we set the denominator equal to 0 and solve:

$y^2 - 15y + 54 = 0$

$(y - 6)(y - 9) = 0$ Factoring

$y - 6 = 0$ or $y - 9 = 0$ Principle of zero products

$y = 6$ or $y = 9$

The meaningful replacements are all real numbers except 6 and 9.

5. $\dfrac{3x}{3x} \cdot \dfrac{x + 1}{x + 3} = \dfrac{3x(x + 1)}{3x(x + 3)}$ Multiplying numerators and multiplying denominators

7. $\dfrac{t - 3}{t + 2} \cdot \dfrac{t + 3}{t + 3} = \dfrac{(t - 3)(t + 3)}{(t + 2)(t + 3)}$ Multiplying numerators and multiplying denominators

9. $\dfrac{9y^2}{15y} = \dfrac{3y \cdot 3y}{3y \cdot 5}$ Factoring the numerator and the denominator

$= \dfrac{3y}{3y} \cdot \dfrac{3y}{5}$ Factoring the rational expression

$= 1 \cdot \dfrac{3y}{5}$ $\qquad \dfrac{3y}{3y} = 1$

$= \dfrac{3y}{5}$ Removing a factor of 1

11. $\dfrac{16p^3}{24p^7} = \dfrac{8p^3 \cdot 2}{8p^3 \cdot 3p^4}$

$= \dfrac{8p^3}{8p^3} \cdot \dfrac{2}{3p^4}$

$= 1 \cdot \dfrac{2}{3p^4}$ $\qquad \dfrac{8p^3}{8p^3} = 1$

$= \dfrac{2}{3p^4}$ Removing a factor of 1

13. $\dfrac{2a - 6}{2} = \dfrac{2(a - 3)}{2 \cdot 1} = \dfrac{2}{2} \cdot \dfrac{a - 3}{1} = a - 3$

16. $\dfrac{6x + 9}{24} = \dfrac{3(2x + 3)}{3 \cdot 8} = \dfrac{3}{3} \cdot \dfrac{2x + 3}{8} = \dfrac{2x + 3}{8}$

17. $\dfrac{4y - 12}{4y + 12} = \dfrac{4(y - 3)}{4(y + 3)} = \dfrac{4}{4} \cdot \dfrac{y - 3}{y + 3} = \dfrac{y - 3}{y + 3}$

19. $\dfrac{t^2 - 16}{t^2 - 8t + 16} = \dfrac{(t + 4)(t - 4)}{(t - 4)(t - 4)} = \dfrac{t + 4}{t - 4} \cdot \dfrac{t - 4}{t - 4} =$

$= \dfrac{t + 4}{t - 4}$

21. $\dfrac{x^2 - 9x + 8}{x^2 + 3x - 4} = \dfrac{(x - 8)(x - 1)}{(x + 4)(x - 1)} = \dfrac{x - 8}{x + 4} \cdot \dfrac{x - 1}{x - 1} =$

$\dfrac{x - 8}{x + 4}$

23. $\dfrac{a^3 - b^3}{a^2 - b^2} = \dfrac{(a - b)(a^2 + ab + b^2)}{(a - b)(a + b)}$

$= \dfrac{a - b}{a - b} \cdot \dfrac{a^2 + ab + b^2}{a + b}$

$= \dfrac{a^2 + ab + b^2}{a + b}$

25. $\dfrac{x^4}{3x + 6} \cdot \dfrac{5x + 10}{5x^7}$

$= \dfrac{x^4(5x + 10)}{(3x + 6)(5x^7)}$ Multiplying the numerators and the denominators

$= \dfrac{x^4(5)(x + 2)}{3(x + 2)(5)(x^4)(x^3)}$ Factoring the numerators and the denominators

$= \dfrac{\cancel{x^4}\cancel{(5)}\cancel{(x + 2)}(1)}{3\cancel{(x + 2)}\cancel{(5)}\cancel{(x^4)}(x^3)}$ Removing a factor of 1: $\dfrac{(x^4)(5)(x + 2)}{(x + 2)(5)(x^4)} = 1$

$= \dfrac{1}{3x^3}$ Simplifying

27. $\dfrac{x^2 - 16}{x^2} \cdot \dfrac{x^2 - 4x}{x^2 - x - 12}$

$= \dfrac{(x^2 - 16)(x^2 - 4x)}{x^2(x^2 - x - 12)}$ Multiplying the numerators and the denominators

$= \dfrac{(x + 4)(x - 4)(x)(x - 4)}{x \cdot x(x - 4)(x + 3)}$ Factoring the numerators and the denominators

$= \dfrac{(x + 4)\cancel{(x - 4)}\cancel{(x)}(x - 4)}{\cancel{x} \cdot x\cancel{(x - 4)}(x + 3)}$ Removing a factor of 1

$= \dfrac{(x + 4)(x - 4)}{x(x + 3)}$

29. $\dfrac{y^2 - 16}{2y + 6} \cdot \dfrac{y + 3}{y - 4} = \dfrac{(y^2 - 16)(y + 3)}{(2y + 6)(y - 4)}$

$= \dfrac{(y + 4)(y - 4)(y + 3)}{2(y + 3)(y - 4)}$

$= \dfrac{(y + 4)\cancel{(y - 4)}\cancel{(y + 3)}}{2\cancel{(y + 3)}\cancel{(y - 4)}}$

$= \dfrac{y + 4}{2}$

31. $\dfrac{x^2 - 2x - 35}{2x^3 - 3x^2} \cdot \dfrac{4x^3 - 9x}{7x - 49}$

$= \dfrac{(x^2 - 2x - 35)(4x^3 - 9x)}{(2x^3 - 3x^2)(7x - 49)}$

$= \dfrac{(x - 7)(x + 5)(x)(2x + 3)(2x - 3)}{x \cdot x(2x - 3)(7)(x - 7)}$

$= \dfrac{\cancel{(x - 7)}(x + 5)\cancel{(x)}(2x + 3)\cancel{(2x - 3)}}{\cancel{x} \cdot x\cancel{(2x - 3)}(7)\cancel{(x - 7)}}$

$= \dfrac{(x + 5)(2x + 3)}{7x}$

33. $\frac{c^3 + 8}{c^2 - 4} \cdot \frac{c^2 - 4c + 4}{c^2 - 2c + 4}$

$= \frac{(c^3 + 8)(c^2 - 4c + 4)}{(c^2 - 4)(c^2 - 2c + 4)}$

$= \frac{(c + 2)(c^2 - 2c + 4)(c - 2)(c - 2)}{(c + 2)(c - 2)(c^2 - 2c + 4)\cdot 1}$

$= \frac{(c + 2)(c^2 - 2c + 4)(c - 2)(c - 2)}{(c + 2)(c - 2)(c^2 - 2c + 4)\cdot 1}$

$= \frac{c - 2}{1}$

$= c - 2$

35. $\frac{x^2-y^2}{x^3-y^3} \cdot \frac{x^2 + xy + y^2}{x^2 + 2xy + y^2} = \frac{(x^2 - y^2)(x^2 + xy + y^2)}{(x^3 - y^3)(x^2 + 2xy + y^2)}$

$= \frac{(x+y)(x-y)(x^2+xy+y^2)}{(x-y)(x^2+xy+y^2)(x+y)(x+y)}$

$= \frac{(x+y)(x-y)(x^2+xy+y^2)\cdot 1}{(x-y)(x^2+xy+y^2)(x+y)(x+y)}$

$= \frac{1}{x + y}$

37. $\frac{12x^8}{3y^4} \div \frac{16x^3}{6y}$

$= \frac{12x^8}{3y^4} \cdot \frac{6y}{16x^3}$ Multiplying by the reciprocal of the divisor

$= \frac{12x^8(6y)}{3y^4(16x^3)}$ Multiplying the numerators and the denominators

$= \frac{3\cdot 4\cdot x^3\cdot x^5\cdot 2\cdot 3\cdot y}{3\cdot y\cdot y^3\cdot 4\cdot 2\cdot 2\cdot x^3}$ Factoring the numerator and the denominator

$= \frac{3\cdot 4\cdot x^3\cdot x^5\cdot 2\cdot 3\cdot y}{3\cdot y\cdot y^3\cdot 4\cdot 2\cdot 2\cdot x^3}$ Removing a factor of 1

$= \frac{3x^5}{2y^3}$

39. $\frac{3y + 15}{y} \div \frac{y + 5}{y} = \frac{3y + 15}{y} \cdot \frac{y}{y + 5}$

$= \frac{(3y + 15)(y)}{y(y + 5)}$

$= \frac{3(y + 5)(y)}{y(y + 5)\cdot 1}$

$= \frac{3(y + 5)(y)}{y(y + 5)\cdot 1}$

$= \frac{3}{1}$

$= 3$

41. $\frac{y^2 - 9}{y} \div \frac{y + 3}{y + 2} = \frac{y^2 - 9}{y} \cdot \frac{y + 2}{y + 3}$

$= \frac{(y^2 - 9)(y + 2)}{y(y + 3)}$

$= \frac{(y + 3)(y - 3)(y + 2)}{y(y + 3)}$

$= \frac{(y + 3)(y - 3)(y + 2)}{y(y + 3)}$

$= \frac{(y - 3)(y + 2)}{y}$

43. $\frac{4a^2 - 1}{a^2 - 4} \div \frac{2a - 1}{a - 2} = \frac{4a^2 - 1}{a^2 - 4} \cdot \frac{a - 2}{2a - 1}$

$= \frac{(4a^2 - 1)(a - 2)}{(a^2 - 4)(2a - 1)}$

$= \frac{(2a + 1)(2a - 1)(a - 2)}{(a + 2)(a - 2)(2a - 1)}$

$= \frac{(2a + 1)(2a - 1)(a - 2)}{(a + 2)(a - 2)(2a - 1)}$

$= \frac{2a + 1}{a + 2}$

45. $\frac{x^2 - 16}{x^2 - 10x + 25} \div \frac{3x - 12}{x^2 - 3x - 10}$

$= \frac{x^2 - 16}{x^2 - 10x + 25} \cdot \frac{x^2 - 3x - 10}{3x - 12}$

$= \frac{(x^2 - 16)(x^2 - 3x - 10)}{(x^2 - 10x + 25)(3x - 12)}$

$= \frac{(x + 4)(x - 4)(x - 5)(x + 2)}{(x - 5)(x - 5)(3)(x - 4)}$

$= \frac{(x + 4)(x - 4)(x - 5)(x + 2)}{(x - 5)(x - 5)(3)(x - 4)}$

$= \frac{(x + 4)(x + 2)}{3(x - 5)}$

47. $\frac{y^3 + 3y}{y^2 - 9} \div \frac{y^2 + 5y - 14}{y^2 + 4y - 21}$

$= \frac{y^3 + 3y}{y^2 - 9} \cdot \frac{y^2 + 4y - 21}{y^2 + 5y - 14}$

$= \frac{(y^3 + 3y)(y^2 + 4y - 21)}{(y^2 - 9)(y^2 + 5y - 14)}$

$= \frac{y(y^2 + 3)(y + 7)(y - 3)}{(y + 3)(y - 3)(y + 7)(y - 2)}$

$= \frac{y(y^2 + 3)(y + 7)(y - 3)}{(y + 3)(y - 3)(y + 7)(y - 2)}$

$= \frac{y(y^2 + 3)}{(y + 3)(y - 2)}$

49. $\frac{x^3 - 64}{x^3 + 64} \div \frac{x^2 - 16}{x^2 - 4x + 16}$

$= \frac{x^3 - 64}{x^3 + 64} \cdot \frac{x^2 - 4x + 16}{x^2 - 16}$

$= \frac{(x^3 - 64)(x^2 - 4x + 16)}{(x^3 + 64)(x^2 - 16)}$

$= \frac{(x - 4)(x^2 + 4x + 16)(x^2 - 4x + 16)}{(x + 4)(x^2 - 4x + 16)(x + 4)(x - 4)}$

$= \frac{(x - 4)(x^2 + 4x + 16)(x^2 - 4x + 16)}{(x + 4)(x^2 - 4x + 16)(x + 4)(x - 4)}$

$= \frac{x^2 + 4x + 16}{(x + 4)(x + 4)}$, or $\frac{x^2 + 4x + 16}{(x + 4)^2}$

51. $\left[\frac{r^2 - 4s^2}{r + 2s} \div (r + 2s)\right] \cdot \frac{2s}{r - 2s}$

$= \left[\frac{r^2 - 4s^2}{r + 2s} \cdot \frac{1}{r + 2s}\right] \cdot \frac{2s}{r - 2s}$

$= \frac{(r + 2s)(r - 2s)(2s)}{(r + 2s)(r + 2s)(r - 2s)}$

$= \frac{(r + 2s)(r - 2s)(2s)}{(r + 2s)(r + 2s)(r - 2s)}$

$= \frac{2s}{r + 2s}$

53. Familiarize. Let x = the number of field goals and y = the number of free throws the player made. Then she scores 2x points for field goals and y points for free throws.

Translate.

The total number of scores is 15.

$x + y = 15$

The total number of points is 27.

$2x + y = 27$

We have a system of equations:

$$x + y = 15,$$
$$2x + y = 27$$

Solve. Solving the system of equations, we get (12,3).

Check. If the player scores 12 field goals and 3 free throws, she has a total of 15 scores. She scores 2·12, or 24 points from field goals and 3 points from free throws, so the total points are 27. These values check.

State. The player scored 12 field goals and 3 free throws.

55. Familiarize. We make a drawing. Let ℓ = the length and w = the width of the rectangle.

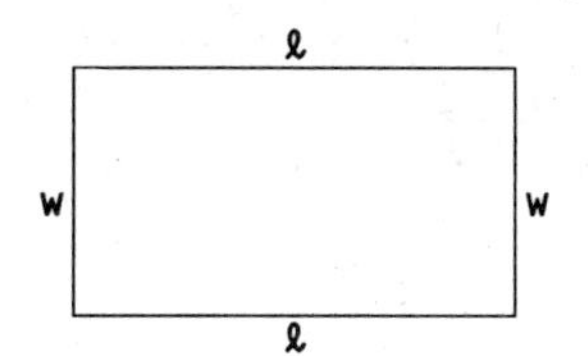

We will first find ℓ and w. Then we will find the area, A = ℓw.

Translate.

The length is 6 m greater than the width.

$\ell = 6 + w$

The perimeter is 628 m.

$2\ell + 2w = 628$

We have a system of equations:

$$\ell = 6 + w,$$
$$2\ell + 2w = 628$$

Solve. Solving the system of equations, we get (160,154).

Check. The length, 160 m, is 6 m greater than 154 m, the width. If the length of a rectangle is 160 m and the width is 154 m, the perimeter is 2·160 + 2·154, = 320 + 308, or 628 m. These values check. The area of the rectangle is 160(154), or 24,640 m².

State. The area of the field is 24,640 m².

57. $$\frac{x(x+1) - 2(x+3)}{(x+1)(x+2)(x+3)} = \frac{x^2 + x - 2x - 6}{(x+1)(x+2)(x+3)}$$
$$= \frac{x^2 - x - 6}{(x+1)(x+2)(x+3)}$$
$$= \frac{(x-3)(x+2)}{(x+1)(x+2)(x+3)}$$
$$= \frac{(x-3)\cancel{(x+2)}}{(x+1)\cancel{(x+2)}(x+3)}$$
$$= \frac{x-3}{(x+1)(x+3)}$$

59. $$\frac{m^2 - t^2}{m^2 + t^2 + m + t + 2mt} = \frac{m^2 - t^2}{(m^2 + 2mt + t^2) + (m + t)}$$
$$= \frac{(m+t)(m-t)}{(m+t)^2 + (m+t)}$$
$$= \frac{(m+t)(m-t)}{(m+t)[(m+t)+1]}$$
$$= \frac{\cancel{(m+t)}(m-t)}{\cancel{(m+t)}(m+t+1)}$$
$$= \frac{m-t}{m+t+1}$$

61. $$\frac{x^3 + x^2 - y^3 - y^2}{x^2 - 2xy + y^2}$$
$$= \frac{(x^3 - y^3) + (x^2 - y^2)}{x^2 - 2xy + y^2}$$
$$= \frac{(x-y)(x^2 + xy + y^2) + (x+y)(x-y)}{(x-y)^2}$$
$$= \frac{(x-y)[(x^2 + xy + y^2) + (x+y)]}{(x-y)(x-y)}$$
$$= \frac{\cancel{(x-y)}(x^2 + xy + y^2 + x + y)}{\cancel{(x-y)}(x-y)}$$
$$= \frac{x^2 + xy + y^2 + x + y}{x - y}$$

Exercise Set 9.2

1. 12 = 2·2·3

18 = 2·3·3

LCM = 2·2·3·3, or 36

3. 18 = 2·3·3

48 = 2·2·2·2·3

LCM = 2·2·2·2·3·3, or 144

5. 24 = 2·2·2·3

36 = 2·2·3·3

LCM = 2·2·2·3·3, or 72

7. 9 = 3·3

15 = 3·5

5 = 5

LCM = 3·3·5, or 45

9. $\frac{5}{6} + \frac{4}{15} = \frac{5}{2\cdot3} + \frac{4}{3\cdot5}$, LCD $= 2\cdot3\cdot5$, or 30

$= \frac{5}{2\cdot3}\cdot\frac{5}{5} + \frac{4}{3\cdot5}\cdot\frac{2}{2}$

$= \frac{25}{2\cdot3\cdot5} + \frac{8}{2\cdot3\cdot5}$

$= \frac{33}{2\cdot3\cdot5} = \frac{\cancel{3}\cdot11}{2\cdot\cancel{3}\cdot5}$

$= \frac{11}{10}$

11. $\frac{7}{12} + \frac{11}{18} = \frac{7}{2\cdot2\cdot3} + \frac{11}{2\cdot3\cdot3}$, LCD $= 2\cdot2\cdot3\cdot3$, or 36

$= \frac{7}{2\cdot2\cdot3}\cdot\frac{3}{3} + \frac{11}{2\cdot3\cdot3}\cdot\frac{2}{2}$

$= \frac{21}{2\cdot2\cdot3\cdot3} + \frac{22}{2\cdot2\cdot3\cdot3}$

$= \frac{43}{2\cdot2\cdot3\cdot3}$

$= \frac{43}{36}$

13. $\frac{3}{4} + \frac{7}{30} + \frac{1}{16}$

$= \frac{3}{2\cdot2} + \frac{7}{2\cdot3\cdot5} + \frac{1}{2\cdot2\cdot2\cdot2}$, LCD $= 2\cdot2\cdot2\cdot2\cdot3\cdot5$

$= \frac{3}{2\cdot2}\cdot\frac{2\cdot2\cdot3\cdot5}{2\cdot2\cdot3\cdot5} + \frac{7}{2\cdot3\cdot5}\cdot\frac{2\cdot2\cdot2}{2\cdot2\cdot2} + \frac{1}{2\cdot2\cdot2\cdot2}\cdot\frac{3\cdot5}{3\cdot5}$

$= \frac{180}{2\cdot2\cdot2\cdot2\cdot3\cdot5} + \frac{56}{2\cdot2\cdot2\cdot2\cdot3\cdot5} + \frac{15}{2\cdot2\cdot2\cdot2\cdot3\cdot5}$

$= \frac{251}{2\cdot2\cdot2\cdot2\cdot3\cdot5}$

$= \frac{251}{240}$

15. $12x^2y = 2\cdot2\cdot3\cdot x\cdot x\cdot y$

$4xy = 2\cdot2\cdot x\cdot y$

LCM $= 2\cdot2\cdot3\cdot x\cdot x\cdot y$, or $12x^2y$

17. $y^2 - 9 = (y + 3)(y - 3)$

$3y + 9 = 3(y + 3)$

LCM $= 3(y + 3)(y - 3)$

19. $15ab^2 = 3\cdot5\cdot a\cdot b\cdot b$

$3ab = 3\cdot a\cdot b$

$10a^3b = 2\cdot5\cdot a\cdot a\cdot a\cdot b$

LCM $= 2\cdot3\cdot5\cdot a\cdot a\cdot a\cdot b\cdot b$, or $30a^3b^2$

21. $5y - 15 = t(y - 3)$

$y^2 - 6y + 9 = (y - 3)(y - 3)$

LCM $= 5(y - 3)(y - 3)$, or $5(y - 3)^2$

23. $x^2 - 4 = (x + 2)(x - 2)$

$2 - x = (2 - x)$

We can use $(x - 2)$ or $(2 - x)$, but we do not use both.

LCM $= (x + 2)(x - 2)$, or $(x + 2)(2 - x)$

25. $2r^2 - 5r - 12 = (2r + 3)(r - 4)$

$3r^2 - 13r + 4 = (3r - 1)(r - 4)$

$r^2 - 16 = (r + 4)(r - 4)$

LCM $= (2r + 3)(3r - 1)(r + 4)(r - 4)$

27. $x^5 + 4x^3 = x^3(x^2 + 4) = x\cdot x\cdot x(x^2 + 4)$

$x^3 - 4x^2 + 4x = x(x^2 - 4x + 4) = x(x - 2)(x - 2)$

LCM $= x\cdot x\cdot x(x - 2)(x - 2)(x^2 + 4)$, or

$x^3(x - 2)(x - 2)(x^2 + 4)$

29. $x^5 - 2x^4 + x^3 = x^3(x^2 - 2x + 1) =$

$x\cdot x\cdot x(x - 1)(x - 1)$

$2x^3 + 2x = 2x(x^2 + 1)$

$5x + 5 = 5(x + 1)$

LCM $= 2\cdot5\cdot x\cdot x\cdot x(x - 1)(x - 1)(x + 1)(x^2 + 1)$, or

$10x^3(x - 1)(x - 1)(x + 1)(x^2 + 1)$

31. $\frac{a - 3b}{a + b} + \frac{a + 5b}{a + b} = \frac{a - 3b + a + 5b}{a + b}$ Adding the numerators

$= \frac{2a + 2b}{a + b} = \frac{2(a + b)}{a + b}$ Simplifying and factoring the numerator

$= \frac{2\cancel{(a + b)}}{1\cancel{(a + b)}}$ Removing a factor of 1

$= \frac{2}{1} = 2$

33. $\frac{4y + 2}{y - 2} - \frac{y - 3}{y - 2} = \frac{4y + 2 - (y - 3)}{y - 2}$ Subtracting the numerators

$= \frac{4y + 2 - y + 3}{y - 2}$

$= \frac{3y + 5}{y - 2}$

35. $\frac{a^2}{a - b} + \frac{b^2}{b - a} = \frac{a^2}{a - b} + \frac{-1}{-1}\cdot\frac{b^2}{b - a}$ Multiplying by $\frac{-1}{-1}$

$= \frac{a^2}{a - b} + \frac{-b^2}{a - b}$

$= \frac{a^2 - b^2}{a - b} = \frac{(a + b)(a - b)}{a - b}$

Adding numerators and factoring

$= \frac{(a + b)\cancel{(a - b)}}{1\cancel{(a - b)}}$ Removing a factor of 1

$= \frac{a + b}{1} = a + b$

37. $\frac{3}{x} - \frac{8}{-x} = \frac{3}{x} - \frac{-1}{-1}\cdot\frac{8}{-x} = \frac{3}{x} + \frac{8}{x} = \frac{11}{x}$

39. $\frac{2x - 10}{x^2 - 25} - \frac{5 - x}{25 - x^2} = \frac{2x - 10}{x^2 - 25} - \frac{-1}{-1}\cdot\frac{5 - x}{25 - x^2}$

$= \frac{2x - 10}{x^2 - 25} + \frac{5 - x}{x^2 - 25}$

$= \frac{x - 5}{x^2 - 25} = \frac{x - 5}{(x + 5)(x - 5)}$

$= \frac{1\cancel{(x - 5)}}{(x + 5)\cancel{(x - 5)}}$

$= \frac{1}{x + 5}$

41. $\dfrac{y-2}{y+4} + \dfrac{y+3}{y-5}$

[LCD is $(y+4)(y-5)$.]

$= \dfrac{y-2}{y+4} \cdot \dfrac{y-5}{y-5} + \dfrac{y+3}{y-5} \cdot \dfrac{y+4}{y+4}$

$= \dfrac{(y^2 - 7y + 10) + (y^2 + 7y + 12)}{(y+4)(y-5)}$

$= \dfrac{2y^2 + 22}{(y+4)(y-5)}$, or $\dfrac{2y^2 + 22}{y^2 - y - 20}$

43. $\dfrac{4xy}{x^2 - y^2} + \dfrac{x-y}{x+y}$

$= \dfrac{4xy}{(x+y)(x-y)} + \dfrac{x-y}{x+y}$

[LCD is $(x+y)(x-y)$.]

$= \dfrac{4xy}{(x+y)(x-y)} + \dfrac{x-y}{x+y} \cdot \dfrac{x-y}{x-y}$

$= \dfrac{4xy + x^2 - 2xy + y^2}{(x+y)(x-y)}$

$= \dfrac{x^2 + 2xy + y^2}{(x+y)(x-y)} = \dfrac{(x+y)(x+y)}{(x+y)(x-y)}$

$= \dfrac{\cancel{(x+y)}(x+y)}{\cancel{(x+y)}(x-y)} = \dfrac{x+y}{x-y}$

45. $\dfrac{9x+2}{3x^2 - 2x - 8} + \dfrac{7}{3x^2 + x - 4}$

$= \dfrac{9x+2}{(3x+4)(x-2)} + \dfrac{7}{(3x+4)(x-1)}$

[LCD is $(3x+4)(x-2)(x-1)$.]

$= \dfrac{9x+2}{(3x+4)(x-2)} \cdot \dfrac{x-1}{x-1} + \dfrac{7}{(3x+4)(x-1)} \cdot \dfrac{x-2}{x-2}$

$= \dfrac{9x^2 - 7x - 2 + 7x - 14}{(3x+4)(x-2)(x-1)}$

$= \dfrac{9x^2 - 16}{(3x+4)(x-2)(x-1)} = \dfrac{(3x+4)(3x-4)}{(3x+4)(x-2)(x-1)}$

$= \dfrac{\cancel{(3x+4)}(3x-4)}{\cancel{(3x+4)}(x-2)(x-1)}$

$= \dfrac{3x-4}{(x-2)(x-1)}$, or $\dfrac{3x-4}{x^2 - 3x + 2}$

47. $\dfrac{4}{x+1} + \dfrac{x+2}{x^2 - 1} + \dfrac{3}{x-1}$

$= \dfrac{4}{x+1} + \dfrac{x+2}{(x+1)(x-1)} + \dfrac{3}{x-1}$

[LCD is $(x+1)(x-1)$.]

$= \dfrac{4}{x+1} \cdot \dfrac{x-1}{x-1} + \dfrac{x+2}{(x+1)(x-1)} + \dfrac{3}{x-1} \cdot \dfrac{x+1}{x+1}$

$= \dfrac{4x - 4 + x + 2 + 3x + 3}{(x+1)(x-1)}$

$= \dfrac{8x+1}{(x+1)(x-1)}$, or $\dfrac{8x+1}{x^2 - 1}$

49. $\dfrac{x-1}{3x+15} - \dfrac{x+3}{5x+25}$

$= \dfrac{x-1}{3(x+5)} - \dfrac{x+3}{5(x+5)}$

[LCD is $3 \cdot 5(x+5)$, or $15(x+5)$.]

$= \dfrac{x-1}{3(x+5)} \cdot \dfrac{5}{5} - \dfrac{x+3}{5(x+5)} \cdot \dfrac{3}{3}$

$= \dfrac{5x - 5 - (3x + 9)}{15(x+5)}$

$= \dfrac{5x - 5 - 3x - 9}{15(x+5)}$

$= \dfrac{2x - 14}{15(x+5)}$, or $\dfrac{2x - 14}{15x + 75}$

51. $\dfrac{5ab}{a^2 - b^2} - \dfrac{a-b}{a+b}$

$= \dfrac{5ab}{(a+b)(a-b)} - \dfrac{a-b}{a+b}$

[LCD is $(a+b)(a-b)$.]

$= \dfrac{5ab}{(a+b)(a-b)} - \dfrac{a-b}{a+b} \cdot \dfrac{a-b}{a-b}$

$= \dfrac{5ab - (a^2 - 2ab + b^2)}{(a+b)(a-b)}$

$= \dfrac{5ab - a^2 + 2ab - b^2}{(a+b)(a-b)}$

$= \dfrac{-a^2 + 7ab - b^2}{(a+b)(a-b)}$, or $\dfrac{-a^2 + 7ab - b^2}{a^2 - b^2}$

53. $\dfrac{3y}{y^2 - 7y + 10} - \dfrac{2y}{y^2 - 8y + 15}$

$= \dfrac{3y}{(y-5)(y-2)} - \dfrac{2y}{(y-5)(y-3)}$

[LCD is $(y-5)(y-2)(y-3)$.]

$= \dfrac{3y}{(y-5)(y-2)} \cdot \dfrac{y-3}{y-3} - \dfrac{2y}{(y-5)(y-3)} \cdot \dfrac{y-2}{y-2}$

$= \dfrac{3y^2 - 9y - (2y^2 - 4y)}{(y-5)(y-2)(y-3)}$

$= \dfrac{3y^2 - 9y - 2y^2 + 4y}{(y-5)(y-2)(y-3)}$

$= \dfrac{y^2 - 5y}{(y-5)(y-2)(y-3)} = \dfrac{y(y-5)}{(y-5)(y-2)(y-3)}$

$= \dfrac{y\cancel{(y-5)}}{\cancel{(y-5)}(y-2)(y-3)}$

$= \dfrac{y}{(y-2)(y-3)}$, or $\dfrac{y}{y^2 - 5y + 6}$

55. $\dfrac{y}{y^2 - y - 20} + \dfrac{2}{y+4}$

$= \dfrac{y}{(y-5)(y+4)} + \dfrac{2}{y+4}$

[LCD is $(y-5)(y+4)$.]

$= \dfrac{y}{(y-5)(y+4)} + \dfrac{2}{y+4} \cdot \dfrac{y-5}{y-5}$

$= \dfrac{y + 2y - 10}{(y-5)(y+4)}$

$= \dfrac{3y - 10}{(y-5)(y+4)}$, or $\dfrac{3y - 10}{y^2 - y - 20}$

57. $\frac{3y + 2}{y^2 + 5y - 24} + \frac{7}{y^2 + 4y - 32}$

$= \frac{3y + 2}{(y + 8)(y - 3)} + \frac{7}{(y + 8)(y - 4)}$

[LCD is $(y + 8)(y - 3)(y - 4)$.]

$= \frac{3y + 2}{(y + 8)(y - 3)} \cdot \frac{y - 4}{y - 4} + \frac{7}{(y + 8)(y - 4)} \cdot \frac{y - 3}{y - 3}$

$= \frac{3y^2 - 10y - 8 + 7y - 21}{(y + 8)(y - 3)(y - 4)}$

$= \frac{3y^2 - 3y - 29}{(y + 8)(y - 3)(y - 4)}$

59. $\frac{3x - 1}{x^2 + 2x - 3} - \frac{x + 4}{x^2 - 9}$

$= \frac{3x - 1}{(x + 3)(x - 1)} - \frac{x + 4}{(x + 3)(x - 3)}$

[LCD is $(x + 3)(x - 1)(x - 3)$.]

$= \frac{3x - 1}{(x + 3)(x - 1)} \cdot \frac{x - 3}{x - 3} - \frac{x + 4}{(x + 3)(x - 3)} \cdot \frac{x - 1}{x - 1}$

$= \frac{3x^2 - 10x + 3 - (x^2 + 3x - 4)}{(x + 3)(x - 1)(x - 3)}$

$= \frac{3x^2 - 10x + 3 - x^2 - 3x + 4}{(x + 3)(x - 1)(x - 3)}$

$= \frac{2x^2 - 13x + 7}{(x + 3)(x - 1)(x - 3)}$

61. $\frac{1}{x + 1} - \frac{x}{x - 2} + \frac{x^2 + 2}{x^2 - x - 2}$

$= \frac{1}{x + 1} - \frac{x}{x - 2} + \frac{x^2 + 2}{(x - 2)(x + 1)}$

[LCD is $(x + 1)(x - 2)$.]

$= \frac{1}{x + 1} \cdot \frac{x - 2}{x - 2} - \frac{x}{x - 2} \cdot \frac{x + 1}{x + 1} + \frac{x^2 + 2}{(x - 2)(x + 1)}$

$= \frac{x - 2 - (x^2 + x) + x^2 + 2}{(x + 1)(x - 2)}$

$= \frac{x - 2 - x^2 - x + x^2 + 2}{(x + 1)(x - 2)}$

$= \frac{0}{(x + 1)(x - 2)} = 0$

63. $\frac{x - 1}{x - 2} - \frac{x + 1}{x + 2} + \frac{x - 6}{x^2 - 4}$

$= \frac{x - 1}{x - 2} - \frac{x + 1}{x + 2} + \frac{x - 6}{(x - 2)(x + 2)}$

[LCD = $(x - 2)(x + 2)$.]

$= \frac{x - 1}{x - 2} \cdot \frac{x + 2}{x + 2} - \frac{x + 1}{x + 2} \cdot \frac{x - 2}{x - 2} + \frac{x - 6}{(x - 2)(x + 2)}$

$= \frac{(x^2 + x - 2) - (x^2 - x - 2) + (x - 6)}{(x - 2)(x + 2)}$

$= \frac{x^2 + x - 2 - x^2 + x + 2 + x - 6}{(x - 2)(x + 2)}$

$= \frac{3x - 6}{(x - 2)(x + 2)}$

$= \frac{3(x - 2)}{(x - 2)(x + 2)}$

$= \frac{3\cancel{(x - 2)}}{\cancel{(x - 2)}(x + 2)}$

$= \frac{3}{x + 2}$

65. $\frac{4x}{x^2 - 1} + \frac{3x}{1 - x} - \frac{4}{x - 1}$

$= \frac{4x}{x^2 - 1} + \frac{-1}{-1} \cdot \frac{3x}{1 - x} - \frac{4}{x - 1}$

$= \frac{4x}{(x + 1)(x - 1)} + \frac{-3x}{x - 1} - \frac{4}{x - 1}$

[LCD = $(x + 1)(x - 1)$.]

$= \frac{4x}{(x + 1)(x - 1)} + \frac{-3x}{x - 1} \cdot \frac{x + 1}{x + 1} - \frac{4}{x - 1} \cdot \frac{x + 1}{x + 1}$

$= \frac{4x - 3x^2 - 3x - 4x - 4}{(x + 1)(x - 1)}$

$= \frac{-3x^2 - 3x - 4}{x^2 - 1}$

67. $\frac{5}{3 - 2x} - \frac{3}{2x - 3} + \frac{x - 3}{2x^2 - x - 3}$

$= \frac{-1}{-1} \cdot \frac{5}{3 - 2x} - \frac{3}{2x - 3} + \frac{x - 3}{2x^2 - x - 3}$

$= \frac{-5}{2x - 3} - \frac{3}{2x - 3} + \frac{x - 3}{(2x - 3)(x + 1)}$

[LCD is $(2x - 3)(x + 1)$.]

$= \frac{-5}{2x - 3} \cdot \frac{x + 1}{x + 1} - \frac{3}{2x - 3} \cdot \frac{x + 1}{x + 1} + \frac{x - 3}{(2x-3)(x+1)}$

$= \frac{-5x - 5 - 3x - 3 + x - 3}{(2x - 3)(x + 1)}$

$= \frac{-7x - 11}{2x^2 - x - 3}$

69. $\frac{3}{2c - 1} - \frac{1}{c + 2} + \frac{5}{2c^2 + 3c - 2}$

$= \frac{3}{2c - 1} - \frac{1}{c + 2} + \frac{5}{(2c - 1)(c + 2)}$

[LCD is $(2c - 1)(c + 2)$.]

$= \frac{3}{2c - 1} \cdot \frac{c + 2}{c + 2} - \frac{1}{c + 2} \cdot \frac{2c - 1}{2c - 1} + \frac{5}{(2c-1)(c+2)}$

$= \frac{3c + 6 - 2c + 1 + 5}{(2c - 1)(c + 2)}$

$= \frac{c + 12}{2c^2 + 3c - 2}$

71. $\frac{1}{x + y} + \frac{1}{y - x} - \frac{2x}{x^2 - y^2}$

$= \frac{1}{x + y} + \frac{-1}{-1} \cdot \frac{1}{y - x} - \frac{2x}{x^2 - y^2}$

$= \frac{1}{x + y} + \frac{-1}{x - y} - \frac{2x}{(x + y)(x - y)}$

[LCD is $(x + y)(x - y)$.]

$= \frac{1}{x + y} \cdot \frac{x - y}{x - y} + \frac{-1}{x - y} \cdot \frac{x + y}{x + y} - \frac{2x}{(x + y)(x - y)}$

$= \frac{x - y - x - y - 2x}{(x + y)(x - y)}$

$= \frac{-2y - 2x}{(x + y)(x - y)} = \frac{-2(y + x)}{(x + y)(x - y)}$

$= \frac{-2\cancel{(y + x)}}{\cancel{(x + y)}(x - y)}$

$= \frac{-2}{x - y}$, or $\frac{2}{y - x}$

73. Graph: $2x - 3y > 6$

We first graph the line $2x - 3y = 6$. The intercepts are (0,-2) and (3,0). We draw the line dashed since the inequality symbol is >. To determine which half-plane to shade, we consider a test point not on the line. We try (0,0):

$2x - 3y > 6$	
$2\cdot 0 - 3\cdot 0$	6
0	FALSE

Since 0 > 6 is false, we shade the half-plane that does not contain (0,0).

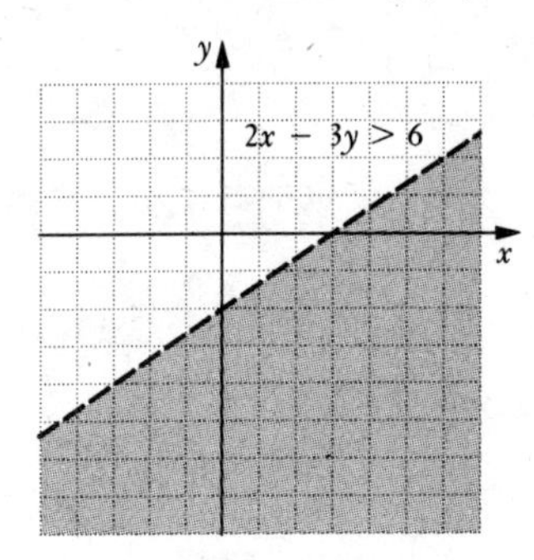

75. $12y^4 - 15y^3 + 3y^2 = 3y^2(4y^2 - 5y + 1)$
$= 3y^2(4y - 1)(y - 1)$

77. $x^2 - 14x + 49 = (x - 7)^2$

79. $x^8 - x^4 = x^4(x^2 + 1)(x + 1)(x - 1)$
$x^5 - x^2 = x^2(x - 1)(x^2 + x + 1)$
$x^5 - x^3 = x^3(x + 1)(x - 1)$
$x^5 + x^2 = x^2(x + 1)(x^2 - x + 1)$
The LCM is
$x^4(x^2 + 1)(x + 1)(x - 1)(x^2 + x + 1)(x^2 - x + 1)$.

81. $\dfrac{b - c}{a - (b - c)} - \dfrac{b - a}{(b - a) - c}$

$= \dfrac{b - c}{a - b + c} - \dfrac{b - a}{b - a - c}$

$= \dfrac{b - c}{a - b + c} - \dfrac{-1}{-1}\cdot\dfrac{b - a}{b - a - c}$

$= \dfrac{b - c}{a - b + c} - \dfrac{a - b}{a - b + c}$

$= \dfrac{b - c - (a - b)}{a - b + c}$

$= \dfrac{b - c - a + b}{a - b + c}$

$= \dfrac{2b - c - a}{a - b + c}$

83. $\dfrac{x^2}{3x^2 - 5x - 2} - \dfrac{2x}{3x + 1}\cdot\dfrac{1}{x - 2}$

$= \dfrac{x^2}{(3x + 1)(x - 2)} - \dfrac{2x}{(3x + 1)(x - 2)}$

$= \dfrac{x^2 - 2x}{(3x + 1)(x - 2)}$

$= \dfrac{x(x - 2)}{(3x + 1)(x - 2)}$

$= \dfrac{x\cancel{(x - 2)}}{(3x + 1)\cancel{(x - 2)}}$

$= \dfrac{x}{3x + 1}$

Exercise Set 9.3

1. $\dfrac{2}{5} + \dfrac{7}{8} = \dfrac{y}{20}$, LCM is 40

$40\left(\dfrac{2}{5} + \dfrac{7}{8}\right) = 40\cdot\dfrac{y}{20}$

$40\cdot\dfrac{2}{5} + 40\cdot\dfrac{7}{8} = 40\cdot\dfrac{y}{20}$

$16 + 35 = 2y$

$51 = 2y$

$\dfrac{51}{2} = y$

Check:

$\dfrac{2}{5} + \dfrac{7}{8} = \dfrac{y}{20}$	
$\dfrac{16}{40} + \dfrac{35}{40}$	$\dfrac{\frac{51}{2}}{20}$
$\dfrac{51}{40}$	$\dfrac{51}{2}\cdot\dfrac{1}{20}$
	$\dfrac{51}{40}$ TRUE

The solution is $\dfrac{51}{2}$.

3. $\dfrac{1}{3} - \dfrac{5}{6} = \dfrac{1}{x}$, LCM is $6x$

$6x\left(\dfrac{1}{3} - \dfrac{5}{6}\right) = 6x\cdot\dfrac{1}{x}$

$6x\cdot\dfrac{1}{3} - 6x\cdot\dfrac{5}{6} = 6x\cdot\dfrac{1}{x}$

$2x - 5x = 6$

$-3x = 6$

$x = -2$

Check:

$\dfrac{1}{3} - \dfrac{5}{6} = \dfrac{1}{x}$	
$\dfrac{2}{6} - \dfrac{5}{6}$	$\dfrac{1}{-2}$
$-\dfrac{3}{6}$	$-\dfrac{1}{2}$
$-\dfrac{1}{2}$	TRUE

The solution is -2.

5. $\dfrac{x}{3} - \dfrac{x}{4} = 12$, LCM is 12

$12\left(\dfrac{x}{3} - \dfrac{x}{4}\right) = 12\cdot 12$

$12\cdot\dfrac{x}{3} - 12\cdot\dfrac{x}{4} = 12\cdot 12$

$4x - 3x = 144$

$x = 144$

Check:

$\dfrac{x}{3} - \dfrac{x}{4} = 12$	
$\dfrac{144}{3} - \dfrac{144}{4}$	12
$48 - 36$	
12	TRUE

The solution is 144.

7. $y + \frac{5}{y} = -6$, LCM is y

$y\left(y + \frac{5}{y}\right) = y(-6)$

$y \cdot y + y \cdot \frac{5}{y} = -6y$

$y^2 + 5 = -6y$

$y^2 + 6y + 5 = 0$

$(y + 5)(y + 1) = 0$

$y + 5 = 0$ or $y + 1 = 0$

$y = -5$ or $y = -1$

Check:

For -5:

$y + \frac{5}{y} = -6$	
$-5 + \frac{5}{-5}$	-6
$-5 - 1$	
-6	TRUE

For -1:

$y + \frac{5}{y} = -6$	
$-1 + \frac{5}{-1}$	-6
$-1 - 5$	
-6	TRUE

The solutions are -5 and -1.

9. $\frac{4}{z} + \frac{2}{z} = 3$, LCM is z

$z\left(\frac{4}{z} + \frac{2}{z}\right) = z \cdot 3$

$z \cdot \frac{4}{z} + z \cdot \frac{2}{z} = 3z$

$4 + 2 = 3z$

$6 = 3z$

$2 = z$

Check:

$\frac{4}{z} + \frac{2}{z} = 3$	
$\frac{4}{2} + \frac{2}{2}$	3
$2 + 1$	
3	TRUE

The solution is 2.

11. $\frac{x - 3}{x + 2} = \frac{1}{5}$, LCM is $5(x + 2)$

$5(x + 2) \cdot \frac{x - 3}{x + 2} = 5(x + 2) \cdot \frac{1}{5}$

$5(x - 3) = x + 2$

$5x - 15 = x + 2$

$4x = 17$

$x = \frac{17}{4}$

Check:

$\frac{x - 3}{x + 2} = \frac{1}{5}$	
$\frac{\frac{17}{4} - \frac{12}{4}}{\frac{17}{4} + \frac{8}{4}}$	$\frac{1}{5}$
$\frac{5}{4} \cdot \frac{4}{25}$	
$\frac{1}{5}$	TRUE

The solution is $\frac{17}{4}$.

13. $\frac{3}{y + 1} = \frac{2}{y - 3}$, LCM is $(y + 1)(y - 3)$

$(y + 1)(y - 3) \cdot \frac{3}{y + 1} = (y + 1)(y - 3) \cdot \frac{2}{y - 3}$

$3(y - 3) = 2(y + 1)$

$3y - 9 = 2y + 2$

$y = 11$

Check:

$\frac{3}{y + 1} = \frac{2}{y - 3}$	
$\frac{3}{11 + 1}$	$\frac{2}{11 - 3}$
$\frac{3}{12}$	$\frac{2}{8}$
$\frac{1}{4}$	$\frac{1}{4}$ TRUE

The solution is 11.

15. $\frac{y - 1}{y - 3} = \frac{2}{y - 3}$, LCM is $y - 3$

$(y - 3) \cdot \frac{y - 1}{y - 3} = (y - 3) \cdot \frac{2}{y - 3}$

$y - 1 = 2$

$y = 3$

Check:

$\frac{y - 1}{y - 3} = \frac{2}{y - 3}$	
$\frac{3 - 1}{3 - 3}$	$\frac{2}{3 - 3}$
$\frac{2}{0}$	$\frac{2}{0}$
	FALSE

We know that 3 is not a solution of the original equation because it results in division by 0. The equation has no solution.

17. $\frac{x + 1}{x} = \frac{3}{2}$, LCM is $2x$

$2x \cdot \frac{x + 1}{x} = 2x \cdot \frac{3}{2}$

$2(x + 1) = x \cdot 3$

$2x + 2 = 3x$

$2 = x$

Check:

$\frac{x + 1}{x} = \frac{3}{2}$	
$\frac{2 + 1}{2}$	$\frac{3}{2}$
$\frac{3}{2}$	TRUE

The solution is 2.

19. $\frac{2}{x} - \frac{3}{x} + \frac{4}{x} = 5$, LCM is x

$x\left(\frac{2}{x} - \frac{3}{x} + \frac{4}{x}\right) = x \cdot 5$

$2 - 3 + 4 = 5x$

$3 = 5x$

$\frac{3}{5} = x$

Check:

$\frac{2}{x} - \frac{3}{x} + \frac{4}{x} = 5$	
$\frac{2}{\frac{3}{5}} - \frac{3}{\frac{3}{5}} + \frac{4}{\frac{3}{5}}$	$\frac{15}{3}$
$\frac{10}{3} - \frac{15}{3} + \frac{20}{3}$	
$\frac{15}{3}$	TRUE

The solution is $\frac{3}{5}$.

21. $\frac{1}{2} - \frac{4}{9x} = \frac{4}{9} - \frac{1}{6x}$, LCM is $18x$

$18x\left(\frac{1}{2} - \frac{4}{9x}\right) = 18x\left(\frac{4}{9} - \frac{1}{6x}\right)$

$9x - 8 = 8x - 3$

$x = 5$

Since 5 checks, it is the solution.

23. $\frac{60}{x} - \frac{60}{x - 5} = \frac{2}{x}$, LCM is $x(x - 5)$

$x(x - 5)\left(\frac{60}{x} - \frac{60}{x - 5}\right) = x(x - 5) \cdot \frac{2}{x}$

$60(x - 5) - 60x = 2(x - 5)$

$60x - 300 - 60x = 2x - 10$

$-300 = 2x - 10$

$-290 = 2x$

$-145 = x$

Since -145 checks, it is the solution.

25. $\frac{7}{5x - 2} = \frac{5}{4x}$, LCM is $4x(5x - 2)$

$$4x(5x - 2) \cdot \frac{7}{5x - 2} = 4x(5x - 2) \cdot \frac{5}{4x}$$
$$4x \cdot 7 = 5(5x - 2)$$
$$28x = 25x - 10$$
$$3x = -10$$
$$x = -\frac{10}{3}$$

Since $-\frac{10}{3}$ checks, it is the solution.

27. $\frac{x}{x - 2} + \frac{x}{x^2 - 4} = \frac{x + 3}{x + 2}$

$\frac{x}{x - 2} + \frac{x}{(x + 2)(x - 2)} = \frac{x + 3}{x + 2}$, LCM is $(x + 2)(x - 2)$

$$(x+2)(x-2)\left[\frac{x}{x-2} + \frac{x}{(x+2)(x-2)}\right] = (x+2)(x-2) \cdot \frac{x+3}{x+2}$$
$$x(x + 2) + x = (x - 2)(x + 3)$$
$$x^2 + 2x + x = x^2 + x - 6$$
$$3x = x - 6$$
$$2x = -6$$
$$x = -3$$

Since -3 checks, it is the solution.

29. $\frac{6}{x^2 - 4x + 3} - \frac{1}{x - 3} = \frac{1}{4x - 4}$

$\frac{6}{(x-3)(x-1)} - \frac{1}{x-3} = \frac{1}{4(x-1)}$, LCM is $4(x-3)(x-1)$

$$4(x-3)(x-1)\left[\frac{6}{(x-3)(x-1)} - \frac{1}{x-3}\right] = 4(x-3)(x-1) \cdot \frac{1}{4(x-1)}$$
$$4 \cdot 6 - 4(x - 1) = x - 3$$
$$24 - 4x + 4 = x - 3$$
$$-5x = -31$$
$$x = \frac{31}{5}$$

Since $\frac{31}{5}$ checks, it is the solution.

31. $\frac{1}{4y^2 - 36} + \frac{2}{y - 3} = \frac{5}{y + 3}$

$\frac{1}{4(y+3)(y-3)} + \frac{2}{y-3} = \frac{5}{y + 3}$, LCM is $4(y+3)(y-3)$

$$4(y+3)(y-3)\left[\frac{1}{4(y+3)(y-3)} + \frac{2}{y-3}\right] = 4(y+3)(y-3) \cdot \frac{5}{y+3}$$
$$1 + 4 \cdot 2(y + 3) = 4 \cdot 5(y - 3)$$
$$1 + 8y + 24 = 20y - 60$$
$$85 = 12y$$
$$\frac{85}{12} = y$$

Since $\frac{85}{12}$ checks, it is the solution.

33. $\frac{a}{2a - 6} - \frac{3}{a^2 - 6a + 9} = \frac{a - 2}{3a - 9}$

$\frac{a}{2(a - 3)} - \frac{3}{(a - 3)(a - 3)} = \frac{a - 2}{3(a - 3)}$

LCM is $2 \cdot 3(a - 3)(a - 3)$

$$6(a-3)(a-3)\left[\frac{a}{2(a-3)} - \frac{3}{(a-3)(a-3)}\right] = 6(a-3)(a-3) \cdot \frac{a-2}{3(a-3)}$$
$$3a(a - 3) - 6 \cdot 3 = 2(a - 3)(a - 2)$$
$$3a^2 - 9a - 18 = 2(a^2 - 5a + 6)$$
$$3a^2 - 9a - 18 = 2a^2 - 10a + 12$$
$$a^2 + a - 30 = 0$$
$$(a + 6)(a - 5) = 0$$
$$a + 6 = 0 \quad \text{or} \quad a - 5 = 0$$
$$a = -6 \quad \text{or} \quad a = 5$$

Both -6 and 5 check. The solutions are -6 and 5.

35. $\frac{2x + 3}{x - 1} = \frac{10}{x^2 - 1} + \frac{2x - 3}{x + 1}$

$\frac{2x + 3}{x - 1} = \frac{10}{(x - 1)(x + 1)} + \frac{2x - 3}{x + 1}$

LCM is $(x - 1)(x + 1)$

$$(x-1)(x+1) \cdot \frac{2x+3}{x-1} = (x-1)(x+1)\left[\frac{10}{(x-1)(x+1)} + \frac{2x-3}{x+1}\right]$$
$$(x + 1)(2x + 3) = 10 + (x - 1)(2x - 3)$$
$$2x^2 + 5x + 3 = 10 + 2x^2 - 5x + 3$$
$$5x + 3 = 13 - 5x$$
$$10x = 10$$
$$x = 1$$

We know that 1 is not a solution of the original equation because it results in division by 0. The equation has no solution.

37. $\frac{4}{x + 3} + \frac{7}{x^2 - 3x + 9} = \frac{108}{x^3 + 27}$

Note: $x^3 + 27 = (x + 3)(x^2 - 3x + 9)$
Thus the LCM is $(x + 3)(x^2 - 3x + 9)$.

$$(x + 3)(x^2 - 3x + 9)\left(\frac{4}{x + 3} + \frac{7}{x^2 - 3x + 9}\right) = (x + 3)(x^2 - 3x + 9)\left(\frac{108}{(x + 3)(x^2 - 3x + 9)}\right)$$
$$4(x^2 - 3x + 9) + 7(x + 3) = 108$$
$$4x^2 - 12x + 36 + 7x + 21 = 108$$
$$4x^2 - 5x + 51 = 0$$
$$(4x - 17)(x + 3) = 0$$

$$4x - 17 = 0 \quad \text{or} \quad x + 3 = 0$$
$$4x = 17 \quad \text{or} \quad x = -3$$
$$x = \frac{17}{4} \quad \text{or} \quad x = -3$$

We know that -3 is not a solution of the original equation because it results in division by 0. Since $\frac{17}{4}$ checks, it is the solution.

39. $\frac{3x}{x+2} + \frac{6}{x} + 4 = \frac{12}{x^2 + 2x}$

Note: $x^2 + 2x = x(x + 2)$
Thus the LCM is $x(x + 2)$.

$$x(x+2)\left(\frac{3x}{x+2} + \frac{6}{x} + 4\right) = x(x+2)\cdot\frac{12}{x(x+2)}$$

$$x\cdot 3x + (x+2)\cdot 6 + x(x+2)\cdot 4 = 12$$
$$3x^2 + 6x + 12 + 4x^2 + 8x = 12$$
$$7x^2 + 14x = 0$$
$$7x(x+2) = 0$$

$x = 0$ or $x + 2 = 0$

$x = 0$ or $x = -2$

Neither 0 nor -2 checks. Each results in division by 0. The equation has no solution.

41. $$\frac{2x-14}{x^2+3x-28} + \frac{2-x}{4-x} - \frac{x+3}{x+7} = 0$$

$$\frac{2x-14}{(x+7)(x-4)} + \frac{-1}{-1}\cdot\frac{2-x}{4-x} - \frac{x+3}{x+7} = 0$$

$$\frac{2x-14}{(x+7)(x-4)} + \frac{x-2}{x-4} - \frac{x+3}{x+7} = 0,$$

LCM is $(x + 7)(x - 4)$

$$(x+7)(x-4)\left[\frac{2x-14}{(x+7)(x-4)} + \frac{x-2}{x-4} - \frac{x+3}{x+7}\right] = (x+7)(x-4)\cdot 0$$

$$2x - 14 + (x+7)(x-2) - (x-4)(x+3) = 0$$
$$2x - 14 + x^2 + 5x - 14 - x^2 + x + 12 = 0$$
$$8x - 16 = 0$$
$$8x = 16$$
$$x = 2$$

Since 2 checks, it is the solution.

43. $4x^2 - 5x - 51 = (4x - 17)(x + 3)$

45. $1 - t^6$
$= (1 + t^3)(1 - t^3)$
$= (1 + t)(1 - t + t^2)(1 - t)(1 + t + t^2)$

47. $\frac{x+3}{x+2} - \frac{x+4}{x+3} = \frac{x+5}{x+4} - \frac{x+6}{x+5}$

After multiplying by the LCM of the denominators, $(x + 2)(x + 3)(x + 4)(x + 5)$, we have:

$$(x+3)(x+4)(x+5)(x+3) - (x+2)(x+4)(x+5)(x+4) =$$
$$(x+2)(x+3)(x+5)(x+5) - (x+2)(x+3)(x+4)(x+6)$$
$$x^4+15x^3+83x^2+201x+180 - (x^4+15x^3+82x^2+192x+160) =$$
$$x^4+15x^3+81x^2+185x+150 - (x^4+15x^3+80x^2+180x+144)$$
$$x^2 + 9x + 20 = x^2 + 5x + 6$$
$$4x = -14$$
$$x = -\frac{7}{2}$$

Since $-\frac{7}{2}$ checks, it is the solution.

49. $$\frac{36}{x+4} - \frac{27}{x+3} = \frac{9x}{x^2+7x+12}$$

$$\frac{36}{x+4} - \frac{27}{x+3} = \frac{9x}{(x+4)(x+3)},\quad \text{LCM is } (x+4)(x+3)$$

$$(x+4)(x+3)\left[\frac{36}{x+4} - \frac{27}{x+3}\right] = (x+4)(x+3)\cdot\frac{9x}{(x+4)(x+3)}$$

$$36(x+3) - 27(x+4) = 9x$$
$$36x + 108 - 27x - 108 = 9x$$
$$9x = 9x$$

We know that -4 and -3 cannot be solutions of the original equation because they result in division by 0. Since $9x = 9x$ is true for all values of x, all real numbers except -4 and -3 are solutions.

Exercise Set 9.4

1. Familiarize. Let x represent the number. Then $\frac{1}{x}$ represents its reciprocal.

Translate.

A number plus 21 times its reciprocal is -10.

$$x + 21\cdot\frac{1}{x} = -10$$

Solve. We solve the equation.

$$x + \frac{21}{x} = -10,\ \text{LCM is } x$$

$$x\left(x + \frac{21}{x}\right) = x(-10)$$

$$x\cdot x + x\cdot\frac{21}{x} = -10x$$
$$x^2 + 21 = -10x$$
$$x^2 + 10x + 21 = 0$$
$$(x+3)(x+7) = 0$$

$x + 3 = 0$ or $x + 7 = 0$

$x = -3$ or $x = -7$

Check. If the number is -3, its reciprocal is $-\frac{1}{3}$ and $-3 + 21\left(-\frac{1}{3}\right) = -3 - 7 = -10$. If the number is -7, its reciprocal is $-\frac{1}{7}$ and $-7 + 21\left(-\frac{1}{7}\right) = -7 - 3 = -10$. Both values check.

State. The number is -3 or -7.

3. Familiarize. Let x represent the number. Then $\frac{1}{x}$ represents its reciprocal.

Translate.

The reciprocal of 5 plus the reciprocal of 6 is the reciprocal of what number?

$$\frac{1}{5} + \frac{1}{6} = \frac{1}{x}$$

3. (continued)

Solve. We solve the equation.

$$\frac{1}{5} + \frac{1}{6} = \frac{1}{x}, \text{ LCM is } 30x$$

$$30x\left(\frac{1}{5} + \frac{1}{6}\right) = 30x\left(\frac{1}{x}\right)$$

$$\frac{30x}{5} + \frac{30x}{6} = \frac{30x}{x}$$

$$6x + 5x = 30$$

$$11x = 30$$

$$x = \frac{30}{11}$$

Check. $\frac{1}{5} + \frac{1}{6} = \frac{6}{30} + \frac{5}{30} = \frac{11}{30}$, which is the reciprocal of $\frac{30}{11}$. The value checks.

State. The number is $\frac{30}{11}$.

5. Familiarize. Let t = the time it takes them, working together, to fill the order.

Translate. Using the work principle, we get the following equation:

$$\frac{t}{5} + \frac{t}{9} = 1$$

Solve. We solve the equation.

$$\frac{t}{5} + \frac{t}{9} = 1, \text{ LCM is } 45$$

$$45\left(\frac{t}{5} + \frac{t}{9}\right) = 45\cdot 1$$

$$45 \cdot \frac{t}{5} + 45 \cdot \frac{t}{9} = 45$$

$$9t + 5t = 45$$

$$14t = 45$$

$$t = \frac{45}{14}, \text{ or } 3\frac{3}{14}$$

Check. We verify the work principle.

$$\frac{\frac{45}{14}}{5} + \frac{\frac{45}{14}}{9} = \frac{45}{14} \cdot \frac{1}{5} + \frac{45}{14} \cdot \frac{1}{9} = \frac{9}{14} + \frac{5}{14} = \frac{14}{14} = 1$$

State. It will take them $3\frac{3}{14}$ hr, working together.

7. Familiarize. Let t = the time it will take to fill the pool using both the pipe and the hose.

Translate. Using the work principle, we get the following equation:

$$\frac{t}{12} + \frac{t}{30} = 1$$

Solve. We solve the equation.

$$\frac{t}{12} + \frac{t}{30} = 1, \text{ LCM is } 60$$

$$60\left(\frac{t}{12} + \frac{t}{30}\right) = 60\cdot 1$$

$$\frac{60t}{12} + \frac{60t}{30} = 60$$

$$5t + 2t = 60$$

$$7t = 60$$

$$t = \frac{60}{7}, \text{ or } 8\frac{4}{7}$$

Check. We verify the work principle.

$$\frac{\frac{60}{7}}{12} + \frac{\frac{60}{7}}{30} = \frac{60}{7} \cdot \frac{1}{12} + \frac{60}{7} \cdot \frac{1}{30} = \frac{5}{7} + \frac{2}{7} = 1$$

State. It will take $8\frac{4}{7}$ hr to fill the pool using both the pipe and the hose.

9. Familiarize. Let t = the time it will take them to clear the lot working together.

Translate. Using the work principle, we get the following equation:

$$\frac{t}{5.5} + \frac{t}{7.5} = 1$$

Solve. We solve the equation.

$$\frac{t}{5.5} + \frac{t}{7.5} = 1$$

$$\frac{1}{10}\left(\frac{t}{5.5} + \frac{t}{7.5}\right) = \frac{1}{10} \cdot 1 \quad \text{Clearing decimals}$$

$$\frac{t}{55} + \frac{t}{75} = \frac{1}{10}, \text{ LCM is } 1650$$

$$1650\left(\frac{t}{55} + \frac{t}{75}\right) = 1650 \cdot \frac{1}{10}$$

$$30t + 22t = 165$$

$$52t = 165$$

$$t = \frac{165}{52}, \text{ or } 3\frac{9}{52}$$

Check. We verify the work principle.

$$\frac{\frac{165}{52}}{5.5} + \frac{\frac{165}{52}}{7.5} = \frac{165}{52} \cdot \frac{1}{5.5} + \frac{165}{52} \cdot \frac{1}{7.5} =$$

$$\frac{30}{52} + \frac{22}{52} = \frac{52}{52} = 1$$

State. It will take them $3\frac{9}{52}$ hr to clear the lot working together.

11. Familiarize. Let a = the number of hours it takes A to paint the house. Then 4a = the number of hours it takes B to paint the house.

Translate. Using the work principle, we get the following equation.

$$\frac{8}{a} + \frac{8}{4a} = 1, \text{ or } \frac{8}{a} + \frac{2}{a} = 1$$

Solve. We solve the equation.

$$\frac{8}{a} + \frac{2}{a} = 1$$

$$\frac{10}{a} = 1$$

$$10 = a$$

Check. If this number checks, the answer will be that A takes 10 hr and B takes 4·10, or 40 hr. Part of the check is obvious: B takes 4 times as long as A, or A is 4 times as fast as B. We complete the check by verifying the work principle.

$$\frac{8}{10} + \frac{8}{40} = \frac{4}{5} + \frac{1}{5} = \frac{5}{5} = 1$$

State. It would take A 10 days to paint the house alone, and it would take B 40 days.

13. Let d = the distance light can travel in 60 sec. We translate to a proportion and solve for d.

$$\text{Distance} \longrightarrow \frac{930{,}000}{5} = \frac{d}{60} \longleftarrow \text{Distance}, \quad \text{Time} \longrightarrow 5, \; 60 \longleftarrow \text{Time}$$

$930{,}000 \cdot 60 = 5 \cdot d$ Using cross products

$$\frac{930{,}000 \cdot 60}{5} = d$$

$$11{,}160{,}000 = d$$

Light can travel 11,160,000 mi in 60 sec.

15. Let h = the number of hits the player will have. We translate to a proportion and solve for h.

$$\text{Hits} \longrightarrow \frac{120}{300} = \frac{h}{500} \longleftarrow \text{Hits}, \quad \text{At bats} \longrightarrow 300, \; 500 \longleftarrow \text{At bats}$$

$120 \cdot 500 = 300 \cdot h$ Using cross products

$$\frac{120 \cdot 500}{300} = h$$

$$200 = h$$

The player will get 200 hits.

17. Let d = the distance the student would travel in 56 days. We translate to a proportion and solve for d.

$$\text{Distance} \longrightarrow \frac{234}{14} = \frac{d}{56} \longleftarrow \text{Distance}, \quad \text{Days} \longrightarrow 14, \; 56 \longleftarrow \text{Days}$$

$234 \cdot 56 = 14 \cdot d$ Using cross products

$$\frac{234 \cdot 56}{14} = d$$

$$936 = d$$

The student would travel 936 km.

19. Let D = the number of deer in the game preserve. We translate to a proportion and solve for D.

Deer tagged originally $\longrightarrow$ 318, Deer in preserve $\longrightarrow$ D:

$$\frac{318}{D} = \frac{56}{168}$$

56 $\longleftarrow$ Tagged deer caught later; 168 $\longleftarrow$ Deer caught later

$318 \cdot 168 = D \cdot 56$ Using cross products

$$\frac{318 \cdot 168}{56} = D$$

$$954 = D$$

There are 954 deer in the game preserve.

21. a) Let w = the number of tons the rocket will weigh on Mars. We translate to a proportion and solve for w.

$$\text{Weight on Mars} \longrightarrow \frac{0.4}{1} = \frac{w}{12} \longleftarrow \text{Weight on Mars}, \quad \text{Weight on earth} \longrightarrow 1, \; 12 \longleftarrow \text{Weight on earth}$$

$0.4 \cdot 12 = 1 \cdot w$ Using cross products

$$4.8 = w$$

A 12-T rocket will weigh 4.8 T on Mars.

b) Let w = the number of pounds the astronaut will weigh on Mars. We translate to a proportion and solve for w.

$$\text{Weight on Mars} \longrightarrow \frac{0.4}{1} = \frac{w}{120} \longleftarrow \text{Weight on Mars}, \quad \text{Weight on earth} \longrightarrow 1, \; 120 \longleftarrow \text{Weight on earth}$$

$0.4 \cdot 120 = 1 \cdot w$ Using cross products

$$48 = w$$

A 120-lb astronaut will weigh 48 lb on Mars.

23. Let x = the number that is added to each of the given numbers. We translate to a proportion and solve for x.

$$\frac{1 + x}{2 + x} = \frac{3 + x}{5 + x}$$

$(1 + x)(5 + x) = (2 + x)(3 + x)$ Using cross products

$$5 + 6x + x^2 = 6 + 5x + x^2$$

$$x = 1$$

The number is 1.

25. Familiarize. We first make a drawing. We let r represent the speed of the boat in still water. Then r - 3 is the speed upstream and r + 3 is the speed downstream.

Upstream 4 miles r - 3 mph

10 miles r + 3 mph Downstream

We organize the information in a table. The time is the same both upstream and downstream so we use t for each time.

25. (continued)

	Distance	Speed	Time
Upstream	4	r - 3	t
Downstream	10	r + 3	t

Translate. Using $t = \frac{d}{r}$ we get two different equations from the rows of the table.

$$t = \frac{4}{r - 3} \text{ and } t = \frac{10}{r + 3}$$

Solve. Since both rational expressions represent the same time, t, we can set them equal to each other and solve.

$$\frac{4}{r - 3} = \frac{10}{r + 3}, \text{ LCM is } (r - 3)(r + 3)$$

$$(r - 3)(r + 3) \cdot \frac{4}{r - 3} = (r - 3)(r + 3) \cdot \frac{10}{r + 3}$$

$$4(r + 3) = 10(r - 3)$$

$$4r + 12 = 10r - 30$$

$$42 = 6r$$

$$7 = r$$

Check. If r = 7 mph, then r - 3 is 4 mph and r + 3 is 10 mph. The time upstream is $\frac{4}{4}$, or 1 hour. The time downstream is $\frac{10}{10}$, or 1 hour. The times are the same. The values check.

State. The speed of the boat in still water is 7 mph.

27. Familiarize. We first make a drawing. We let t represent the time for the first train. Then t - 2 represents the time for the second train. Since the distances are the same, we use d for each distance.

75 km/h
t hours
d km
First train

125 km/h
t - 2 hours
d km
Second train

We organize the information in a table.

Train	Distance	Speed	Time
First	d	75	t
Second	d	125	t - 2

Translate. From the rows of the table we get two different equations.

$d = 75t$ and $d = 125(t - 2)$.

Since the distances are the same, we get

$75t = 125(t - 2)$

27. (continued)

Solve. We solve the equation.

$75t = 125t - 250$

$250 = 50t$

$5 = t$

Check. If t is 5 hours, then the time for the second train is 5 - 2, or 3 hours. The first train travels 75·5, or 375 km. The second train travels 125·3, or 375 km. The distances are the same.

State. The second train will overtake the first train 375 km from the station.

29. Familiarize. We let r represent the speed of Train B. Then r - 12 represents the speed of Train A. The times are the same. We use t for each. We organize the information in a table.

	Distance	Rate	Time
Train A	230	r - 12	t
Train B	290	r	t

Translate. Using $t = \frac{d}{r}$, we get two equations.

$$t = \frac{230}{r - 12} \text{ and } t = \frac{290}{r}$$

Solve. Set the rational expressions equal to each other and solve.

$$\frac{230}{r - 12} = \frac{290}{r}, \text{ LCM is } r(r - 12)$$

$$r(r - 12) \cdot \frac{230}{r - 12} = r(r - 12) \cdot \frac{290}{r}$$

$$230r = 290(r - 12)$$

$$230r = 290r - 3480$$

$$-60r = -3480$$

$$r = 58$$

Check. If the speed of Train B is 58 mph, then the speed of Train A is 58 - 12, or 46 mph. The time for Train A is $\frac{230}{46}$, or 5 hours. The time for Train B is $\frac{290}{58}$, or 5 hours. The times are the same. The values check.

State. The speed of Train A is 46 mph; the speed of Train B is 58 mph.

31. Familiarize. We first make a drawing. We let r represent the speed of the river. Then 15 + r is the speed downstream and 15 - r is the speed upstream. The times are the same.

140 km 15 + r t hours Downstream

35 km 15 - r t hours Upstream

We organize the information in a table.

	Distance	Speed	Time
Downstream	140	15 + r	t
Upstream	35	15 - r	t

Translate. Using $t = \frac{d}{r}$, we get two equations from the rows of the table.

$t = \frac{140}{15 + r}$ and $t = \frac{35}{15 - r}$

Since the times are the same, we have

$$\frac{140}{15 + r} = \frac{35}{15 - r}.$$

Solve. We solve the equation. The LCM is $(15 + r)(15 - r)$.

$$(15 + r)(15 - r)\cdot\frac{140}{15 + r} = (15 + r)(15 - r)\cdot\frac{35}{15 - r}$$

$$140(15 - r) = 35(15 + r)$$

$$2100 - 140r = 525 + 35r$$

$$1575 = 175r$$

$$9 = r$$

Check. If r = 9, then the speed downstream is 15 + 9, or 24 km/h and the speed upstream is 15 - 9, or 6 km/h. The time for the trip downstream is $\frac{140}{24}$, or $5\frac{5}{6}$ hours. The time for the trip upstream is $\frac{35}{6}$, or $5\frac{5}{6}$ hours. The times are the same. The values check.

State. The speed of the river is 9 km/h.

33. Familiarize. Let x represent the number of city-miles driven and let y represent the number of highway-miles. Then $\frac{x}{22.5}$ and $\frac{y}{30}$ represent the number of gallons of gasoline used in city and highway driving, respectively.

Translate.

The total number of miles is 465.

$$x + y = 465$$

The total number of gallons is 18.4.

$$\frac{x}{22.5} + \frac{y}{30} = 18.4$$

We have a system of equations:

$$x + y = 465,$$

$$\frac{x}{22.5} + \frac{y}{30} = 18.4$$

33. (continued)

Solve. Solving the system we get (261,204).

Check. The total number of miles is 261 + 204, or 465. The total number of gallons of gasoline used is $\frac{261}{22.5} + \frac{204}{30} = 11.6 + 6.8 = 18.4$. The answer checks.

State. 261 miles were driven in the city, and 204 miles were driven on the highway.

35. Familiarize. It helps to first make a drawing.

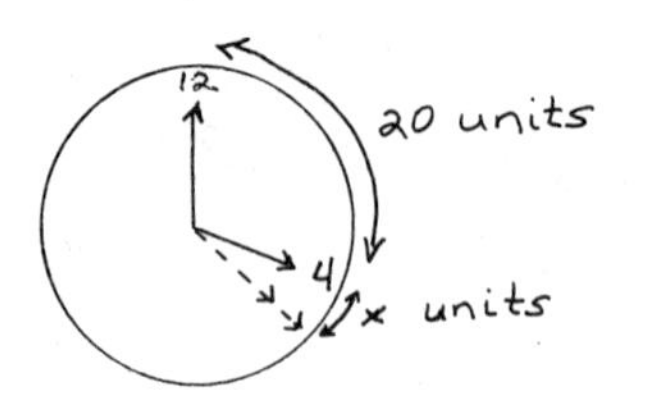

The minute hand moves 60 units per hour while the hour hand moves 5 units per hour, where one unit represents one minute on the face of the clock. When the hands are in the same position the first time, the hour hand will have moved x units and the minute hand will have moved x + 20 units. The times are the same. We use t for time.

	Distance	Speed	Time
Minute	x + 20	60	t
Hour	x	5	t

Translate. Using $t = \frac{d}{r}$, we get two equations from the table.

$t = \frac{x + 20}{60}$ and $t = \frac{x}{5}$

Solve. Set the rational expressions equal to each other and solve.

$$\frac{x + 20}{60} = \frac{x}{5}, \text{ LCM is } 60$$

$$60\cdot\frac{x + 20}{60} = 60\cdot\frac{x}{5}$$

$$x + 20 = 12x$$

$$20 = 11x$$

$$\frac{20}{11} = x$$

$$\text{or } x = 1\frac{9}{11}$$

Check. If the hour hand moves $1\frac{9}{11}$ units, then the minute hand moves $1\frac{9}{11} + 20$, or $21\frac{9}{11}$ units $\left[21\frac{9}{11} \text{ minutes after } 4\right]$. The time for the hour hand is $\frac{20}{11} \div 5$, or $\frac{4}{11}$ hour. The time for the minute hand is $\frac{240}{11} \div 60$, or $\frac{4}{11}$ hour. The times are the same. The values check.

State. At $21\frac{9}{11}$ minutes after 4:00, the hands will be in the same position.

37. Familiarize. We first make a drawing. We let x represent the speed of the boat in still water and r represent the speed of the stream. Then x + r represents the speed downstream and x - r represents the speed upstream.

96 km 4 hr x + r km/h

Downstream

28 km 7 hr x - r km/h

Upstream

We organize the information in a table.

	Distance	Speed	Time
Downstream	96	x + r	4
Upstream	28	x - r	7

Translate. Using d = rt, we get a system of equations from the table.

$96 = (x + r)4$ or $x + r = 24$

$28 = (x - r)7$ or $x - r = 4$

Solve. Solving the system we get (10,14).

Check. The speed downstream is 14 + 10, or 24 km/h. The distance downstream is 24·4, or 96 km. The speed upstream is 14 - 10, or 4 km/h. The distance upstream is 4·7, or 28 km. The values check.

State. The speed of the boat is 14 km/h; the speed of the stream is 10 km/h.

39. Familiarize. We let x represent the speed of the current and 3x represent the speed of the boat. Then the speed up the river is 3x - x, or 2x, and the speed down the river is 3x + x, or 4x. The total distance is 100 km; thus the distance each way is 50 km. Using $t = \frac{d}{r}$, we can use $\frac{50}{2x}$ for the time up the river and $\frac{50}{4x}$ for the time down the river.

Translate. Since the total of the times is 10 hours, we have the following equation.

$$\frac{50}{2x} + \frac{50}{4x} = 10$$

Solve. We solve the equation. The LCM is 4x.

$$4x\left[\frac{50}{2x} + \frac{50}{4x}\right] = 4x\cdot 10$$

$$100 + 50 = 40x$$

$$150 = 40x$$

$$\frac{15}{5} = x$$

$$\text{or } x = 3\frac{3}{4}$$

39. (continued)

Check. If the speed of the current is $\frac{15}{4}$ km/h, then the speed of the boat is $3 \cdot \frac{15}{4}$, or $\frac{45}{4}$. The speed up the river is $\frac{45}{4} - \frac{15}{4}$, or $\frac{15}{2}$ km/h, and the time traveling up the river is $50 \div \frac{15}{2}$, or $6\frac{2}{3}$ hr. The speed down the river is $\frac{45}{4} + \frac{15}{4}$, or 15 km/h, and the time traveling down the river is $50 \div 15$, or $3\frac{1}{3}$ hr. The total time for the trip is $6\frac{2}{3} + 3\frac{1}{3}$, or 10 hr. The value checks.

State. The speed of the current is $3\frac{3}{4}$ km/h.

41. Familiarize. We make a drawing and organize the information in a table. Remember: $t = \frac{d}{r}$.

200 km

100 km 100 km

40 km/h 60 km/h

	Distance	Speed	Time
1st part	100	40	$\frac{100}{40}$, or $\frac{5}{2}$
2nd part	100	60	$\frac{100}{60}$, or $\frac{5}{3}$

The total distance is 200 km.

The total time is $\frac{5}{2} + \frac{5}{3}$, or $\frac{25}{6}$ hr.

Translate. $\text{Average speed} = \frac{\text{Total distance}}{\text{Total time}}$

$$= \frac{200 \text{ km}}{\frac{25}{6} \text{ hr}}$$

Solve. We simplify.

$$\text{Average speed} = \frac{200 \text{ km}}{1} \cdot \frac{6}{25 \text{ hr}} = 48 \text{ km/h}$$

Check. Calculate the total time at 48 km/h. $\frac{200}{48} = \frac{25}{6}$ hr. The answer checks.

State. The average speed was 48 km/h.

43. Familiarize. Trucks A, B, and C, working together, move a load of sand in t hours. Thus, together, they do $\frac{1}{t}$ of the job in 1 hour. Truck A, alone, can move a load of sand in t + 1 hours. Thus, Truck A does $\frac{1}{t + 1}$ of the job in 1 hour. Truck B, alone, can move a load of sand in t + 6 hours. Thus, Truck B does $\frac{1}{t + 6}$ of the job in 1 hour. Truck C, alone, can move a load of sand in t + t, or 2t hours. Thus, Truck C does $\frac{1}{2t}$ of the job in 1 hour. Working together, they can do $\frac{1}{t + 1} + \frac{1}{t + 6} + \frac{1}{2t}$ of the job in 1 hour.

43. (continued)

Translate. We want to find t such that

$$t\left(\frac{1}{t+1}+\frac{1}{t+6}+\frac{1}{2t}\right)=1\text{, or}$$

$$\frac{t}{t+1}+\frac{t}{t+6}+\frac{1}{2}=1$$

Solve. We solve the equation.

$$\frac{t}{t+1}+\frac{t}{t+6}=\frac{1}{2}\quad\text{Adding } -\frac{1}{2}$$

$$2(t+1)(t+6)\left[\frac{t}{t+1}+\frac{t}{t+6}\right]=2(t+1)(t+6)\cdot\frac{1}{2}$$

$$2t(t+6)+2t(t+1)=(t+1)(t+6)$$

$$2t^2+12t+2t^2+2t=t^2+7t+6$$

$$4t^2+14t=t^2+7t+6$$

$$3t^2+7t-6=0$$

$$(3t-2)(t+3)=0$$

$$3t-2=0\quad\text{or}\quad t+3=0$$

$$3t=2\quad\text{or}\quad t=-3$$

$$t=\frac{2}{3}\quad\text{or}\quad t=-3$$

Check. Since time cannot be negative, we need only check $\frac{2}{3}$. If $t=\frac{2}{3}$, then $\frac{1}{t}=\frac{3}{2}$. If $t=\frac{2}{3}$, then working alone it takes A $\frac{2}{3}+1$, or $\frac{5}{3}$ hr, B $\frac{2}{3}+6$, or $\frac{20}{3}$ hr, and C $2\cdot\frac{2}{3}$, or $\frac{4}{3}$ hr. We calculate.

$$\frac{1}{\frac{5}{3}}+\frac{1}{\frac{20}{3}}+\frac{1}{\frac{4}{3}}$$

$$=\frac{3}{5}+\frac{3}{20}+\frac{3}{4}=\frac{12}{20}+\frac{3}{20}+\frac{15}{20}=\frac{30}{20}=\frac{3}{2}$$

State. Working together, it takes $\frac{2}{3}$ hour.

45. Let x = the number of operations the computer can do in one minute. We translate to a proportion and solve for x. We will express 1 min as 60 sec.

$$\begin{array}{l}\text{Operations}\longrightarrow\\ \text{Seconds}\longrightarrow\end{array}\frac{1}{1.23\times10^{-6}}=\frac{x}{60}\begin{array}{l}\longleftarrow\text{Operations}\\ \longleftarrow\text{Seconds}\end{array}$$

$$1\cdot60=1.23\times10^{-6}\cdot x$$

Using cross products

$$\frac{1\cdot60}{1.23\times10^{-6}}=x$$

$$48{,}780{,}488\approx x$$

The computer can do about 48,780,488 operations in one minute.

Exercise Set 9.5

1. $$\frac{W_1}{W_2}=\frac{d_1}{d_2}$$

$$\frac{d_2W_1}{W_2}=d_1\qquad\text{Multiplying by } d_2$$

3. $$s=\frac{(v_1+v_2)t}{2}$$

$$2s=(v_1+v_2)t\qquad\text{Multiplying by 2}$$

$$\frac{2s}{v_1+v_2}=t\qquad\text{Dividing by } v_1+v_2$$

5. $$\frac{1}{R}=\frac{1}{r_1}+\frac{1}{r_2}$$

$$Rr_1r_2\cdot\frac{1}{R}=Rr_1r_2\left(\frac{1}{r_1}+\frac{1}{r_2}\right)\qquad\text{Multiplying by } Rr_1r_2$$

$$r_1r_2=Rr_2+Rr_1$$

$$r_1r_2-Rr_2=Rr_1\qquad\text{Subtracting } Rr_2$$

$$r_2(r_1-R)=Rr_1\qquad\text{Factoring}$$

$$r_2=\frac{Rr_1}{r_1-R}\qquad\text{Dividing by } r_1-R$$

7. $$R=\frac{gs}{g+s}$$

$$R(g+s)=gs\qquad\text{Multiplying by } g+s$$

$$Rg+Rs=gs\qquad\text{Removing parentheses}$$

$$Rg=gs-Rs\qquad\text{Subtracting Rs}$$

$$Rg=s(g-r)\qquad\text{Factoring}$$

$$\frac{Rg}{g-R}=s\qquad\text{Dividing } g-R$$

9. $$\frac{1}{p}+\frac{1}{q}=\frac{1}{f}$$

$$pqf\left(\frac{1}{p}+\frac{1}{q}\right)=pqf\cdot\frac{1}{f}\qquad\text{Multiplying by pqf}$$

$$qf+pf=pq$$

$$qf=pq-pf\qquad\text{Subtracting pf}$$

$$qf=p(q-f)\qquad\text{Factoring}$$

$$\frac{qf}{q-f}=p\qquad\text{Dividing by } q-f$$

11. $$I=\frac{nE}{E+nr}$$

$$(E+nr)I=(E+nr)\cdot\frac{nE}{E+nr}\qquad\text{Multiplying by } E+nr$$

$$EI+nrI=nE$$

$$nrI=nE-EI\qquad\text{Subtracting EI}$$

$$nrI=E(n-I)\qquad\text{Factoring}$$

$$\frac{nrI}{n-I}=E\qquad\text{Dividing by } n-I$$

13. $$S=\frac{H}{m(t_1-t_2)}$$

$$Sm(t_1-t_2)=H\qquad\text{Multiplying by } m(t_1-t_2)$$

15. $$\frac{E}{e}=\frac{R+r}{r}$$

$$er\cdot\frac{E}{e}=er\cdot\frac{R+r}{r}\qquad\text{Multiplying by er}$$

$$rE=e(R+r)$$

$$\frac{rE}{R+r}=e\qquad\text{Dividing by } R+r$$

17. $A = P(1 + rt)$

$A = P + Prt$ Removing parentheses

$A - P = Prt$ Subtracting P

$\frac{A - P}{Pt} = r$ Dividing by Pt

19. Graph: $6x - y < 6$

First graph the line $6x - y = 6$. The intercepts are (0,-6) and (1,0). We draw the line dashed since the inequality is <. Since the ordered pair (0,0) is a solution of the inequality ($6 \cdot 0 - 0 < 0$ is true), we shade the half-plane containing (0,0).

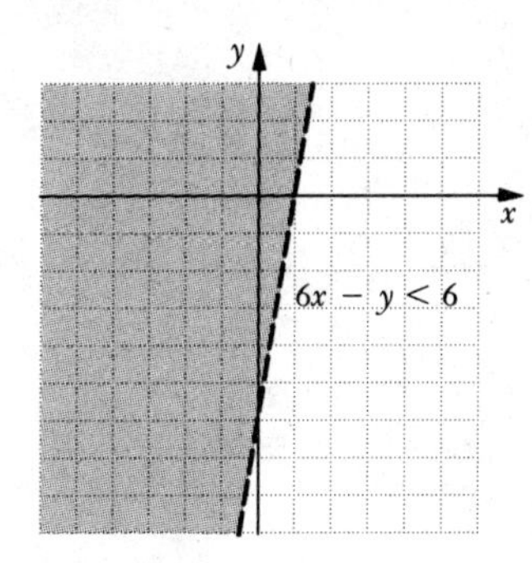

21. $t^3 + 8b^3 = t^3 + (2b)^3 = (t + 2b)(t^2 - 2bt + 4b^2)$

23. Using the work principle, we write the formula:

$\frac{t}{a} + \frac{t}{b} = 1$

25. $\frac{t}{a} + \frac{t}{b} = 1$

$ab\left(\frac{t}{a} + \frac{t}{b}\right) = ab \cdot 1$

$bt + at = ab$

$bt = ab - at$

$bt = a(b - t)$

$\frac{bt}{b - t} = a$

Exercise Set 9.6

1. $\frac{\frac{1}{x} + 4}{\frac{1}{x} - 3} = \frac{\frac{1}{x} + 4}{\frac{1}{x} - 3} \cdot \frac{x}{x}$ Using the LCM

$= \frac{\frac{1}{x} \cdot x + 4 \cdot x}{\frac{1}{x} \cdot x - 3 \cdot x}$

$= \frac{1 + 4x}{1 - 3x}$

3. $\frac{x - \frac{1}{x}}{x + \frac{1}{x}} = \frac{x - \frac{1}{x}}{x + \frac{1}{x}} \cdot \frac{x}{x}$ Using the LCM

$= \frac{x \cdot x - \frac{1}{x} \cdot x}{x \cdot x + \frac{1}{x} \cdot x}$

$= \frac{x^2 - 1}{x^2 + 1}$

5. $\frac{\frac{3}{x} + \frac{4}{y}}{\frac{4}{x} - \frac{3}{y}} = \frac{\frac{3}{x} + \frac{4}{y}}{\frac{4}{x} - \frac{3}{y}} \cdot \frac{xy}{xy}$ Using the LCM

$= \frac{\frac{3}{x} \cdot xy + \frac{4}{y} \cdot xy}{\frac{4}{x} \cdot xy - \frac{3}{y} \cdot xy}$

$= \frac{3y + 4x}{4y - 3x}$

7. $\frac{\frac{x^2 - y^2}{xy}}{\frac{x - y}{y}} = \frac{x^2 - y^2}{xy} \cdot \frac{y}{x - y}$ Multiplying by the reciprocal of the divisor

$= \frac{(x + y)(x - y)}{xy} \cdot \frac{y}{x - y}$

$= \frac{y(x - y)}{y(x - y)} \cdot \frac{x + y}{x}$

$= \frac{x + y}{x}$

9. $\frac{a - \frac{3a}{b}}{b - \frac{b}{a}} = \frac{a - \frac{3a}{b}}{b - \frac{b}{a}} \cdot \frac{ab}{ab}$ Using the LCM

$= \frac{a(ab) - \frac{3a}{b} \cdot ab}{b(ab) - \frac{b}{a} \cdot ab}$

$= \frac{a^2b - 3a^2}{ab^2 - b^2}$

$= \frac{a^2(b - 3)}{b^2(a - 1)}$

11. $\frac{\frac{1}{a} + \frac{1}{b}}{\frac{a^2 - b^2}{ab}} = \frac{\frac{1}{a} + \frac{1}{b}}{\frac{a^2 - b^2}{ab}} \cdot \frac{ab}{ab}$ Using the LCM

$= \frac{\frac{1}{a} \cdot ab + \frac{1}{b} \cdot ab}{\frac{a^2 - b^2}{ab} \cdot ab}$

$= \frac{b + a}{a^2 - b^2} = \frac{b + a}{(a + b)(a - b)}$

$= \frac{a + b}{a + b} \cdot \frac{1}{a - b}$ $(b + a = a + b)$

$= \frac{1}{a - b}$

13. $\dfrac{\frac{1}{x+h} - \frac{1}{x}}{h} = \dfrac{\frac{1}{x+h}\cdot\frac{x}{x} - \frac{1}{x}\cdot\frac{x+h}{x+h}}{h}$ Adding in the numerator

$= \dfrac{\frac{x - x - h}{x(x+h)}}{h} = \dfrac{\frac{-h}{x(x+h)}}{h}$

$= \dfrac{-h}{x(x+h)}\cdot\dfrac{1}{h}$ Multiplying by the reciprocal of the divisor

$= \dfrac{h}{h}\cdot\dfrac{-1}{x(x+h)}$

$= \dfrac{-1}{x(x+h)}$, or $-\dfrac{1}{x(x+h)}$

15. $\dfrac{\frac{y^2 - y - 6}{y^2 - 5y - 14}}{\frac{y^2 + 6y + 5}{y^2 - 6y - 7}} = \dfrac{y^2 - y - 6}{y^2 - 5y - 14}\cdot\dfrac{y^2 - 6y - 7}{y^2 + 6y + 5}$

$= \dfrac{(y-3)(y+2)(y-7)(y+1)}{(y-7)(y+2)(y+5)(y+1)}$

$= \dfrac{(y+2)(y-7)(y+1)}{(y+2)(y-7)(y+1)}\cdot\dfrac{y-3}{y+5}$

$= \dfrac{y-3}{y+5}$

17. $\dfrac{\frac{1}{x+2} + \frac{4}{x-3}}{\frac{2}{x-3} - \frac{7}{x+2}}$

$= \dfrac{\frac{1}{x+2} + \frac{4}{x-3}}{\frac{2}{x-3} - \frac{7}{x+2}}\cdot\dfrac{(x+2)(x-3)}{(x+2)(x-3)}$ Using the LCM

$= \dfrac{\frac{1}{x+2}\cdot(x+2)(x-3) + \frac{4}{x-3}\cdot(x+2)(x-3)}{\frac{2}{x-3}\cdot(x+2)(x-3) - \frac{7}{x+2}\cdot(x+2)(x-3)}$

$= \dfrac{x - 3 + 4(x+2)}{2(x+2) - 7(x-3)}$

$= \dfrac{x - 3 + 4x + 8}{2x + 4 - 7x + 21}$

$= \dfrac{5x + 5}{-5x + 25}$

$= \dfrac{5(x+1)}{5(-x+5)}$ Removing a factor of 1

$= \dfrac{x+1}{-x+5}$, or $\dfrac{x+1}{5-x}$

19. $\dfrac{\frac{6}{x^2 - 4} - \frac{5}{x+2}}{\frac{7}{x^2 - 4} - \frac{4}{x-2}}$

$= \dfrac{\frac{6}{(x+2)(x-2)} - \frac{5}{x+2}\cdot\frac{x-2}{x-2}}{\frac{7}{(x+2)(x-2)} - \frac{4}{x-2}\cdot\frac{x+2}{x+2}}$ Find the LCM and multiplying by 1

$= \dfrac{\frac{6}{(x+2)(x-2)} - \frac{5x - 10}{(x+2)(x-2)}}{\frac{7}{(x+2)(x-2)} - \frac{4x + 8}{(x+2)(x-2)}}$

$= \dfrac{\frac{6 - 5x + 10}{(x+2)(x-2)}}{\frac{7 - 4x - 8}{(x+2)(x-2)}}$ Subtracting in the numerator and in the denominator

$= \dfrac{16 - 5x}{(x+2)(x-2)}\cdot\dfrac{(x+2)(x-2)}{-1 - 4x}$ Multiplying by the reciprocal of the denominator

$= \dfrac{(16 - 5x)(x+2)(x-2)}{(x+2)(x-2)(-1 - 4x)}$ Removing a factor of 1

$= \dfrac{16 - 5x}{-1 - 4x}$

$= \dfrac{-1\cdot(5x - 16)}{-1\cdot(4x + 1)}$ Removing a factor of 1

$= \dfrac{5x - 16}{4x + 1}$

21. $\dfrac{\frac{1}{a^2} - \frac{1}{b^2}}{\frac{1}{a^3} + \frac{1}{b^3}} = \dfrac{\frac{1}{a^2} - \frac{1}{b^2}}{\frac{1}{a^3} + \frac{1}{b^3}}\cdot\dfrac{a^3b^3}{a^3b^3}$ Using the LCM

$= \dfrac{\frac{1}{a^2}\cdot a^3b^3 - \frac{1}{b^2}\cdot a^3b^3}{\frac{1}{a^3}\cdot a^3b^3 + \frac{1}{b^3}\cdot a^3b^3}$

$= \dfrac{ab^3 - a^3b}{b^3 + a^3}$

$= \dfrac{ab(b^2 - a^2)}{(b+a)(b^2 - ab + a^2)}$

$= \dfrac{ab(b+a)(b-a)}{(b+a)(b^2 - ab + a^2)}$ Removing a factor of 1

$= \dfrac{ab(b-a)}{b^2 - ab + a^2}$

23. $\dfrac{5x^{-1} - 5y^{-1} + 10x^{-1}y^{-1}}{6x^{-1} - 6y^{-1} + 12x^{-1}y^{-1}} = \dfrac{\frac{5}{x} - \frac{5}{y} + \frac{10}{xy}}{\frac{6}{x} - \frac{6}{y} + \frac{12}{xy}}$

$= \dfrac{\frac{5}{x} - \frac{5}{y} + \frac{10}{xy}}{\frac{6}{x} - \frac{6}{y} + \frac{12}{xy}}\cdot\dfrac{xy}{xy}$

$= \dfrac{5y - 5x + 10}{6y - 6x + 12} = \dfrac{5(y - x + 2)}{6(y - x + 2)} = \dfrac{5}{6}$

25. $2 + \cfrac{2}{2 + \cfrac{2}{2 + \cfrac{2}{2 + \frac{2}{x}}}} = 2 + \cfrac{2}{2 + \cfrac{2}{2 + \cfrac{2}{\frac{2x + 2}{x}}}}$

$= 2 + \cfrac{2}{2 + \cfrac{2}{2 + \frac{2x}{2x + 2}}} = 2 + \cfrac{2}{2 + \cfrac{2}{\frac{6x + 4}{2x + 2}}}$

$= 2 + \cfrac{2}{2 + \frac{4x + 4}{6x + 4}} = 2 + \cfrac{2}{\frac{16x + 12}{6x + 4}}$

$= 2 + \frac{12x + 8}{16x + 12} = \frac{44x + 32}{16x + 12}$

$= \frac{4(11x + 8)}{4(4x + 3)} = \frac{11x + 8}{4x + 3}$

27. $\frac{(a^2b^{-1} + b^2a^{-1})(a^{-2} - b^{-2})}{(a^2 - ab + b^2)(a^{-2} + 2a^{-1}b^{-1} + b^{-2})}$

$= \frac{\left(\frac{a^2}{b} + \frac{b^2}{a}\right)\left(\frac{1}{a^2} - \frac{1}{b^2}\right)}{(a^2 - ab + b^2)\left(\frac{1}{a^2} + \frac{2}{ab} + \frac{1}{b^2}\right)}$

$= \frac{\left(\frac{a^3 + b^3}{ab}\right)\left(\frac{b^2 - a^2}{a^2b^2}\right)}{(a^2 - ab + b^2)\left(\frac{b^2 + 2ab + a^2}{a^2b^2}\right)}$

$= \frac{(a+b)(a^2-ab+b^2)(b+a)(b-a)}{a^3b^3} \cdot \frac{a^2b^2}{(a^2-ab+b^2)(b+a)^2}$

$= \frac{b - a}{ab}$

29. $\frac{a - 1}{1 - \frac{1}{a}} = \frac{a - 1}{\frac{a - 1}{a}} = \frac{a - 1}{1} \cdot \frac{a}{a - 1} = a$

Exercise Set 9.7

1. $y = kx$

$24 = k \cdot 3$ Substituting

$8 = k$

The variation constant is 8.
The equation of variation is $y = 8x$.

3. $y = kx$

$3.6 = k \cdot 1$ Substituting

$3.6 = k$

The variation constant is 3.6.
The equation of variation is $y = 3.6x$.

5. $y = kx$

$0.8 = k(0.5)$ Substituting

$\frac{0.8}{0.5} = k$

$\frac{8}{5} = k$

The variation constant is $\frac{8}{5}$.
The equation of variation is $y = \frac{8}{5}x$.

7. $I = kV$ I varies directly as V

$4 = k \cdot 12$ Substituting

$\frac{1}{3} = k$ Variation constant

$I = \frac{1}{3}V$ Equation of variation

$I = \frac{1}{3} \cdot 18$ Substituting

$I = 6$

The current is 6 amperes.

9. $A = kG$ A varies directly as G

$\$9.66 = k \cdot 9$ Substituting

$\$\frac{161}{150} = k$ Variation constant

$A = \$\frac{161}{150}G$ Equation of variation

$A = \$\frac{161}{150} \cdot 4$ Substituting

$A \approx \$4.29$

The average weekly allowance of a 4th grade student is about \$4.29.

11. Let T represent the person's weight.

$W = kT$ W varies directly as T

$64 = k \cdot 96$ Substituting

$\frac{64}{96} = k$ Variation constant

$\frac{2}{3} = k$

$W = \frac{2}{3}T$ Equation of variation

$W = \frac{2}{3} \cdot 75$ Substituting

$W = 50$

A person weighing 75 kg contains 50 kg of water.

13. $y = \frac{k}{x}$

$6 = \frac{k}{10}$ Substituting

$60 = k$

The variation constant is 60.
The equation of variation is $y = \frac{60}{x}$.

15. $y = \frac{k}{x}$

$12 = \frac{k}{3}$ Substituting

$36 = k$

The variation constant is 36.
The equation of variation is $y = \frac{36}{x}$.

17. $y = \frac{k}{x}$

$0.4 = \frac{k}{0.8}$ Substituting

$0.32 = k$

The variation constant is 0.32.

The equation of variation is $y = \frac{0.32}{x}$.

19. $I = \frac{k}{R}$ I varies inversely as R

$\frac{1}{2} = \frac{k}{240}$ Substituting

$\frac{240}{2} = k$

$120 = k$ Variation constant

$I = \frac{120}{R}$ Equation of variation

$I = \frac{120}{540}$ Substituting

$I = \frac{2}{9}$

The current is $\frac{2}{9}$ ampere.

21. $W = \frac{k}{L}$ W varies inversely as L

$1200 = \frac{k}{8}$ Substituting

$9600 = k$ Variation constant

$W = \frac{9600}{L}$ Equation of variation

$W = \frac{9600}{14}$ Substituting

$W = 685\frac{5}{7}$

A 14-meter beam can support $685\frac{5}{7}$ kg.

23. $y = kx^2$

We first find k.

$0.15 = k(0.1)^2$ Substituting 0.15 for y and 0.1 for x

$0.15 = 0.01k$

$\frac{0.15}{0.01} = k$

$15 = k$

The equation of variation is $y = 15x^2$.

25. $y = \frac{k}{x^2}$

We first find k.

$0.15 = \frac{k}{(0.1)^2}$ Substituting 0.15 for y and 0.1 for x

$0.15 = \frac{k}{0.01}$

$0.15(0.01) = k$

$0.0015 = k$

The equation of variation is $y = \frac{0.0015}{x^2}$.

27. $y = kxz$

We first find k.

$56 = k \cdot 7 \cdot 8$ Substituting 56 for y, 7 for x, and 8 for z

$56 = 56k$

$1 = k$

The equation of variation is $y = xz$.

29. $y = kxz^2$

We first find k.

$105 = k \cdot 14 \cdot 5^2$ Substituting 105 for y, 14 for x, and 5 for z

$105 = 350k$

$\frac{105}{350} = k$

$\frac{3}{10} = k$

The equation of variation is $y = \frac{3}{10}xz^2$.

31. $y = k\,\frac{xz}{wp}$

We first find k.

$\frac{3}{28} = k\,\frac{3 \cdot 10}{7 \cdot 8}$ Substituting $\frac{3}{28}$ for y, 3 for x, 10 for z, 7 for w, and 8 for p

$\frac{3}{28} = k \cdot \frac{30}{56}$

$\frac{3}{28} \cdot \frac{56}{30} = k$

$\frac{1}{5} = k$

The equation of variation is $y = \frac{xz}{5wp}$.

33. $d = kr^2$

We first find k.

$200 = k \cdot 60^2$ (Substituting 200 for d and 60 for r)

$200 = 3600k$

$\frac{200}{3600} = k$

$\frac{1}{18} = k$

The equation of variation is $d = \frac{1}{18}r^2$.

33. (continued)

Substitute 80 for r and solve for d.

$d = \frac{1}{18} \cdot 80^2 = \frac{6400}{18} = 355\frac{5}{9}$

It will take $355\frac{5}{9}$ ft to stop when traveling 80 mph.

35. $W = \frac{k}{d^2}$

We first find k.

$220 = \frac{k}{(3978)^2}$ Substituting 220 for W and 3978 for d

$220 = \frac{k}{15{,}824{,}484}$

$3{,}481{,}386{,}480 = k$

The equation of variation is $W = \frac{3{,}481{,}386{,}480}{d^2}$.

Substitute 3978 + 200, or 4178 for d and solve for W.

$W = \frac{3{,}481{,}386{,}480}{(4178)^2} = \frac{3{,}481{,}386{,}480}{17{,}455{,}684} \approx 199.4$

When the astronaut is 200 km above the surface of the earth, his weight is about 199.4 lb.

37. $A = k\frac{R}{I}$

We first find k.

$2.92 = k \cdot \frac{85}{262}$ Substituting 2.92 for A, 85 for R, and 262 for I

$2.92(\frac{262}{85}) = k$

$9 \approx k$

The equation of variation is $A = \frac{9R}{I}$.

Substitute 2.92 for A and 300 for I and solve for R.

$2.92 = \frac{9R}{300}$

$876 = 9R$

$\frac{876}{9} = R$

$97 \approx R$

For an earned run average of 2.92 in 300 innings, the pitcher would have to give up 97 earned runs.

39. $Q = kd^2$

We first find k.

$225 = k \cdot 5^2$

$225 = 25k$

$9 = k$

The equation of variation is $Q = 9d^2$.

39. (continued)

Substitute 9 for d and compute Q.

$Q = 9 \cdot 9^2$

$Q = 9 \cdot 81$

$Q = 729$

729 gallons of water are emptied by a pipe that is 9 in. in diameter.

41. Familiarize. Let x = the number of correct answers and y = the number of incorrect answers. Then 2x points are awarded for the correct answers, and $\frac{1}{2} \cdot y$, or $\frac{y}{2}$, points are deducted for the incorrect answers.

Translate.

The total number of answers	is	75.
$x + y$	=	75

The total score	is	100.
$2x - \frac{y}{2}$	=	100

We have a system of equations:

$x + y = 75,$

$2x - \frac{y}{2} = 100$

Solve. Solving the system, we get (55,20).

Check. If there are 55 correct answers and 20 incorrect answers, the total number of answers is 75. Also, $2 \cdot 55$, or 110, points are awarded for the correct answers, and $\frac{1}{2} \cdot 20$, or 10, points are deducted for the incorrect answers. The score is 110 - 10, or 100. The solution checks.

State. There were 55 correct answers and 20 incorrect answers.

43. $I = kP$ I varies directly as P

$1665 = k \cdot 9000$ Substituting

$\frac{1665}{9000} = k$

$0.185 = k$ Variation constant

The equation of variation is $I = 0.185P$.

45. We are told $A = kd^2$, and we know $A = \pi r^2$ so we have:

$kd^2 = \pi r^2$

$kd^2 = \pi\left(\frac{d}{2}\right)^2$ $r = \frac{d}{2}$

$kd^2 = \frac{\pi d^2}{4}$

$k = \frac{\pi}{4}$ Variation constant

47. $Q = \frac{kp^2}{q^3}$

Q varies directly as the square of p and inversely as the cube of q.

Exercise Set 10.1

1. The square roots of 16 are 4 and -4, because $4^2 = 16$ and $(-4)^2 = 16$.

3. The square roots of 144 are 12 and -12, because $12^2 = 144$ and $(-12)^2 = 144$.

5. The square roots of 400 are 20 and -20, because $20^2 = 400$ and $(-20)^2 = 400$.

7. $-\sqrt{\frac{49}{36}} = -\frac{7}{6}$ Since $\sqrt{\frac{49}{36}} = \frac{7}{6}$, $-\sqrt{\frac{49}{36}} = -\frac{7}{6}$.

9. $\sqrt{196} = 14$ Remember, $\sqrt{\ }$ indicates the principal square root.

11. $-\sqrt{\frac{16}{81}} = -\frac{4}{9}$ Since $\sqrt{\frac{16}{81}} = \frac{4}{9}$, $-\sqrt{\frac{16}{81}} = -\frac{4}{9}$.

13. $\sqrt{0.09} = 0.3$

15. $-\sqrt{0.0049} = -0.07$

17. $5\sqrt{p^2 + 4}$

The radical is the expression written under the radical sign, $p^2 + 4$.

19. $x^2y^2\sqrt{\frac{x}{y + 4}}$

The radicand is the expression written under the radical sign, $\frac{x}{y + 4}$.

21. $\sqrt{16x^2} = \sqrt{(4x)^2} = |4x| = 4|x|$

Since x might be negative, absolute-value notation is necessary.

23. $\sqrt{(-7c)^2} = |-7c| = |-7|\cdot|c| = 7|c|$

Since c might be negative, absolute-value notation is necessary.

25. $\sqrt{(a + 1)^2} = |a + 1|$

Since a + 1 might be negative, absolute-value notation is necessary.

27. $\sqrt{x^2 - 4x + 4} = \sqrt{(x - 2)^2} = |x - 2|$

Since x - 2 might be negative, absolute-value notation is necessary.

29. $\sqrt{4x^2 + 28x + 49} = \sqrt{(2x + 7)^2} = |2x + 7|$

Since 2x + 7 might be negative, absolute-value notation is necessary.

31. $\sqrt[3]{27} = 3$ $[3^3 = 27]$

33. $\sqrt[3]{-64x^3} = -4x$ $[(-4x)^3 = -64x^3]$

35. $\sqrt[3]{-216} = -6$ $[(-6)^3 = -216]$

37. $\sqrt[3]{0.343(x + 1)^3} = 0.7(x + 1)$

$[(0.7(x + 1))^3 = 0.343(x + 1)^3]$

39. $\sqrt[4]{625} = 5$ Since $5^4 = 625$

41. $\sqrt[5]{-1} = -1$ Since $(-1)^5 = -1$

43. $\sqrt[5]{-\frac{32}{243}} = -\frac{2}{3}$ Since $\left(-\frac{2}{3}\right)^5 = -\frac{32}{243}$

45. $\sqrt[6]{x^6} = |x|$

The index is even. Use absolute-value notation since x could have a negative value.

47. $\sqrt[4]{(5a)^4} = |5a| = 5|a|$

The index is even. Use absolute-value notation since a could have a negative value.

49. $\sqrt[10]{(-6)^{10}} = |-6| = 6$

51. $\sqrt[414]{(a + b)^{414}} = |a + b|$

The index is even. Use absolute-value notation since a + b could have a negative value.

53. $\sqrt[7]{y^7} = y$

We do not use absolute-value notation when the index is odd.

55. $\sqrt[5]{(x - 2)^5} = x - 2$

We do not use absolute-value notation when the index is odd.

57. $x^2 + x - 2 = 0$

$(x + 2)(x - 1) = 0$ Factoring

$x + 2 = 0$ or $x - 1 = 0$ Principle of zero products

$x = -2$ or $x = 1$

The solutions are -2 and 1.

59. $4x^2 - 49 = 0$

$(2x + 7)(2x - 7) = 0$ Factoring

$2x + 7 = 0$ or $2x - 7 = 0$ Principle of zero products

$2x = -7$ or $2x = 7$

$x = -\frac{7}{2}$ or $x = \frac{7}{2}$

The solutions are $-\frac{7}{2}$ and $\frac{7}{2}$.

61. $N = 2.5\sqrt{A}$

a) $N = 2.5\sqrt{25} = 2.5(5) = 12.5 \approx 13$

b) $N = 2.5\sqrt{36} = 2.5(6) = 15$

c) $N = 2.5\sqrt{49} = 2.5(7) = 17.5 \approx 18$

d) $N = 2.5\sqrt{64} = 2.5(8) = 20$

Exercise Set 10.2

1. $\sqrt{8} = \sqrt{4\cdot 2} = \sqrt{4}\,\sqrt{2} = 2\sqrt{2}$

3. $\sqrt{24} = \sqrt{4\cdot 6} = \sqrt{4}\,\sqrt{6} = 2\sqrt{6}$

5. $\sqrt{40} = \sqrt{4\cdot 10} = \sqrt{4}\,\sqrt{10} = 2\sqrt{10}$

7. $\sqrt{180x^4} = \sqrt{36\cdot 5\cdot x^4} = \sqrt{36x^4}\,\sqrt{5} = 6x^2\,\sqrt{5}$

9. $\sqrt[3]{54x^8} = \sqrt[3]{27\cdot 2\cdot x^6\cdot x^2} = \sqrt[3]{27x^6}\,\sqrt[3]{2x^2} = 3x^2\,\sqrt[3]{2x^2}$

11. $\sqrt[3]{80t^8} = \sqrt[3]{8\cdot 10\cdot t^6\cdot t^2} = \sqrt[3]{8t^6}\,\sqrt[3]{10t^2} = 2t^2\,\sqrt[3]{10t^2}$

13. $\sqrt[4]{80} = \sqrt[4]{16\cdot 5} = \sqrt[4]{16}\,\sqrt[4]{5} = 2\,\sqrt[4]{5}$

15. $\sqrt[4]{243x^8y^{10}} = \sqrt[4]{81\cdot 3\cdot x^8\cdot y^8\cdot y^2} = \sqrt[4]{81x^8y^8}\,\sqrt[4]{3y^2} =$
$3x^2y^2\,\sqrt[4]{3y^2}$

17. $\sqrt[3]{(x+y)^4} = \sqrt[3]{(x+y)^3\cdot(x+y)} =$
$\sqrt[3]{(x+y)^3}\,\sqrt[3]{x+y} = (x+y)\,\sqrt[3]{x+y}$

19. $\sqrt[3]{-24x^4y^5} = \sqrt[3]{-8\cdot 3\cdot x^3\cdot x\cdot y^3\cdot y^2} = \sqrt[3]{-8x^3y^3}\,\sqrt[3]{3xy^2} =$
$-2xy\,\sqrt[3]{3xy^2}$

21. $\sqrt[5]{96x^7y^{15}} = \sqrt[5]{32\cdot 3\cdot x^5\cdot x^2\cdot y^{15}} = \sqrt[5]{32x^5y^{15}}\,\sqrt[5]{3x^2} =$
$2xy^3\,\sqrt[5]{3x^2}$

23. $\sqrt{15}\,\sqrt{6} = \sqrt{15\cdot 6} = \sqrt{90}$
$= \sqrt{9\cdot 10} = \sqrt{9}\,\sqrt{10} = 3\sqrt{10}$

25. $\sqrt[3]{3}\,\sqrt[3]{18} = \sqrt[3]{3\cdot 18} = \sqrt[3]{54}$
$= \sqrt[3]{27\cdot 2} = \sqrt[3]{27}\,\sqrt[3]{2} = 3\sqrt[3]{2}$

27. $\sqrt{45}\,\sqrt{60} = \sqrt{45\cdot 60} = \sqrt{2700}$
$= \sqrt{900\cdot 3} = \sqrt{900}\,\sqrt{3} = 30\sqrt{3}$

29. $\sqrt{5b^3}\,\sqrt{10c^4}$
$= \sqrt{5b^3\cdot 10c^4}$
$= \sqrt{50b^3c^4}$
$= \sqrt{25\cdot 2\cdot b^2\cdot b\cdot c^4}$
$= \sqrt{25b^2c^4}\,\sqrt{2b}$
$= 5bc^2\,\sqrt{2b}$

31. $\sqrt[3]{y^4}\,\sqrt[3]{16y^5}$
$= \sqrt[3]{y^4\cdot 16y^5}$
$= \sqrt[3]{16y^9}$
$= \sqrt[3]{8\cdot 2\cdot y^9}$
$= \sqrt[3]{8y^9}\,\sqrt[3]{2}$
$= 2y^3\,\sqrt[3]{2}$

33. $\sqrt[3]{(b+3)^4}\,\sqrt[3]{(b+3)^2}$
$= \sqrt[3]{(b+3)^4(b+3)^2}$
$= \sqrt[3]{(b+3)^6}$
$= (b+3)^2$

35. $\sqrt{12a^3b}\,\sqrt{8a^4b^2}$
$= \sqrt{12a^3b\cdot 8a^4b^2}$
$= \sqrt{96a^7b^3}$
$= \sqrt{16\cdot 6\cdot a^6\cdot a\cdot b^2\cdot b}$
$= \sqrt{16a^6b^2}\,\sqrt{6ab}$
$= 4a^3b\sqrt{6ab}$

37. $\sqrt[4]{16}\cdot\sqrt[4]{64} = \sqrt[4]{16\cdot 64} = \sqrt[4]{1024} = \sqrt[4]{256\cdot 4} =$
$\sqrt[4]{256}\,\sqrt[4]{4} = 4\,\sqrt[4]{4}$

39. $\sqrt[4]{10a^3}\cdot\sqrt[4]{8a^2} = \sqrt[4]{10a^3\cdot 8a^2} = \sqrt[4]{80a^5} =$
$\sqrt[4]{16\cdot 5\cdot a^4\cdot a} = \sqrt[4]{16a^4}\,\sqrt[4]{5a} = 2a\,\sqrt[4]{5a}$

41. $\sqrt{30x^3y^4}\,\sqrt{18x^2y^5} = \sqrt{30x^3y^4\cdot 18x^2y^5} = \sqrt{540x^5y^9} =$
$\sqrt{36\cdot 15\cdot x^4\cdot x\cdot y^8\cdot y} = \sqrt{36x^4y^8}\,\sqrt{15xy} = 6x^2y^4\,\sqrt{15xy}$

43. $\sqrt[5]{a^3(b-c)^7}\,\sqrt[5]{a^8(b-c)^{11}} =$
$\sqrt[5]{a^3(b-c)^7\cdot a^8(b-c)^{11}} = \sqrt[5]{a^{11}(b-c)^{18}} =$
$\sqrt[5]{a^{10}\cdot a\cdot(b-c)^{15}\cdot(b-c)^3} =$
$\sqrt[5]{a^{10}(b-c)^{15}}\,\sqrt[5]{a(b-c)^3} = a^2(b-c)^3\,\sqrt[5]{a(b-c)^3}$

45. Using a calculator,
$\sqrt{180} \approx 13.41640787 \approx 13.416$

Using Table 1,
$\sqrt{180} = \sqrt{36\cdot 5} = \sqrt{36}\,\sqrt{5} = 6\sqrt{5} \approx 6(2.236)$
≈ 13.416

47. Using a calculator,
$\frac{8+\sqrt{480}}{4} \approx \frac{8+21.9089023}{4} = \frac{29.9089023}{4} \approx$
$7.477225575 \approx 7.477$

Using Table 1,
$\frac{8+\sqrt{480}}{4} = \frac{8+\sqrt{16\cdot 30}}{4} = \frac{8+4\sqrt{30}}{4}$
$= \frac{4(2+\sqrt{30})}{4}$
$= 2+\sqrt{30}$
$\approx 2+5.477$
$= 7.477$

49. Using a calculator,

$\frac{16 - \sqrt{48}}{20} \approx \frac{16 - 6.92820323}{20} = \frac{9.07179677}{20} \approx$

$0.453589838 \approx 0.454$

Using Table 1,

$\frac{16 - \sqrt{48}}{20} \approx \frac{16 - 6.928}{20} = \frac{9.072}{20} = 0.4536 \approx 0.454$

51. Using a calculator,

$\frac{24 + \sqrt{128}}{8} \approx \frac{24 + 11.3137085}{8} = \frac{35.3137085}{8} \approx$

$\approx 4.414213562 \approx 4.414$

Using Table 1,

$\frac{24 + \sqrt{128}}{8} = \frac{24 + \sqrt{64 \cdot 2}}{8} = \frac{24 + 8\sqrt{2}}{8} =$

$\frac{8(3 + \sqrt{2})}{8} = 3 + \sqrt{2} \approx 3 + 1.414 = 4.414$

53. $r = 2\sqrt{5L}$

a) $r = 2\sqrt{5 \cdot 20} = 2\sqrt{100} = 2 \cdot 10 = 20$ mph

b) $r = 2\sqrt{5 \cdot 70} = 2\sqrt{350}$

$\approx 2 \times 18.708$ Using Table 1 or a calculator

≈ 37.4 mph Multiplying and rounding

c) $r = 2\sqrt{5 \cdot 90} = 2\sqrt{450}$

$\approx 2 \times 21.213$ Using Table 1 or a calculator

≈ 42.4 mph Multiplying and rounding

55. $x^2 - \frac{2}{3}x = 0$

$x\left(x - \frac{2}{3}\right) = 0$ Factoring

$x = 0$ or $x - \frac{2}{3} = 0$ Principle of zero products

$x = 0$ or $x = \frac{2}{3}$

The solutions are 0 and $\frac{2}{3}$.

57. $\sqrt{1.6 \times 10^3}\sqrt{36 \times 10^{-8}}$

$= \sqrt{1.6 \times 36 \times 10^3 \times 10^{-8}}$

$= \sqrt{57.6 \times 10^{-5}}$

$= \sqrt{5.76 \times 10^{-4}}$

$= \sqrt{5.76} \times \sqrt{10^{-4}}$

$= 2.4 \times 10^{-2}$

59. $\sqrt[3]{48}\ \sqrt[3]{63}\ \sqrt[3]{196}$

$= \sqrt[3]{48 \times 63 \times 196}$

$= \sqrt[3]{2 \cdot 2 \cdot 2 \cdot 2 \cdot 3 \times 7 \cdot 3 \cdot 3 \times 2 \cdot 2 \cdot 7 \cdot 7}$

$= \sqrt[3]{2^6 \cdot 3^3 \cdot 7^3}$

$= 2^2 \cdot 3 \cdot 7$

$= 84$

Exercise Set 10.3

1. $\frac{\sqrt{21a}}{\sqrt{3a}} = \sqrt{\frac{21a}{3a}} = \sqrt{7}$

3. $\frac{\sqrt[3]{54}}{\sqrt[3]{2}} = \sqrt[3]{\frac{54}{2}} = \sqrt[3]{27} = 3$

5. $\frac{\sqrt{40xy^3}}{\sqrt{8x}} = \sqrt{\frac{40xy^3}{8x}} = \sqrt{5y^3} = \sqrt{y^2 \cdot 5y} = \sqrt{y^2}\sqrt{5y} =$

$y\sqrt{5y}$

7. $\frac{\sqrt[3]{96a^4b^2}}{\sqrt[3]{12a^2b}} = \sqrt[3]{\frac{96a^4b^2}{12a^2b}} = \sqrt[3]{8a^2b} = \sqrt[3]{8}\ \sqrt[3]{a^2b} =$

$2\sqrt[3]{a^2b}$

9. $\frac{\sqrt{144xy}}{2\sqrt{2}} = \frac{1}{2} \cdot \sqrt{\frac{144xy}{2}} = \frac{1}{2}\sqrt{72xy} = \frac{1}{2}\sqrt{36 \cdot 2 \cdot x \cdot y} =$

$\frac{1}{2}\sqrt{36}\sqrt{2xy} = \frac{1}{2} \cdot 6\sqrt{2xy} = 3\sqrt{2xy}$

11. $\frac{\sqrt[4]{48x^9y^{13}}}{\sqrt[4]{3xy^5}} = \sqrt[4]{\frac{48x^9y^{13}}{3xy^5}} = \sqrt[4]{16x^8y^8} = 2x^2y^2$

13. $\frac{\sqrt{x^3 - y^3}}{\sqrt{x - y}} = \sqrt{\frac{x^3 - y^3}{x - y}} = \sqrt{\frac{(x - y)(x^2 + xy + y^2)}{x - y}} =$

$\sqrt{\frac{\cancel{(x - y)}(x^2 + xy + y^2)}{\cancel{(x - y)}(1)}} = \sqrt{x^2 + xy + y^2}$

15. $\sqrt{\frac{16}{25}} = \frac{\sqrt{16}}{\sqrt{25}} = \frac{4}{5}$

17. $\sqrt[3]{\frac{64}{27}} = \frac{\sqrt[3]{64}}{\sqrt[3]{27}} = \frac{4}{3}$

19. $\sqrt{\frac{49}{y^2}} = \frac{\sqrt{49}}{\sqrt{y^2}} = \frac{7}{y}$

21. $\sqrt{\frac{25y^3}{x^4}} = \frac{\sqrt{25y^3}}{\sqrt{x^4}} = \frac{\sqrt{25y^2 \cdot y}}{\sqrt{x^4}} = \frac{\sqrt{25y^2}\sqrt{y}}{\sqrt{x^4}} = \frac{5y\sqrt{y}}{x^2}$

23. $\sqrt[3]{\frac{8x^5}{27y^3}} = \frac{\sqrt[3]{8x^5}}{\sqrt[3]{27y^3}} = \frac{\sqrt[3]{8x^3 \cdot x^2}}{\sqrt[3]{27y^3}} = \frac{\sqrt[3]{8x^3}\ \sqrt[3]{x^2}}{\sqrt[3]{27y^3}} = \frac{2x\sqrt[3]{x^2}}{3y}$

25. $\sqrt[4]{\frac{81x^4}{16}} = \frac{\sqrt[4]{81x^4}}{\sqrt[4]{16}} = \frac{3x}{2}$

27. $\sqrt[4]{\frac{p^5q^8}{r^{12}}} = \frac{\sqrt[4]{p^5q^8}}{\sqrt[4]{r^{12}}} = \frac{\sqrt[4]{p^4\cdot p\cdot q^8}}{\sqrt[4]{r^{12}}} = \frac{\sqrt[4]{p^4q^8}\sqrt[4]{p}}{\sqrt[4]{r^{12}}} = \frac{pq^2\sqrt[4]{p}}{r^3}$

29. $\sqrt[5]{\frac{32x^8}{y^{10}}} = \frac{\sqrt[5]{32x^8}}{\sqrt[5]{y^{10}}} = \frac{\sqrt[5]{32\cdot x^5\cdot x^3}}{\sqrt[5]{y^{10}}} = \frac{\sqrt[5]{32x^5}\sqrt[5]{x^3}}{\sqrt[5]{y^{10}}} =$

$\frac{2x\sqrt[5]{x^3}}{y^2}$

31. $\sqrt[6]{\frac{x^{13}}{y^6z^{12}}} = \frac{\sqrt[6]{x^{13}}}{\sqrt[6]{y^6z^{12}}} = \frac{\sqrt[6]{x^{12}\cdot x}}{\sqrt[6]{y^6z^{12}}} = \frac{\sqrt[6]{x^{12}}\sqrt[6]{x}}{\sqrt[6]{y^6z^{12}}} = \frac{x^2\sqrt[6]{x}}{yz^2}$

33. a) $\sqrt{(6a)^3} = \sqrt{6^3a^2} = \sqrt{6^2a^2}\sqrt{6a} = 6a\sqrt{6a}$

b) $(\sqrt{6a})^3 = \sqrt{6a}\sqrt{6a}\sqrt{6a} = 6a\sqrt{6a}$

35. a) $(\sqrt[3]{16b^2})^2 = \sqrt[3]{16b^2}\sqrt[3]{16b^2} = \sqrt[3]{256b^4} =$

$\sqrt[3]{64\cdot 4\cdot b^3\cdot b} = 4b\sqrt[3]{4b}$

b) $\sqrt[3]{(16b^2)^2} = \sqrt[3]{256b^4} = 4b\sqrt[3]{4b}$, as in part (a)

37. a) $\sqrt{(18a^2b)^3} = \sqrt{5832a^6b^3} = \sqrt{2916a^6b^2\cdot 2b} =$

$54a^3b\sqrt{2b}$

b) $(\sqrt{18a^2b})^3 = \sqrt{18a^2b}\sqrt{18a^2b}\sqrt{18a^2b} =$

$18a^2b\sqrt{18a^2b} = 18a^2b\sqrt{9a^2\cdot 2b} = 18a^2b\cdot 3a\sqrt{2b} =$

$54a^3b\sqrt{2b}$

39. a) $(\sqrt[3]{12c^2d})^2 = \sqrt[3]{12c^2d}\cdot\sqrt[3]{12c^2d} = \sqrt[3]{144c^4d^2} =$

$\sqrt[3]{8\cdot 18\cdot c^3\cdot c\cdot d^2} = 2c\sqrt[3]{18cd^2}$

b) $\sqrt[3]{(12c^2d)^2} = \sqrt[3]{144c^4d^2} = 2c\sqrt[3]{18cd^2}$, as in part (a)

41. a) $\sqrt[3]{(7x^2y)^2} = \sqrt[3]{49x^4y^2} = \sqrt[3]{49\cdot x^3\cdot x\cdot y^2} = x\sqrt[3]{49xy^2}$

b) $(\sqrt[3]{7x^2y})^2 = \sqrt[3]{7x^2y}\cdot\sqrt[3]{7x^2y} = \sqrt[3]{49x^4y} =$

$x\sqrt[3]{49xy^2}$, as in part (a)

43. a) $(\sqrt[4]{81xy^5})^2 = \sqrt[4]{81xy^5}\cdot\sqrt[4]{81xy^5} = \sqrt[4]{6561x^2y^{10}} =$

$\sqrt[4]{6561\cdot x^2\cdot y^8\cdot y^2} = 9y^2\sqrt[4]{x^2y^2}$

b) $\sqrt[4]{(81xy^5)^2} = \sqrt[4]{6561x^2y^{10}} = 9y^2\sqrt[4]{x^2y^2}$, as in part (a)

45. $\frac{12x}{x-4} - \frac{3x^2}{x+4} = \frac{384}{x^2-16}$

$\frac{12x}{x-4} - \frac{3x^2}{x+4} = \frac{384}{(x+4)(x-4)}$

The LCM is $(x + 4)(x - 4)$.

$(x+4)(x-4)\left[\frac{12x}{x-4} - \frac{3x^2}{x+4}\right] = (x+4)(x-4)\cdot\frac{384}{(x+4)(x-4)}$

$12x(x + 4) - 3x^2(x - 4) = 384$

$12x^2 + 48x - 3x^3 + 12x^2 = 384$

$-3x^3 + 24x^2 + 48x - 384 = 0$

$-3x^2(x - 8) + 48(x - 8) = 0$

$(x - 8)(-3x^2 + 48) = 0$

$x - 8 = 0$ or $-3x^2 + 48 = 0$

$x = 8$ or $-3(x^2 - 16) = 0$

$x = 8$ or $-3(x + 4)(x - 4) = 0$

$x = 8$ or $x + 4 = 0$ or $x - 4 = 0$

$x = 8$ or $x = -4$ or $x = 4$

Check: For $x = 8$:

$\frac{12x}{x-4} - \frac{3x^2}{x+4}$	$= \frac{384}{x^2-16}$
$\frac{12\cdot 8}{8-4} - \frac{3\cdot 8^2}{8+4}$	$\frac{384}{8^2-16}$
$\frac{96}{4} - \frac{192}{12}$	$\frac{384}{48}$
$24 - 16$	8
8	

8 is a solution.

For $x = -4$:

$\frac{12x}{x-4} - \frac{3x^2}{x+4}$	$= \frac{384}{x^2-16}$
$\frac{12(-4)}{-4-4} - \frac{3(-4)^2}{-4+4}$	$\frac{384}{(-4)^2-16}$
$\frac{-48}{-8} - \frac{48}{0}$	$\frac{384}{16-16}$

-4 is not a solution.

For $x = 4$:

$\frac{12x}{x-4} - \frac{3x^2}{x+4}$	$= \frac{384}{x^2-16}$
$\frac{12\cdot 4}{4-4} - \frac{3\cdot 4^2}{4+4}$	$\frac{384}{4^2-16}$
$\frac{48}{0} - \frac{48}{8}$	$\frac{384}{16-16}$

4 is not a solution.

The solution is 8.

47. Familiarize. Let x and y represent the width and length of the rectangle, respectively.

Translate. We write two equations.

The width is one-fourth the length.

$x = \frac{1}{4}\cdot y$

The area is twice the perimeter.

$xy = 2\cdot(2x + 2y)$

47. (continued)

Carry out. Solving the system of equations we get (5,20).

Check. The width, 5, is one-fourth the length, 20. The area is $5\cdot20$, or 100. The perimeter is $2\cdot5 + 2\cdot20$, or 50. Since $100 = 2\cdot50$, the area is twice the perimeter. The values check.

State. The width is 5, and the length is 20.

49. $\dfrac{7\sqrt{a^2b}\,\sqrt{25xy}}{5\sqrt{a^{-4}b^{-1}}\,\sqrt{49x^{-1}y^{-3}}} = \dfrac{7\sqrt{25a^2bxy}}{5\sqrt{49a^{-4}b^{-1}x^{-1}y^{-3}}} =$

$\dfrac{7}{5}\sqrt{\dfrac{25a^2bxy}{49a^{-4}b^{-1}x^{-1}y^{-3}}} = \dfrac{7}{5}\sqrt{\dfrac{25}{49}\cdot a^6b^2x^2y^4} =$

$\dfrac{7}{5}\cdot\dfrac{5}{7}\cdot a^3bxy^2 = a^3bxy^2$

51. $\dfrac{\sqrt{44x^2y^9z}\,\sqrt{22y^9z^6}}{(\sqrt{11xy^8z^2})^2} = \dfrac{\sqrt{44\cdot22x^2y^{18}z^7}}{\sqrt{11\cdot11x^2y^{16}z^4}} =$

$\sqrt{\dfrac{44\cdot22x^2y^{18}z^7}{11\cdot11x^2y^{16}z^4}} = \sqrt{4\cdot2y^2z^3} = \sqrt{4y^2z^2\cdot2z} = 2yz\sqrt{2z}$

Exercise Set 10.4

1. $6\sqrt{3} + 2\sqrt{3} = (6 + 2)\sqrt{3} = 8\sqrt{3}$

3. $9\sqrt[3]{5} - 6\sqrt[3]{5} = (9 - 6)\sqrt[3]{5} = 3\sqrt[3]{5}$

5. $4\sqrt[3]{y} + 9\sqrt[3]{y} = (4 + 9)\sqrt[3]{y} = 13\sqrt[3]{y}$

7. $8\sqrt{2} - 6\sqrt{2} + 5\sqrt{2} = (8 - 6 + 5)\sqrt{2} = 7\sqrt{2}$

9. $4\sqrt[3]{3} - \sqrt{5} + 2\sqrt[3]{3} + \sqrt{5} =$

$(4 + 2)\sqrt[3]{3} + (-1 + 1)\sqrt{5} = 6\sqrt[3]{3}$

11. $8\sqrt{27} - 3\sqrt{3} = 8\sqrt{9\cdot3} - 3\sqrt{3}$ } Factoring the first radical

$= 8\sqrt{9}\cdot\sqrt{3} - 3\sqrt{3}$

$= 8\cdot3\sqrt{3} - 3\sqrt{3}$ Taking the square root

$= 24\sqrt{3} - 3\sqrt{3}$

$= (24 - 3)\sqrt{3}$ Factoring out $\sqrt{3}$

$= 21\sqrt{3}$

13. $8\sqrt{45} + 7\sqrt{20} = 8\sqrt{9\cdot5} + 7\sqrt{4\cdot5}$ } Factoring the radicals

$= 8\sqrt{9}\cdot\sqrt{5} + 7\sqrt{4}\cdot\sqrt{5}$

$= 8\cdot3\sqrt{5} + 7\cdot2\sqrt{5}$ Taking the square roots

$= 24\sqrt{5} + 14\sqrt{5}$

$= (24 + 14)\sqrt{5}$ Factoring out $\sqrt{5}$

$= 38\sqrt{5}$

15. $18\sqrt{72} + 2\sqrt{98} = 18\sqrt{36\cdot2} + 2\sqrt{49\cdot2} =$

$18\sqrt{36}\cdot\sqrt{2} + 2\sqrt{49}\cdot\sqrt{2} = 18\cdot6\sqrt{2} + 2\cdot7\sqrt{2} =$

$108\sqrt{2} + 14\sqrt{2} = (108 + 14)\sqrt{2} = 122\sqrt{2}$

17. $3\sqrt[3]{16} + \sqrt[3]{54} = 3\sqrt[3]{8\cdot2} + \sqrt[3]{27\cdot2} =$

$3\sqrt[3]{8}\cdot\sqrt[3]{2} + \sqrt[3]{27}\cdot\sqrt[3]{2} = 3\cdot2\sqrt[3]{2} + 3\sqrt[3]{2} =$

$6\sqrt[3]{2} + 3\sqrt[3]{2} = (6 + 3)\sqrt[3]{2} = 9\sqrt[3]{2}$

19. $2\sqrt{128} - \sqrt{18} + 4\sqrt{32} =$

$2\sqrt{64\cdot2} - \sqrt{9\cdot2} + 4\sqrt{16\cdot2} =$

$2\sqrt{64}\cdot\sqrt{2} - \sqrt{9}\cdot\sqrt{2} + 4\sqrt{16}\cdot\sqrt{2} =$

$2\cdot8\sqrt{2} - 3\sqrt{2} + 4\cdot4\sqrt{2} = 16\sqrt{2} - 3\sqrt{2} + 16\sqrt{2} =$

$(16 - 3 + 16)\sqrt{2} = 29\sqrt{2}$

21. $\sqrt{5a} + 2\sqrt{45a^3} = \sqrt{5a} + 2\sqrt{9a^2\cdot5a} =$

$\sqrt{5a} + 2\sqrt{9a^2}\cdot\sqrt{5a} = \sqrt{5a} + 2\cdot3a\sqrt{5a} =$

$\sqrt{5a} + 6a\sqrt{5a} = (1 + 6a)\sqrt{5a}$

23. $\sqrt[3]{24x} - \sqrt[3]{3x^4} = \sqrt[3]{8\cdot3x} - \sqrt[3]{x^3\cdot3x} =$

$\sqrt[3]{8}\cdot\sqrt[3]{3x} - \sqrt[3]{x^3}\cdot\sqrt[3]{3x} = 2\sqrt[3]{3x} - x\sqrt[3]{3x} =$

$(2 - x)\sqrt[3]{3x}$

25. $\sqrt{8y - 8} + \sqrt{2y - 2} = \sqrt{4(2y - 2)} + \sqrt{2y - 2} =$

$\sqrt{4}\cdot\sqrt{2y - 2} + \sqrt{2y - 2} = 2\sqrt{2y - 2} + \sqrt{2y - 2} =$

$(2 + 1)\sqrt{2y - 2} = 3\sqrt{2y - 2}$

27. $\sqrt{x^3 - x^2} + \sqrt{9x - 9} = \sqrt{x^2(x - 1)} + \sqrt{9(x - 1)} =$

$\sqrt{x^2}\cdot\sqrt{x - 1} + \sqrt{9}\cdot\sqrt{x - 1} = x\sqrt{x - 1} + 3\sqrt{x - 1} =$

$(x + 3)\sqrt{x - 1}$

29. $5\sqrt[3]{32} - \sqrt[3]{108} + 2\sqrt[3]{256} =$

$5\sqrt[3]{8\cdot4} - \sqrt[3]{27\cdot4} + 2\sqrt[3]{64\cdot4} =$

$5\sqrt[3]{8}\cdot\sqrt[3]{4} - \sqrt[3]{27}\cdot\sqrt[3]{4} + 2\sqrt[3]{64}\cdot\sqrt[3]{4} =$

$5\cdot2\sqrt[3]{4} - 3\sqrt[3]{4} + 2\cdot4\sqrt[3]{4} = 10\sqrt[3]{4} - 3\sqrt[3]{4} + 8\sqrt[3]{4} =$

$(10 - 3 + 8)\sqrt[3]{4} = 15\sqrt[3]{4}$

31. $\sqrt{6}(2 - 3\sqrt{6}) = \sqrt{6}\cdot2 - \sqrt{6}\cdot3\sqrt{6} = 2\sqrt{6} - 3\cdot6 =$

$2\sqrt{6} - 18$

33. $\sqrt{2}(\sqrt{3} - \sqrt{5}) = \sqrt{2}\cdot\sqrt{3} - \sqrt{2}\cdot\sqrt{5} = \sqrt{6} - \sqrt{10}$

35. $\sqrt{3}(2\sqrt{5} - 3\sqrt{4}) = \sqrt{3}(2\sqrt{5} - 3\cdot2) =$

$\sqrt{3}\cdot2\sqrt{5} - \sqrt{3}\cdot6 = 2\sqrt{15} - 6\sqrt{3}$

37. $\sqrt[3]{2}(\sqrt[3]{4} - 2\sqrt[3]{32}) = \sqrt[3]{2}\cdot\sqrt[3]{4} - \sqrt[3]{2}\cdot 2\sqrt[3]{32} =$

$\sqrt[3]{8} - 2\sqrt[3]{64} = 2 - 2\cdot 4 = 2 - 8 = -6$

39. $\sqrt[3]{a}(\sqrt[3]{2a^2} + \sqrt[3]{16a^2}) = \sqrt[3]{a}\cdot\sqrt[3]{2a^2} + \sqrt[3]{a}\cdot\sqrt[3]{16a^2} =$

$\sqrt[3]{2a^3} + \sqrt[3]{16a^3} = \sqrt[3]{a^3\cdot 2} + \sqrt[3]{8a^3\cdot 2} = a\sqrt[3]{2} + 2a\sqrt[3]{2} =$

$3a\sqrt[3]{2}$

41. $(\sqrt{3} - \sqrt{2})(\sqrt{3} + \sqrt{2}) = (\sqrt{3})^2 - (\sqrt{2})^2 =$

$3 - 2 = 1$

43. $(\sqrt{8} + 2\sqrt{5})(\sqrt{8} - 2\sqrt{5}) = (\sqrt{8})^2 - (2\sqrt{5})^2 =$

$8 - 4\cdot 5 = 8 - 20 = -12$

45. $(7 + \sqrt{5})(7 - \sqrt{5}) = 7^2 - (\sqrt{5})^2 = 49 - 5 = 44$

47. $(2 - \sqrt{3})(2 + \sqrt{3}) = 2^2 - (\sqrt{3})^2 = 4 - 3 = 1$

49. $(\sqrt{8} + \sqrt{5})(\sqrt{8} - \sqrt{5}) = (\sqrt{8})^2 - (\sqrt{5})^2 =$

$8 - 5 = 3$

51. $(3 + 2\sqrt{7})(3 - 2\sqrt{7}) = 3^2 - (2\sqrt{7})^2 =$

$9 - 4\cdot 7 = 9 - 28 = -19$

53. $(\sqrt{a} + \sqrt{b})(\sqrt{a} - \sqrt{b}) = (\sqrt{a})^2 - (\sqrt{b})^2 = a - b$

55. $(3 - \sqrt{5})(2 + \sqrt{5})$

$= 3\cdot 2 + 3\sqrt{5} - 2\sqrt{5} - (\sqrt{5})^2$ Using FOIL

$= 6 + 3\sqrt{5} - 2\sqrt{5} - 5$

$= 1 + \sqrt{5}$ Simplifying

57. $(\sqrt{3} + 1)(2\sqrt{3} + 1)$

$= \sqrt{3}\cdot 2\sqrt{3} + \sqrt{3}\cdot 1 + 1\cdot 2\sqrt{3} + 1^2$ Using FOIL

$= 2\cdot 3 + \sqrt{3} + 2\sqrt{3} + 1$

$= 7 + 3\sqrt{3}$ Simplifying

59. $(2\sqrt{7} - 4\sqrt{2})(3\sqrt{7} + 6\sqrt{2}) =$

$2\sqrt{7}\cdot 3\sqrt{7} + 2\sqrt{7}\cdot 6\sqrt{2} - 4\sqrt{2}\cdot 3\sqrt{7} - 4\sqrt{2}\cdot 6\sqrt{2} =$

$6\cdot 7 + 12\sqrt{14} - 12\sqrt{14} - 24\cdot 2 =$

$42 + 12\sqrt{14} - 12\sqrt{14} - 48 = -6$

61. $(\sqrt{a} + \sqrt{2})(\sqrt{a} + \sqrt{3}) =$

$(\sqrt{a})^2 + \sqrt{a}\cdot\sqrt{3} + \sqrt{2}\cdot\sqrt{a} + \sqrt{2}\cdot\sqrt{3} =$

$a + \sqrt{3a} + \sqrt{2a} + \sqrt{6}$

63. $(2\sqrt[3]{3} + \sqrt[3]{2})(\sqrt[3]{3} - 2\sqrt[3]{2}) =$

$2\sqrt[3]{3}\cdot\sqrt[3]{3} - 2\sqrt[3]{3}\cdot 2\sqrt[3]{2} + \sqrt[3]{2}\cdot\sqrt[3]{3} - \sqrt[3]{2}\cdot 2\sqrt[3]{2} =$

$2\sqrt[3]{9} - 4\sqrt[3]{6} + \sqrt[3]{6} - 2\sqrt[3]{4} = 2\sqrt[3]{9} - 3\sqrt[3]{6} - 2\sqrt[3]{4}$

65. $(2 + \sqrt{3})^2 = 2^2 + 4\sqrt{3} + (\sqrt{3})^2$ Squaring a binomial

$= 4 + 4\sqrt{3} + 3$

$= 7 + 4\sqrt{3}$

67. $(\sqrt[5]{9} - \sqrt[5]{3})(\sqrt[5]{8} + \sqrt[5]{27})$

$= \sqrt[5]{9}\cdot\sqrt[5]{8} + \sqrt[5]{9}\cdot\sqrt[5]{27} - \sqrt[5]{3}\cdot\sqrt[5]{8} - \sqrt[5]{3}\cdot\sqrt[5]{27}$ Using FOIL

$= \sqrt[5]{72} + \sqrt[5]{243} - \sqrt[5]{24} - \sqrt[5]{81}$

$= \sqrt[5]{72} + 3 - \sqrt[5]{24} - \sqrt[5]{81}$

69. $\sqrt{9 + 3\sqrt{5}}\,\sqrt{9 - 3\sqrt{5}} = \sqrt{(9 + 3\sqrt{5})(9 - 3\sqrt{5})} =$

$\sqrt{81 - 9\cdot 5} = \sqrt{81 - 45} = \sqrt{36} = 6$

71. $(\sqrt{3} + \sqrt{5} - \sqrt{6})^2 = [(\sqrt{3} + \sqrt{5}) - \sqrt{6}]^2 =$

$(\sqrt{3} + \sqrt{5})^2 - 2(\sqrt{3} + \sqrt{5})(\sqrt{6}) + (\sqrt{6})^2 =$

$3 + 2\sqrt{15} + 5 - 2\sqrt{18} - 2\sqrt{30} + 6 =$

$14 + 2\sqrt{15} - 2\sqrt{9\cdot 2} - 2\sqrt{30} =$

$14 + 2\sqrt{15} - 6\sqrt{2} - 2\sqrt{30}$

73. $(\sqrt[3]{9} - 2)(\sqrt[3]{9} + 4)$

$= \sqrt[3]{81} + 2\sqrt[3]{9} - 8$

$= \sqrt[3]{27\cdot 3} + 2\sqrt[3]{9} - 8$

$= 3\sqrt[3]{3} + 2\sqrt[3]{9} - 8$

Exercise Set 10.5

1. $\sqrt{\frac{6}{5}} = \sqrt{\frac{6}{5}\cdot\frac{5}{5}} = \sqrt{\frac{30}{25}} = \frac{\sqrt{30}}{\sqrt{25}} = \frac{\sqrt{30}}{5}$

3. $\sqrt{\frac{10}{3}} = \sqrt{\frac{10}{3}\cdot\frac{3}{3}} = \sqrt{\frac{30}{9}} = \frac{\sqrt{30}}{\sqrt{9}} = \frac{\sqrt{30}}{3}$

5. $\frac{6\sqrt{5}}{5\sqrt{3}} = \frac{6\sqrt{5}}{5\sqrt{3}}\cdot\frac{\sqrt{3}}{\sqrt{3}} = \frac{6\sqrt{15}}{5\cdot 3} = \frac{2\sqrt{15}}{5}$

7. $\sqrt[3]{\frac{16}{9}} = \sqrt[3]{\frac{16}{9}\cdot\frac{3}{3}} = \sqrt[3]{\frac{48}{27}} = \frac{\sqrt[3]{8\cdot 6}}{\sqrt[3]{27}} = \frac{2\sqrt[3]{6}}{3}$

9. $\frac{\sqrt[3]{3a}}{\sqrt[3]{5c}} = \frac{\sqrt[3]{3a}}{\sqrt[3]{5c}}\cdot\frac{\sqrt[3]{5^2c^2}}{\sqrt[3]{5^2c^2}} = \frac{\sqrt[3]{75ac^2}}{\sqrt[3]{5^3c^3}} = \frac{\sqrt[3]{75ac^2}}{5c}$

11. $\frac{\sqrt[3]{2y^4}}{\sqrt[3]{6x^4}} = \frac{\sqrt[3]{2y^4}}{\sqrt[3]{6x^4}}\cdot\frac{\sqrt[3]{6^2x^2}}{\sqrt[3]{6^2x^2}} = \frac{\sqrt[3]{72x^2y^4}}{\sqrt[3]{6^3x^6}}$

$= \frac{\sqrt[3]{8y^3\cdot 9x^2y}}{6x^2} = \frac{2y\sqrt[3]{9x^2y}}{6x^2} = \frac{y\sqrt[3]{9x^2y}}{3x^2}$

13. $\frac{1}{\sqrt[3]{xy}} = \frac{1}{\sqrt[3]{xy}}\cdot\frac{\sqrt[3]{x^2y^2}}{\sqrt[3]{x^2y^2}} = \frac{\sqrt[3]{x^2y^2}}{\sqrt[3]{x^3y^3}} = \frac{\sqrt[3]{x^2y^2}}{xy}$

15. $\sqrt{\frac{5x}{18}} = \sqrt{\frac{5x}{18}\cdot\frac{2}{2}} = \sqrt{\frac{10x}{36}} = \frac{\sqrt{10x}}{\sqrt{36}} = \frac{\sqrt{10x}}{6}$

17. $\sqrt[3]{\frac{4}{5x^5y^2}} = \sqrt[3]{\frac{4}{5x^5y^2}\cdot\frac{25xy}{25xy}} = \sqrt[3]{\frac{100xy}{125x^6y^3}} = \frac{\sqrt[3]{100xy}}{\sqrt[3]{125x^6y^3}} =$
$\frac{\sqrt[3]{100xy}}{5x^2y}$

19. $\sqrt[4]{\frac{1}{8x^7y^3}} = \sqrt[4]{\frac{1}{8x^7y^3}\cdot\frac{2xy}{2xy}} = \sqrt[4]{\frac{2xy}{16x^8y^4}} = \frac{\sqrt[4]{2xy}}{2x^2y}$

21. $\frac{2x}{\sqrt[5]{18x^8y^6}} = \frac{2x}{\sqrt[5]{18x^8y^6}}\cdot\frac{\sqrt[5]{432x^2y^4}}{\sqrt[5]{432x^2y^4}} = \frac{2x\sqrt[5]{432x^2y^4}}{\sqrt[5]{7776x^{10}y^{10}}} =$
$\frac{2x\sqrt[5]{432x^2y^4}}{6x^2y^2} = \frac{2x\sqrt[5]{432x^2y^4}}{2\cdot3\cdot x\cdot x\cdot y\cdot y} = \frac{\cancel{2x}\sqrt[5]{432x^2y^4}}{\cancel{2}\cdot3\cdot\cancel{x}\cdot x\cdot y\cdot y} =$
$\frac{\sqrt[5]{432x^2y^4}}{3xy^2}$

23. $\frac{\sqrt{7}}{\sqrt{3x}} = \frac{\sqrt{7}}{\sqrt{3x}}\cdot\frac{\sqrt{7}}{\sqrt{7}} = \frac{7}{\sqrt{21x}}$

25. $\sqrt{\frac{14}{21}} = \sqrt{\frac{2}{3}\cdot\frac{2}{2}} = \sqrt{\frac{4}{6}} = \frac{\sqrt{4}}{\sqrt{6}} = \frac{2}{\sqrt{6}}$

27. $\frac{4\sqrt{13}}{3\sqrt{7}} = \frac{4\sqrt{13}}{3\sqrt{7}}\cdot\frac{\sqrt{13}}{\sqrt{13}} = \frac{4\cdot13}{3\sqrt{91}} = \frac{52}{3\sqrt{91}}$

29. $\frac{\sqrt[3]{7}}{\sqrt[3]{2}} = \frac{\sqrt[3]{7}}{\sqrt[3]{2}}\cdot\frac{\sqrt[3]{7^2}}{\sqrt[3]{7^2}} = \frac{\sqrt[3]{7^3}}{\sqrt[3]{98}} = \frac{7}{\sqrt[3]{98}}$

31. $\sqrt{\frac{7x}{3y}} = \sqrt{\frac{7x}{3y}\cdot\frac{7x}{7x}} = \frac{\sqrt{(7x)^2}}{\sqrt{21xy}} = \frac{7x}{\sqrt{21xy}}$

33. $\frac{\sqrt[3]{5y^4}}{\sqrt[3]{6x^5}} = \frac{\sqrt[3]{5y^4}}{\sqrt[3]{6x^5}}\cdot\frac{\sqrt[3]{5^2y^2}}{\sqrt[3]{5^2y^2}} = \frac{\sqrt[3]{5^3y^6}}{\sqrt[3]{150x^5y^2}} = \frac{5y^2}{x\sqrt[3]{150x^2y^2}}$

35. $\frac{\sqrt{ab}}{3} = \frac{\sqrt{ab}}{3}\cdot\frac{\sqrt{ab}}{\sqrt{ab}} = \frac{ab}{3\sqrt{ab}}$

37. $\frac{5}{8-\sqrt{6}} = \frac{5}{8-\sqrt{6}}\cdot\frac{8+\sqrt{6}}{8+\sqrt{6}} = \frac{5(8+\sqrt{6})}{8^2-(\sqrt{6})^2} =$
$\frac{5(8+\sqrt{6})}{64-6} = \frac{5(8+\sqrt{6})}{58} = \frac{40+5\sqrt{6}}{58}$

39. $\frac{-4\sqrt{7}}{\sqrt{5}-\sqrt{3}} = \frac{-4\sqrt{7}}{\sqrt{5}-\sqrt{3}}\cdot\frac{\sqrt{5}+\sqrt{3}}{\sqrt{5}+\sqrt{3}} =$
$\frac{-4\sqrt{7}(\sqrt{5}+\sqrt{3})}{5-3} = \frac{-4\sqrt{7}(\sqrt{5}+\sqrt{3})}{2} =$
$-2\sqrt{7}(\sqrt{5}+\sqrt{3}) = -2\sqrt{35}-2\sqrt{21}$

41. $\frac{\sqrt{5}-2\sqrt{6}}{\sqrt{3}-4\sqrt{5}} = \frac{\sqrt{5}-2\sqrt{6}}{\sqrt{3}-4\sqrt{5}}\cdot\frac{\sqrt{3}+4\sqrt{5}}{\sqrt{3}+4\sqrt{5}} =$
$\frac{\sqrt{15}+4\cdot5-2\sqrt{18}-8\sqrt{30}}{(\sqrt{3})^2-(4\sqrt{5})^2} =$
$\frac{\sqrt{15}+20-2\sqrt{9\cdot2}-8\sqrt{30}}{3-16\cdot5} =$
$\frac{\sqrt{15}+20-6\sqrt{2}-8\sqrt{30}}{-77}$

43. $\frac{\sqrt{x}-\sqrt{y}}{\sqrt{x}+\sqrt{y}} = \frac{\sqrt{x}-\sqrt{y}}{\sqrt{x}+\sqrt{y}}\cdot\frac{\sqrt{x}-\sqrt{y}}{\sqrt{x}-\sqrt{y}} =$
$\frac{x-\sqrt{xy}-\sqrt{xy}+y}{x-y} = \frac{x-2\sqrt{xy}+y}{x-y}$

45. $\frac{5\sqrt{3}-3\sqrt{2}}{3\sqrt{2}-2\sqrt{3}} = \frac{5\sqrt{3}-3\sqrt{2}}{3\sqrt{2}-2\sqrt{3}}\cdot\frac{3\sqrt{2}+2\sqrt{3}}{3\sqrt{2}+2\sqrt{3}} =$
$\frac{15\sqrt{6}+10\cdot3-9\cdot2-6\sqrt{6}}{9\cdot2-4\cdot3} = \frac{12+9\sqrt{6}}{6} =$
$\frac{3(4+3\sqrt{6})}{3\cdot2} = \frac{4+3\sqrt{6}}{2}$

47. $\frac{\sqrt{x}-2\sqrt{y}}{2\sqrt{x}+\sqrt{y}} = \frac{\sqrt{x}-2\sqrt{y}}{2\sqrt{x}+\sqrt{y}}\cdot\frac{2\sqrt{x}-\sqrt{y}}{2\sqrt{x}-\sqrt{y}} =$
$\frac{2(\sqrt{x})^2-\sqrt{xy}-4\sqrt{xy}+2(\sqrt{y})^2}{(2\sqrt{x})^2-(\sqrt{y})^2} = \frac{2x-5\sqrt{xy}+2y}{4x-y}$

49. $\frac{\sqrt{3}+5}{8} = \frac{\sqrt{3}+5}{8}\cdot\frac{\sqrt{3}-5}{\sqrt{3}-5} = \frac{3-25}{8(\sqrt{3}-5)} =$
$\frac{-22}{8(\sqrt{3}-5)} = \frac{2(-11)}{2\cdot4(\sqrt{3}-5)} = \frac{2}{2}\cdot\frac{-11}{4(\sqrt{3}-5)} =$
$\frac{-11}{4(\sqrt{3}-5)} = \frac{-11}{4\sqrt{3}-20}$

51. $\frac{\sqrt{3}-5}{\sqrt{2}+5} = \frac{\sqrt{3}-5}{\sqrt{2}+5}\cdot\frac{\sqrt{3}+5}{\sqrt{3}+5} =$
$\frac{3-25}{\sqrt{6}+5\sqrt{2}+5\sqrt{3}+25} = \frac{-22}{\sqrt{6}+5\sqrt{2}+5\sqrt{3}+25}$

53. $\frac{\sqrt{x}-\sqrt{y}}{\sqrt{x}+\sqrt{y}} = \frac{\sqrt{x}-\sqrt{y}}{\sqrt{x}+\sqrt{y}}\cdot\frac{\sqrt{x}+\sqrt{y}}{\sqrt{x}+\sqrt{y}} =$
$\frac{x-y}{x+\sqrt{xy}+\sqrt{xy}+y} = \frac{x-y}{x+2\sqrt{xy}+y}$

55. $\frac{4\sqrt{6}-5\sqrt{3}}{2\sqrt{3}+7\sqrt{6}} = \frac{4\sqrt{6}-5\sqrt{3}}{2\sqrt{3}+7\sqrt{6}}\cdot\frac{4\sqrt{6}+5\sqrt{3}}{4\sqrt{6}+5\sqrt{3}} =$
$\frac{16\cdot6-25\cdot3}{8\sqrt{18}+10\cdot3+28\cdot6+35\sqrt{18}} = \frac{96-75}{43\sqrt{18}+30+168} =$
$\frac{21}{43\sqrt{9\cdot2}+198} = \frac{21}{43\cdot3\sqrt{2}+198} = \frac{3\cdot7}{3(43\sqrt{2}+66)} =$
$\frac{7}{43\sqrt{2}+66}$

57. $\frac{\sqrt{2} + 3\sqrt{x}}{\sqrt{2} - \sqrt{x}} = \frac{\sqrt{2} + 3\sqrt{x}}{\sqrt{2} - \sqrt{x}} \cdot \frac{\sqrt{2} - 3\sqrt{x}}{\sqrt{2} - 3\sqrt{x}} =$

$\frac{(\sqrt{2})^2 - (3\sqrt{x})^2}{(\sqrt{2})^2 - 3\sqrt{2x} - \sqrt{2x} + 3(\sqrt{x})^2} = \frac{2 - 9x}{2 - 4\sqrt{2x} + 3x}$

59. $\frac{a\sqrt{b} + c}{\sqrt{b} + c} = \frac{a\sqrt{b} + c}{\sqrt{b} + c} \cdot \frac{a\sqrt{b} - c}{a\sqrt{b} - c} =$

$\frac{(a\sqrt{b})^2 - c^2}{a(\sqrt{b})^2 - c\sqrt{b} + ac\sqrt{b} - c^2} =$

$\frac{a^2b - c^2}{ab - c\sqrt{b} + ac\sqrt{b} - c^2}$

61. $\frac{1}{2} - \frac{1}{3} = \frac{1}{t}$ LCM is 6t

$6t\left(\frac{1}{2} - \frac{1}{3}\right) = 6t\left(\frac{1}{t}\right)$

$3t - 2t = 6$

$t = 6$

Check:

$\frac{1}{2} - \frac{1}{3}$	$\frac{1}{t}$
$\frac{1}{2} - \frac{1}{3}$	$\frac{1}{6}$
$\frac{3}{6} - \frac{2}{6}$	
$\frac{1}{6}$	

The solution is 6.

63. $\frac{1}{x^3 - y^2} \div \frac{1}{(x - y)(x^2 + xy + y^2)}$

$= \frac{1}{(x - y)(x^2 + xy + y^2)} \cdot \frac{(x - y)(x^2 + xy + y^2)}{1}$

$= \frac{(x - y)(x^2 + xy + y^2)}{(x - y)(x^2 + xy + y^2)}$

$= 1$

65. $\frac{\sqrt{5} + \sqrt{10} - \sqrt{6}}{\sqrt{50}} = \frac{\sqrt{5} + \sqrt{10} - \sqrt{6}}{\sqrt{50}} \cdot \frac{\sqrt{2}}{\sqrt{2}}$

$= \frac{\sqrt{10} + \sqrt{20} - \sqrt{12}}{\sqrt{100}}$

$= \frac{\sqrt{10} + 2\sqrt{5} - 2\sqrt{3}}{10}$

67. $\frac{b + \sqrt{b}}{1 + b + \sqrt{b}} = \frac{b + \sqrt{b}}{(1 + b) + \sqrt{b}} \cdot \frac{(1 + b) - \sqrt{b}}{(1 + b) - \sqrt{b}}$

$= \frac{(b + \sqrt{b})(1 + b - \sqrt{b})}{(1 + b)^2 - (\sqrt{b})^2}$

$= \frac{b + b^2 - b\sqrt{b} + \sqrt{b} + b\sqrt{b} - b}{1 + 2b + b^2 - b}$

$= \frac{b^2 + \sqrt{b}}{b^2 + b + 1}$

69. $\frac{36a^2b}{\sqrt[3]{6a^2b}} = \frac{36a^2b}{\sqrt[3]{6a^2b}} \cdot \frac{\sqrt[3]{6^2ab^2}}{\sqrt[3]{6^2ab^2}}$

$= \frac{36a^2b \sqrt[3]{6^2ab^2}}{\sqrt[3]{6^3a^3b^3}}$

$= \frac{36a^2b \sqrt[3]{6^2ab^2}}{6ab}$

$= 6a \sqrt[3]{36ab^2}$

71. $\frac{\sqrt{x + 6} - 5}{\sqrt{x + 6} + 5} = \frac{\sqrt{x + 6} - 5}{\sqrt{x + 6} + 5} \cdot \frac{\sqrt{x + 6} + 5}{\sqrt{x + 6} + 5}$

$= \frac{(x + 6) - 25}{(x + 6) + 10\sqrt{x + 6} + 25}$

$= \frac{x - 19}{x + 10\sqrt{x + 6} + 31}$

73. $\sqrt{a^2 - 3} - \frac{a^2}{\sqrt{a^2 - 3}} =$

$\sqrt{a^2 - 3} \cdot \frac{\sqrt{a^2 - 3}}{\sqrt{a^2 - 3}} - \frac{a^2}{\sqrt{a^2 - 3}} =$

$\frac{a^2 - 3}{\sqrt{a^2 - 3}} - \frac{a^2}{\sqrt{a^2 - 3}} = \frac{a^2 - 3 - a^2}{\sqrt{a^2 - 3}} =$

$\frac{-3}{\sqrt{a^2 - 3}} = \frac{-3}{\sqrt{a^2 - 3}} \cdot \frac{\sqrt{a^2 - 3}}{\sqrt{a^2 - 3}} = \frac{-3\sqrt{a^2 - 3}}{a^2 - 3}$

75. $\frac{\frac{1}{\sqrt{w}} - \sqrt{w}}{\frac{\sqrt{w} + 1}{\sqrt{w}}} = \frac{\frac{1}{\sqrt{w}} - \sqrt{w}}{\frac{\sqrt{w} + 1}{\sqrt{w}}} \cdot \frac{\sqrt{w}}{\sqrt{w}} = \frac{1 - w}{\sqrt{w} + 1} =$

$\frac{1 - w}{\sqrt{w} + 1} \cdot \frac{\sqrt{w} - 1}{\sqrt{w} - 1} = \frac{\sqrt{w} - 1 - w\sqrt{w} + w}{w - 1} =$

$\frac{(w - 1) - \sqrt{w}(w - 1)}{w - 1} = \frac{(w - 1)(1 - \sqrt{w})}{w - 1} = 1 - \sqrt{w}$

Exercise Set 10.6

1. $x^{1/4} = \sqrt[4]{x}$

3. $(8)^{1/3} = \sqrt[3]{8} = 2$

5. $(a^2b^2)^{1/5} = \sqrt[5]{a^2b^2}$

7. $16^{3/4} = \sqrt[4]{16^3} = (\sqrt[4]{16})^3 = 2^3 = 8$

9. $\sqrt[3]{20} = 20^{1/3}$

11. $\sqrt[5]{xy^2z} = (xy^2z)^{1/5}$

13. $(\sqrt{3mn})^3 = (3mn)^{3/2}$

15. $(\sqrt[7]{8x^2y})^5 = (8x^2y)^{5/7}$

17. $x^{-1/3} = \frac{1}{x^{1/3}}$

19. $\frac{1}{x^{-2/3}} = x^{2/3}$

21. $5^{3/4} \cdot 5^{1/8} = 5^{3/4+1/8} = 5^{6/8+1/8} = 5^{7/8}$

23. $\frac{7^{5/8}}{7^{3/8}} = 7^{5/8-3/8} = 7^{2/8} = 7^{1/4}$

25. $\frac{8.3^{3/4}}{8.3^{2/5}} = 8.3^{3/4-2/5} = 8.3^{15/20-8/20} = 8.3^{7/20}$

27. $(10^{3/8})^{2/5} = 10^{(3/8)(2/5)} = 10^{6/40} = 10^{3/20}$

29. $a^{2/3} \cdot a^{5/6} = a^{2/3+5/6} = a^{4/6+5/6} = a^{9/6} = a^{3/2}$

31. $(a^{2/3} \cdot b^{5/8})^4 = (a^{2/3})^4(b^{5/8})^4 = a^{8/3}b^{20/8} =$
$a^{8/3}b^{5/2}$

33. $\sqrt[6]{a^4} = a^{4/6} = a^{2/3} = \sqrt[3]{a^2}$

35. $\sqrt[3]{8y^6} = (2^3y^6)^{1/3} = 2^{3/3}y^{6/3} = 2y^2$

37. $\sqrt[4]{32} = \sqrt[4]{2^5} = 2^{5/4} = 2^{4/4} \cdot 2^{1/4} = 2\sqrt[4]{2}$

39. $\sqrt[6]{4x^2} = (2^2x^2)^{1/6} = 2^{2/6}x^{2/6}$
$= 2^{1/3}x^{1/3} = (2x)^{1/3} = \sqrt[3]{2x}$

41. $\sqrt[5]{32c^{10}d^{15}} = (2^5c^{10}d^{15})^{1/5} = 2^{5/5}c^{10/5}d^{15/5}$
$= 2c^2d^3$

43. $\sqrt[6]{\frac{m^{12}n^{24}}{64}} = \left[\frac{m^{12}n^{24}}{2^6}\right]^{1/6} = \frac{m^{12/6}n^{24/6}}{2^{6/6}} = \frac{m^2n^4}{2}$

45. $\sqrt[8]{r^4s^2} = (r^4s^2)^{1/8} = r^{4/8}s^{2/8} = r^{2/4}s^{1/4}$
$= (r^2s)^{1/4} = \sqrt[4]{r^2s}$

47. $\sqrt[3]{27a^3b^9} = (3^3a^3b^9)^{1/3} = 3^{3/3}a^{3/3}b^{9/3} = 3ab^3$

49. $\sqrt[5]{32x^{15}y^{40}} = (2^5x^{15}y^{40})^{1/5} = 2^{5/5}x^{15/5}y^{40/5} =$
$2x^3y^8$

51. $\sqrt[4]{64p^{12}q^{32}} = (2^6p^{12}q^{32})^{1/4} = 2^{6/4}p^{12/4}q^{32/4} =$
$2^{3/2}p^3q^8 = 2 \cdot 2^{1/2}p^3q^8 = 2\sqrt{2}p^3q^8$

53. $\sqrt[3]{3}\sqrt{3} = 3^{1/3}3^{1/2} = 3^{2/6}3^{3/6} = (3^2 \cdot 3^3)^{1/6} =$
$(9 \cdot 27)^{1/6} = (243)^{1/6} = \sqrt[6]{243}$

55. $\sqrt{x}\sqrt[3]{2x} = x^{1/2} \cdot (2x)^{1/3} = x^{3/6} \cdot (2x)^{2/6}$
$= [x^3(2x)^2]^{1/6}$
$= (x^3 \cdot 4x^2)^{1/6}$
$= (4x^5)^{1/6}$
$= \sqrt[6]{4x^5}$

57. $\sqrt{x}\sqrt[3]{x-2} = x^{1/2} \cdot (x-2)^{1/3}$
$= x^{3/6} \cdot (x-2)^{2/6}$
$= [x^3(x-2)^2]^{1/6}$
$= [x^3(x^2 - 4x + 4)]^{1/6}$
$= (x^5 - 4x^4 + 4x^3)^{1/6}$
$= \sqrt[6]{x^5 - 4x^4 + 4x^3}$

59. $\frac{\sqrt[3]{(a+b)^2}}{\sqrt{(a+b)}} = \frac{(a+b)^{2/3}}{(a+b)^{1/2}} = \frac{(a+b)^{4/6}}{(a+b)^{3/6}}$
$= (a+b)^{4/6-3/6}$
$= (a+b)^{1/6}$
$= \sqrt[6]{a+b}$

61. $a^{2/3} \cdot b^{3/4} = a^{8/12} \cdot b^{9/12}$
$= (a^8b^9)^{1/12}$
$= \sqrt[12]{a^8b^9}$

63. $\frac{s^{7/12} \cdot t^{7/6}}{x^{1/3} \cdot t^{-1/6}} = \frac{s^{7/12} \cdot t^{7/6}}{s^{4/12} \cdot t^{-1/6}} = s^{7/12-4/12} \cdot t^{7/6-(-1/6)}$
$= s^{3/12} \cdot t^{8/6}$
$= s^{3/12} \cdot t^{16/12}$
$= (s^3t^{16})^{1/12}$
$= \sqrt[12]{s^3t^{16}}$, or $t\sqrt[12]{s^3t^4}$

65. $\sqrt[4]{x^3y^5} \cdot \sqrt{xy} = (x^3y^5)^{1/4}(xy)^{1/2}$
$= x^{3/4}y^{5/4}x^{1/2}y^{1/2}$
$= x^{3/4+1/2}y^{5/4+1/2}$
$= x^{5/4}y^{7/4}$
$= (x^5y^7)^{1/4}$
$= \sqrt[4]{x^5y^7}$, or $xy\sqrt[4]{xy^3}$

67. $\sqrt{a^4b^3c^4}\sqrt[3]{ab^2c} = (a^4b^3c^4)^{1/2}(ab^2c)^{1/3}$
$= a^{4/2}b^{3/2}c^{4/2}a^{1/3}b^{2/3}c^{1/3}$
$= a^{4/2+1/3}b^{3/2+2/3}c^{4/2+1/3}$
$= a^{14/6}b^{13/6}c^{14/6}$
$= (a^{14}b^{13}c^{14})^{1/6}$
$= \sqrt[6]{a^{14}b^{13}c^{14}}$, or
$a^2b^2c^2\sqrt[6]{a^2bc^2}$

69. $\left(\frac{p^{3/4}q^{-2/3}}{p^{7/8}q^{-3/4}}\right)^2 = \left[p^{3/4-7/8}q^{-2/3-(-3/4)}\right]^2$

$= (p^{6/8-7/8}q^{-8/12+9/12})^2$

$= (p^{-1/8}q^{1/12})^2$

$= p^{-2/8}q^{2/12}$

$= p^{-1/4}q^{1/6}$

$= p^{-3/12}q^{2/12}$

$= (p^{-3}q^2)^{1/12}$

$= \left(\frac{q^2}{p^3}\right)^{1/12}$

$= \sqrt[12]{\frac{q^2}{p^3}}$

71. $\sqrt{x^5 \sqrt[3]{x^4}} = \sqrt{x^5x^{4/3}} = \sqrt{x^{19/3}} = (x^{19/3})^{1/2} = x^{19/6} =$

$\sqrt[6]{x^{19}} = x^3\sqrt[6]{x}$

73. $\sqrt[4]{\sqrt[3]{8x^3y^6}} = \sqrt[4]{(2^3x^3y^6)^{1/3}} = \sqrt[4]{2^{3/3}x^{3/3}y^{6/3}} = \sqrt[4]{2xy^2}$

75. $\sqrt[12]{p^2 + 2pq + q^2} = \sqrt[12]{(p + q)^2} = [(p + q)^2]^{1/12} =$

$(p + q)^{2/12} = (p + q)^{1/6} = \sqrt[6]{p + q}$

77. $\left[\sqrt[10]{\sqrt[5]{x^{15}}}\right]^5 \left[\sqrt[5]{\sqrt[10]{x^{15}}}\right]^5$

$= \left[\left[(x^{15})^{1/5}\right]^{1/10}\right]^5 \left[\left[(x^{15})^{1/10}\right]^{1/5}\right]^5$

$= x^{3/2}x^{3/2}$ Multiplying exponents

$= x^3$

Exercise Set 10.7

1. $\sqrt{4x - 5} = 1$

$(\sqrt{4x - 5})^2 = 1^2$ Principle of powers

$4x - 5 = 1$

$4x = 6$

$x = \frac{6}{4}$

$x = \frac{3}{2}$

Check:

$\sqrt{4x - 5} = 1$	
$\sqrt{4 \cdot \frac{3}{2} - 5}$	1
$\sqrt{6 - 5}$	
$\sqrt{1}$	
1	TRUE

The solution is $\frac{3}{2}$.

3. $\sqrt{5x} + 2 = 9$

$\sqrt{5x} = 7$ Subtracting to isolate the radical

$(\sqrt{5x})^2 = 7^2$ Principle of powers

$5x = 49$

$x = \frac{49}{5}$

Check:

$\sqrt{5x} + 2 = 9$	
$\sqrt{5 \cdot \frac{49}{5}} + 2$	9
$\sqrt{49} + 2$	
$7 + 2$	
9	TRUE

The solution is $\frac{49}{5}$.

5. $\sqrt{y + 1} - 5 = 8$

$\sqrt{y + 1} = 13$

$(\sqrt{y + 1})^2 = 13^2$

$y + 1 = 169$

$y = 168$

Check:

$\sqrt{y + 1} - 5 = 8$	
$\sqrt{168 + 1} - 5$	8
$\sqrt{169} - 5$	
$13 - 5$	
8	TRUE

The solution is 168.

7. $\sqrt{3y + 1} = 9$

$(\sqrt{3y + 1})^2 = 9^2$

$3y + 1 = 81$

$3y = 80$

$y = \frac{80}{3}$

Check:

$\sqrt{3y + 1} = 9$	
$\sqrt{3 \cdot \frac{80}{3} + 1}$	9
$\sqrt{80 + 1}$	
$\sqrt{81}$	
9	TRUE

The solution is $\frac{80}{3}$.

9. $\sqrt[3]{x} = -3$

$(\sqrt[3]{x})^3 = (-3)^3$

$x = -27$

Check:

$\sqrt[3]{x} = -3$	
$\sqrt[3]{-27}$	-3
-3	TRUE

The solution is -27.

11. $\sqrt{x + 2} = -4$

$(\sqrt{x + 2})^2 = (-4)^2$

$x + 2 = 16$

$x = 14$

Check:

$\sqrt{x + 2} = -4$	
$\sqrt{14 + 2}$	-4
$\sqrt{16}$	
4	FALSE

The number 14 does not check. The equation has no solution. We might have observed at the outset that this equation has no solution because the principle square root of a number is never negative.

13. $\sqrt[3]{x + 5} = 2$

$(\sqrt[3]{x + 5})^3 = 2^3$

$x + 5 = 8$

$x = 3$

Check:

$\sqrt[3]{x + 5} = 2$	
$\sqrt[3]{3 + 5}$	2
$\sqrt[3]{8}$	
2	TRUE

The solution is 3.

15. $\sqrt[4]{y - 3} = 2$

$(\sqrt[4]{y - 3})^4 = 2^4$

$y - 3 = 16$

$y = 19$

Check:

$\sqrt[4]{y - 3} = 2$	
$\sqrt[4]{19 - 3}$	2
$\sqrt[4]{16}$	
2	TRUE

The solution is 19.

17. $\sqrt[3]{6x + 9} + 8 = 5$

$\sqrt[3]{6x + 9} = -3$

$(\sqrt[3]{6x + 9})^3 = (-3)^3$

$6x + 9 = -27$

$6x = -36$

$x = -6$

Check:

$\sqrt[3]{6x + 9} + 8 = 5$	
$\sqrt[3]{6(-6) + 9} + 8$	5
$\sqrt[3]{-27} + 8$	
$-3 + 8$	
5	TRUE

The solution is -6.

19. $8 = \frac{1}{\sqrt{x}}$

$8 \cdot \sqrt{x} = \frac{1}{\sqrt{x}} \cdot \sqrt{x}$

$8\sqrt{x} = 1$

$(8\sqrt{x})^2 = 1^2$

$64x = 1$

$x = \frac{1}{64}$

Check:

$8 = \frac{1}{\sqrt{x}}$	
8	$\frac{1}{\sqrt{\frac{1}{64}}}$
	$\frac{1}{\frac{1}{8}}$
	8 TRUE

The solution is $\frac{1}{64}$.

21. $\sqrt{3y + 1} = \sqrt{2y + 6}$

$(\sqrt{3y + 1})^2 = (\sqrt{2y + 6})^2$

$3y + 1 = 2y + 6$

$y = 5$

Check:

$\sqrt{3y + 1} = \sqrt{2y + 6}$	
$\sqrt{3 \cdot 5 + 1}$	$\sqrt{2 \cdot 5 + 6}$
$\sqrt{16}$	$\sqrt{16}$ TRUE

The solution is 5.

23. $\sqrt{y - 5} + \sqrt{y} = 5$

$\sqrt{y - 5} = 5 - \sqrt{y}$

$(\sqrt{y - 5})^2 = (5 - \sqrt{y})^2$

$y - 5 = 25 - 10\sqrt{y} + y$

$10\sqrt{y} = 30$

$\sqrt{y} = 3$

$(\sqrt{y})^2 = 3^2$

$y = 9$

The number 9 checks, so it is the solution.

25. $3 + \sqrt{z - 6} = \sqrt{z + 9}$

$(3 + \sqrt{z - 6})^2 = (\sqrt{z + 9})^2$

$9 + 6\sqrt{z - 6} + z - 6 = z + 9$

$6\sqrt{z - 6} = 6$

$\sqrt{z - 6} = 1$

$(\sqrt{z - 6})^2 = 1^2$

$z - 6 = 1$

$z = 7$

The number 7 checks, so it is the solution.

27. $\sqrt{20 - x} + 8 = \sqrt{9 - x} + 11$

$\sqrt{20 - x} = \sqrt{9 - x} + 3$

$(\sqrt{20 - x})^2 = (\sqrt{9 - x} + 3)^2$

$20 - x = 9 - x + 6\sqrt{9 - x} + 9$

$2 = 6\sqrt{9 - x}$

$1 = 3\sqrt{9 - x}$

$1^2 = (3\sqrt{9 - x})^2$

$1 = 9(9 - x)$

$1 = 81 - 9x$

$9x = 80$

$x = \frac{80}{9}$

The number $\frac{80}{9}$ checks, so it is the solution.

29. $\sqrt{4y + 1} - \sqrt{y - 2} = 3$

$$\sqrt{4y + 1} = 3 + \sqrt{y - 2}$$
$$(\sqrt{4y + 1})^2 = (3 + \sqrt{y - 2})^2$$
$$4y + 1 = 9 + 6\sqrt{y - 2} + y - 2$$
$$3y - 6 = 6\sqrt{y - 2}$$
$$y - 2 = 2\sqrt{y - 2}$$
$$(y - 2)^2 = (2\sqrt{y - 2})^2$$
$$y^2 - 4y + 4 = 4(y - 2)$$
$$y^2 - 4y + 4 = 4y - 8$$
$$y^2 - 8y + 12 = 0$$
$$(y - 6)(y - 2) = 0$$

$y - 6 = 0$ or $y - 2 = 0$

$y = 6$ or $y = 2$

The numbers 6 and 2 check, so they are the solutions.

31. $\sqrt{x + 2} + \sqrt{3x + 4} = 2$

$\sqrt{x + 2} = 2 - \sqrt{3x + 4}$ Isolating one radical

$$(\sqrt{x + 2})^2 = (2 - \sqrt{3x + 4})^2$$
$$x + 2 = 4 - 4\sqrt{3x + 4} + 3x + 4$$

$-2x - 6 = -4\sqrt{3x + 4}$ Isolating the remaining radical

$x + 3 = 2\sqrt{3x + 4}$ Dividing by 2

$$(x + 3)^2 = (2\sqrt{3x + 4})^2$$
$$x^2 + 6x + 9 = 4(3x + 4)$$
$$x^2 + 6x + 9 = 12x + 16$$
$$x^2 - 6x - 7 = 0$$
$$(x - 7)(x + 1) = 0$$

$x - 7 = 0$ or $x + 1 = 0$

$x = 7$ or $x = -1$

Check: For 7:

$\sqrt{x + 2} + \sqrt{3x + 4} = 2$	
$\sqrt{7 + 2} + \sqrt{3 \cdot 7 + 4}$	2
$\sqrt{9} + \sqrt{25}$	
8	FALSE

For -1:

$\sqrt{x + 2} + \sqrt{3x + 4} = 2$	
$\sqrt{-1 + 2} + \sqrt{3(-1) + 4}$	2
$\sqrt{1} + \sqrt{1}$	
2	TRUE

Since -1 checks but 7 does not, the solution is -1.

33. $\sqrt{3x - 5} + \sqrt{2x + 3} + 1 = 0$

$$\sqrt{3x - 5} + 1 = -\sqrt{2x + 3}$$
$$(\sqrt{3x - 5} + 1)^2 = (-\sqrt{2x + 3})^2$$
$$3x - 5 + 2\sqrt{3x - 5} + 1 = 2x + 3$$
$$2\sqrt{3x - 5} = -x + 7$$
$$(2\sqrt{3x - 5})^2 = (-x + 7)^2$$
$$4(3x - 5) = x^2 - 14x + 49$$
$$12x - 20 = x^2 - 14x + 49$$
$$0 = x^2 - 26x + 69$$
$$0 = (x - 23)(x - 3)$$

$x - 23 = 0$ or $x - 3 = 0$

$x = 23$ or $x = 3$

Neither number checks. There is no solution.

35. $2\sqrt{t - 1} - \sqrt{3t - 1} = 0$

$$2\sqrt{t - 1} = \sqrt{3t - 1}$$
$$(2\sqrt{t - 1})^2 = (\sqrt{3t - 1})^2$$
$$4(t - 1) = 3t - 1$$
$$4t - 4 = 3t - 1$$
$$t = 3$$

Since 3 checks, it is the solution.

37. $x^2 + 2.8x = 0$

$x(x + 2.8) = 0$

$x = 0$ or $x + 2.8 = 0$

$x = 0$ or $x = -2.8$

39. $\sqrt[3]{\frac{z}{4}} - 10 = 2$

$$\sqrt[3]{\frac{z}{4}} = 12$$
$$\left(\sqrt[3]{\frac{z}{4}}\right)^3 = 12^3$$
$$\frac{z}{4} = 1728$$
$$z = 6912$$

The number 6912 checks, so it is the solution.

41. $\sqrt{\sqrt{y + 49} - \sqrt{y}} = \sqrt{7}$

$(\sqrt{\sqrt{y + 49} - \sqrt{y}})^2 = (\sqrt{7})^2$

$\sqrt{y + 49} - \sqrt{y} = 7$

$\sqrt{y + 49} = 7 + \sqrt{y}$

$(\sqrt{y + 49})^2 = (7 + \sqrt{y})^2$

$y + 49 = 49 + 14\sqrt{y} + y$

$0 = 14\sqrt{y}$

$0 = \sqrt{y}$

$0 = y$

The number 0 checks and is the solution.

43. $\sqrt{\sqrt{x^2 + 9x + 34}} = 2$

$(\sqrt{\sqrt{x^2 + 9x + 34}})^2 = 2^2$

$\sqrt{x^2 + 9x + 34} = 4$

$(\sqrt{x^2 + 9x + 34})^2 = 4^2$

$x^2 + 9x + 34 = 16$

$x^2 + 9x + 18 = 0$

$(x + 6)(x + 3) = 0$

$x + 6 = 0$ or $x + 3 = 0$

$x = -6$ or $x = -3$

Both values check. The solutions are -6 and -3.

45. $\sqrt{x - 2} - \sqrt{x + 2} + 2 = 0$

$\sqrt{x - 2} + 2 = \sqrt{x + 2}$

$(\sqrt{x - 2} + 2)^2 = (\sqrt{x + 2})^2$

$(x - 2) + 4\sqrt{x - 2} + 4 = x + 2$

$4\sqrt{x - 2} = 0$

$\sqrt{x - 2} = 0$

$(\sqrt{x - 2})^2 = 0^2$

$x - 2 = 0$

$x = 2$

The number 2 checks, so it is the solution.

47. $\sqrt{a^2 + 30a} = a + \sqrt{5a}$

$(\sqrt{a^2 + 30a})^2 = (a + \sqrt{5a})^2$

$a^2 + 30a = a^2 + 2a\sqrt{5a} + 5a$

$25a = 2a\sqrt{5a}$

$(25a)^2 = (2a\sqrt{5a})^2$

$625a^2 = 4a^2 \cdot 5a$

$625a^2 = 20a^3$

$0 = 20a^3 - 625a^2$

$0 = a^2(20a - 625)$

47. (continued)

$a^2 = 0$ or $20a - 625 = 0$

$a = 0$ or $20a = 625$

$a = 0$ or $a = \frac{125}{4}$

Both values check. The solutions are 0 and $\frac{125}{4}$.

49. $\frac{x - 1}{\sqrt{x^2 + 3x + 6}} = \frac{1}{4}$, LCM $= 4\sqrt{x^2 + 3x + 6}$

$4x - 4 = \sqrt{x^2 + 3x + 6}$

$16x^2 - 32x + 16 = x^2 + 3x + 6$

$15x^2 - 35x + 10 = 0$

$3x^2 - 7x + 2 = 0$

$(3x - 1)(x - 2) = 0$

$3x - 1 = 0$ or $x - 2 = 0$

$3x = 1$ or $x = 2$

$x = \frac{1}{3}$ or $x = 2$

The number 2 does check but $\frac{1}{3}$ does not. The solution is 2.

51. $\sqrt{y^2 + 6} + y - 3 = 0$

$\sqrt{y^2 + 6} = 3 - y$

$(\sqrt{y^2 + 6})^2 = (3 - y)^2$

$y^2 + 6 = 9 - 6y + y^2$

$-3 = -6y$

$\frac{1}{2} = y$

The number $\frac{1}{2}$ checks and is the solution.

53. $\sqrt{y + 1} - \sqrt{2y - 5} = \sqrt{y - 2}$

$(\sqrt{y + 1} - \sqrt{2y - 5})^2 = (\sqrt{y - 2})^2$

$y + 1 - 2\sqrt{(y + 1)(2y - 5)} + 2y - 5 = y - 2$

$-2\sqrt{2y^2 - 3y - 5} = -2y + 2$

$\sqrt{2y^2 - 3y - 5} = y - 1$ Dividing by -2

$(\sqrt{2y^2 - 3y - 5})^2 = (y - 1)^2$

$2y^2 - 3y - 5 = y^2 - 2y + 1$

$y^2 - y - 6 = 0$

$(y - 3)(y + 2) = 0$

$y - 3 = 0$ or $y + 2 = 0$

$y = 3$ or $y = -2$

The number 3 checks but -2 does not. The solution is 3.

Exercise Set 10.8

1. $a = 3, \quad b = 5$

Find c.

$c^2 = a^2 + b^2$ Pythagorean equation

$c^2 = 3^2 + 5^2$ Substituting

$c^2 = 9 + 25$

$c^2 = 34$

$c = \sqrt{34}$ Exact answer

$c \approx 5.831$ Approximation

3. $a = 12, \quad b = 12$

Find c.

$c^2 = a^2 + b^2$ Pythagorean equation

$c^2 = 12^2 + 12^2$ Substituting

$c^2 = 144 + 144$

$c^2 = 288$

$c = \sqrt{288}$, or $12\sqrt{2}$ Exact answer

$c \approx 16.971$ Approximation

5. $b = 12, \quad c = 13$

Find a.

$a^2 + b^2 = c^2$

$a^2 + 12^2 = 13^2$ Substituting

$a^2 + 144 = 169$

$a^2 = 25$

$a = 5$

7. $c = 6, \quad a = \sqrt{5}$

Find b.

$a^2 + b^2 = c^2$

$(\sqrt{5})^2 + b^2 = 6^2$

$5 + b^2 = 36$

$b^2 = 31$

$b = \sqrt{31}$ Exact answer

$b \approx 5.568$ Approximation

9. $b = 1, \quad c = \sqrt{13}$

Find a.

$a^2 + b^2 = c^2$

$a^2 + 1^2 = (\sqrt{13})^2$

$a^2 + 1 = 13$

$a^2 = 12$

$a = \sqrt{12}$, or $2\sqrt{3}$ Exact answer

$a \approx 3.464$ Approximation

11. $a = 1, \quad c = \sqrt{n}$

Find b.

$a^2 + b^2 = c^2$

$1^2 + b^2 = (\sqrt{n})^2$

$1 + b^2 = n$

$b^2 = n - 1$

$b = \sqrt{n - 1}$

13. $d^2 = 10^2 + 15^2$

$d^2 = 100 + 225$

$d^2 = 325$

$d = \sqrt{325}$

$d \approx 18.028$

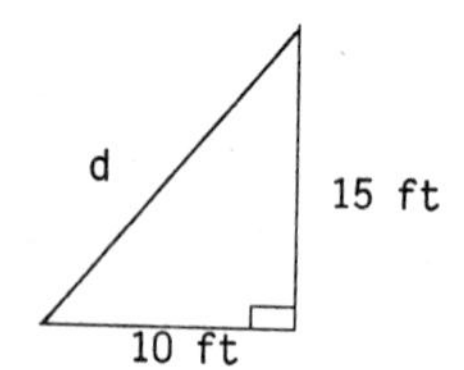

The wire is $\sqrt{325}$, or 18.028 ft long.

15.

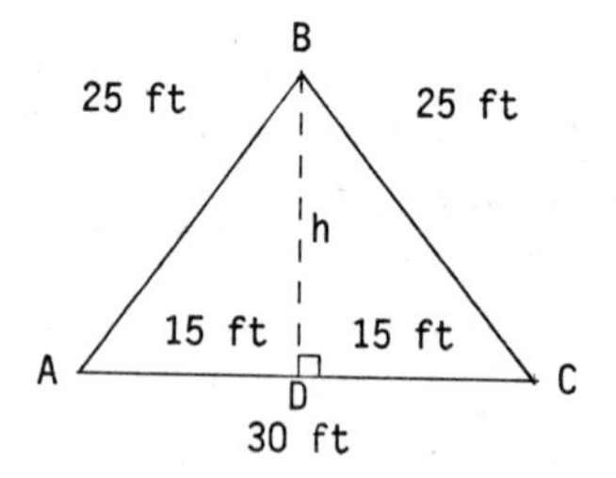

Δ ABC is an isosceles triangle. The height from B to $\overline{AC}$ bisects $\overline{AC}$. Thus, $\overline{DC}$ measures 15 ft.

$h^2 + 15^2 = 25^2$

$h^2 + 225 = 625$

$h^2 = 400$

$h = \sqrt{400}$

$h = 20$

If the height of Δ ABC is 20 ft and the base is 30 ft, the area is $\frac{1}{2} \cdot 30 \cdot 20$, or 300 ft^2.

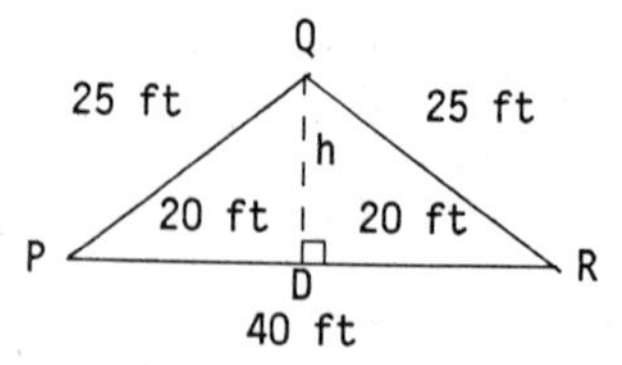

Δ PQR is an isosceles triangle. The height from Q to $\overline{PR}$ bisects $\overline{PR}$. Thus, $\overline{SR}$ measures 20 ft.

$h^2 + 20^2 = 25^2$

$h^2 + 400 = 625$

$h^2 = 225$

$h = \sqrt{225}$

$h = 15$

If the height of Δ PQR is 15 ft and the base is 40 ft, the area is $\frac{1}{2} \cdot 40 \cdot 15$, or 300 ft^2.

The areas of the two triangles are the same.

17.

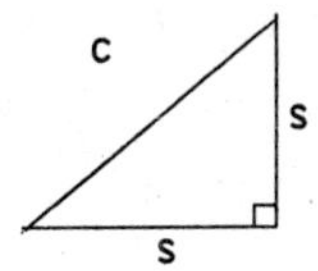

$c^2 = s^2 + s^2$

$c^2 = 2s^2$

$c = \sqrt{2s^2}$

$c = s\sqrt{2}$

19. $L = \dfrac{0.000169d^{2.27}}{h}$

$L = \dfrac{0.000169(200)^{2.27}}{4}$

≈ 7.1

The length of the letters should be about 7.1 ft.

21.

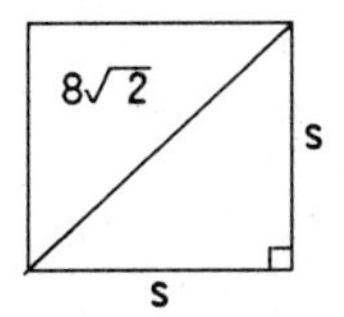

$s^2 + s^2 = (8\sqrt{2})^2$

$2s^2 = 128$

$s^2 = 64$

$s = 8$

The length of a side of the square is 8 ft.

23.

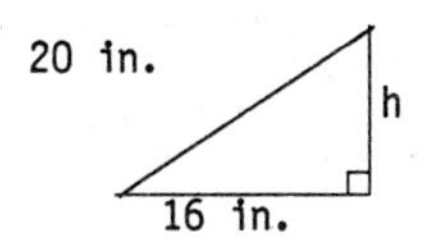

$h^2 + 16^2 = 20^2$

$h^2 + 256 = 400$

$h^2 = 144$

$h = 12$

The height is 12 in.

25.

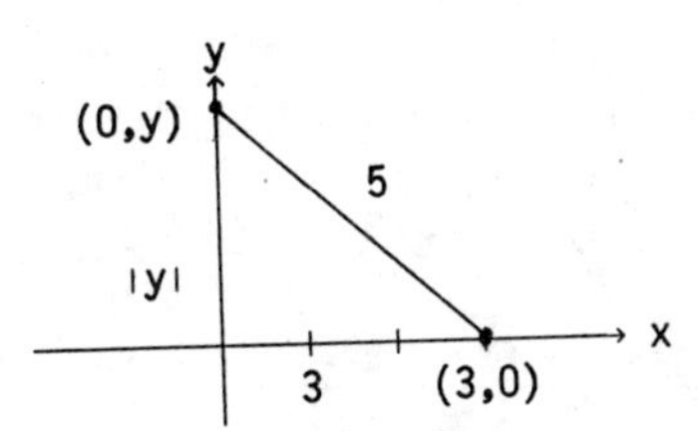

$|y|^2 + 3^2 = 5^2$

$y^2 + 9 = 25$

$y^2 = 16$

$y = \pm 4$

The points are (0,4) and (0,-4).

27. $x^2 - 11x + 24 = 0$

$(x - 8)(x - 3) = 0$

$x - 8 = 0$ or $x - 3 = 0$

$x = 8$ or $x = 3$

29. If the area of square PQRS is 100 ft^2, then each side measures 10 ft. If A, B, C, and D are midpoints, then each of the segments PB, BQ, QC, CS, SD, DR, RA, and AP measures 5 ft. We can label the figure with additional information.

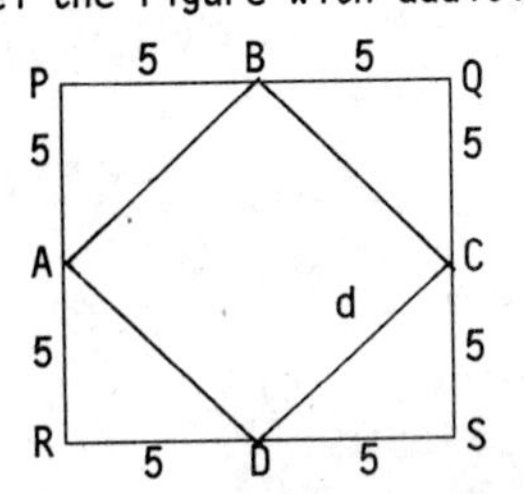

We label a side of the square ABCD with d. Then we use the Pythagorean property.

$5^2 + 5^2 = d^2$

$25 + 25 = d^2$

$50 = d^2$

$\sqrt{50} = d$

If a side of square ABCD is $\sqrt{50}$, then its area is $\sqrt{50}\cdot\sqrt{50}$, or 50 ft^2.

Exercise Set 10.9

1. $\sqrt{-15} = \sqrt{-1\cdot 15} = \sqrt{-1}\cdot\sqrt{15} = i\sqrt{15}$, or $\sqrt{15}i$

3. $\sqrt{-16} = \sqrt{-1\cdot 16} = \sqrt{-1}\cdot\sqrt{16} = 4i$

5. $-\sqrt{-12} = -\sqrt{-1\cdot 12} = -\sqrt{-1}\cdot\sqrt{12} = -i\cdot 2\sqrt{3} = -2\sqrt{3}i$, or $-2i\sqrt{3}$

7. $\sqrt{-3} = \sqrt{-1\cdot 3} = \sqrt{-1}\cdot\sqrt{3} = i\sqrt{3}$, or $\sqrt{3}i$

9. $\sqrt{-81} = \sqrt{-1\cdot 81} = \sqrt{-1}\cdot\sqrt{81} = i\cdot 9 = 9i$

11. $\sqrt{-98} = \sqrt{-1\cdot 98} = \sqrt{-1}\cdot\sqrt{98} = i\cdot 7\sqrt{2} = 7\sqrt{2}i$, or $7i\sqrt{2}$

13. $-\sqrt{-49} = -\sqrt{-1\cdot 49} = -\sqrt{-1}\cdot\sqrt{49} = -i\cdot 7 = -7i$

15. $4 - \sqrt{-60} = 4 - \sqrt{-1\cdot 60} = 4 - \sqrt{-1}\cdot\sqrt{60} =$

$4 - i\cdot 2\sqrt{15} = 4 - 2\sqrt{15}i$, or $4 - 2i\sqrt{15}$

17. $\sqrt{-4} + \sqrt{-12} = \sqrt{-1\cdot 4} + \sqrt{-1\cdot 12} =$

$\sqrt{-1}\cdot\sqrt{4} + \sqrt{-1}\cdot\sqrt{12} = i\cdot 2 + i\cdot 2\sqrt{3} =$

$(2 + 2\sqrt{3})i$

19. $(3 + 2i) + (5 - i) = (3 + 5) + (2 - 1)i$ Collecting the real and the imaginary parts
$= 8 + i$

21. $(4 - 3i) + (5 - 2i) = (4 + 5) + (-3 - 2)i$ Collecting the real and the imaginary parts
$= 9 - 5i$

23. $(9 - i) + (-2 + 5i) = (9 - 2) + (-1 + 5)i$
$= 7 + 4i$

25. $(3 - i) - (5 + 2i) = (3 - 5) + (-1 - 2)i$
$= -2 - 3i$

27. $(4 - 2i) - (5 - 3i) = (4 - 5) + [-2 - (-3)]i$
$= -1 + i$

29. $(9 + 5i) - (-2 - i) = [9 - (-2)] + [5 - (-1)]i$
$= 11 + 6i$

31. $\sqrt{-36} \cdot \sqrt{-9} = \sqrt{-1} \cdot \sqrt{36} \cdot \sqrt{-1} \cdot \sqrt{9}$
$= i \cdot 6 \cdot i \cdot 3$
$= i^2 \cdot 18$
$= -1 \cdot 18$ $\quad i^2 = -1$
$= -18$

33. $\sqrt{-5} \cdot \sqrt{-2} = \sqrt{-1} \cdot \sqrt{5} \cdot \sqrt{-1} \cdot \sqrt{2}$
$= i \cdot \sqrt{5} \cdot i \cdot \sqrt{2}$
$= i^2(\sqrt{10})$
$= -1(\sqrt{10})$ $\quad i^2 = -1$
$= -\sqrt{10}$

35. $-3i \cdot 7i = -21 \cdot i^2$
$= -21(-1)$ $\quad i^2 = -1$
$= 21$

37. $5i(4 - 7i) = 5i \cdot 4 - 5i \cdot 7i$
$= 20i - 35i^2$
$= 20i - 35(-1)$ $\quad i^2 = -1$
$= 20i + 35$
$= 35 + 20i$

39. $(3 + 2i)(1 + i) = 3 + 3i + 2i + 2i^2$ Using FOIL
$= 3 + 3i + 2i - 2$ $\quad i^2 = -1$
$= 1 + 5i$

41. $(2 + 3i)(6 - 2i) = 12 - 4i + 18i - 6i^2$ Using FOIL
$= 12 - 4i + 18i + 6$ $\quad i^2 = -1$
$= 18 + 14i$

43. $(6 - 5i)(3 + 4i) = 18 + 24i - 15i - 20i^2 =$
$18 + 24i - 15i + 20 = 38 + 9i$

45. $(7 - 2i)(2 - 6i) = 14 - 42i - 4i + 12i^2 =$
$14 - 42i - 4i - 12 = 2 - 46i$

47. $(3 - 2i)^2 = 3^2 - 2 \cdot 3 \cdot 2i + (2i)^2$ Squaring a binomial
$= 9 - 12i + 4i^2$
$= 9 - 12i - 4$ $\quad i^2 = -1$
$= 5 - 12i$

49. $(2 + 3i)^2 = 2^2 + 2 \cdot 2 \cdot 3i + (3i)^2$ Squaring a binomial
$= 4 + 12i + 9i^2$
$= 4 + 12i - 9$
$= -5 + 12i$

51. $(-2 + 3i)^2 = 4 - 12i + 9i^2 = 4 - 12i - 9 =$
$-5 - 12i$

53. $i^7 = i^6 \cdot i = (i^2)^3 \cdot i = (-1)^3 \cdot i = -1 \cdot i = -i$

55. $i^{24} = (i^2)^{12} = (-1)^{12} = 1$

57. $i^{42} = (i^2)^{21} = (-1)^{21} = -1$

59. $i^9 = (i^2)^4 \cdot i = (-1)^4 \cdot i = 1 \cdot i = i$

61. $i^6 = (i^2)^3 = (-1)^3 = -1$

63. $(5i)^3 = 5^3 \cdot i^3 = 125 \cdot i^2 \cdot i = 125(-1)(i) = -125i$

65. $7 + i^4 = 7 + (i^2)^2 = 7 + (-1)^2 = 7 + 1 = 8$

67. $i^4 - 26i = (i^2)^2 - 26i = (-1)^2 - 26i = 1 - 26i$

69. $i^2 + i^4 = -1 + (i^2)^2 = -1 + (-1)^2 = -1 + 1 = 0$

71. $i^5 + i^7 = i^4 \cdot i + i^6 \cdot i = (i^2)^2 \cdot i + (i^2)^3 \cdot i =$
$(-1)^2 \cdot i + (-1)^3 \cdot i = 1 \cdot i + (-1)i = i - i = 0$

73. $1 + i + i^2 + i^3 + i^4 = 1 + i + i^2 + i^2 \cdot i + (i^2)^2$
$= 1 + i + (-1) + (-1) \cdot i + (-1)^2$
$= 1 + i - 1 - i + 1$
$= 1$

75. $5 - \sqrt{-64} = 5 - \sqrt{-1} \cdot \sqrt{64} = 5 - i \cdot 8 = 5 - 8i$

77. $\dfrac{8 - \sqrt{-24}}{4} = \dfrac{8 - \sqrt{-1} \cdot \sqrt{24}}{4} = \dfrac{8 - i \cdot 2\sqrt{6}}{4} =$
$\dfrac{2(4 - i\sqrt{6})}{2 \cdot 2} = \dfrac{\cancel{2}(4 - i\sqrt{6})}{\cancel{2} \cdot 2} = \dfrac{4 - i\sqrt{6}}{2} = 2 - \dfrac{\sqrt{6}}{2}i$

79. $\frac{3+2i}{2+i} = \frac{3+2i}{2+i} \cdot \frac{2-i}{2-i}$

$= \frac{6-3i+4i-2i^2}{4-i^2}$

$= \frac{6-3i+4i+2}{4+1}$

$= \frac{8+i}{5}$

$= \frac{8}{5} + \frac{1}{5}i$

81. $\frac{5-2i}{2+5i} = \frac{5-2i}{2+5i} \cdot \frac{2-5i}{2-5i}$

$= \frac{10-25i-4i+10i^2}{4-25i^2}$

$= \frac{10-25i-4i-10}{4+25}$

$= \frac{-29i}{29}$

$= -i$

83. $\frac{8-3i}{7i} = \frac{8-3i}{7i} \cdot \frac{-7i}{-7i}$

$= \frac{-56i+21i^2}{-49i^2}$

$= \frac{-21-56i}{49}$

$= -\frac{3}{7} - \frac{8}{7}i$

85. $\frac{4}{3+i} = \frac{4}{3+i} \cdot \frac{3-i}{3-i}$

$= \frac{12-4i}{9-i^2}$

$= \frac{12-4i}{9-(-1)}$

$= \frac{12-4i}{10}$

$= \frac{12}{10} - \frac{4}{10}i$

$= \frac{6}{5} - \frac{2}{5}i$

87. $\frac{2i}{5-4i} = \frac{2i}{5-4i} \cdot \frac{5+4i}{5+4i}$

$= \frac{10i+8i^2}{25-16i^2}$

$= \frac{10i+8(-1)}{25-16(-1)}$

$= \frac{-8+10i}{41}$

$= -\frac{8}{41} + \frac{10}{41}i$

89. $\frac{4}{3i} = \frac{4}{3i} \cdot \frac{-3i}{-3i}$

$= \frac{-12i}{-9i^2}$

$= \frac{-12i}{-9(-1)}$

$= \frac{-12i}{9}$

$= -\frac{4}{3}i$

91. $\frac{2-4i}{8i} = \frac{2-4i}{8i} \cdot \frac{-8i}{-8i}$

$= \frac{-16i+32i^2}{-64i^2}$

$= \frac{-16i+32(-1)}{-64(-1)}$

$= \frac{-32-16i}{64}$

$= -\frac{32}{64} - \frac{16}{64}i$

$= -\frac{1}{2} - \frac{1}{4}i$

93. $\frac{6+3i}{6-3i} = \frac{6+3i}{6-3i} \cdot \frac{6+3i}{6+3i}$

$= \frac{36+18i+18i+9i^2}{36-9i^2}$

$= \frac{36+36i-9}{36-9(-1)}$

$= \frac{27+36i}{45}$

$= \frac{27}{45} + \frac{36}{45}i$

$= \frac{3}{5} + \frac{4}{5}i$

95. Substitute $1 + 2i$ for x in the equation.

$x^2 - 2x + 5 = 0$	
$(1+2i)^2 - 2(1+2i) + 5$	0
$1 + 4i + 4i^2 - 2 - 4i + 5$	
$1 - 4 - 2 + 5$	
0	

$1 + 2i$ is a solution.

97. Substitute $1 - i$ for x in the equation.

$x^2 + 2x + 2 = 0$	
$(1-i)^2 + 2(1-i) + 2$	0
$1 - 2i + i^2 + 2 - 2i + 2$	
$4 - 4i$	

$1 - i$ is not a solution.

99. $\frac{196}{x^2 - 7x + 49} - \frac{2x}{x + 7} = \frac{2058}{x^3 + 343}$

Note: $x^3 + 343 = (x + 7)(x^2 - 7x + 49)$

The LCM = $(x + 7)(x^2 - 7x + 49)$.

$$(x + 7)(x^2 - 7x + 49)\left(\frac{196}{x^2 - 7x + 49} - \frac{2x}{x + 7}\right) = (x + 7)(x^2 - 7x + 49) \cdot \frac{2058}{x^3 + 343}$$

$$196(x + 7) - 2x(x^2 - 7x + 49) = 2058$$

$$196x + 1372 - 2x^3 + 14x^2 - 98x = 2058$$

$$98x - 686 - 2x^3 + 14x^2 = 0$$

$$49x - 343 - x^3 + 7x^2 = 0$$

$$49(x - 7) - x^2(x - 7) = 0$$

$$(49 - x^2)(x - 7) = 0$$

$$(7 - x)(7 + x)(x - 7) = 0$$

$7 - x = 0$ or $7 + x = 0$ or $x - 7 = 0$

$x = 7$ or $x = -7$ or $x = 7$

Only 7 checks. It is the solution.

101. $\frac{(2i)^4 - (2i)^2}{2i - 1} = \frac{16i^4 - 4i^2}{-1 + 2i} = \frac{20}{-1 + 2i} =$

$\frac{20}{-1 + 2i} \cdot \frac{-1 - 2i}{-1 - 2i} = \frac{-20 - 40i}{5} = -4 - 8i$

103. $\frac{1}{8}(-24 - \sqrt{-1024}) = \frac{1}{8}(-24 - 32i) = -3 - 4i$

105. $7\sqrt{-64} - 9\sqrt{-256} = 7 \cdot 8i - 9 \cdot 16i = 56i - 144i = -88i$

107. $(1 - i)^3(1 + i)^3 =$

$(1 - i)(1 + i) \cdot (1 - i)(1 + i) \cdot (1 - i)(1 + i) =$

$(1 - i^2)(1 - i^2)(1 - i^2) = (1 + 1)(1 + 1)(1 + 1) =$

$2 \cdot 2 \cdot 2 = 8$

109. $\frac{6}{1 + \frac{3}{i}} = \frac{6}{\frac{i + 3}{i}} = \frac{6i}{i - 3} = \frac{6i}{i + 3} \cdot \frac{-i + 3}{-i + 3} =$

$\frac{6i^2 + 18i}{-i^2 + 9} = \frac{6 + 18i}{10} = \frac{6}{10} + \frac{18}{10}i = \frac{3}{5} + \frac{9}{5}i$

111. $\frac{i - i^{38}}{1 + i} = \frac{i - (i^2)^{19}}{1 + i} = \frac{i - (-1)^{19}}{1 + i} = \frac{i - (-1)}{1 + i} =$

$\frac{i + 1}{1 + i} = 1$

41. $x^2 + 7x - 2 = 0$

$x^2 + 7x = 2$ Adding 2

$x^2 + 7x + \frac{49}{4} = 2 + \frac{49}{4}$ $\left(\frac{7}{2}\right)^2 = \frac{49}{4}$

$\left(x + \frac{7}{2}\right)^2 = \frac{57}{4}$

$x + \frac{7}{2} = \frac{\sqrt{57}}{2}$ or $x + \frac{7}{2} = -\frac{\sqrt{57}}{2}$

$x = \frac{-7 + \sqrt{57}}{2}$ or $x = \frac{-7 - \sqrt{57}}{2}$

The solutions are $\frac{-7 \pm \sqrt{57}}{2}$.

43. $x^2 - 3x - 28 = 0$

$x^2 - 3x = 28$

$x^2 - 3x + \frac{9}{4} = 28 + \frac{9}{4}$ $\left(\frac{-3}{2}\right)^2 = \frac{9}{4}$

$\left(x - \frac{3}{2}\right)^2 = \frac{121}{4}$

$x - \frac{3}{2} = \frac{11}{2}$ or $x - \frac{3}{2} = -\frac{11}{2}$

$x = \frac{14}{2}$ or $x = -\frac{8}{2}$

$x = 7$ or $x = -4$

The solutions are 7 and -4.

45. $x^2 - \frac{3}{2}x - \frac{1}{2} = 0$

$x^2 - \frac{3}{2}x = \frac{1}{2}$

$x^2 - \frac{3}{2}x + \frac{9}{16} = \frac{1}{2} + \frac{9}{16}$ $\left[\frac{1}{2}\left(-\frac{3}{2}\right)\right]^2 = \left(-\frac{3}{4}\right)^2 = \frac{9}{16}$

$\left(x - \frac{3}{4}\right)^2 = \frac{17}{16}$

$x - \frac{3}{4} = \frac{\sqrt{17}}{4}$ or $x - \frac{3}{4} = -\frac{\sqrt{17}}{4}$

$x = \frac{3 + \sqrt{17}}{4}$ or $x = \frac{3 - \sqrt{17}}{4}$

The solutions are $\frac{3 \pm \sqrt{17}}{4}$.

47. $2x^2 - 3x - 17 = 0$

$\frac{1}{2}(2x^2 - 3x - 17) = \frac{1}{2} \cdot 0$ Multiplying by $\frac{1}{2}$ to make the x^2-coefficient 1

$x^2 - \frac{3}{2}x - \frac{17}{2} = 0$

$x^2 - \frac{3}{2}x = \frac{17}{2}$ Adding $\frac{17}{2}$

$x^2 - \frac{3}{2}x + \frac{9}{16} = \frac{17}{2} + \frac{9}{16}$ $\left[\frac{1}{2}\left(-\frac{3}{2}\right)\right]^2 = \left(-\frac{3}{4}\right)^2 = \frac{9}{16}$

$\left(x - \frac{3}{4}\right)^2 = \frac{145}{16}$

$x - \frac{3}{4} = \frac{\sqrt{145}}{4}$ or $x - \frac{3}{4} = -\frac{\sqrt{145}}{4}$

$x = \frac{3 + \sqrt{145}}{4}$ or $x = \frac{3 - \sqrt{145}}{4}$

The solutions are $\frac{3 \pm \sqrt{145}}{4}$.

49. $3x^2 - 4x - 1 = 0$

$\frac{1}{3}(3x^2 - 4x - 1) = \frac{1}{3} \cdot 0$ Multiplying to make the x^2-coefficient 1

$x^2 - \frac{4}{3}x - \frac{1}{3} = 0$

$x^2 - \frac{4}{3}x = \frac{1}{3}$ Adding $\frac{1}{3}$

$x^2 - \frac{4}{3}x + \frac{4}{9} = \frac{1}{3} + \frac{4}{9}$ $\left[\frac{1}{2}\left(-\frac{4}{3}\right)\right]^2 = \left(-\frac{2}{3}\right)^2 = \frac{4}{9}$

$\left(x - \frac{2}{3}\right)^2 = \frac{7}{9}$

$x - \frac{2}{3} = \frac{\sqrt{7}}{3}$ or $x - \frac{2}{3} = -\frac{\sqrt{7}}{3}$

$x = \frac{2 + \sqrt{7}}{3}$ or $x = \frac{2 - \sqrt{7}}{3}$

The solutions are $\frac{2 \pm \sqrt{7}}{3}$.

51. $x^2 + x + 2 = 0$

$x^2 + x = -2$ Subtracting 2

$x^2 + x + \frac{1}{4} = -2 + \frac{1}{4}$ $\left(\frac{1}{2}\right)^2 = \frac{1}{4}$

$\left(x + \frac{1}{2}\right)^2 = -\frac{7}{4}$

$x + \frac{1}{2} = \sqrt{-\frac{7}{4}}$ or $x + \frac{1}{2} = -\sqrt{-\frac{7}{4}}$

$x + \frac{1}{2} = \frac{i\sqrt{7}}{2}$ or $x + \frac{1}{2} = -\frac{i\sqrt{7}}{2}$

$x = \frac{-1 + i\sqrt{7}}{2}$ or $x = \frac{-1 - i\sqrt{7}}{2}$

The solutions are $\frac{-1 \pm i\sqrt{7}}{2}$.

53. $x^2 - 4x + 13 = 0$

$x^2 - 4x = -13$ Subtracting 13

$x^2 - 4x + 4 = -13 + 4$ $\left(\frac{-4}{2}\right)^2 = (-2)^2 = 4$

$(x - 2)^2 = -9$

$x - 2 = \sqrt{-9}$ or $x - 2 = -\sqrt{-9}$

$x - 2 = 3i$ or $x - 2 = -3i$

$x = 2 + 3i$ or $x = 2 - 3i$

The solutions are $2 \pm 3i$.

55. $A = P(1 + r)^t$

$2890 = 2560(1 + r)^2$ Substituting

$\frac{2890}{2560} = (1 + r)^2$

$\frac{289}{256} = (1 + r)^2$

$\sqrt{\frac{289}{256}} = 1 + r$ or $-\sqrt{\frac{289}{256}} = 1 + r$ Principle of square roots

$\frac{17}{16} = 1 + r$ or $-\frac{17}{16} = 1 + r$

$\frac{1}{16} = r$ or $-\frac{33}{16} = r$

Since the interest rate cannot be negative, we have

$\frac{1}{16} = r$

$r = 0.0625$, or 6.25%.

57. $A = P(1 + r)^2$

$1440 = 1000(1 + r)^2$ Substituting

$\frac{1440}{1000} = (1 + r)^2$

$\frac{36}{25} = (1 + r)^2$ Simplifying

$\sqrt{\frac{36}{25}} = 1 + r$ or $-\sqrt{\frac{36}{25}} = 1 + r$

$\frac{6}{5} = 1 + r$ or $-\frac{6}{5} = 1 + r$

$\frac{1}{5} = r$ or $-\frac{11}{5} = r$

Since the interest rate cannot be negative, we have

$\frac{1}{5} = r$

$r = 0.2$, or 20%.

59. $A = P(1 + r)^t$

$6760 = 6250(1 + r)^2$ Substituting

$\frac{6760}{6250} = (1 + r)^2$

$\frac{676}{625} = (1 + r)^2$

$\sqrt{\frac{676}{625}} = 1 + r$ or $-\sqrt{\frac{676}{625}} = 1 + r$

$\frac{26}{25} = 1 + r$ or $-\frac{26}{25} = 1 + r$

$\frac{1}{25} = r$ or $-\frac{51}{25} = r$

Since the interest rate cannot be negative, we have

$\frac{1}{25} = r$

$r = 0.04$, or 4%.

61. $s = 16t^2$

$1377 = 16t^2$ Substituting

$\frac{1377}{16} = t^2$

$86.0625 = t^2$

$\sqrt{86.0625} = t$ Principle of square roots; rejecting the negative square root

$9.3 \approx t$

It takes about 9.3 sec.

63. Graph: $y = 2x + 1$

x	y
0	1
2	5
-3	-5

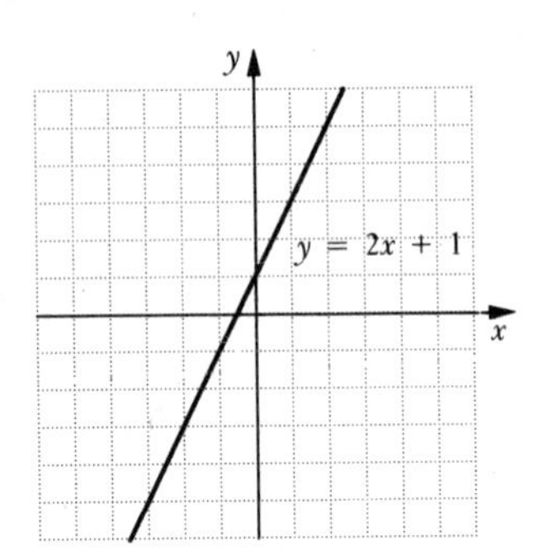

65. $14 - \sqrt{88} \approx 14 - 9.3808 \approx 4.6$

67. In order for $x^2 + bx + 64$ to be a trinomial square, the following must be true:

$\left(\frac{b}{2}\right)^2 = 64$

$\frac{b^2}{4} = 64$

$b^2 = 256$

$b = 16$ or $b = -16$

69. $x(2x^2 + 9x - 56)(3x + 10) = 0$

$x(2x - 7)(x + 8)(3x + 10) = 0$

$x = 0$ or $2x - 7 = 0$ or $x + 8 = 0$ or $3x + 10 = 0$

$x = 0$ or $x = \frac{7}{2}$ or $x = -8$ or $x = -\frac{10}{3}$

The solutions are -8, $-\frac{10}{3}$, 0, and $\frac{7}{2}$.

71. Familiarize. It is helpful to list information in a chart and make a drawing. Let r represent the speed of Boat A. Then r - 7 represents the speed of Boat B.

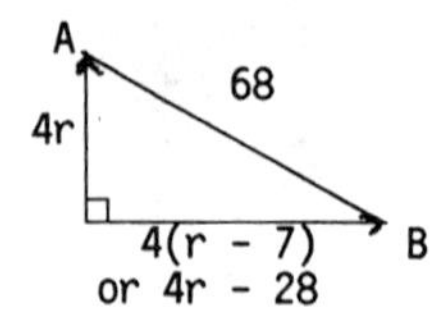

Boat	r	t	d
A	r	4	4r
B	r - 7	4	4(r - 7)

71. (continued)

Translate. We use the Pythagorean equation: $a^2 + b^2 = c^2$

$(4r - 28)^2 + (4r)^2 = 68^2$

Solve.

$(4r - 28)^2 + (4r)^2 = 68^2$

$16r^2 - 224r + 784 + 16r^2 = 4624$

$32r^2 - 224r - 3840 = 0$

$r^2 - 7r - 120 = 0$

$(r + 8)(r - 15) = 0$

$r + 8 = 0$ or $r - 15 = 0$

$r = -8$ or $r = 15$

Check. We only check $r = 15$ since the speeds of the boats cannot be negative. If the speed of Boat A is 15 km/h, then the speed of Boat B is 15 - 7, or 8 km/h, and the distances they travel are 4·15 (or 60) and 4·8 (or 32).

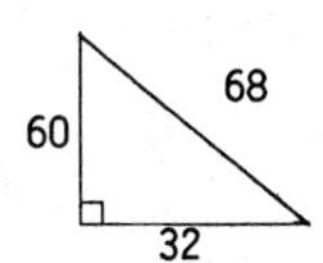

$60^2 + 32^2 = 68^2$

$3600 + 1024 = 4624$

$4624 = 4624$

The values check.

State. The speed of Boat A is 15 km/h, and the speed of Boat B is 8 km/h.

Exercise Set 11.2

1. $x^2 + 6x + 4 = 0$

$a = 1,\ b = 6,\ c = 4$

$x = \frac{-b \pm \sqrt{b^2 - 4ac}}{2a}$

$x = \frac{-6 \pm \sqrt{6^2 - 4\cdot1\cdot4}}{2\cdot1} = \frac{-6 \pm \sqrt{36 - 16}}{2}$

$x = \frac{-6 \pm \sqrt{20}}{2} = \frac{-6 \pm 2\sqrt{5}}{2}$

$x = \frac{2(-3 \pm \sqrt{5})}{2} = -3 \pm \sqrt{5}$

The solutions are $-3 + \sqrt{5}$ and $-3 - \sqrt{5}$.

3. $3p^2 = -8p - 5$

$3p^2 + 8p + 5 = 0$

$(3p + 5)(p + 1) = 0$

$3p + 5 = 0$ or $p + 1 = 0$

$p = -\frac{5}{3}$ or $p = -1$

The solutions are $-\frac{5}{3}$ and -1.

5. $x^2 - x + 1 = 0$

$a = 1,\ b = -1,\ c = 1$

$x = \frac{-(-1) \pm \sqrt{(-1)^2 - 4\cdot1\cdot1}}{2\cdot1} = \frac{1 \pm \sqrt{1 - 4}}{2}$

$x = \frac{1 \pm \sqrt{-3}}{2} = \frac{1 \pm i\sqrt{3}}{2}$

The solutions are $\frac{1 + i\sqrt{3}}{2}$ and $\frac{1 - i\sqrt{3}}{2}$.

7. $x^2 + 13 = 4x$

$x^2 - 4x + 13 = 0$ Finding standard form

$a = 1,\ b = -4,\ c = 13$

$x = \frac{-(-4) \pm \sqrt{(-4)^2 - 4\cdot1\cdot13}}{2\cdot1} = \frac{4 \pm \sqrt{16 - 52}}{2}$

$x = \frac{4 \pm \sqrt{-36}}{2} = \frac{4 \pm 6i}{2} = 2 \pm 3i$

The solutions are $2 + 3i$ and $2 - 3i$.

9. $r^2 + 3r = 8$

$r^2 + 3r - 8 = 0$ Finding standard form

$a = 1,\ b = 3,\ c = -8$

$r = \frac{-3 \pm \sqrt{3^2 - 4\cdot1\cdot(-8)}}{2\cdot1} = \frac{-3 \pm \sqrt{9 + 32}}{2}$

$r = \frac{-3 \pm \sqrt{41}}{2}$

The solutions are $\frac{-3 + \sqrt{41}}{2}$ and $\frac{-3 - \sqrt{41}}{2}$.

11. $1 + \frac{2}{x} + \frac{5}{x^2} = 0$

$x^2 + 2x + 5 = 0$ Multiplying by x^2, the LCM of the denominators

$a = 1,\ b = 2,\ c = 5$

$x = \frac{-2 \pm \sqrt{2^2 - 4\cdot1\cdot5}}{2\cdot1} = \frac{-2 \pm \sqrt{4 - 20}}{2}$

$x = \frac{-2 \pm \sqrt{-16}}{2} = \frac{-2 \pm 4i}{2} = -1 \pm 2i$

The solutions are $-1 + 2i$ and $-1 - 2i$.

13. $3x + x(x - 2) = 0$

$3x + x^2 - 2x = 0$

$x^2 + x = 0$

$x(x + 1) = 0$

$x = 0$ or $x + 1 = 0$

$x = 0$ or $x = -1$

The solutions are 0 and -1.

15. $14x^2 + 9x = 0$

$x(14x + 9) = 0$

$x = 0$ or $14x + 9 = 0$

$x = 0$ or $14x = -9$

$x = 0$ or $x = -\frac{9}{14}$

The solutions are 0 and $-\frac{9}{14}$.

17. $25x^2 - 20x + 4 = 0$

$(5x - 2)(5x - 2) = 0$

$5x - 2 = 0$ or $5x - 2 = 0$

$5x = 2$ or $5x = 2$

$x = \frac{2}{5}$ or $x = \frac{2}{5}$

The solution is $\frac{2}{5}$.

19. $4x(x - 2) - 5x(x - 1) = 2$

$4x^2 - 8x - 5x^2 + 5x = 2$

$-x^2 - 3x = 2$

$-x^2 - 3x - 2 = 0$

$x^2 + 3x + 2 = 0$ Multiplying by -1

$(x + 2)(x + 1) = 0$

$x + 2 = 0$ or $x + 1 = 0$

$x = -2$ or $x = -1$

The solutions are -2 and -1.

21. $14(x - 4) - (x + 2) = (x + 2)(x - 4)$

$14x - 56 - x - 2 = x^2 - 2x - 8$ Removing parentheses

$13x - 58 = x^2 - 2x - 8$

$0 = x^2 - 15x + 50$

$0 = (x - 10)(x - 5)$

$x - 10 = 0$ or $x - 5 = 0$

$x = 10$ or $x = 5$

The solutions are 10 and 5.

23. $5x^2 = 13x + 17$

$5x^2 - 13x - 17 = 0$

$a = 5,\ b = -13,\ c = -17$

$$x = \frac{-(-13) \pm \sqrt{(-13)^2 - 4(5)(-17)}}{2\cdot5}$$

$$x = \frac{13 \pm \sqrt{169 + 340}}{10} = \frac{13 \pm \sqrt{509}}{10}$$

The solutions are $\frac{13 + \sqrt{509}}{10}$ and $\frac{13 - \sqrt{509}}{10}$.

25. $x^2 + 5 = 2x$

$x^2 - 2x + 5 = 0$

$a = 1,\ b = -2,\ c = 5$

$$x = \frac{-(-2) \pm \sqrt{(-2)^2 - 4\cdot1\cdot5}}{2\cdot1} = \frac{2 \pm \sqrt{4 - 20}}{2}$$

$$x = \frac{2 \pm \sqrt{-16}}{2} = \frac{2 \pm 4i}{2}$$

$$x = \frac{2(1 \pm 2i)}{2} = 1 \pm 2i$$

The solutions are $1 + 2i$ and $1 - 2i$.

27. $x + \frac{1}{x} = \frac{13}{6}$, LCM is $6x$

$6x\left(x + \frac{1}{x}\right) = 6x \cdot \frac{13}{6}$

$6x^2 + 6 = 13x$

$6x^2 - 13x + 6 = 0$

$(2x - 3)(3x - 2) = 0$

$2x - 3 = 0$ or $3x - 2 = 0$

$2x = 3$ or $3x = 2$

$x = \frac{3}{2}$ or $x = \frac{2}{3}$

The solutions are $\frac{3}{2}$ and $\frac{2}{3}$.

29. $\frac{1}{x} + \frac{1}{x + 3} = \frac{1}{2}$, LCM is $2x(x + 3)$

$2x(x + 3)\left(\frac{1}{x} + \frac{1}{x + 3}\right) = 2x(x + 3) \cdot \frac{1}{2}$

$2(x + 3) + 2x = x(x + 3)$

$2x + 6 + 2x = x^2 + 3x$

$0 = x^2 - x - 6$

$0 = (x - 3)(x + 2)$

$x - 3 = 0$ or $x + 2 = 0$

$x = 3$ or $x = -2$

The solutions are 3 and -2.

31. $(2t - 3)^2 + 17t = 15$

$4t^2 - 12t + 9 + 17t = 15$

$4t^2 + 5t - 6 = 0$

$(4t - 3)(t + 2) = 0$

$4t - 3 = 0$ or $t + 2 = 0$

$t = \frac{3}{4}$ or $t = -2$

The solutions are $\frac{3}{4}$ and -2.

33. $(x - 2)^2 + (x + 1)^2 = 0$

$x^2 - 4x + 4 + x^2 + 2x + 1 = 0$

$2x^2 - 2x + 5 = 0$

$a = 2,\quad b = -2,\quad c = 5$

$$x = \frac{-(-2) \pm \sqrt{(-2)^2 - 4\cdot2\cdot5}}{2\cdot2} = \frac{2 \pm \sqrt{4 - 40}}{4}$$

$$x = \frac{2 \pm \sqrt{-36}}{4} = \frac{2 \pm 6i}{4}$$

$$x = \frac{2(1 \pm 3i)}{2\cdot2} = \frac{1 \pm 3i}{2}$$

The solutions are $\frac{1 + 3i}{2}$ and $\frac{1 - 3i}{2}$.

35. $x^3 - 8 = 0$

$x^3 - 2^3 = 0$

$(x - 2)(x^2 + 2x + 4) = 0$

$x - 2 = 0$ or $x^2 + 2x + 4 = 0$

$x = 2$ or $x = \frac{-2 \pm \sqrt{2^2 - 4\cdot1\cdot4}}{2\cdot1}$

$x = 2$ or $x = \frac{-2 \pm \sqrt{-12}}{2} = \frac{-2 \pm 2i\sqrt{3}}{2}$

$x = 2$ or $x = \frac{-2 \pm 2i\sqrt{3}}{2}$

$x = 2$ or $x = -1 \pm i\sqrt{3}$

The solutions are 2, $-1 + i\sqrt{3}$, and $-1 - i\sqrt{3}$.

37. $x^2 + 4x - 7 = 0$

$a = 1, \quad b = 4, \quad c = -7$

$x = \frac{-4 \pm \sqrt{4^2 - 4\cdot1\cdot(-7)}}{2\cdot1} = \frac{-4 \pm \sqrt{16 + 28}}{2}$

$x = \frac{-4 \pm \sqrt{44}}{2} = \frac{-4 \pm 2\sqrt{11}}{2} = -2 \pm \sqrt{11}$

Using a calculator we find that $\sqrt{11} \approx 3.317$.

$-2 + \sqrt{11} \approx -2 + 3.317 \approx 1.317 \approx 1.3$

$-2 - \sqrt{11} \approx -2 - 3.317 \approx -5.317 \approx -5.3$

The solutions are approximately 1.3 and -5.3.

39. $x^2 - 6x + 4 = 0$

$a = 1, \quad b = -6, \quad c = 4$

$x = \frac{-(-6) \pm \sqrt{(-6)^2 - 4\cdot1\cdot4}}{2\cdot1} = \frac{6 \pm \sqrt{36 - 16}}{2}$

$x = \frac{6 \pm \sqrt{20}}{2} = \frac{6 \pm 2\sqrt{5}}{2} = 3 \pm \sqrt{5}$

Using a calculator we find that $\sqrt{5} \approx 2.236$.

$3 + \sqrt{5} \approx 3 + 2.236 \approx 5.236 \approx 5.2$

$3 - \sqrt{5} \approx 3 - 2.236 \approx 0.764 \approx 0.8$

The solutions are approximately 5.2 and 0.8.

41. $2x^2 - 3x - 7 = 0$

$a = 2, \quad b = -3, \quad c = -7$

$x = \frac{-(-3) \pm \sqrt{(-3)^2 - 4\cdot2\cdot(-7)}}{2\cdot2} = \frac{3 \pm \sqrt{9 + 56}}{4}$

$x = \frac{3 \pm \sqrt{65}}{4}$

Using a calculator we find that $\sqrt{65} \approx 8.062$.

$\frac{3 + \sqrt{65}}{4} \approx \frac{3 + 8.062}{4} \approx \frac{11.062}{4} \approx 2.7655 \approx 2.8$

$\frac{3 - \sqrt{65}}{4} \approx \frac{3 - 8.062}{4} \approx \frac{-5.062}{4} \approx -1.2655 \approx -1.3$

The solutions are approximately 2.8 and -1.3.

43. $5x^2 = 3 + 8x$

$5x^2 - 8x - 3 = 0$

$a = 5, b = -8, c = -3$

$x = \frac{-(-8) \pm \sqrt{(-8)^2 - 4\cdot5\cdot(-3)}}{2\cdot5} = \frac{8 \pm \sqrt{64 + 60}}{10}$

$x = \frac{8 \pm \sqrt{124}}{10} = \frac{8 \pm 2\sqrt{31}}{10} = \frac{4 \pm \sqrt{31}}{5}$

Using a calculator we find that $\sqrt{31} \approx 5.568$.

$\frac{4 + \sqrt{31}}{5} \approx \frac{4 + 5.568}{5} \approx \frac{9.568}{5} \approx 1.9136 \approx 1.9$

$\frac{4 - \sqrt{31}}{5} \approx \frac{4 - 5.568}{5} \approx \frac{-1.568}{5} \approx -0.3136 \approx -0.3$

The solutions are approximately 1.9 and -0.3.

45. Familiarize. Let x represent the number of pounds of coffee A to be used, and let y represent the number of pounds of coffee B. We organize the information in a table.

Coffee	Price per pound	Number of pounds	Total cost
A	\$1.50	x	\$1.50x
B	\$2.50	y	\$2.50y
Blend	\$1.90	50	1.90 × 50, or \$95

Translate. From the last two columns of the table we get a system of equations.

$x + y = 50,$

$1.50x + 2.50y = 95$

Solve. Solving the system of equations, we get (30,20).

Check. The total number of pounds in the blend is 30 + 20, or 50. The total cost of the blend is 1.50(30) + 2.50(20), or 45 + 50, or \$95. The values check.

State. The blend should consist of 30 pounds of coffee A and 20 pounds of coffee B.

47. $5.33x^2 - 8.23x - 3.24 = 0$

$533x^2 - 823x - 324 = 0$ Clearing decimals

$x = \frac{-(-823) \pm \sqrt{(-823)^2 - 4\cdot533\cdot(-324)}}{2\cdot533}$

$x = \frac{823 \pm \sqrt{1,368,097}}{1066} \approx \frac{823 \pm 1169.656787}{1066}$

$x \approx \frac{823 + 1169.656787}{1066} \approx 1.8692840$

$x \approx \frac{823 - 1169.656787}{1066} \approx -0.3251940$

The solutions are approximately 1.8692840 and -0.3251940.

49. $2x^2 - x - \sqrt{5} = 0$

$a = 2, \quad b = -1, \quad c = -\sqrt{5}$

$$x = \frac{-(-1) \pm \sqrt{(-1)^2 - 4\cdot 2\cdot(-\sqrt{5})}}{2\cdot 2} = \frac{1 \pm \sqrt{1 + 8\sqrt{5}}}{4}$$

The solutions are $\frac{1 + \sqrt{1 + 8\sqrt{5}}}{4}$ and $\frac{1 - \sqrt{1 + 8\sqrt{5}}}{4}$.

51. $ix^2 - x - 1 = 0$

$a = i, \quad b = -1, \quad c = -1$

$$x = \frac{-(-1) \pm \sqrt{(-1)^2 - 4\cdot i\cdot(-1)}}{2\cdot i} = \frac{1 \pm \sqrt{1 + 4i}}{2i}$$

$$x = \frac{i \pm i\sqrt{1 + 4i}}{2i^2} = \frac{i \pm i\sqrt{1 + 4i}}{-2}$$

$$x = \frac{-i \pm i\sqrt{1 + 4i}}{2}$$

The solutions are $\frac{-i + i\sqrt{1 + 4i}}{2}$ and $\frac{-i - i\sqrt{1 + 4i}}{2}$.

53. $$\frac{x}{x + 1} = 4 + \frac{1}{3x^2 - 3}$$

LCM is $3(x + 1)(x - 1)$

$$3(x+1)(x-1)\ \frac{x}{x+1} = 3(x+1)(x-1)\left[4 + \frac{1}{3(x+1)(x-1)}\right]$$

$$3x(x - 1) = 12(x + 1)(x - 1) + 1$$

$$3x^2 - 3x = 12x^2 - 12 + 1$$

$$0 = 9x^2 + 3x - 11$$

$a = 9, b = 3, c = -11$

$$x = \frac{-3 \pm \sqrt{3^2 - 4\cdot 9\cdot(-11)}}{2\cdot 9} = \frac{-3 \pm \sqrt{9 + 396}}{18}$$

$$x = \frac{-3 \pm \sqrt{405}}{18} = \frac{-3 \pm 9\sqrt{5}}{18}$$

$$x = \frac{3(-1 \pm 3\sqrt{5})}{3\cdot 6} = \frac{-1 \pm 3\sqrt{5}}{6}$$

The solutions are $\frac{-1 + 3\sqrt{5}}{6}$ and $\frac{-1 - 3\sqrt{5}}{6}$.

Exercise Set 11.3

1. $x^2 - 6x + 9 = 0$

$a = 1, \quad b = -6, \quad c = 9$

We compute the discriminant.

$b^2 - 4ac = (-6)^2 - 4\cdot 1\cdot 9$
$= 36 - 36$
$= 0$

Since $b^2 - 4ac = 0$, there is just one solution, and it is a real number.

3. $x^2 + 7 = 0$

$a = 1, \quad b = 0, \quad c = 7$

We compute the discriminant.

$b^2 - 4ac = 0^2 - 4\cdot 1\cdot 7$
$= -28$

Since $b^2 - 4ac < 0$, there are two nonreal solutions.

5. $x^2 - 2 = 0$

$a = 1, \quad b = 0, \quad c = -2$

We compute the discriminant.

$b^2 - 4ac = 0^2 - 4\cdot 1\cdot(-2)$
$= 8$

Since $b^2 - 4ac > 0$, there are two real solutions.

7. $4x^2 - 12x + 9 = 0$

$a = 4, \quad b = -12, \quad c = 9$

We compute the discriminant.

$b^2 - 4ac = (-12)^2 - 4\cdot 4\cdot 9$
$= 144 - 144$
$= 0$

Since $b^2 - 4ac = 0$, there is just one solution, and it is a real number.

9. $x^2 - 2x + 4 = 0$

$a = 1, \quad b = -2, \quad c = 4$

We compute the discriminant.

$b^2 - 4ac = (-2)^2 - 4\cdot 1\cdot 4$
$= 4 - 16$
$= -12$

Since $b^2 - 4ac < 0$, there are two nonreal solutions.

11. $9t^2 - 3t = 0$

$a = 9, \quad b = -3, \quad c = 0$

We compute the discriminant.

$b^2 - 4ac = (-3)^2 - 4\cdot 9\cdot 0$
$= 9 - 0$
$= 9$

Since $b^2 - 4ac > 0$, there are two real solutions.

13. $y^2 = \frac{1}{2}y + \frac{3}{5}$

$y^2 - \frac{1}{2}y - \frac{3}{5} = 0$ Standard form

$a = 1, \quad b = -\frac{1}{2}, \quad c = -\frac{3}{5}$

We compute the discriminant.

13. (continued)

$$b^2 - 4ac = (-\tfrac{1}{2})^2 - 4\cdot1\cdot(-\tfrac{3}{5})$$
$$= \tfrac{1}{4} + \tfrac{12}{5}$$
$$= \tfrac{53}{20}$$

Since $b^2 - 4ac > 0$, there are two real solutions.

15. $4x^2 - 4\sqrt{3}x + 3 = 0$

$a = 4, \quad b = -4\sqrt{3}, \quad c = 3$

We compute the discriminant.

$$b^2 - 4ac = (-4\sqrt{3})^2 - 4\cdot4\cdot3$$
$$= 48 - 48$$
$$= 0$$

Since $b^2 - 4ac = 0$, there is just one solution, and it is a real number.

17. The solutions are -11 and 9.

$x = -11$ or $x = 9$

$x + 11 = 0$ or $x - 9 = 0$

$(x + 11)(x - 9) = 0$ Principle of zero products

$x^2 + 2x - 99 = 0$ FOIL

19. The only solution is 7. It must be a double solution.

$x = 7$ or $x = 7$

$x - 7 = 0$ or $x - 7 = 0$

$(x - 7)(x - 7) = 0$ Principle of zero products

$x^2 - 14x + 49 = 0$ FOIL

21. The solutions are $-\frac{2}{5}$ and $\frac{6}{5}$.

$x = -\frac{2}{5}$ or $x = \frac{6}{5}$

$x + \frac{2}{5} = 0$ or $x - \frac{6}{5} = 0$

$(x + \frac{2}{5})(x - \frac{6}{5}) = 0$ Principle of zero products

$x^2 - \frac{4}{5}x - \frac{12}{25} = 0$ FOIL

$25x^2 - 20x - 12 = 0$ Multiplying by 25

23. The solutions are $\frac{c}{2}$ and $\frac{d}{2}$.

$x = \frac{c}{2}$ or $x = \frac{d}{2}$

$x - \frac{c}{2} = 0$ or $x - \frac{d}{2} = 0$

$(x - \frac{c}{2})(x - \frac{d}{2}) = 0$

$x^2 - \frac{d}{2}x - \frac{c}{2}x + \frac{cd}{4} = 0$

or $x^2 - \frac{d + c}{2}x + \frac{cd}{4} = 0$

or $4x^2 - 2(c + d)x + cd = 0$ Multiplying by 4

25. The solutions are $\sqrt{2}$ and $3\sqrt{2}$.

$x = \sqrt{2}$ or $x = 3\sqrt{2}$

$x - \sqrt{2} = 0$ or $x - 3\sqrt{2} = 0$

$(x - \sqrt{2})(x - 3\sqrt{2}) = 0$

$x^2 - 4\sqrt{2}x + 6 = 0$

27. Familiarize. Let x and y represent the number of 30-sec and 60-sec commercials, respectively. Then the amount of time for the 30-sec commercials was 30x sec, or $\frac{30x}{60} = \frac{x}{2}$ min. The amount of time for the 60-sec commercials was 60x sec, or $\frac{60x}{60} = x$ min.

Translate. Rewording, we write two equations. We will express time in minutes.

Total number of commercials is 12.

$$x + y = 12$$

Time for 30-sec commercials is total commercial time less 6 min.

$$\frac{x}{2} = \frac{x}{2} + x - 6$$

Solve. Solving the system of equations we get (6,6).

Check. If there are six 30-sec and six 60-sec commercials, the total number of commercials is 12. The amount of time for six 30-sec commercials is 180 sec, or 3 min, and for six 60-sec commercials is 360 sec, or 6 min. The total commercial time is 9 min, and the amount of time for 30-sec commercials is 6 min less than this. The numbers check.

State. There were six 30-sec and six 60-sec commercials.

29. $ax^2 + bx + c = 0$

The solutions are $x = \frac{-b \pm \sqrt{b^2 - 4ac}}{2a}$.

The product of the solutions is

$$\left(\frac{-b + \sqrt{b^2 - 4ac}}{2a}\right)\left(\frac{-b - \sqrt{b^2 - 4ac}}{2a}\right)$$

$$= \frac{(-b)^2 - (\sqrt{b^2 - 4ac})^2}{(2a)^2}$$

$$= \frac{b^2 - (b^2 - 4ac)}{4a^2}$$

$$= \frac{4ac}{4a^2}$$

$$= \frac{c}{a}$$

31. a) $kx^2 - 2x + k = 0$; one solution is -3

We first find k by substituting -3 for x.

$k(-3)^2 - 2(-3) + k = 0$

$9k + 6 + k = 0$

$10k = -6$

$k = -\frac{6}{10}$

$k = -\frac{3}{5}$

b) $-\frac{3}{5}x^2 - 2x + (-\frac{3}{5}) = 0$ Substituting $-\frac{3}{5}$ for k

$3x^2 + 10x + 3 = 0$ Multiplying by -5

$(3x + 1)(x + 3) = 0$

$3x + 1 = 0$ or $x + 3 = 0$

$3x = -1$ or $x = -3$

$x = -\frac{1}{3}$ or $x = -3$

The other solution is $-\frac{1}{3}$.

33. For $ax^2 + bx + c = 0$, $-\frac{b}{a}$ is the sum of the solutions and $\frac{c}{a}$ is the product of the solutions.

Thus

$-\frac{b}{a} = \sqrt{3}$ and $\frac{c}{a} = 8$

$ax^2 + bx + c = 0$

$x^2 + \frac{b}{a}x + \frac{c}{a} = 0$ Multiplying by $\frac{1}{a}$

$x^2 - (-\frac{b}{a})x + \frac{c}{a} = 0$

$x^2 - \sqrt{3}x + 8 = 0$ Substituting $\sqrt{3}$ for $-\frac{b}{a}$ and 8 for $\frac{c}{a}$

Exercise Set 11.4

1. $x^4 - 10x^2 + 25 = 0$

Let $u = x^2$ and think of x^4 as $(x^2)^2$.

$u^2 - 10u + 25 = 0$ Substituting u for x^2

$(u - 5)(u - 5) = 0$

$u - 5 = 0$ or $u - 5 = 0$

$u = 5$ or $u = 5$

Now we substitute x^2 for u and solve the equation.

$x^2 = 5$

$x = \pm\sqrt{5}$

Both $\sqrt{5}$ and $-\sqrt{5}$ check. They are the solutions.

3. $x - 10\sqrt{x} + 9 = 0$

Let $u = \sqrt{x}$ and think of x as $(\sqrt{x})^2$.

$u^2 - 10u + 9 = 0$ Substituting u for $\sqrt{x}$

$(u - 9)(u - 1) = 0$

$u - 9 = 0$ or $u - 1 = 0$

$u = 9$ or $u = 1$

Now we substitute $\sqrt{x}$ for u and solve these equations:

$\sqrt{x} = 9$ or $\sqrt{x} = 1$

$x = 81$ $x = 1$

The numbers 81 and 1 both check. They are the solutions.

5. $(x^2 - 6x) - 2(x^2 - 6x) - 35 = 0$

Let $u = x^2 - 6x$.

$u^2 - 2u - 35 = 0$ Substituting u for $x^2 - 6x$

$(u - 7)(u + 5) = 0$

$u - 7 = 0$ or $u + 5 = 0$

$u = 7$ or $u = -5$

$x^2 - 6x = 7$ or $x^2 - 6x = -5$

Substituting $x^2 - 6x$ for u

$x^2 - 6x - 7 = 0$ or $x^2 - 6x + 5 = 0$

$(x - 7)(x + 1) = 0$ or $(x - 5)(x - 1) = 0$

$x = 7$ or $x = -1$ or $x = 5$ or $x = 1$

The numbers -1, 1, 5, and 7 check. They are the solutions.

7. $x^{-2} - x^{-1} - 6 = 0$

Let $u = x^{-1}$ and think of x^{-2} as $(x^{-1})^2$.

$u^2 - u - 6 = 0$ Substituting u for x^{-1}

$(u - 3)(u + 2) = 0$

$u = 3$ or $u = -2$

Now we substitute x^{-1} for u and solve these equations:

$x^{-1} = 3$ or $x^{-1} = -2$

$\frac{1}{x} = 3$ or $\frac{1}{x} = -2$

$\frac{1}{3} = x$ or $-\frac{1}{2} = x$

Both $\frac{1}{3}$ and $-\frac{1}{2}$ check. They are the solutions.

9. $(1 + \sqrt{x})^2 + (1 + \sqrt{x}) - 6 = 0$

Let $u = 1 + \sqrt{x}$.

$u^2 + u - 6 = 0$ Substituting u for $1 + \sqrt{x}$

$(u + 3)(u - 2) = 0$

$u = -3$ or $u = 2$

$1 + \sqrt{x} = -3$ or $1 + \sqrt{x} = 2$ Substituting $1 + \sqrt{x}$ for u

$\sqrt{x} = -4$ or $\sqrt{x} = 1$

No real solution $\quad x = 1$

The number 1 checks. It is the solution.

11. $(y^2 - 5y)^2 - 2(y^2 - 5y) - 24 = 0$

Let $u = y^2 - 5y$.

$u^2 - 2u - 24 = 0$ Substituting u for $y^2 - 5y$

$(u - 6)(u + 4) = 0$

$u = 6$ or $u = -4$

$y^2 - 5y = 6$ or $y^2 - 5y = -4$ Substituting $y^2 - 5y$ for u

$y^2 - 5y - 6 = 0$ or $y^2 - 5y + 4 = 0$

$(y - 6)(y + 1) = 0$ or $(y - 4)(y - 1) = 0$

$y = 6$ or $y = -1$ or $y = 4$ or $y = 1$

The numbers -1, 1, 4, and 6 check. They are the solutions.

13. $w^4 - 4w^2 - 2 = 0$

Let $u = w^2$.

$u^2 - 4u - 2 = 0$ Substituting u for w^2

$u = \frac{-(-4) \pm \sqrt{(-4)^2 - 4\cdot1\cdot(-2)}}{2\cdot1}$

$u = \frac{4 \pm \sqrt{24}}{2}$

$u = \frac{4 \pm 2\sqrt{6}}{2}$

$u = 2 \pm \sqrt{6}$

13. (continued)

Now we substitute w^2 for u and solve these equations:

$w^2 = 2 + \sqrt{6}$ or $w^2 = 2 - \sqrt{6}$

$w = \pm\sqrt{2 + \sqrt{6}}$ or $w = \pm\sqrt{2 - \sqrt{6}}$

All four numbers check. They are the solutions.

15. $2x^{-2} + x^{-1} - 1 = 0$

Let $u = x^{-1}$.

$2u^2 + u - 1 = 0$ Substituting u for x^{-1}

$(2u - 1)(u + 1) = 0$

$2u = 1$ or $u = -1$

$u = \frac{1}{2}$ or $u = -1$

$x^{-1} = \frac{1}{2}$ or $x^{-1} = -1$ Substituting x^{-1} for u

$\frac{1}{x} = \frac{1}{2}$ or $\frac{1}{x} = -1$

$x = 2$ or $x = -1$

Both 2 and -1 check. They are the solutions.

17. $6x^4 - 19x^2 + 15 = 0$

Let $u = x^2$.

$6u^2 - 19u + 15 = 0$ Substituting u for x^2

$(3u - 5)(2u - 3) = 0$

$3u = 5$ or $2u = 3$

$u = \frac{5}{3}$ or $u = \frac{3}{2}$

$x^2 = \frac{5}{3}$ or $x^2 = \frac{3}{2}$ Substituting x^2 for u

$x = \pm\sqrt{\frac{5}{3}}$ or $x = \pm\sqrt{\frac{3}{2}}$

$x = \pm\frac{\sqrt{15}}{3}$ or $x = \pm\frac{\sqrt{6}}{2}$ Rationalizing denominators

All four numbers check. They are the solutions.

19. $x^{2/3} - 4x^{1/3} - 5 = 0$

Let $u = x^{1/3}$.

$u^2 - 4u - 5 = 0$ Substituting u for $x^{1/3}$

$(u - 5)(u + 1) = 0$

$u = 5$ or $u = -1$

$x^{1/3} = 5$ or $x^{1/3} = -1$ Substituting $x^{1/3}$ for u

$(x^{1/3})^3 = 5^3$ or $(x^{1/3})^3 = (-1)^3$ Principle of powers

$x = 125$ or $x = -1$

Both 125 and -1 check. They are the solutions.

21. $\left(\frac{x + 3}{x - 3}\right)^2 - \left(\frac{x + 3}{x - 3}\right) - 6 = 0$

Let $u = \frac{x + 3}{x - 3}$.

$u^2 - u - 6 = 0$ Substituting u for $\frac{x + 3}{x - 3}$

$(u - 3)(u + 2) = 0$

$u = 3$ or $u = -2$

$\frac{x + 3}{x - 3} = 3$ or $\frac{x + 3}{x - 3} = -2$ Substituting $\frac{x + 3}{x - 3}$ for u

$x + 3 = 3(x - 3)$ or $x + 3 = -2(x - 3)$ Multiplying by $(x - 3)$

$x + 3 = 3x - 9$ or $x + 3 = -2x + 6$

$-2x = -12$ or $3x = 3$

$x = 6$ or $x = 1$

Both 6 and 1 check. They are the solutions.

23. $9\left(\frac{x + 2}{x + 3}\right)^2 - 6\left(\frac{x + 2}{x + 3}\right) + 1 = 0$

Let $u = \frac{x + 2}{x + 3}$.

$9u^2 - 6u + 1 = 0$ Substituting u for $\frac{x + 2}{x + 3}$

$(3u - 1)(3u - 1) = 0$

$3u = 1$

$u = \frac{1}{3}$

$\frac{x + 2}{x + 3} = \frac{1}{3}$ Substituting $\frac{x + 2}{x + 3}$ for u

$3(x + 2) = x + 3$ Multiplying by $3(x + 3)$

$3x + 6 = x + 3$

$2x = -3$

$x = -\frac{3}{2}$

The number $-\frac{3}{2}$ checks. It is the solution.

25. $\left(\frac{y^2 - 1}{y}\right)^2 - 4\left(\frac{y^2 - 1}{y}\right) - 12 = 0$

Let $u = \frac{y^2 - 1}{y}$.

$u^2 - 4u - 12 = 0$ Substituting u for $\frac{y^2 - 1}{y}$

$(u - 6)(u + 2) = 0$

$u = 6$ or $u = -2$

$\frac{y^2 - 1}{y} = 6$ or $\frac{y^2 - 1}{y} = -2$ Substituting $\frac{y^2 - 1}{y}$ for u

$y^2 - 1 = 6y$ or $y^2 - 1 = -2y$ Multiplying by y

$y^2 - 6y - 1 = 0$ or $y^2 + 2y - 1 = 0$

$y = \frac{-(-6)\pm\sqrt{(-6)^2-4\cdot1\cdot(-1)}}{2\cdot1}$ or $y = \frac{-2\pm\sqrt{2^2-4\cdot1\cdot(-1)}}{2\cdot1}$

$y = \frac{6 \pm \sqrt{40}}{2}$ or $y = \frac{-2 \pm \sqrt{8}}{2}$

25. (continued)

$y = \frac{6 \pm 2\sqrt{10}}{2}$ or $y = \frac{-2 \pm 2\sqrt{2}}{2}$

$y = 3 \pm \sqrt{10}$ or $y = -1 \pm \sqrt{2}$

All four numbers check. They are the solutions.

27. $\sqrt{3x^2}\sqrt{3x^3} = \sqrt{3x^2\cdot3x^3} = \sqrt{9x^5} = \sqrt{9x^4\cdot x} = 3x^2\sqrt{x}$

29. $\frac{x + 1}{x - 1} - \frac{x + 1}{x^2 + x + 1}$, LCM is $(x - 1)(x^2 + x + 1)$

$= \frac{x + 1}{x - 1}\cdot\frac{x^2 + x + 1}{x^2 + x + 1} - \frac{x + 1}{x^2 + x + 1}\cdot\frac{x - 1}{x - 1}$

$= \frac{(x^3 + 2x^2 + 2x + 1) - (x^2 - 1)}{(x - 1)(x^2 + x + 1)}$

$= \frac{x^3 + x^2 + 2x + 2}{x^3 - 1}$

31. $6.75x - 35\sqrt{x} - 5.36 = 0$

Let $u = \sqrt{x}$. (Since we will use a calculator, we do not clear the decimals.)

$6.75u^2 - 35u - 5.36 = 0$

$u = \frac{-(-35) \pm \sqrt{(-35)^2 - 4(6.75)(-5.36)}}{2(6.75)}$

$u = \frac{35 \pm \sqrt{1225 + 144.72}}{13.5}$

$u \approx \frac{35 \pm 37.01}{13.5}$

$u \approx \frac{72.01}{13.5}$ or $u \approx \frac{-2.01}{13.5}$

$u \approx 5.334$ or $u \approx -0.149$

$\sqrt{x} \approx 5.334$ or $\sqrt{x} \approx -0.149$

$x \approx 28.5$ No real solution

The number 28.5 checks, so it is the solution.

33. $\frac{x}{x - 1} - 6\sqrt{\frac{x}{x - 1}} - 40 = 0$

Let $u = \sqrt{\frac{x}{x - 1}}$.

$u^2 - 6u - 40 = 0$ Substituting for $\sqrt{\frac{x}{x - 1}}$

$(u - 10)(u + 4) = 0$

$u = 10$ or $u = -4$

$\sqrt{\frac{x}{x - 1}} = 10$ or $\sqrt{\frac{x}{x - 1}} = -4$ Substituting for u

$\frac{x}{x - 1} = 100$ No real solution

$x = 100x - 100$ Multiplying by $(x - 1)$

$100 = 99x$

$\frac{100}{99} = x$

This number checks. It is the solution.

35. $\sqrt{x - 3} - \sqrt[4]{x - 3} = 12$

$(x - 3)^{1/2} - (x - 3)^{1/4} - 12 = 0$

Let $u = (x - 3)^{1/4}$.

$u^2 - u - 12 = 0$ Substituting for $(x - 3)^{1/4}$

$(u - 4)(u + 3) = 0$

$u = 4$ or $u = -3$

$(x - 3)^{1/4} = 4$ or $(x - 3)^{1/4} = -3$ Substituting for u

$x - 3 = 4^4$ No real solution

$x - 3 = 256$

$x = 259$

This number checks. It is the solution.

37. $x^6 - 28x^3 + 27 = 0$

Let $u = x^3$.

$u^2 - 28u + 27 = 0$ Substituting for x^3

$(u - 27)(u - 1) = 0$

$u = 27$ or $u = 1$

$x^3 = 27$ or $x^3 = 1$ Substituting for u

$x = 3$ or $x = 1$

Both 3 and 1 check. They are the solutions.

Exercise Set 11.5

1. $A = 6s^2$

$\frac{A}{6} = s^2$ Dividing by 6

$\sqrt{\frac{A}{6}} = s$ Taking the positive square root

3. $F = \frac{Gm_1m_2}{r^2}$

$Fr^2 = Gm_1m_2$ Multiplying by r^2

$r^2 = \frac{Gm_1m_2}{F}$ Dividing by F

$r = \sqrt{\frac{Gm_1m_2}{F}}$ Taking the positive square root

5. $E = mc^2$

$\frac{E}{m} = c^2$

$\sqrt{\frac{E}{m}} = c$

7. $a^2 + b^2 = c^2$

$b^2 = c^2 - a^2$

$b = \sqrt{c^2 - a^2}$

9. $N = \frac{k^2 - 3k}{2}$

$2N = k^2 - 3k$

$0 = k^2 - 3k - 2N$ Standard form

$a = 1,\quad b = -3,\quad c = -2N$

$k = \frac{-(-3) \pm \sqrt{(-3)^2 - 4\cdot 1\cdot(-2N)}}{2\cdot 1}$ Using the quadratic formula

$= \frac{3 \pm \sqrt{9 + 8N}}{2}$

Since taking the negative square root would result in a negative answer, we take the positive one.

$k = \frac{3 + \sqrt{9 + 8N}}{2}$

11. $A = 2\pi r^2 + 2\pi rh$

$0 = 2\pi r^2 + 2\pi rh - A$ Standard form

$a = 2\pi,\quad b = 2\pi h,\quad c = -A$

$r = \frac{-2\pi h \pm \sqrt{(2\pi h)^2 - 4\cdot 2\pi\cdot(-A)}}{2\cdot 2\pi}$ Using the quadratic formula

$= \frac{-2\pi h \pm \sqrt{4\pi^2h^2 + 8\pi A}}{4\pi}$

$= \frac{-2\pi h \pm 2\sqrt{\pi^2h^2 + 2\pi A}}{4\pi}$

$= \frac{-\pi h \pm \sqrt{\pi^2h^2 + 2\pi A}}{2\pi}$

Since taking the negative square root would result in a negative answer, we take the positive one.

$r = \frac{-\pi h + \sqrt{\pi^2h^2 + 2\pi A}}{2\pi}$

13. $T = 2\pi\sqrt{\frac{L}{g}}$

$\frac{T}{2\pi} = \sqrt{\frac{L}{g}}$ g Dividing by 2π

$\frac{T^2}{4\pi^2} = \frac{L}{g}$ Squaring

$gT^2 = 4\pi^2L$ Multiplying by $4\pi^2g$

$g = \frac{4\pi^2L}{T^2}$ Dividing by T^2

15. $P_1 - P_2 = \frac{32LV}{gD^2}$

$gD^2(P_1 - P_2) = 32LV$

$D^2 = \frac{32LV}{g(P_1 - P_2)}$

$D = \sqrt{\frac{32LV}{g(P_1 - P_2)}}$

17.
$$m = \frac{m_0}{\sqrt{1 - \frac{v^2}{c^2}}}$$
$$m^2 = \frac{m_0{}^2}{1 - \frac{v^2}{c^2}}$$
$$m^2(1 - \frac{v^2}{c^2}) = m_0{}^2$$
$$m^2 - \frac{m^2v^2}{c^2} = m_0{}^2$$
$$m^2 - m_0{}^2 = \frac{m^2v^2}{c^2}$$
$$c^2(m^2 - m_0{}^2) = m^2v^2$$
$$\frac{c^2(m^2 - m_0{}^2)}{m^2} = v^2$$
$$\sqrt{\frac{c^2(m^2 - m_0{}^2)}{m^2}} = v$$
$$\frac{c}{m}\sqrt{m^2 - m_0{}^2} = v$$

19. Familiarize. We make a drawing and label it. We let x represent the length of the rectangle. Then x - 4 represents the width.

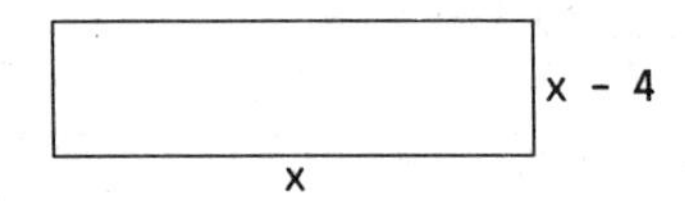

Translate.

$A = \ell \cdot w$

$12 = x(x - 4)$ Substituting

Solve. We solve the equation.

$12 = x^2 - 4x$

$0 = x^2 - 4x - 12$

$0 = (x - 6)(x + 2)$

$x - 6 = 0$ or $x + 2 = 0$

$x = 6$ or $x = -2$

Check. We only check 6 since the length of a rectangle cannot be negative. If x = 6, then x - 4 = 2 and the area is 6·2, or 12. The value checks.

State. The length is 6 ft, and the width is 2 ft.

21. Familiarize. We first make a drawing and label it. We let x represent the width of the rectangle. Then 2x represents the length.

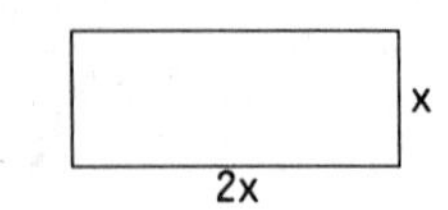

Translate.

$A = \ell \cdot w$

$288 = 2x \cdot x$ Substituting

21. (continued)

Solve. We solve the equation.

$288 = 2x^2$

$144 = x^2$

$x = 12$ or $x = -12$

Check. We only check 12 since the width of a rectangle cannot be negative. If x = 12, then 2x = 24 and the area is 24·12, or 288. The value checks.

State. The length is 24 yd, and the width is 12 yd.

23. Familiarize. We make a drawing and label it with both known and unknown information. We let x represent the width of the frame.

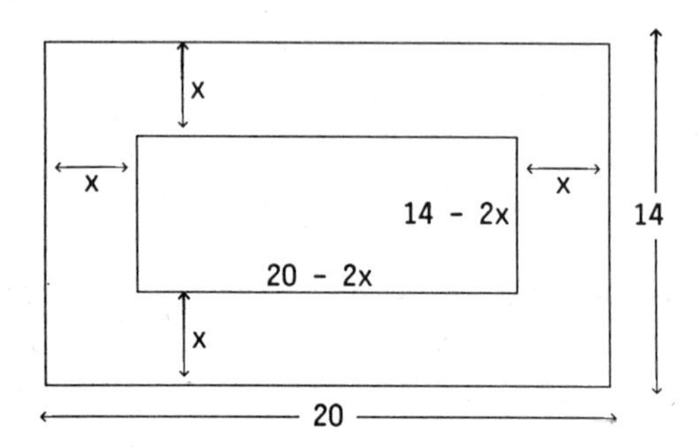

The length and width of the picture that shows are represented by 20 - 2x and 14 - 2x. The area of the picture that shows is 160 cm^2.

Translate. Using the formula for the area of a rectangle, $A = \ell \cdot w$, we have

$160 = (20 - 2x)(14 - 2x)$

Solve. We solve the equation.

$160 = (20 - 2x)(14 - 2x)$

$160 = 280 - 68x + 4x^2$

$0 = 120 - 68x + 4x^2$

$0 = 4x^2 - 68x + 120$

$0 = x^2 - 17x + 30$ Dividing by 4

$0 = (x - 15)(x - 2)$

$x - 15 = 0$ or $x - 2 = 0$

$x = 15$ or $x = 2$

Check. We see that 15 is not a solution because when x = 15, 20 - 2x = -10 and 14 - 2x = -16, and the length and width of the frame cannot be negative. Let's check 2. When x = 2, 20 - 2x = 16, 14 - 2x = 10 and 16·10 = 160. The area is 160. The value checks.

State. The width of the frame is 2 in.

25. Familiarize. We first make a drawing. We let x represent the shorter leg. Then x + 14 represents the longer leg.

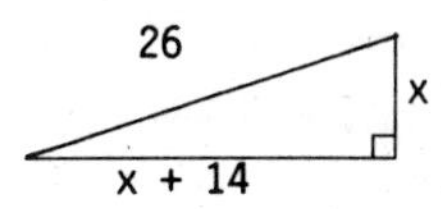

Translate. We use the Pythagorean equation.

$$a^2 + b^2 = c^2$$

$$(x + 14)^2 + x^2 = 26^2 \quad \text{Substituting}$$

Solve. We solve the equation.

$$x^2 + 28x + 196 + x^2 = 676$$
$$2x^2 + 28x - 480 = 0$$
$$x^2 + 14x - 240 = 0$$
$$(x - 10)(x + 24) = 0$$

$$x - 10 = 0 \text{ or } x + 24 = 0$$
$$x = 10 \text{ or } x = -24$$

Check. We only check 10 since the length of a leg cannot be negative. If x = 10, then x + 14 = 24.

$a^2 + b^2$	$= c^2$
$24^2 + 10^2$	26^2
$576 + 100$	676
676	

The lengths check.

State. The lengths of the legs are 24 ft and 10 ft.

27. Familiarize. We first make a drawing. We let x represent the longer leg. Then x - 1 represents the shorter leg.

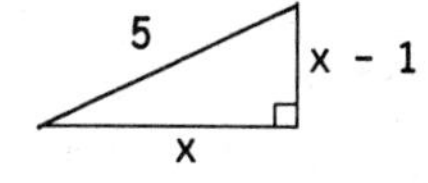

Translate. We use the Pythagorean equation.

$$a^2 + b^2 = c^2$$

$$x^2 + (x - 1)^2 = 5^2 \quad \text{Substituting}$$

Solve. We solve the equation.

$$x^2 + x^2 - 2x + 1 = 25$$
$$2x^2 - 2x - 24 = 0$$
$$x^2 - x - 12 = 0$$
$$(x - 4)(x + 3) = 0$$

$$x - 4 = 0 \text{ or } x + 3 = 0$$
$$x = 4 \text{ or } x = -3$$

Check. We only check 4 since the length of a leg cannot be negative. If x = 4, then x - 1 = 3.

$a^2 + b^2$	$= c^2$
$4^2 + 3^2$	5^2
$16 + 9$	25
25	

The lengths check.

State. The lengths of the legs are 4 ft and 3 ft.

29. Familiarize. The page numbers on facing pages are consecutive integers. Let x represent the number on the left-hand page. Then x + 1 represents the number on the right-hand page.

Translate.

$$\underbrace{\text{The product of the page numbers}} \text{ is } 2756.$$

$$x(x + 1) = 2756$$

Solve. We solve the equation.

$$x^2 + x = 2756$$
$$x^2 + x - 2756 = 0$$
$$(x + 53)(x - 52) = 0$$

$$x + 53 = 0 \quad \text{or} \quad x - 52 = 0$$
$$x = -53 \quad \text{or} \quad x = 52$$

Check. We only check 52 since a page number cannot be negative. If the left-hand page number is 52, then the right-hand page number is 53 and the product of the numbers is 52·53, or 2756. The numbers check.

State. The page numbers are 52 and 53.

31. Using the same reasoning that we did in Exercise 19, we translate the problem to the equation

$$10 = x(x - 4).$$

We solve as follows:

$$10 = x^2 - 4x$$
$$0 = x^2 - 4x - 10 \quad \text{Standard form}$$

$$x = \frac{-b \pm \sqrt{b^2 - 4ac}}{2a} = \frac{-(-4) \pm \sqrt{(-4)^2 - 4\cdot1\cdot(-10)}}{2\cdot1}$$

$$x = \frac{4 \pm \sqrt{16 + 40}}{2} = \frac{4 \pm \sqrt{56}}{2} = \frac{4 \pm \sqrt{4\cdot14}}{2}$$

$$x = \frac{4 \pm 2\sqrt{14}}{2} = 2 \pm \sqrt{14}$$

Since $\sqrt{14} > 2$, the use of $-\sqrt{14}$ would give a negative length. Then x is given by $2 + \sqrt{14}$ and $x - 4$ is given by $(2 + \sqrt{14}) - 4$, or $\sqrt{14} - 2$. Using a calculator we find that the length is $2 + \sqrt{14} \approx 2 + 3.7 \approx 5.7$ ft and the width is $\sqrt{14} - 2 \approx 3.7 - 2 \approx 1.7$ ft.

33. Using the same reasoning that we did in Exercise 21, we translate the problem to the equation

$$256 = 2x^2, \text{ or}$$
$$128 = x^2.$$

We solve by taking the positive square root, since the use of the negative square root would give a negative length:

$$x = \sqrt{128} = \sqrt{64\cdot2} = 8\sqrt{2}$$

Then $2x = 2(8\sqrt{2}) = 16\sqrt{2}$.

It follows that the width is $8\sqrt{2} \approx 8(1.414) \approx 11.312 \approx 11.3$ yd and the length is $16\sqrt{2} \approx 16(1.414) \approx 22.624 \approx 22.6$ yd.

35. Using the same reasoning that we did in Exercise 23, we translate the problem to the equation

$$100 = (20 - 2x)(14 - 2x).$$

We solve as follows:

$$100 = 280 - 68x + 4x^2$$
$$0 = 4x^2 - 68x + 180$$
$$0 = x^2 - 17x + 45 \quad \text{Dividing by 4}$$

$$x = \frac{-b \pm \sqrt{b^2 - 4ac}}{2a} = \frac{-(-17) \pm \sqrt{(-17)^2 - 4\cdot1\cdot45}}{2\cdot1}$$

$$x = \frac{17 \pm \sqrt{289 - 180}}{2} = \frac{17 \pm \sqrt{109}}{2}$$

Since $10 < \sqrt{109} < 11$, the use of the positive square root would give a frame width between 13.5 in. and 14 in. This is not possible since the outside dimensions of the frame are 14 in. by 20 in. Thus we use $-\sqrt{109}$. It follows that

$x = \frac{17 - \sqrt{109}}{2} \approx \frac{17 - 10.440}{2} \approx \frac{6.560}{2} \approx 3.280 \approx$ 3.3 in.

37. Using the same reasoning that we did in Exercise 25, we translate the problem to the equation

$$(x + 14)^2 + x^2 = 24^2.$$

We solve as follows:

$$x^2 + 28x + 196 + x^2 = 576$$
$$2x^2 + 28x - 380 = 0$$
$$x^2 + 14x - 190 = 0 \quad \text{Dividing by 2}$$

$$x = \frac{-b \pm \sqrt{b^2 - 4ac}}{2a} = \frac{-14 \pm \sqrt{14^2 - 4\cdot1\cdot(-190)}}{2\cdot1}$$

$$x = \frac{-14 \pm \sqrt{196 + 760}}{2} = \frac{-14 \pm \sqrt{956}}{2} = \frac{-14 \pm \sqrt{4\cdot239}}{2}$$

$$x = \frac{-14 \pm 2\sqrt{239}}{2} = -7 \pm \sqrt{239}$$

Since the use of $-\sqrt{239}$ would give a negative length, we use the positive square root. Then $x = -7 + \sqrt{239}$ and $x + 14 = (-7 + \sqrt{239}) + 14 = 7 + \sqrt{239}$. The length of one leg is $-7 + \sqrt{239} \approx -7 + 15.5 \approx 8.5$ ft and the length of the other leg is $7 + \sqrt{239} \approx 7 + 15.5 \approx 22.5$ ft.

39. $\sqrt{-20} = \sqrt{-1\cdot4\cdot5} = 2\sqrt{5}i$

41. Solve: $\sqrt{x^2} = -20$

There is no solution since $\sqrt{x^2}$ is always nonnegative.

Exercise Set 11.6

1. $\frac{1}{x} = \frac{x - 2}{24}$, LCM is $24x$

$$24x \cdot \frac{1}{x} = 24x \cdot \frac{x - 2}{24} \quad \text{Multiplying by the LCM}$$
$$24 = x(x - 2)$$
$$24 = x^2 - 2x$$
$$0 = x^2 - 2x - 24 \quad \text{Standard form}$$
$$0 = (x - 6)(x + 4) \quad \text{Factoring}$$

$x = 6$ or $x = -4$ Principle of zero products

Check: For 6:

$$\begin{array}{c|c} \frac{1}{x} & \frac{x - 2}{24} \\ \hline \frac{1}{6} & \frac{6 - 2}{24} \\ & \frac{4}{24} \\ & \frac{1}{6} \end{array}$$

For -4:

$$\begin{array}{c|c} \frac{1}{x} & \frac{x - 2}{24} \\ \hline \frac{1}{-4} & \frac{-4 - 2}{24} \\ -\frac{1}{4} & \frac{-6}{24} \\ & -\frac{1}{4} \end{array}$$

The solutions are 6 and -4.

3. $$\frac{1}{2x - 1} - \frac{1}{2x + 1} = \frac{1}{4}$$

LCM is $4(2x - 1)(2x + 1)$

$$4(2x-1)(2x+1)\left(\frac{1}{2x-1} - \frac{1}{2x+1}\right) = 4(2x-1)(2x+1) \cdot \frac{1}{4}$$
$$4(2x + 1) - 4(2x - 1) = (2x - 1)(2x + 1)$$
$$8x + 4 - 8x + 4 = 4x^2 - 1$$
$$8 = 4x^2 - 1$$
$$9 = 4x^2$$
$$\frac{9}{4} = x^2$$
$$x = \frac{3}{2} \text{ or } x = -\frac{3}{2}$$

Both numbers check. The solutions are $\frac{3}{2}$ and $-\frac{3}{2}$, or $\pm\frac{3}{2}$.

5. $\frac{50}{x} - \frac{50}{x - 5} = -\frac{1}{2}$, LCM is $2x(x - 5)$

$$2x(x - 5)\left(\frac{50}{x} - \frac{50}{x - 5}\right) = 2x(x - 5)\left(-\frac{1}{2}\right)$$
$$100(x - 5) - 100x = -x(x - 5)$$
$$100x - 500 - 100x = -x^2 + 5x$$
$$x^2 - 5x - 500 = 0$$
$$(x - 25)(x + 20) = 0$$

$x - 25 = 0$ or $x + 20 = 0$

$x = 25$ or $x = -20$

Both numbers check. The solutions are 25 and -20.

7. $\frac{x + 2}{x} = \frac{x - 1}{2}$, LCM is 2x

$$2x \cdot \frac{x + 2}{x} = 2x \cdot \frac{x - 1}{2}$$
$$2(x + 2) = x(x - 1)$$
$$2x + 4 = x^2 - x$$
$$0 = x^2 - 3x - 4$$
$$0 = (x - 4)(x + 1)$$

$x - 4 = 0$ or $x + 1 = 0$

$x = 4$ or $x = -1$

Both numbers check. The solutions are 4 and -1.

9. $x - 6 = \frac{1}{x + 6}$, LCM is $x + 6$

$$(x + 6)(x - 6) = (x + 6) \cdot \frac{1}{x + 6}$$
$$x^2 - 36 = 1$$
$$x^2 = 37$$

$x = \sqrt{37}$ or $x = -\sqrt{37}$

Both numbers check. The solutions are $\pm \sqrt{37}$.

11. $\frac{2}{x} = \frac{x + 3}{5}$, LCM is 5x

$$5x \cdot \frac{2}{x} = 5x \cdot \frac{x + 3}{5}$$
$$5 \cdot 2 = x(x + 3)$$
$$10 = x^2 + 3x$$
$$0 = x^2 + 3x - 10$$
$$0 = (x + 5)(x - 2)$$

$x + 5 = 0$ or $x - 2 = 0$

$x = -5$ or $x = 2$

Both numbers check. The solutions are -5 and 2.

13. $\frac{40}{x} - \frac{20}{x - 3} = \frac{8}{7}$, LCM is $7x(x - 3)$

$$7x(x - 3)\left[\frac{40}{x} - \frac{20}{x - 3}\right] = 7x(x - 3) \cdot \frac{8}{7}$$
$$280(x - 3) - 140x = 8x(x - 3)$$
$$280x - 840 - 140x = 8x^2 - 24x$$
$$0 = 8x^2 - 164x + 840$$
$$0 = 2x^2 - 41x + 210$$
$$0 = (2x - 21)(x - 10)$$

$2x - 21 = 0$ or $x - 10 = 0$

$x = \frac{21}{2}$ or $x = 10$

Both numbers check. The solutions are $\frac{21}{2}$ and 10.

15. $\frac{x + 1}{3x + 2} = \frac{2x - 3}{3x - 2} - 1 - \frac{36}{4 - 9x^2}$

$$\frac{x + 1}{3x + 2} = \frac{2x - 3}{3x - 2} - 1 - \frac{-1}{-1} \cdot \frac{36}{4 - 9x^2}$$
$$\frac{x + 1}{3x + 2} = \frac{2x - 3}{3x - 2} - 1 - \frac{-36}{9x^2 - 4},$$

LCM is $(3x + 2)(3x - 2)$

$$(3x+2)(3x-2) \cdot \frac{x+1}{3x+2} =$$
$$(3x+2)(3x-2)\left[\frac{2x-3}{3x-2} - 1 - \frac{-36}{(3x+2)(3x-2)}\right]$$
$$(3x - 2)(x + 1) =$$
$$(3x+2)(2x-3) - (3x+2)(3x-2) - (-36)$$
$$3x^2 + x - 2 = 6x^2 - 5x - 6 - (9x^2-4) + 36$$
$$3x^2 + x - 2 = 6x^2 - 5x - 6 - 9x^2 + 4 + 36$$
$$6x^2 + 6x - 36 = 0$$
$$x^2 + x - 6 = 0 \quad \text{Dividing by 6}$$
$$(x + 3)(x - 2) = 0$$

$x + 3 = 0$ or $x - 2 = 0$

$x = -3$ or $x = 2$

Both numbers check. The solutions are -3 and 2.

17. $\frac{x - 2}{x + 2} + \frac{x + 2}{x - 2} = \frac{10x - 8}{4 - x^2}$

$$\frac{x - 2}{x + 2} + \frac{x + 2}{x - 2} = \frac{-1}{-1} \cdot \frac{10x - 8}{4 - x^2}$$
$$\frac{x - 2}{x + 2} + \frac{x + 2}{x - 2} = \frac{8 - 10x}{x^2 - 4},$$

LCM is $(x + 2)(x - 2)$

$$(x+2)(x-2)\left[\frac{x-2}{x+2} + \frac{x+2}{x-2}\right] = (x+2)(x-2) \cdot \frac{8-10x}{(x+2)(x-2)}$$
$$(x-2)(x-2) + (x+2)(x+2) = 8 - 10x$$
$$x^2 - 4x + 4 + x^2 + 4x + 4 = 8 - 10x$$
$$2x^2 + 10x = 0$$
$$2x(x + 5) = 0$$

$2x = 0$ or $x + 5 = 0$

$x = 0$ or $x = -5$

Both numbers check. The solutions are 0 and -5.

19. $\frac{13}{7x - 5} + \frac{11x}{3} = \frac{39}{21x - 15}$,

LCM is $3(7x - 5)$

$$3(7x-5)\left[\frac{13}{7x-5} + \frac{11x}{3}\right] = 3(7x - 5) \cdot \frac{39}{3(7x - 5)}$$
$$3 \cdot 13 + (7x - 5)(11x) = 39$$
$$39 + 77x^2 - 55x = 39$$
$$77x^2 - 55x = 0$$
$$11x(7x - 5) = 0$$

$11x = 0$ or $7x - 5 = 0$

$x = 0$ or $7x = 5$

$x = 0$ or $x = \frac{5}{7}$

The number 0 checks, but $\frac{5}{7}$ produces a zero denominator $\left[7 \cdot \frac{5}{7} - 5 = 5 - 5 = 0\right]$. The solution is 0.

21.

$$\frac{12}{x^2 - 9} = 1 + \frac{3}{x - 3},$$

LCM is $(x + 3)(x - 3)$

$$(x + 3)(x - 3)\frac{12}{x^2 - 9} = (x + 3)(x - 3)\left[1 + \frac{3}{x - 3}\right]$$

$$12 = (x + 3)(x - 3) + 3(x + 3)$$

$$12 = x^2 - 9 + 3x + 9$$

$$0 = x^2 + 3x - 12$$

$$x = \frac{-b \pm \sqrt{b^2 - 4ac}}{2a} = \frac{-3 \pm \sqrt{3^2 - 4\cdot 1\cdot(-12)}}{2\cdot 1}$$

$$x = \frac{-3 \pm \sqrt{9 + 48}}{2} = \frac{-3 \pm \sqrt{57}}{2}$$

Both numbers check. The solutions are

$\frac{-3 + \sqrt{57}}{2}$ and $\frac{-3 - \sqrt{57}}{2}$, or $\frac{-3 \pm \sqrt{57}}{2}$.

23. Familiarize. We first make a drawing, labeling it with the known and unknown information. We can also organize the information in a table. We let r represent the speed and t the time for the first part of the trip.

r mph t hr — 80 mi; r - 5 mph 3 - t hr — 35 mi

Canoe trip	Distance	Speed	Time
1st part	80	r	t
2nd part	35	r - 5	3 - t

Translate. Using $r = \frac{d}{t}$, we get two equations from the table, $r = \frac{80}{t}$ and $r - 5 = \frac{35}{3 - t}$.

Solve. We substitute $\frac{80}{t}$ for r in the second equation and solve for t.

$$\frac{80}{t} - 5 = \frac{35}{3 - t}, \text{ LCM is } t(3 - t)$$

$$t(3 - t)\left[\frac{80}{t} - 5\right] = t(3 - t)\cdot\frac{35}{3 - t}$$

$$80(3 - t) - 5t(3 - t) = 35t$$

$$5t^2 - 130t + 240 = 0 \quad \text{Standard form}$$

$$t^2 - 26t + 48 = 0 \quad \text{Multiplying by } \frac{1}{5}$$

$$(t - 24)(t - 2) = 0$$

$t = 24$ or $t = 2$

Check. Since the time cannot be negative (If $t = 24$, $3 - t = -21$.), we only check 2 hr. If $t = 2$, then $3 - t = 1$. The speed of the first part is $\frac{80}{2}$, or 40 mph. The speed of the second part is $\frac{35}{1}$, or 35 mph. The speed of the second part is 5 mph slower than the first part. The value checks.

State. The speed of the first part was 40 mph, and the speed of the second part was 35 mph.

25. Familiarize. We first make a drawing. We also organize the information in a table. We let r represent the speed and t the time of the slower trip.

280 mi r mph t hr

280 mi r + 5 mph t - 1 hr

Trip	Distance	Speed	Time
Slower	280	r	t
Faster	280	r + 5	t - 1

Translate.

Using $t = \frac{d}{r}$, we get two equations from the table, $t = \frac{280}{r}$, and $t - 1 = \frac{280}{r + 5}$.

Solve. We substitute $\frac{280}{r}$ for t in the second equation and solve for r.

$$\frac{280}{r} - 1 = \frac{280}{r + 5}, \text{ LCM is } r(r + 5)$$

$$r(r + 5)\left[\frac{280}{r} - 1\right] = r(r + 5)\cdot\frac{280}{r + 5}$$

$$280(r + 5) - r(r + 5) = 280r$$

$$0 = r^2 + 5r - 1400$$

$$0 = (r - 35)(r + 40)$$

$r = 35$ or $r = -40$

Check. Since negative speed has no meaning in this problem, we only check 35. If $r = 35$, then the time for the slow trip is $\frac{280}{35}$, or 8 hours. If $r = 35$, then $r + 5 = 40$ and the time for the fast trip is $\frac{280}{40}$, or 7 hours. This is 1 hour less time than the slow trip took, so we have an answer to the problem.

State. The speed is 35 mph.

27. Familiarize. We make a drawing and then organize the information in a table. We let r represent the speed and t the time of plane A.

2800 km r km/h t hr

2000 km r + 50 km/h t - 3 hr

Plane	Distance	Rate	Time
A	2800	r	t
B	2000	r + 50	t - 3

Translate. Using $r = \frac{d}{t}$, we get two equations from the table,

$$r = \frac{2800}{t} \text{ and } r + 50 = \frac{2000}{t - 3}.$$

27. (continued)

Solve. We substitute $\frac{2800}{t}$ for r in the second equation and solve for t.

$$\frac{2800}{t} + 50 = \frac{2000}{t - 3}, \text{ LCM is } t(t - 3)$$

$$t(t - 3)\left[\frac{2800}{t} + 50\right] = t(t - 3) \cdot \frac{2000}{t - 3}$$

$$2800(t - 3) + 50t(t - 3) = 2000t$$

$$50t^2 + 650t - 8400 = 0$$

$$t^2 + 13t - 168 = 0$$

$$(t + 21)(t - 8) = 0$$

$t = -21$ or $t = 8$

Check. Since negative time has no meaning in this problem, we only check 8 hours. If $t = 8$, then $t - 3 = 5$. The speed of plane A is $\frac{2800}{8}$, or 350 km/h. The speed of plane B is $\frac{2000}{5}$, or 400 km/h. Since the speed of plane B is 50 km/h faster than the speed of plane A, the value checks.

State. The speed of plane A is 350 km/h; the speed of plane B is 400 km/h.

29. Familiarize. Let x represent the time it takes the smaller pipe to fill the tank. Then $x - 3$ represents the time it takes the larger pipe to fill the tank.

Translate. Using the work formula, we write an equation.

$$\frac{2}{x} + \frac{2}{x - 3} = 1$$

Solve. We solve the equation.

We multiply by the LCM which is $x(x - 3)$.

$$x(x - 3)\left[\frac{2}{x} + \frac{2}{x - 3}\right] = x(x - 3)\cdot 1$$

$$2(x - 3) + 2x = x(x - 3)$$

$$2x - 6 + 2x = x^2 - 3x$$

$$0 = x^2 - 7x + 6$$

$$0 = (x - 6)(x - 1)$$

$x = 6$ or $x = 1$

Check. Since negative time has no meaning in this problem, 1 is not a solution $(1 - 3 = -2)$. We only check 6 hr. This is the time it would take the smaller pipe working alone. Then the larger pipe would take 6 - 3, or 3 hr working alone. The larger pipe would fill $2\left[\frac{1}{3}\right]$, or $\frac{2}{3}$, of the tank in 2 hr, and the smaller pipe would fill $2\left[\frac{1}{6}\right]$, or $\frac{1}{3}$, or the tank in 2 hr. Thus in 2 hr they would fill $\frac{2}{3} + \frac{1}{3}$ of the tank. This is all of it, so the numbers check.

State. It takes the smaller pipe, working alone, 6 hr to fill the tank.

31. Familiarize. We make a drawing and then organize the information in a table. We let r represent the speed of the boat in still water. Then $r - 2$ is the speed upstream and $r + 2$ is the speed downstream. Using $t = \frac{d}{r}$, we let $\frac{1}{r - 2}$ represent the time upstream and $\frac{1}{r + 2}$ represent the time downstream.

1 km — r - 2 km/h → Upstream

Downstream ← 1 km — r + 2 km/h

Trip	Distance	Speed	Time
Upstream	1	$r - 2$	$\frac{1}{r - 2}$
Downstream	1	$r + 2$	$\frac{1}{r + 2}$

Translate. The time for the round trip is 1 hour. We now have an equation.

$$\frac{1}{r - 2} + \frac{1}{r + 2} = 1$$

Solve. We solve the equation. We multiply by the LCM, $(r - 2)(r + 2)$.

$$(r - 2)(r + 2)\left[\frac{1}{r - 2} + \frac{1}{r + 2}\right] = (r - 2)(r + 2)\cdot 1$$

$$(r + 2) + (r - 2) = (r - 2)(r + 2)$$

$$2r = r^2 - 4$$

$$0 = r^2 - 2r - 4$$

$$a = 1,\ b = -2,\ c = -4$$

$$r = \frac{-(-2) \pm \sqrt{(-2)^2 - 4\cdot 1(-4)}}{2\cdot 1}$$

$$r = \frac{2 \pm \sqrt{4 + 16}}{2} = \frac{2 \pm \sqrt{20}}{2}$$

$$r = \frac{2 \pm 2\sqrt{5}}{2} = 1 \pm \sqrt{5}$$

$$1 + \sqrt{5} \approx 1 + 2.236 \approx 3.24$$

$$1 - \sqrt{5} \approx 1 - 2.236 \approx -1.24$$

Check. Since negative speed has no meaning in this problem, we only check 3.24 km/h. If $r \approx 3.24$, then $r - 2 \approx 1.24$ and $r + 2 \approx 5.24$. The time it takes to travel upstream is approximately $\frac{1}{1.24}$, or 0.806 hr, and the time it takes to travel downstream is approximately $\frac{1}{5.24}$, or 0.191 hr. The total time is 0.997 which is approximately 1 hour. The value checks.

State. The speed of the boat in still water is approximately 3.24 km/h.

33. Familarize. We make a drawing and then organize the information in a table. We let r represent the speed on the first part of the trip. Then r - 5 represents the speed on the second part of the trip. Using $t = \frac{d}{r}$, we let $\frac{80}{r}$ represent the time for the first part of the trip and $\frac{25}{r - 5}$ represent the speed on the second part of the trip.

r km/h — r - 5 km/h

80 mi — 25 mi

Trip	Distance	Speed	Time
First part	80	r	$\frac{80}{r}$
Second part	25	r - 5	$\frac{25}{r - 5}$

Translate. The total time for the trip is 3 hr. We now have an equation.

$$\frac{80}{r} + \frac{25}{r - 5} = 3$$

Solve. We solve the equation. We multiply by the LCM, r(r - 5).

$$r(r - 5)\left[\frac{80}{r} + \frac{25}{r - 5}\right] = r(r - 5)\cdot 3$$

$$80(r - 5) + 25r = 3r(r - 5)$$

$$80r - 400 + 25r = 3r^2 - 15r$$

$$0 = 3r^2 - 120r + 400$$

$$a = 3,\ b = -120,\ c = 400$$

$$r = \frac{-(-120) \pm \sqrt{(-120)^2 - 4\cdot 3\cdot 400}}{2\cdot 3}$$

$$r = \frac{120 \pm \sqrt{14,400 - 4800}}{6} = \frac{120 \pm \sqrt{9600}}{6}$$

$$r = \frac{120 \pm 40\sqrt{6}}{6} = \frac{60 \pm 20\sqrt{6}}{3}$$

$$\frac{60 + 20\sqrt{6}}{3} \approx \frac{60 + 20(2.449)}{3} \approx 36.33$$

$$\frac{60 - 20\sqrt{6}}{3} \approx \frac{60 - 20(2.449)}{3} \approx 3.67$$

Check. If $r \approx 3.67$, then $r - 5 \approx -1.33$. Since negative time has no meaning in this problem, 3.67 is not a solution. If $r \approx 36.33$, then $r - 5 \approx 36.33 - 5 \approx 31.33$. The time for the first part of the trip is $\frac{80}{36.33} \approx 2.2$ hr, and the time for the second part of the trip is $\frac{25}{31.33} \approx 0.8$ hr. The total time is 2.2 + 0.8, or 3 hr. The value checks.

State. The speed on the first part of the trip was approximately 36.33 mph, and the speed on the second part of the trip was approximately 31.33 mph.

35.

$$\sqrt{3x + 1} = \sqrt{2x - 1} + 1$$

$$3x + 1 = 2x - 1 + 2\sqrt{2x - 1} + 1 \quad \text{Squaring both sides}$$

$$x + 1 = 2\sqrt{2x - 1}$$

$$x^2 + 2x + 1 = 4(2x - 1) \quad \text{Squaring both sides again}$$

$$x^2 + 2x + 1 = 8x - 4$$

$$x^2 - 6x + 5 = 0$$

$$(x - 1)(x - 5) = 0$$

$$x = 1 \text{ or } x = 5$$

Both numbers check. The solutions are 1 and 5.

37. $\sqrt[3]{18y^3}\,\sqrt[3]{4x^2} = \sqrt[3]{72x^2y^3} = \sqrt[3]{8y^3\cdot 9x^2} = 2y\sqrt[3]{9x^2}$

39. Familiarize. Let t represent the number of hours it takes Chester to make the crusts. Then it takes Ron t - 1.2 hr.

Translate. Using the work formula, we write an equation.

$$\frac{1.8}{t} + \frac{1.8}{t - 1.2} = 1$$

Solve. We solve the equation. The LCM is t(t - 1.2).

$$t(t - 1.2)\left[\frac{1.8}{t} + \frac{1.8}{t - 1.2}\right] = t(t - 1.2)\cdot 1$$

$$1.8(t - 1.2) + 1.8t = t(t - 1.2)$$

$$1.8t - 2.16 + 1.8t = t^2 - 1.2t$$

$$0 = t^2 - 4.8t + 2.16$$

Use the quadratic formula.

$$t = \frac{-(-4.8) \pm \sqrt{(-4.8)^2 - 4\cdot 1\cdot (2.16)}}{2.1}$$

$$t \approx 4.3 \text{ or } t \approx 0.5$$

Check. Since negative time has no meaning in this problem (if $t \approx 0.5$, then $t - 1.2 \approx -0.7$), we only check 4.3. This is the time for Chester to do the job alone. Then it would take Ron about 4.3 - 1.2, or 3.1 hr. In 1.8 hr, Chester would do about $\frac{1.8}{4.3}$ of the job, and Ron would do about $\frac{1.8}{3.1}$ of it. Together in 1.8 hr, they would do $\frac{1.8}{4.3} + \frac{1.8}{3.1}$, or 0.419 + 0.581, of the job. This is all of it so the numbers check.

State. It takes Chester about 4.3 hr and Ron about 3.1 hr to do the job alone.

41.

$$\frac{1}{a - 1} = a + 1, \text{ LCM is } a - 1$$

$$(a - 1)\cdot\frac{1}{a - 1} = (a - 1)(a + 1)$$

$$1 = a^2 - 1$$

$$2 = a^2$$

$$\pm\sqrt{2} = a$$

Both values check. The solution is $\pm\sqrt{2}$.

Exercise Set 11.7

1. Graph: $y = 5x^2$

We choose some values of x and compute y. Then we plot these ordered pairs and connect them with a smooth curve.

x	y
0	0
1	5
-1	5
2	20
-2	20

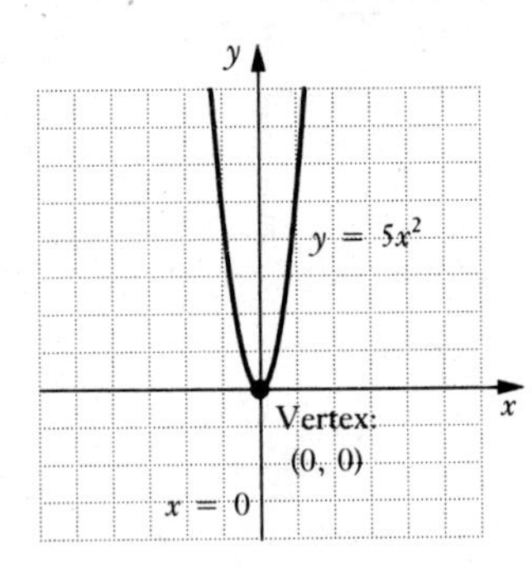

3. Graph: $y = \frac{1}{4}x^2$

We choose some values of x and compute y. Then we plot these ordered pairs and connect them with a smooth curve.

x	y
0	0
2	1
-2	1
4	4
-4	4

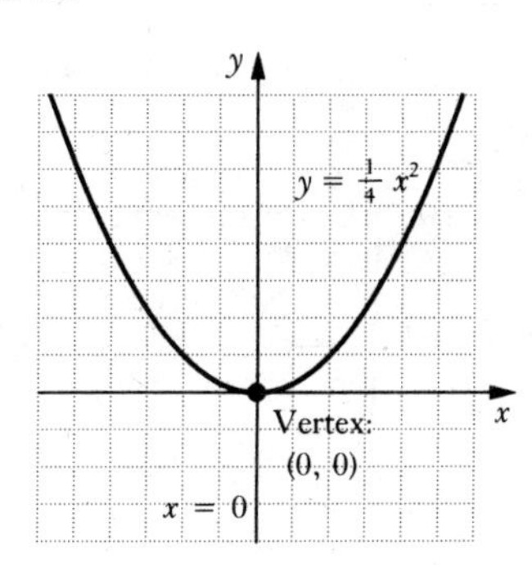

5. Graph: $y = -\frac{1}{2}x^2$

We choose some values of x and compute y. Then we plot these ordered pairs and connect them with a smooth curve.

x	y
0	0
2	-2
-2	-2
3	$-\frac{9}{2}$
-3	$-\frac{9}{2}$

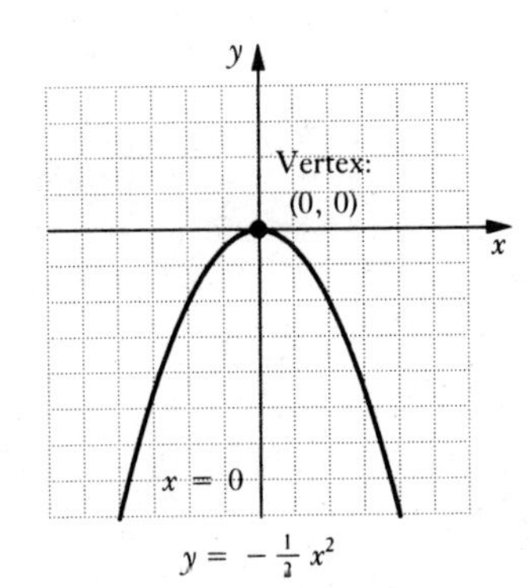

7. Graph: $y = -4x^2$

We choose some values of x and compute y. Then we plot these ordered pairs and connect them with a smooth curve.

x	y
0	0
1	-4
-1	4
2	-16
-2	16

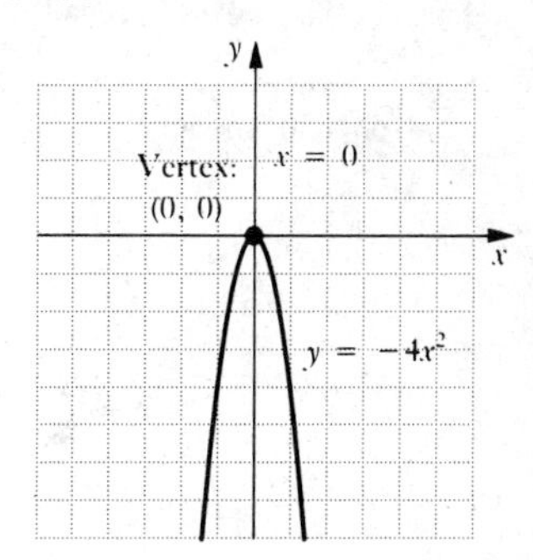

9. Graph: $y = (x - 3)^2$

We choose some values of x and compute y. Then we plot these ordered pairs and connect them with a smooth curve.

x	y
3	0
4	1
2	1
5	4
1	4

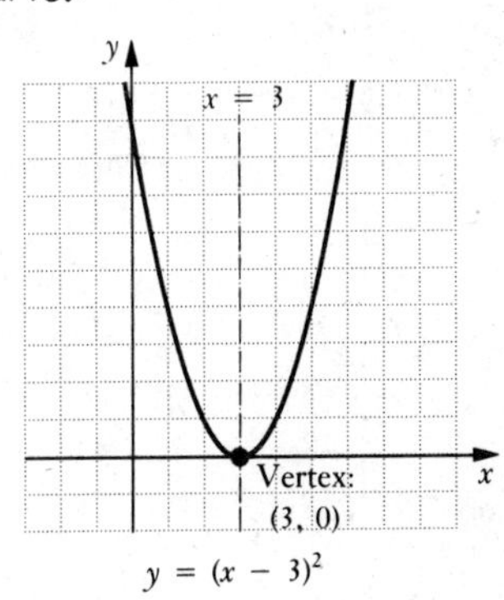

The graph of $y = (x - 3)^2$ looks like the graph of $y = x^2$ except that it is moved three units to the right. The vertex is (3,0), and the line of symmetry is $x = 3$.

11. Graph: $y = 2(x - 4)^2$

We choose some values of x and compute y. Then we plot these ordered pairs and connect them with a smooth curve.

x	y
4	0
5	2
3	2
6	8
2	8

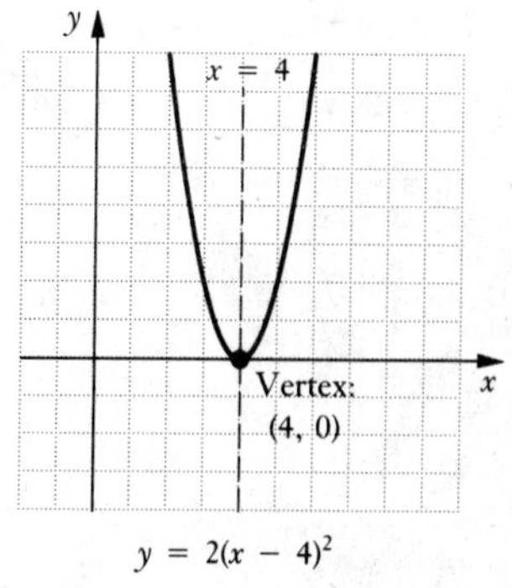

The graph of $y = 2(x - 4)^2$ looks like the graph of $y = 2x^2$ except that it is moved four units to the right. The vertex is (4,0), and the line of symmetry is $x = 4$.

13. Graph: $y = -2(x + 2)^2$

We can express the equation in the equivalent form $y = -2[x - (-2)]^2$. Then we know that the graph looks like that of $y = 2x^2$ translated 2 units to the left, and it will also open downward since $-2 < 0$. The vertex is $(-2,0)$, and the line of symmetry is $x = -2$. Plotting points as needed, we obtain the graph.

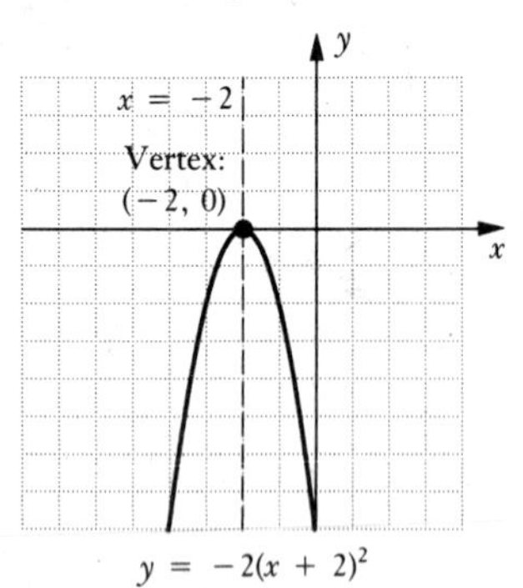

$y = -2(x + 2)^2$

15. Graph: $y = 3(x - 1)^2$

We choose some values of x and compute y. Then we plot these ordered pairs and connect them with a smooth curve.

x	y
1	0
2	3
0	3
3	12
-1	12

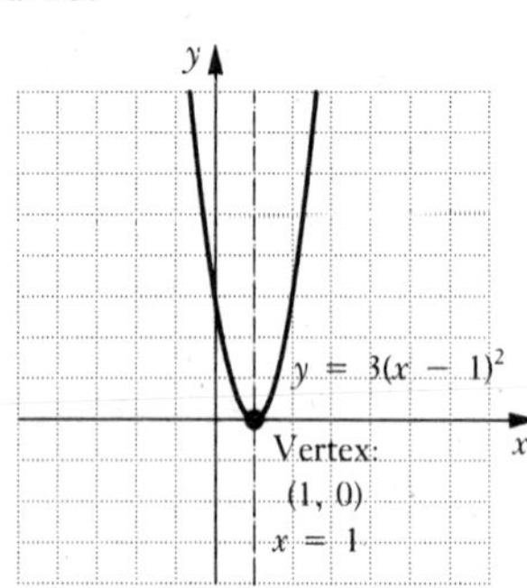

The graph of $y = 3(x - 1)^2$ looks like the graph of $y = 3x^2$ except that it is moved one unit to the right. The vertex is $(1,0)$, and the line of symmetry is $x = 1$.

17. Graph: $y = (x - 3)^2 + 1$

We choose some values of x and compute y. Then we plot these ordered pairs and connect them with a smooth curve.

x	y
3	1
4	2
2	2
5	5
1	5

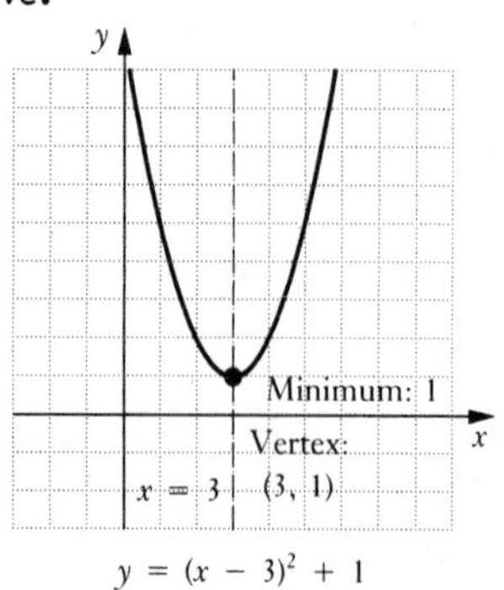

$y = (x - 3)^2 + 1$

The graph of $y = (x - 3)^2 + 1$ looks like the graph of $y = x^2$ except that it is moved three units right and one unit up.

19. Graph: $y = -3(x + 4)^2 + 1$

$y = -3[x - (-4)]^2 + 1$

We choose some values of x and compute y. Then we plot these ordered pairs and connect them with a smooth curve.

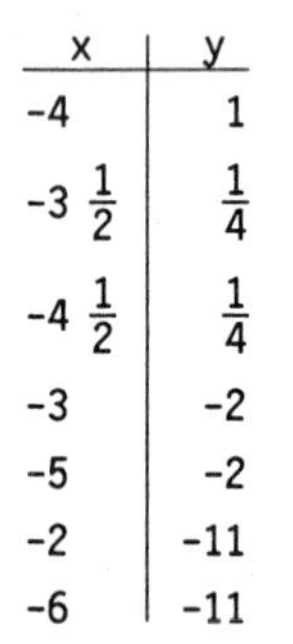

x	y
-4	1
$-3\frac{1}{2}$	$\frac{1}{4}$
$-4\frac{1}{2}$	$\frac{1}{4}$
-3	-2
-5	-2
-2	-11
-6	-11

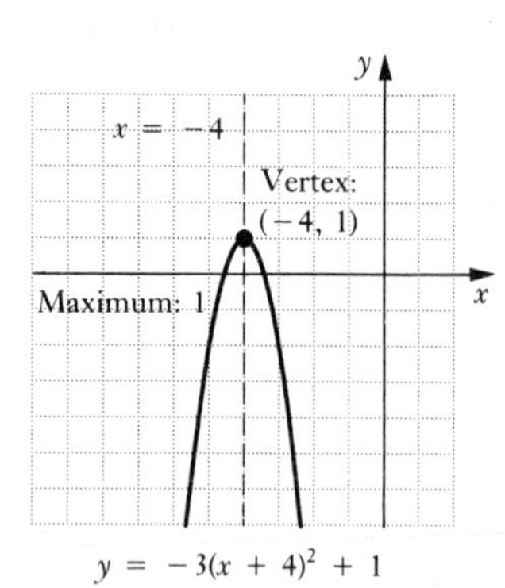

$y = -3(x + 4)^2 + 1$

The graph of $y = -3(x + 4)^2 + 1$ looks like the graph of $y = 3x^2$ except that it is moved four units left and one unit up and opens downward.

21. Graph: $y = \frac{1}{2}(x + 1)^2 + 4$

$y = \frac{1}{2}[x - (-1)]^2 + 4$

We choose some values of x and compute y. Then we plot these ordered pairs and connect them with a smooth curve.

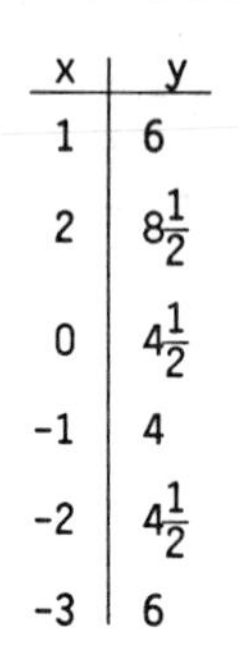

x	y
1	6
2	$8\frac{1}{2}$
0	$4\frac{1}{2}$
-1	4
-2	$4\frac{1}{2}$
-3	6

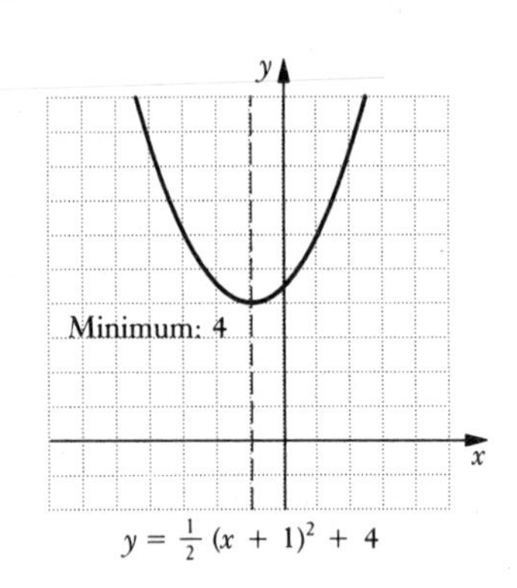

$y = \frac{1}{2}(x + 1)^2 + 4$

The graph of $y = \frac{1}{2}(x + 1)^2 + 4$ looks like the graph of $y = \frac{1}{2}x^2$ except that it is moved one unit left and four units up.

23. Graph: $y = -2(x + 2)^2 - 3$

$y = -2[x - (-2)]^2 + (-3)$

We choose some values of x and compute y. Then we plot these ordered pairs and connect them with a smooth curve.

x	y
-2	-3
-1	-5
-3	-5
0	-11
-4	-11

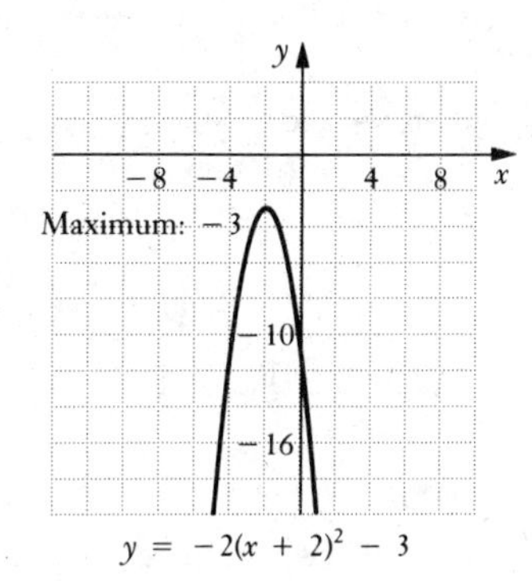

$y = -2(x + 2)^2 - 3$

The graph of $y = -2(x + 2)^2 - 3$ looks like the graph of $y = 2x^2$ except that it is moved two units left and three units down and opens downward.

25. Graph: $y = -(x + 1)^2 - 2$

$y = -[x - (-1)]^2 + (-2)$

We choose some values of x and compute y. Then we plot these ordered pairs and connect them with a smooth curve.

x	y
-1	-2
0	-3
-2	-3
1	-6
-3	-6

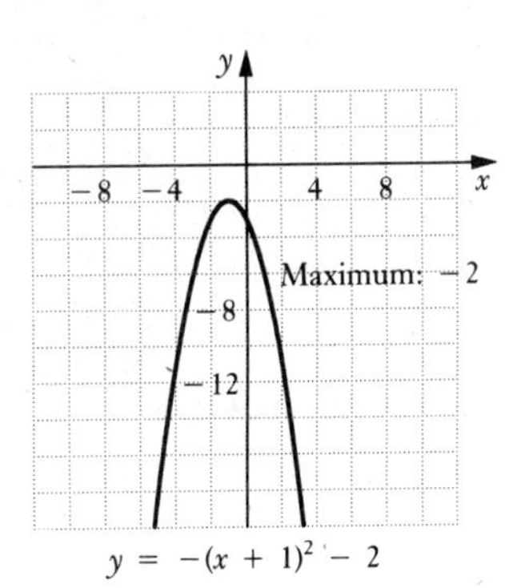

$y = -(x + 1)^2 - 2$

The graph of $y = -(x + 1)^2 - 2$ looks like the graph of $y = x^2$ except that it is moved one unit left and two units down and opens downward.

27. $\sqrt{x} = -7$

The equation has no solution, because the principal square root of a number is always nonnegative.

29.

$x - 5 = \sqrt{x + 7}$

$(x - 5)^2 = (\sqrt{x + 7})^2$ Principle of powers (squaring)

$x^2 - 10x + 25 = x + 7$

$x^2 - 11x + 18 = 0$

$(x - 9)(x - 2) = 0$

$x - 9 = 0$ or $x - 2 = 0$

$x = 9$ or $x = 2$

Check: For x = 9,

$x - 5$	$= \sqrt{x + 7}$
$9 - 5$	$\sqrt{9 + 7}$
4	$\sqrt{16}$
	4 TRUE

For x = 2,

$x - 5$	$= \sqrt{x + 7}$
$2 - 5$	$\sqrt{2 + 7}$
-3	$\sqrt{9}$
	3 FALSE

Only 9 checks. It is the solution.

Exercise Set 11.8

1. $y = x^2 - 2x - 3$

$= (x^2 - 2x) - 3$

We complete the square inside parentheses. We take half the x-coefficient and square it.

$\frac{1}{2} \cdot (-2) = -1 \longrightarrow (-1)^2 = 1$

Then we add 1 - 1 inside the parentheses.

$y = (x^2 - 2x + 1 - 1) - 3$

$= (x^2 - 2x + 1) - 1 - 3$

$= (x - 1)^2 - 4$

$= (x - 1)^2 + (-4)$

Vertex: (1,-4)
Line of symmetry: x = 1

We plot a few points and draw the curve.

x	y
1	-4
2	-3
0	-3
3	0
-1	0
4	5
-2	5

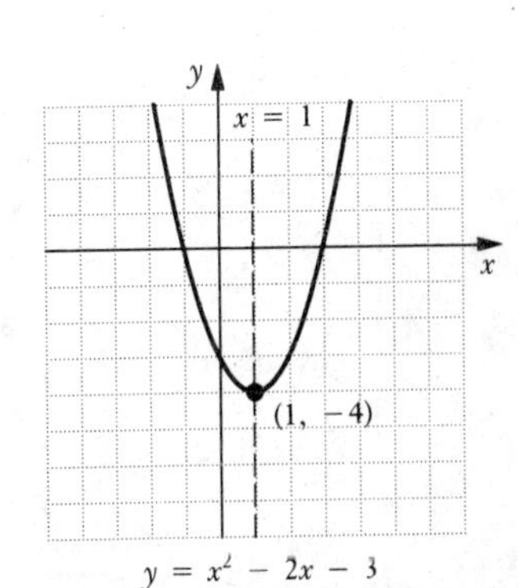

$y = x^2 - 2x - 3$

3. $y = -x^2 + 4x + 1$
$= -(x^2 - 4x) + 1$

We complete the square inside parentheses.
We take half the x-coefficient and square it.

$\frac{1}{2} \cdot (-4) = -2 \longrightarrow (-2)^2 = 4$

Then we add 4 - 4 inside parentheses.

$y = -(x^2 - 4x + 4 - 4) + 1$
$= -(x^2 - 4x + 4) + 4 + 1$
$= -(x - 2)^2 + 5$

Vertex: (2,5)
Line of symmetry: x = 2

We plot a few points and draw the curve.

x	y
2	5
3	4
1	4
4	1
0	1
5	-4
-1	-4

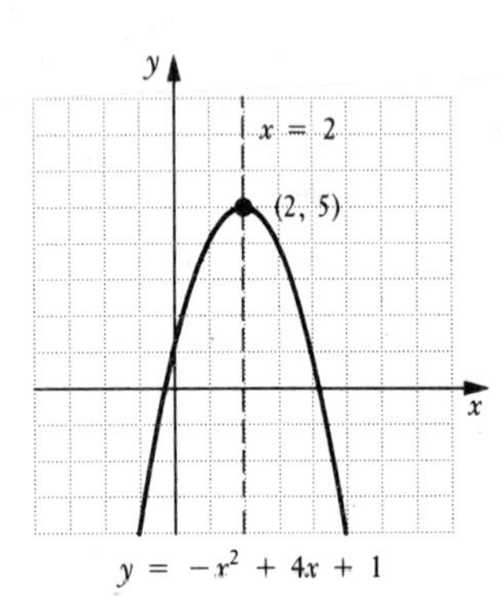

5. $y = 3x^2 - 24x + 50$
$= 3(x^2 - 8x) + 50$

We complete the square inside parentheses.
We take half the x-coefficient and square it.

$\frac{1}{2} \cdot (-8) = -4 \longrightarrow (-4)^2 = 16$

Then we add 16 - 16 inside parentheses.

$y = 3(x^2 - 8x + 16 - 16) + 50$
$= 3(x^2 - 8x + 16) - 48 + 50$
$= 3(x - 4)^2 + 2$

Vertex: (4,2)
Line of symmetry: x = 4

We plot a few points and draw the curve.

x	y
4	2
5	5
3	5
6	14
2	14

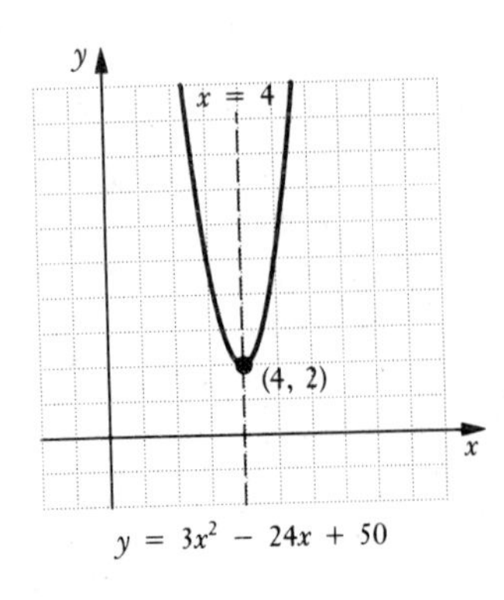

7. $y = -2x^2 + 2x + 1$
$= -2(x^2 - x) + 1$

We complete the square inside parentheses.
We take half the x-coefficient and square it.

$\frac{1}{2} \cdot (-1) = -\frac{1}{2} \longrightarrow (-\frac{1}{2})^2 = \frac{1}{4}$

Then we add $\frac{1}{4} - \frac{1}{4}$ inside parentheses.

$y = -2(x^2 - x + \frac{1}{4} - \frac{1}{4}) + 1$

$= -2(x^2 - x + \frac{1}{4}) + \frac{1}{2} + 1$

$= -2(x - \frac{1}{2})^2 + \frac{3}{2}$

Vertex: $(\frac{1}{2},\frac{3}{2})$
Line of symmetry: $x = \frac{1}{2}$

We plot a few points and draw the curve.

x	y
$\frac{1}{2}$	$\frac{3}{2}$
1	1
0	1
2	-3
-1	-3

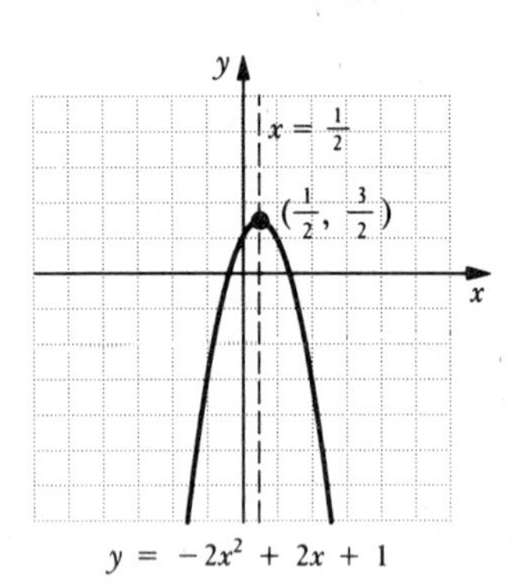

9. $y = 5 - x^2$
$= -x^2 + 5$
$= -(x - 0)^2 + 5$

Vertex: (0,5)
Line of symmetry: x = 0

We plot a few points and draw the curve.

x	y
0	5
1	4
-1	4
2	1
-2	1
3	-4
-3	-4

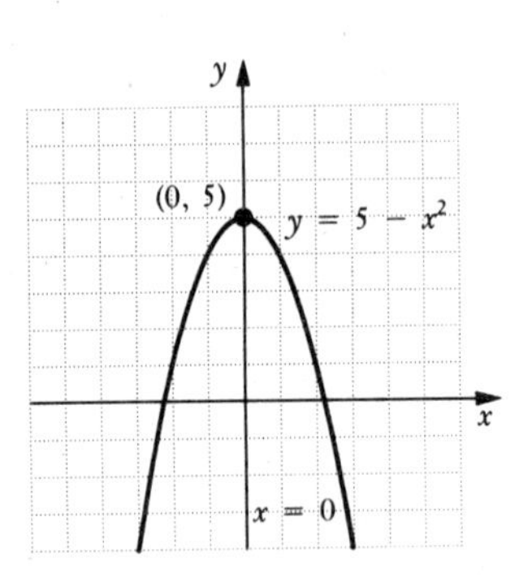

11. $y = x^2 + 6x + 10$

We solve the equation $0 = x^2 + 6x + 10$ using the quadratic formula.

$$x = \frac{-6 \pm \sqrt{6^2 - 4\cdot1\cdot10}}{2\cdot1}$$

$$x = \frac{-6 \pm \sqrt{36 - 40}}{2}$$

$$x = \frac{-6 \pm \sqrt{-4}}{2}$$

$$x = \frac{-6 \pm 2i}{2}$$

$$x = -3 \pm i$$

The solutions are not real. Thus, there are no x-intercepts.

13. $y = -x^2 + 3x + 4$

We solve the following equation.

$0 = -x^2 + 3x + 4$

$0 = x^2 - 3x - 4$

$0 = (x - 4)(x + 1)$

$x - 4 = 0$ or $x + 1 = 0$

$x = 4$ or $x = -1$

The x-intercepts are (4,0) and (-1,0).

15. $y = 3x^2 - 6x + 1$

We solve the equation $0 = 3x^2 - 6x + 1$ using the quadratic formula.

$$x = \frac{-(-6) \pm \sqrt{(-6)^2 - 4\cdot3\cdot1}}{2\cdot3}$$

$$x = \frac{6 \pm \sqrt{36 - 12}}{6}$$

$$x = \frac{6 \pm \sqrt{24}}{6}$$

$$x = \frac{6 \pm 2\sqrt{6}}{6}$$

$$x = \frac{2(3 \pm \sqrt{6})}{2\cdot3}$$

$$x = \frac{3 \pm \sqrt{6}}{3}$$

The x-intercepts are $(\frac{3 + \sqrt{6}}{3},0)$ and $(\frac{3 - \sqrt{6}}{3},0)$.

17. $y = 2x^2 + 4x - 1$

We solve the equation $0 = 2x^2 + 4x - 1$ using the quadratic formula.

$$x = \frac{-4 \pm \sqrt{4^2 - 4\cdot2\cdot(-1)}}{2\cdot2}$$

$$x = \frac{-4 \pm \sqrt{16 + 8}}{4}$$

$$x = \frac{-4 \pm 2\sqrt{6}}{4}$$

$$x = \frac{-2 \pm \sqrt{6}}{2}$$

The x-intercepts are $(\frac{-2 + \sqrt{6}}{2},0)$ and $(\frac{-2 - \sqrt{6}}{2},0)$.

19. We make a drawing and label it.

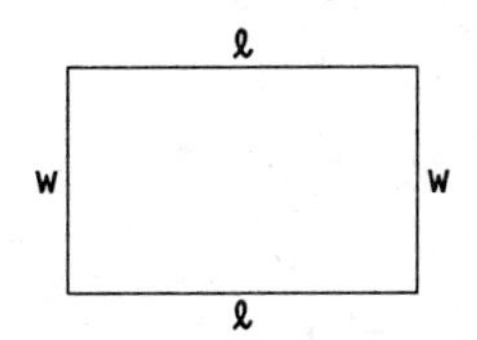

The perimeter must be 68 ft, so we have

$2\ell + 2w = 68.$ (1)

We wish to find the maximum area. We know

$A = \ell w.$ (2)

Solving (1) for ℓ, we get $\ell = 34 - w$. Substituting in (2), we get a quadratic equation:

$A = (34 - w)w$

$A = -w^2 + 34w$

The first coordinate of the vertex is

$$w = -\frac{b}{2a} = -\frac{34}{2(-1)} = 17.$$

Note that when $w = 17$, $\ell = 34 - 17 = 17$. We substitute to find the second coordinate of the vertex:

$A = -(17)^2 + 34(17) = 289$

The coefficient of w^2 is negative, so we know that 289 is a maximum. The maximum area occurs when the dimensions are 17 ft by 17 ft. It is 289 ft^2.

21. Let x and y represent the numbers. Their sum is 45, so we have

$x + y = 45.$ (1)

We wish to find the maximum product P. We know

$P = xy.$ (2)

Solving (1) for y, we get $y = 45 - x$. Substituting in (2), we get a quadratic equation:

$P = x(45 - x)$

$P = -x^2 + 45x$

The first coordinate of the vertex is

$$x = -\frac{b}{2a} = -\frac{45}{2(-1)} = 22.5.$$

21. (continued)

Note that when $x = 22.5$, $y = 45 - 22.5 = 22.5$. We substitute to find the second coordinate of the vertex:

$$P = -(22.5)^2 + 45(22.5) = 506.25$$

The coefficient of x^2 is negative, so we know that 506.25 is a maximum. The maximum product is 506.25. The numbers 22.5 and 22.5 yield this product.

23. Let x and y represent the numbers. Their difference is 6, so we have

$$x - y = 6. \quad (1)$$

We wish to find the minimum product P. We know

$$P = xy. \quad (2)$$

Solving (1) for x, we get $x = y + 6$. Substituting in (2), we get a quadratic equation:

$$P = (y + 6)y$$

$$P = y^2 + 6y$$

The first coordinate of the vertex is

$$y = -\frac{b}{2a} = -\frac{6}{2\cdot 1} = -3.$$

Note that when $y = -3$, $x = -3 + 6 = 3$. We substitute to find the second coordinate of the vertex:

$$P = (-3)^2 + 6(-3) = -9$$

The coefficient of y^2 is positive, so -9 is a minimum. The minimum product is -9. The numbers that yield this product are -3 and 3.

25. Let x and y represent the numbers. Their difference is 7, so we hve

$$x - y = 7. \quad (1)$$

We wish to find the minimum product P. We know

$$P = xy. \quad (2)$$

Solving (1) for x, we get $x = y + 7$. Substituting in (2), we get a quadratic equation:

$$P = (y + 7)y$$

$$P = y^2 + 7y$$

The first coordinate of the vertex is

$$y = -\frac{b}{2a} = -\frac{7}{2\cdot 1} = -\frac{7}{2}.$$

Note that when $y = -\frac{7}{2}$, $x = -\frac{7}{2} + 7 = \frac{7}{2}$. We substitute to find the second coordinate of the vertex:

$$P = \left(-\frac{7}{2}\right)^2 + 7\left(-\frac{7}{2}\right) = -\frac{49}{4}$$

The coefficient of y^2 is positive, so $-\frac{49}{4}$ is a minimum. The minimum product is $-\frac{49}{4}$. The numbers that yield this product are $-\frac{7}{2}$ and $\frac{7}{2}$.

27. $h = -16t^2 + 64t + 2240$

a) Since the coefficient of t^2 is negative, the value of h at the vertex is a maximum. The t-coordinate of the vertex is

$$t = -\frac{b}{2a} = -\frac{64}{2(-16)} = 2.$$

The h-coordinate of the vertex is found by substitution:

$$h = -16(2)^2 + 64(2) + 2240 = 2304$$

Thus, the rocket attains its maximum height of 2304 ft 2 sec after blastoff.

b) When the rocket reaches the ground, $h = 0$.

$$0 = -16t^2 + 64t + 2240$$

$$0 = t^2 - 4t - 140 \quad \text{Dividing by -16}$$

$$0 = (t - 14)(t + 10)$$

$$t - 14 = 0 \quad \text{or} \quad t + 10 = 0$$

$$t = 14 \quad \text{or} \quad t = -10$$

Negative time has no meaning in this problem. Thus, the rocket reaches the ground 14 sec after blastoff.

29. $P = -x^2 + 980x - 3000$

Since the coefficient of x^2 is negative, the value of P at the vertex is a maximum. The x-coordinate of the vertex is

$$x = -\frac{b}{2a} = -\frac{980}{2(-1)} = 490.$$

Thus, 490 units should be produced and sold in order to maximize profit.

31. We look for an equation of the form $y = ax^2 + bx + c$. We use the three data points to find a, b, and c.

$$4 = a(1)^2 + b(1) + c,$$

$$6 = a(-1)^2 + b(-1) + c,$$

$$16 = a(-2)^2 + b(-2) + c, \text{ or}$$

$$4 = a + b + c,$$

$$6 = a - b + c,$$

$$16 = 4a - 2b + c$$

Solving, we get $a = 3$, $b = -1$, and $c = 2$. Then the equation is $y = 3x^2 - x + 2$.

33. We look for an equation of the form $y = ax^2 + bx + c$. We use the three data points to find a, b, and c.

$$-4 = a(1)^2 + b(1) + c,$$

$$-6 = a(2)^2 + b(2) + c,$$

$$-6 = a(3)^2 + b(3) + c, \text{ or}$$

$$-4 = a + b + c,$$

$$-6 = 4a + 2b + c,$$

$$-6 = 9a + 3b + c$$

Solving, we get $a = 1$, $b = -5$, and $c = 0$. Then the equation is $y = x^2 - 5x$.

35. a) We look for an equation of the form $y = ax^2 + bx + c$. We use the three data points to find a, b, and c.

$$1000 = a(1)^2 + b(1) + c,$$
$$2000 = a(2)^2 + b(2) + c,$$
$$8000 = a(3)^2 + b(3) + c, \text{ or}$$

$$1000 = a + b + c,$$
$$2000 = 4a + 2b + c,$$
$$8000 = 9a + 3b + c$$

Solving, we get $a = 2500$, $b = -6500$, and $c = 5000$. Then the equation is $y = 2500x^2 - 6500x + 5000$.

b) To predict the earnings for the fourth month, we substitute 4 for x and compute y:

$$y = 2500(4)^2 - 6500(4) + 5000 = 19{,}000$$

Thus, the earnings are $19,000.

37. a) We look for an equation of the form $y = ax^2 + bx + c$. We use the three data points to find a, b, and c.

$$3 = a(8)^2 + b(8) + c,$$
$$4.25 = a(12)^2 + b(12) + c,$$
$$5.75 = a(16)^2 + b(16) + c, \text{ or}$$

$$3 = 64a + 8b + c,$$
$$4.25 = 144a + 12b + c,$$
$$5.75 = 256a + 16b + c$$

Solving, we get $a = 0.0078125$, $b = 0.15625$, and $c = 1.25$. Then the equation is $y = 0.0078125x^2 + 0.15625x + 1.25$.

b) We substitute 14 for x and compute y:

$y = 0.0078125(14)^2 + 0.15625(14) + 1.25 = 4.96875$. The price of a 14-in. pizza is about $4.97.

39. $\sqrt{9a^3}\sqrt{16ab^4} = \sqrt{144a^4b^4} = 12a^2b^2$

41. $\sqrt{5x - 4} + \sqrt{13 - x} = 7$

$$\sqrt{5x - 4} = 7 - \sqrt{13 - x} \quad \text{Isolating one radical}$$
$$(\sqrt{5x - 4})^2 = (7 - \sqrt{13 - x})^2 \quad \text{Principle of powers}$$
$$5x - 4 = 49 - 14\sqrt{13 - x} + (13 - x)$$
$$6x - 66 = -14\sqrt{13 - x}$$
$$3x - 33 = -7\sqrt{13 - x} \quad \text{Dividing by 2}$$
$$(3x - 33)^2 = (-7\sqrt{13 - x})^2$$
$$9x^2 - 198x + 1089 = 49(13 - x)$$
$$9x^2 - 198x + 1089 = 637 - 49x$$
$$9x^2 - 149x + 452 = 0$$
$$(9x - 113)(x - 4) = 0$$

$$9x - 113 = 0 \quad \text{or} \quad x - 4 = 0$$
$$9x = 113 \quad \text{or} \quad x = 4$$
$$x = \frac{113}{9} \quad \text{or} \quad x = 4$$

The number 4 checks, but $\frac{113}{9}$ does not. The solution is 4.

43. Graph $y = |x^2 + 6x + 4|$

$$y = |(x^2 + 6x + 9 - 9) + 4|$$
$$y = |(x^2 + 6x + 9) - 9 + 4|$$
$$y = |(x + 3)^2 - 5|$$

The graph will lie entirely on or above the x-axis since absolute value must be nonnegative. For values of x for which $(x + 3)^2 - 5 \geqslant 0$, the graph will be the same as the graph of $y = (x + 3)^2 - 5$. For values of x for which $(x + 3)^2 - 5 < 0$, the graph will be the reflection of the graph of $y = (x + 3)^2 - 5$ across the x-axis. We first graph $y = (x + 3)^2 - 5$:

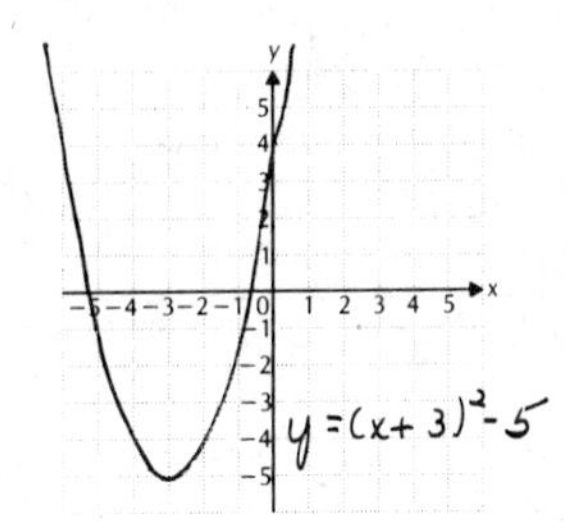

Then, reflecting the points below the axis across the x-axis we get the graph of $y = |(x + 3)^2 - 5|$, or $y = |x^2 + 6x + 4|$:

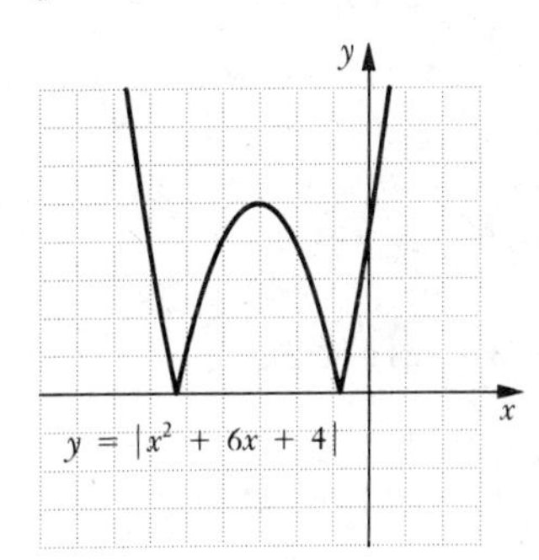

45. We made a drawing.

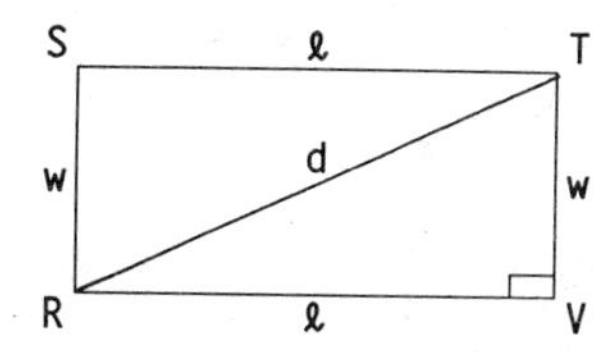

The perimeter is 44, so we have

$$2\ell + 2w = 44. \quad (1)$$

Then we use the Pythagorean equation:

$$d^2 = \ell^2 + w^2 \quad (2)$$

Solving (1) for w, we get $w = 22 - \ell$. Substituting in (2), we get

$$d^2 = \ell^2 + (22 - \ell)^2$$
$$d^2 = \ell^2 + 484 - 44\ell + \ell^2$$
$$d^2 = 2\ell^2 - 44\ell + 484$$

Since the coefficient of ℓ^2 is positive, we know the value of d^2 at the vertex is a minimum. The ℓ-coordinate of the vertex is

$$\ell = -\frac{b}{2a} = -\frac{-44}{2\cdot 2} = 11.$$

45. (continued)

We substitute to find the second coordinate of the vertex:

$d^2 = 2(11)^2 - 44(11) + 484 = 242$

The minimum value of d^2 is 242, so the minimum value of $d = \sqrt{242} = 11\sqrt{2}$ ft.

Exercise Set 11.9

1. $(x - 5)(x + 3) > 0$

The solutions of $(x - 5)(x + 3) = 0$ are 5 and -3. They divide the real-number line into three intervals as shown:

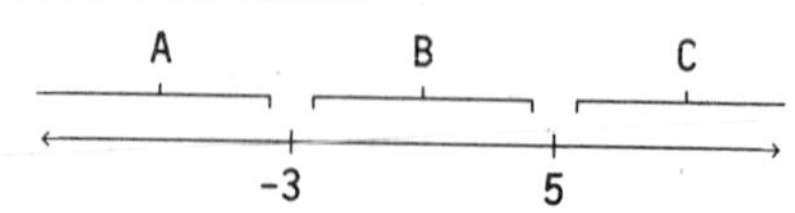

We try test numbers in each interval.

A: Test -4, $y = (-4 - 5)(-4 + 3) = 9$

B: Test 0, $y = (0 - 5)(0 + 3) = -15$

C: Test 6, $y = (6 - 5)(6 + 3) = 9$

The expression is positive for all values of x in intervals A and C. The solution set is $\{x | x < -3$ or $x > 5\}$.

3. $(x + 1)(x - 2) \leqslant 0$

The solutions of $(x + 1)(x - 2) = 0$ are -1 and 2. They divide the real-number line into three intervals as shown:

A B C

-1 2

We try test numbers in each interval.

A: Test -2, $y = (-2 + 1)(-2 - 2) = 4$;

B: Test 0, $y = (0 + 1)(0 - 2) = -2$;

C: Test 3, $y = (3 + 1)(3 - 2) = 4$

The expression is negative for all numbers in interval B. The inequality symbol is $\leqslant$, so we need to include the intercepts.

The solution set is $\{x | -1 \leqslant x \leqslant 2\}$.

5. $x^2 - x - 2 < 0$

$(x + 1)(x - 2) < 0$ Factoring

See the diagram and test numbers in Exercise 3. The solution set is $\{x | -1 < x < 2\}$.

7. $9 - x^2 \leqslant 0$

$(3 + x)(3 - x) \leqslant 0$

The solutions of $(3 + x)(3 - x) = 0$ are -3 and 3. They divide the real-number line into three intervals as shown:

A B C

-3 3

We try test numbers in each interval.

A: Test -4, $y = 9 - (-4)^2 = -7$

B: Test 0, $y = 9 - (0)^2 = 9$

C: Test 4, $y = 9 - (4)^2 = -7$

The expression is negative for all values of x in intervals A and C. Since the inequality symbol is $\leqslant 0$, we need to include the intercepts. The solution set is $\{x | x \leqslant -3$ or $x \geqslant 3\}$.

9. $x^2 - 2x + 1 \geqslant 0$

$(x - 1)^2 \geqslant 0$

The solution of $(x - 1)^2 = 0$ is 1. For all real-number values of x except 1, $(x - 1)^2$ will be positive. Thus the solution set is $\{x | x$ is a real number$\}$.

11. $x^2 + 8 < 6x$

$x^2 - 6x + 8 < 0$

$(x - 4)(x - 2) < 0$

The solutions of $(x - 4)(x - 2) = 0$ are 4 and 2. They divide the real-number line into three intervals as shown:

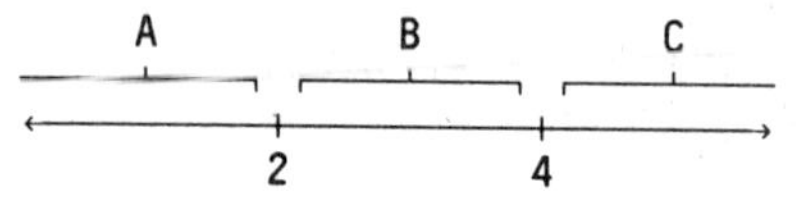

We try test numbers in each interval.

A: Test 0, $y = (0 - 4)(0 - 2) = 8$

B: Test 3, $y = (3 - 4)(3 - 2) = -1$

C: Test 5, $y = (5 - 4)(5 - 2) = 3$

The expression is negative for all values of x in interval B. The solution set is $\{x | 2 < x < 4\}$.

13. $3x(x + 2)(x - 2) < 0$

The solutions of $3x(x + 2)(x - 2) = 0$ are 0, -2, and 2. They divide the real-number line into four intervals as shown.

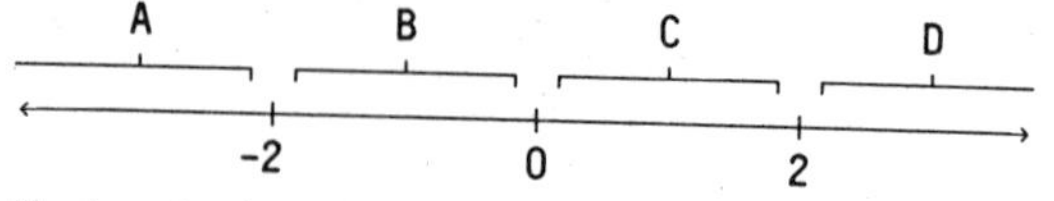

We try test numbers in each interval.

A: Test -3, $y = 3(-3)(-3 + 2)(-3 - 2) = -45$

B: Test -1, $y = 3(-1)(-1 + 2)(-1 - 2) = 9$

C: Test 1, $y = 3(1)(1 + 2)(1 - 2) = -9$

D: Test 3, $y = 3(3)(3 + 2)(3 - 2) = 45$

The expression is negative for all numbers in intervals A and C. The solution set is $\{x | x < -2$ or $0 < x < 2\}$.

15. $(x + 3)(x - 2)(x + 1) > 0$

The solutions of $(x + 3)(x - 2)(x + 1) = 0$ are -3, 2, and -1. They divide the real-number line into four intervals as shown:

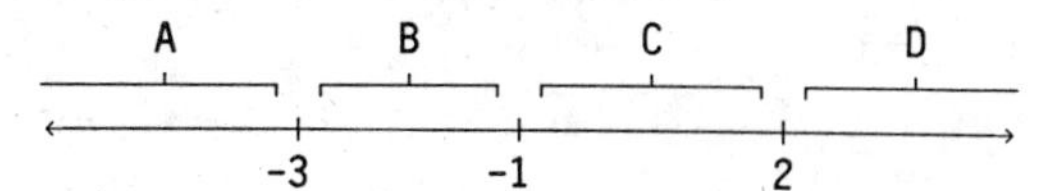

We try test numbers in each interval.

A: Test -4, $y = (-4 + 3)(-4 - 2)(-4 + 1) = -18$

B: Test -2, $y = (-2 + 3)(-2 - 2)(-2 + 1) = 4$

C: Test 0, $y = (0 + 3)(0 - 2)(0 + 1) = -6$

D: Test 3, $y = (3 + 3)(3 - 2)(3 + 1) = 24$

The expression is positive for all values of x intervals B and D. The solution set is $\{x \mid -3 < x < -1 \text{ or } x > 2\}$.

17. $(x + 3)(x + 2)(x - 1) < 0$

The solutions of $(x + 3)(x + 2)(x - 1) = 0$ are -3, -2, and 1. They divide the number line into four intervals as shown:

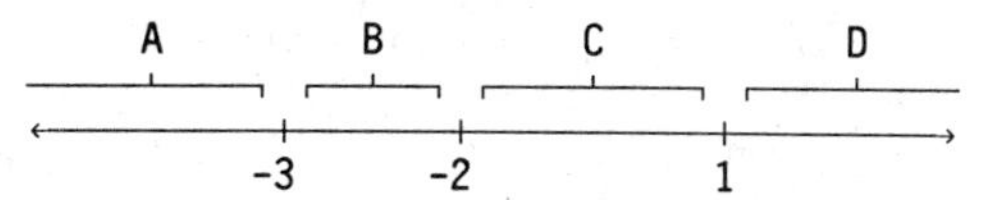

We try test numbers in each interval.

A: Test -4, $y = (-4 + 3)(-4 + 2)(-4 - 1) = -10$

B: Test $-\frac{5}{2}$, $y = \left(-\frac{5}{2} + 3\right)\left(-\frac{5}{2} + 2\right)\left(-\frac{5}{2} - 1\right) = \frac{7}{8}$

C: Test 0, $y = (0 + 3)(0 + 2)(0 - 1) = -6$

D: Test 2, $y = (2 + 3)(2 + 2)(2 - 1) = 20$

The expression is negative for all numbers in intervals A and C. The solution set is $\{x \mid x < -3 \text{ or } -2 < x < 1\}$.

19. $\frac{1}{x - 4} < 0$

We write the related equation by changing the $<$ symbol to $=$:

$$\frac{1}{x - 4} = 0$$

We solve the related equation.

$$(x - 4) \cdot \frac{1}{x - 4} = (x - 4) \cdot 0$$

$$1 = 0$$

The related equation has no solution.

Next we find the replacements that are not meaningful by setting the denominator equal to 0 and solving:

$$x - 4 = 0$$

$$x = 4$$

We use 4 to divide the number line into two intervals as shown:

A B

4

We try test numbers in each interval.

19. (continued)

A: Test 0,

$\frac{1}{x - 4} < 0$	
$\frac{1}{0 - 4}$	0
$-\frac{1}{4}$	TRUE

The number 0 is a solution of the inequality, so the interval A is part of the solution set.

B: Test 5,

$\frac{1}{x - 4} < 0$	
$\frac{1}{5 - 4}$	0
1	FALSE

The number 5 is not a solution of the inequality, so the interval B is not part of the solution set. The solution set is $\{x \mid x < 4\}$.

21. $\frac{x + 1}{x - 3} > 0$

Solve the related equation.

$$\frac{x + 1}{x - 3} = 0$$

$$x + 1 = 0$$

$$x = -1$$

Find replacements that are not meaningful.

$$x - 3 = 0$$

$$x = 3$$

Use the numbers -1 and 3 to divide the number line into intervals as shown:

A B C

-1 3

Try test numbers in each interval.

A: Test -2,

$\frac{x + 1}{x - 3} > 0$	
$\frac{-2 + 1}{-2 - 3}$	0
$\frac{-1}{-5}$	
$\frac{1}{5}$	TRUE

The number -2 is a solution of the inequality so the interval A is part of the solution set.

B: Test 0,

$\frac{x + 1}{x - 3} > 0$	
$\frac{0 + 1}{0 - 3}$	0
$-\frac{1}{3}$	FALSE

The number 0 is not a solution of the inequality, so the interval B is not part of the solution set.

21. (continued)

C: Test 4,

$$\begin{array}{c|c} \frac{x+1}{x-3} > 0 & \\ \hline \frac{4+1}{4-3} & 0 \\ \frac{5}{1} & \\ 5 & \text{TRUE} \end{array}$$

The number 4 is a solution of the inequality, so the interval C is part of the solution set. The solution set is $\{x | x < -1 \text{ or } x > 3\}$.

23. $\frac{3x+2}{x-3} \leqslant 0$

Solve the related equation.

$$\frac{3x+2}{x-3} = 0$$

$$3x + 2 = 0$$

$$3x = -2$$

$$x = -\frac{2}{3}$$

Find replacements that are not meaningful.

$$x - 3 = 0$$

$$x = 3$$

Use the numbers $-\frac{2}{3}$ and 3 to divide the number line into intervals as shown.

A B C

$-\frac{2}{3}$ 3

Try test numbers in each interval.

A: Test -1,

$$\begin{array}{c|c} \frac{3x+2}{x-3} \leqslant 0 & \\ \hline \frac{3(-1)+2}{-1-3} & 0 \\ \frac{-1}{-4} & \\ \frac{1}{4} & \text{FALSE} \end{array}$$

The number -1 is not a solution of the inequality, so the interval A is not part of the solution set.

B: Test 0,

$$\begin{array}{c|c} \frac{3x+2}{x-3} \leqslant 0 & \\ \hline \frac{3\cdot 0+2}{0-3} & 0 \\ \frac{2}{-3} & \\ -\frac{2}{3} & \text{TRUE} \end{array}$$

The number 0 is a solution of the inequality, so the interval B is part of the solution set.

C: Test 4,

$$\begin{array}{c|c} \frac{3x+2}{x-3} \leqslant 0 & \\ \hline \frac{3\cdot 4+2}{4-3} & 0 \\ 14 & \text{FALSE} \end{array}$$

23. (continued)

The number 4 is not a solution of the inequality, so the interval C is not part of the solution set. The solution set includes the interval B. The number $-\frac{2}{3}$ is also included since the inequality symbol is $\leqslant$ and $-\frac{2}{3}$ is the solution of the related equation. The number 3 is not included since it not a meaningful replacement. The solution set is $\left\{x \middle| -\frac{2}{3} \leqslant x < 3\right\}$.

25. $\frac{x-1}{x-2} > 3$

Solve the related equation.

$$\frac{x-1}{x-2} = 3$$

$$x - 1 = 3(x - 2)$$

$$x - 1 = 3x - 6$$

$$5 = 2x$$

$$\frac{5}{2} = x$$

Find replacements that are not meaningful.

$$x - 2 = 0$$

$$x = 2$$

Use the numbers $\frac{5}{2}$ and 2 to divide the number line into intervals as shown:

A B C

2 $\frac{5}{2}$

Try test numbers in each interval.

A: Test 0,

$$\begin{array}{c|c} \frac{x-1}{x-2} > 3 & \\ \hline \frac{0-1}{0-2} & 3 \\ \frac{1}{2} & \text{FALSE} \end{array}$$

The number 0 is not a solution of the inequality, so the interval A is not part of the solution set.

B: Test $\frac{9}{4}$,

$$\begin{array}{c|c} \frac{x-1}{x-2} > 3 & \\ \hline \frac{\frac{9}{4}-1}{\frac{9}{4}-2} & 3 \\ \frac{\frac{5}{4}}{\frac{1}{4}} & \\ 5 & \text{TRUE} \end{array}$$

The number $\frac{9}{4}$ is a solution of the inequality, so the interval B is part of the solution set.

25. (continued)

C: Test 3,

$\frac{x - 1}{x - 2} > 3$	
$\frac{3 - 1}{3 - 2}$	3
2	FALSE

The number 3 is not a solution of the inequality, so the interval C is not part of the solution set. The solution set is $\left\{x \middle| 2 < x < \frac{5}{2}\right\}$.

27. $\frac{(x - 2)(x + 1)}{x - 5} < 0$

Solve the related equation.

$\frac{(x - 2)(x + 1)}{x - 5} = 0$

$(x - 2)(x + 1) = 0$

$x = 2$ or $x = -1$

Find replacements that are not meaningful.

$x - 5 = 0$

$x = 5$

Use the numbers 2, -1, and 5 to divide the number line into intervals as shown:

A B C D

-1 2 5

Try test numbers in each interval.

A: Test -2,

$\frac{(x - 2)(x + 1)}{x - 5} < 0$	
$\frac{(-2 - 2)(-2 + 1)}{-2 - 5}$	0
$\frac{-4(-1)}{-7}$	
$-\frac{4}{7}$	TRUE

Interval A is part of the solution set.

B: Test 0,

$\frac{(x - 2)(x + 1)}{x - 5} < 0$	
$\frac{(0 - 2)(0 + 1)}{0 - 5}$	0
$\frac{-2 \cdot 1}{-5}$	
$\frac{2}{5}$	FALSE

Interval B is not part of the solution set.

C: Test 3,

$\frac{(x - 2)(x + 1)}{x - 5} < 0$	
$\frac{(3 - 2)(3 + 1)}{3 - 5}$	0
$\frac{1 \cdot 4}{-2}$	
-2	TRUE

Interval C is part of the solution set.

27. (continued)

D: Test 6,

$\frac{(x - 2)(x + 1)}{x - 5} < 0$	
$\frac{(6 - 2)(6 + 1)}{6 - 5}$	0
$\frac{4 \cdot 7}{1}$	
28	FALSE

Interval D is not part of the solution set.

The solution set is $\{x \mid x < -1 \text{ or } 2 < x < 5\}$.

29. $\frac{x}{x - 2} \geqslant 0$

Solve the related equation.

$\frac{x}{x - 2} = 0$

$x = 0$

Find replacements that are not meaningful.

$x - 2 = 0$

$x = 2$

Use the numbers 0 and 2 to divide the number line into intervals as shown.

A B C

0 2

Try test numbers in each interval.

A: Test -1,

$\frac{x}{x - 2} \geqslant 0$	
$\frac{-1}{-1 - 2}$	0
$\frac{1}{3}$	TRUE

Interval A is part of the solution set.

B: Test 1,

$\frac{x}{x - 2} \geqslant 0$	
$\frac{1}{1 - 2}$	0
-1	FALSE

Interval B is not part of the solution set.

C: Test 3,

$\frac{x}{x - 2} \geqslant 0$	
$\frac{3}{3 - 2}$	0
3	TRUE

Interval C is part of the solution set.

The solution set includes intervals A and C. The number 0 is also included since the inequality symbol is $\geqslant$ and 0 is the solution of the related equation. The number 2 is not included since it is not a meaningful replacement. The solution set is $\{x \mid x \leqslant 0 \text{ or } x > 2\}$.

31. $\frac{x - 5}{x} < 1$

Solve the related equation.

$$\frac{x - 5}{x} = 1$$

$$x - 5 = x$$

$$-5 = 0$$

The related equation has no solution.

Find replacements that are not meaningful.

$$x = 0$$

Use the number 0 to divide the number line into two intervals as shown.

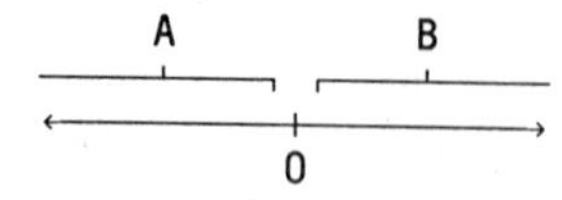

Try test numbers in each interval.

A: Test -1,

$$\frac{x - 5}{x} < 1$$

$\frac{-1 - 5}{-1}$	1
6	FALSE

Interval A is not part of the solution set.

B: Test 1,

$$\frac{x - 5}{x} < 1$$

$\frac{1 - 5}{1}$	1
-4	TRUE

Interval B is part of the solution set.

The solution set is $\{x|x > 0\}$.

33. $\frac{x - 1}{(x - 3)(x + 4)} < 0$

Solve the related equation.

$$\frac{x - 1}{(x - 3)(x + 4)} = 0$$

$$x - 1 = 0$$

$$x = 1$$

Find replacements that are not meaningful.

$$(x - 3)(x + 4) = 0$$

$$x = 3 \quad \text{or} \quad x = -4$$

Use the numbers 1, 3, and -4 to divide the number line into intervals as shown:

A B C D

-4 1 3

Try test numbers in each interval.

A: Test -5,

$$\frac{x - 1}{(x - 3)(x + 4)} < 0$$

$\frac{-5 - 1}{(-5 - 3)(-5 + 4)}$	0
$\frac{-6}{-8(-1)}$	
$-\frac{3}{4}$	TRUE

Interval A is part of the solution set.

33. (continued)

B: Test 0,

$$\frac{x - 1}{(x - 3)(x + 4)} < 0$$

$\frac{0 - 1}{(0 - 3)(0 + 4)}$	0
$\frac{-1}{-3 \cdot 4}$	
$\frac{1}{12}$	FALSE

Interval B is not part of the solution set.

C: Test 2,

$$\frac{x - 1}{(x - 3)(x + 4)} < 0$$

$\frac{2 - 1}{(2 - 3)(2 + 4)}$	0
$\frac{1}{-1 \cdot 6}$	
$-\frac{1}{6}$	TRUE

Interval C is part of the solution set.

D: Test 4,

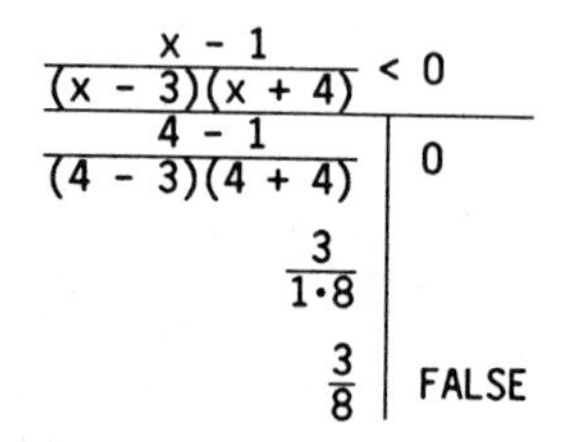

$$\frac{x - 1}{(x - 3)(x + 4)} < 0$$

$\frac{4 - 1}{(4 - 3)(4 + 4)}$	0
$\frac{3}{1 \cdot 8}$	
$\frac{3}{8}$	FALSE

Interval D is not part of the solution set.

The solution set is $\{x|x < -4 \text{ or } 1 < x < 3\}$.

35. $2 < \frac{1}{x}$

Solve the related equation.

$$2 = \frac{1}{x}$$

$$x = \frac{1}{2}$$

Find replacements that are not meaningful.

$$x = 0$$

Use the numbers $\frac{1}{2}$ and 0 to divide the number line into intervals as shown.

A B C

0 $\frac{1}{2}$

Try test numbers in each interval.

A: Test -1,

$$2 < \frac{1}{x}$$

2	$\frac{1}{-1}$
	-1 FALSE

Interval A is not part of the solution set.

B: Test $\frac{1}{4}$,

$$2 < \frac{1}{x}$$

2	$\frac{1}{\frac{1}{4}}$
	4 TRUE

Interval B is part of the solution set.

35. (continued)

C: Test 1,

$$\begin{array}{c|l} 2 < \frac{1}{x} & \\ \hline 2 & \frac{1}{1} \\ & 1 \quad \text{FALSE} \end{array}$$

Interval C is not part of the solution set.

The solution set is $\left\{x \middle| 0 < x < \frac{1}{2}\right\}$.

37. $\dfrac{(x - 1)(x + 2)}{(x + 3)(x - 4)} > 0$

Solve the related equation.

$$\frac{(x - 1)(x + 2)}{(x + 3)(x - 4)} = 0$$

$(x - 1)(x + 2) = 0$

$x = 1$ or $x = -2$

Find replacements that are not meaningful.

$(x + 3)(x - 4) = 0$

$x = -3$ or $x = 4$

Use the numbers 1, -2, -3, and 4 to divide the number line into intervals as shown.

A B C D E

-3 -2 1 4

Try test numbers in each interval.

A: Test -4,

$$\begin{array}{c|l} \frac{(x - 1)(x + 2)}{(x + 3)(x - 4)} > 0 & \\ \hline \frac{(-4 - 1)(-4 + 2)}{(-4 + 3)(-4 - 4)} & 0 \\ \frac{-5(-2)}{-1(-8)} & \\ \frac{5}{4} & \text{TRUE} \end{array}$$

Interval A is part of the solution set.

B: Test $-\frac{5}{2}$,

$$\begin{array}{c|l} \frac{(x - 1)(x + 2)}{(x + 3)(x - 4)} > 0 & \\ \hline \frac{\left(-\frac{5}{2} - 1\right)\left(-\frac{5}{2} + 2\right)}{\left(-\frac{5}{2} + 3\right)\left(-\frac{5}{2} - 4\right)} & 0 \\ \frac{-\frac{7}{2}\left(-\frac{1}{2}\right)}{\frac{1}{2}\left(-\frac{13}{2}\right)} & \\ -\frac{7}{13} & \text{FALSE} \end{array}$$

Interval B is not part of the solution set.

C: Test 0,

$$\begin{array}{c|l} \frac{(x - 1)(x + 2)}{(x + 3)(x - 4)} > 0 & \\ \hline \frac{(0 - 1)(0 + 2)}{(0 + 3)(0 - 4)} & 0 \\ \frac{-1 \cdot 2}{3(-4)} & \\ \frac{1}{6} & \text{TRUE} \end{array}$$

Interval C is part of the solution set.

37. (continued)

D: Test 2,

$$\begin{array}{c|l} \frac{(x - 1)(x + 2)}{(x + 3)(x - 4)} > 0 & \\ \hline \frac{(2 - 1)(2 + 2)}{(2 + 3)(2 - 4)} & 0 \\ \frac{1 \cdot 4}{5 \cdot (-2)} & \\ -\frac{2}{5} & \text{FALSE} \end{array}$$

Interval D is not part of the solution set.

E: Test 5,

$$\begin{array}{c|l} \frac{(x - 1)(x + 2)}{(x + 3)(x - 4)} > 0 & \\ \hline \frac{(5 - 4)(5 + 2)}{(5 + 3)(5 - 4)} & 0 \\ \frac{1 \cdot 7}{8 \cdot 1} & \\ \frac{7}{8} & \text{TRUE} \end{array}$$

Interval E is part of the solution set.

The solution set is $\{x \mid x < -3 \text{ or } -2 < x < 1 \text{ or } x > 4\}$.

39. $\dfrac{x^2 + 3x - 10}{x^2 - x - 56} \leqslant 0$

$\dfrac{(x + 5)(x - 2)}{(x - 8)(x + 7)} \leqslant 0$

Solve the related equation.

$$\frac{(x + 5)(x - 2)}{(x - 8)(x + 7)} = 0$$

$(x + 5)(x - 2) = 0$

$x = -5$ or $x = 2$

Find replacements that are not meaningful.

$(x - 8)(x + 7) = 0$

$x = 8$ or $x = -7$

Use the numbers -5, 2, 8, and -7 to divide the number line into intervals as shown.

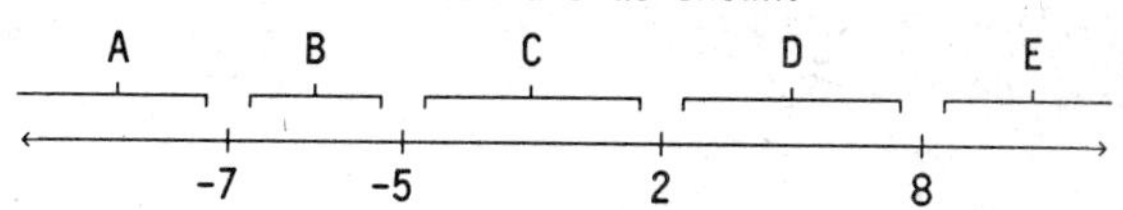

Try test numbers in each interval.

A: Test -8,

$$\begin{array}{c|l} \frac{(x + 5)(x - 2)}{(x - 8)(x + 7)} \leqslant 0 & \\ \hline \frac{(-8 + 5)(-8 - 2)}{(-8 - 8)(-8 + 7)} & 0 \\ \frac{-3 \cdot (-10)}{-16 \cdot (-1)} & \\ \frac{15}{8} & \text{FALSE} \end{array}$$

Interval A is not part of the solution set.

B: Test -6,

$$\begin{array}{c|l} \frac{(x + 5)(x - 2)}{(x - 8)(x + 7)} \leqslant 0 & \\ \hline \frac{(-6 + 5)(-6 - 2)}{(-6 - 8)(-6 + 7)} & 0 \\ \frac{-1 \cdot (-8)}{-14 \cdot 1} & \\ -\frac{4}{7} & \text{TRUE} \end{array}$$

Interval B is part of the solution set.

39. (continued)

C: Test 0,

$$\frac{(x + 5)(x - 2)}{(x - 8)(x + 7)} \leqslant 0$$

$\frac{(0 + 5)(0 - 2)}{(0 - 8)(0 + 7)}$	0
$\frac{5\cdot(-2)}{-8\cdot 7}$	
$\frac{5}{28}$	FALSE

Interval C is not part of the solution set.

D: Test 3,

$$\frac{(x + 5)(x - 2)}{(x - 8)(x + 7)} \leqslant 0$$

$\frac{(3 + 5)(3 - 2)}{(3 - 8)(3 + 7)}$	0
$\frac{8\cdot 1}{-5\cdot 10}$	
$-\frac{4}{25}$	TRUE

Interval D is part of the solution set.

E: Test 9,

$$\frac{(x + 5)(x - 2)}{(x - 8)(x + 7)} \leqslant 0$$

$\frac{(9 + 5)(9 - 2)}{(9 - 8)(9 + 7)}$	0
$\frac{14\cdot 7}{1\cdot 16}$	
$\frac{49}{8}$	FALSE

Interval E is not part of the solution set.

The solution set includes B and D. The numbers -5 and 2 are also included since the inequality symbol is $\leqslant 0$ and -5 and 2 are solutions of the related equation. The numbers -7 and 8 are not included since they are not meaningful replacements. The solution set is $\{x \mid -7 < x \leqslant -5 \text{ or } 2 \leqslant x < 8\}$.

41. $x^2 - 2x \leqslant 2$

$x^2 - 2x - 2 \leqslant 0$

The solutions of $x^2 - 2x - 2 = 0$ are found using the quadratic formula. They are $1 \pm \sqrt{3}$, or about 2.7 and -0.7. These numbers divide the number line into three intervals as shown:

A B C

$1 - \sqrt{3}$ $1 + \sqrt{3}$

We try test numbers in each interval.

A: Test -1, $y = (-1)^2 - 2(-1) - 2 = 1$

B: Test 0, $y = 0^2 - 2\cdot 0 - 2 = -2$

C: Test 3, $y = 3^2 - 2\cdot 3 - 2 = 1$

The expression is negative for all values of x in interval B. The inequality symbol is $\leqslant 0$, so we must also include the intercepts. The solution set is $\{x \mid 1 - \sqrt{3} \leqslant x \leqslant 1 + \sqrt{3}\}$.

43. $x^4 + 2x^2 > 0$

$x^2(x^2 + 2) > 0$

$x^2 > 0$ for all $x \neq 0$, and $x^2 + 2 > 0$ for all values of x. Then $x^2(x^2 + 2) > 0$ for all $x \neq 0$. The solution set is $\{x \mid x \neq 0\}$, or the set of all real numbers except 0.

45. $\left|\frac{x + 2}{x - 1}\right| < 3$

$-3 < \frac{x + 2}{x - 1} < 3$

Case I: $x - 1 > 0$, or $x > 1$. Then we have

$-3(x - 1) < x + 2 < 3(x - 1)$

$-3x + 3 < x + 2 < 3x - 3$

$-3x + 3 < x + 2$ and $x + 2 < 3x - 3$

$-4x < -1$ and $-2x < -5$

$x > \frac{1}{4}$ and $x > \frac{5}{2}$

The solution set for Case I is $\left\{x \middle| x > 1 \text{ and } x > \frac{1}{4} \text{ and } x > \frac{5}{2}\right\}$, or $\left\{x \middle| x > \frac{5}{2}\right\}$.

Case II: $x - 1 < 0$, or $x < 1$. Then we have

$-3(x - 1) > x + 2 > 3(x - 1)$

$-3x + 3 > x + 2 > 3x - 3$

$-3x + 3 > x + 2$ and $x + 2 > 3x - 3$

$-4x > -1$ and $-2x > -5$

$x < \frac{1}{4}$ and $x < \frac{5}{2}$

The solution set for Case II is $\left\{x \middle| x < 1 \text{ and } x < \frac{1}{4} \text{ and } x < \frac{5}{2}\right\}$, or $\left\{x \middle| x < \frac{1}{4}\right\}$.

The solution set for the original inequality is $\left\{x \middle| x > \frac{5}{2}\right\} \cup \left\{x \middle| x < \frac{1}{4}\right\}$, or $\left\{x \middle| x < \frac{1}{4} \text{ or } x > \frac{5}{2}\right\}$.

47. a) Solve: $-16t^2 + 32t + 1920 > 1920$

$-16t^2 + 32t > 0$

$t^2 - 2t < 0$

$t(t - 2) < 0$

The solutions of $t(t - 2) = 0$ are 0 and 2. They divide the real-number line into 3 intervals as shown:

A B C

0 2

A: Test -1, $y = (-1)^2 - 2(-1) = 3$

B: Test 1, $y = 1^2 - 2\cdot 1 = -1$

C: Test 3, $y = 3^2 - 2\cdot 3 = 3$

The expression is negative for all values of t in interval B. The solution set is $\{t \mid 0 < t < 2\}$.

47. (continued)

b) Solve: $-16t^2 + 32t + 1920 < 640$

$$-16t^2 + 32t + 1280 < 0$$

$$t^2 - 2t - 80 > 0$$

$$(t - 10)(t + 8) > 0$$

The solutions of $(t - 10)(t + 8)$ are 10 and -8. They divide the real-number line into three intervals as shown:

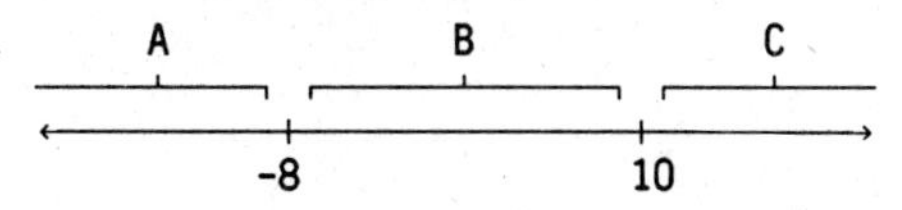

A: Test -10, $y = (-10)^2 - 2(-10) - 80 = 40$

B: Test 0, $y = 0^2 - 2\cdot0 - 80 = -80$

C: Test 20, $y = 20^2 - 2\cdot20 - 80 = 280$

The expression is positive for all values of t in intervals A and C. However, since negative values of t have no meaning in this problem, we disregard interval A. Thus, the solution set is $\{t | t > 10\}$.

Exercise Set 12.1

1. Graph: $y = x^2$

The graph is a parabola. The vertex is (0,0); the line of symmetry is $x = 0$. The curve opens upward. We choose some x-values on both sides of the vertex and compute the corresponding y-values. Then we plot the points and graph the parabola.

x	y
0	0
1	1
-1	1
2	4
-2	4

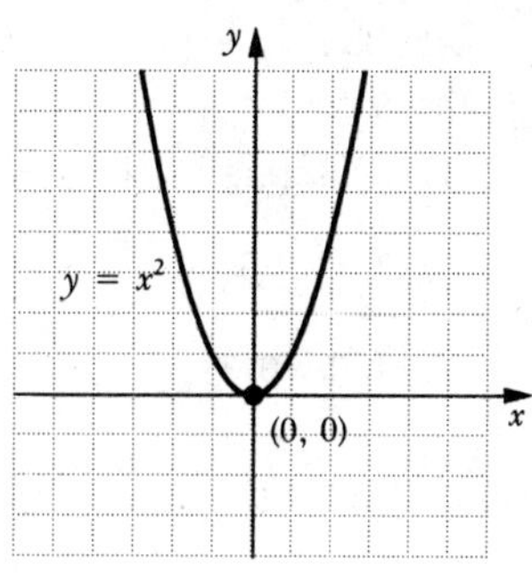

3. Graph: $x = y^2 + 4y + 1$

We complete the square.

$x = (y^2 + 4y + 4 - 4) + 1$
$= (y^2 + 4y + 4) - 4 + 1$
$= (y + 2)^2 - 3$, or
$= [y - (-2)]^2 + (-3)$

The graph is a parabola. The vertex is (-3,-2); the line of symmetry is $y = -2$. The curve opens to the right.

x	y
-3	-2
-2	-3
-2	-1
1	-4
1	0

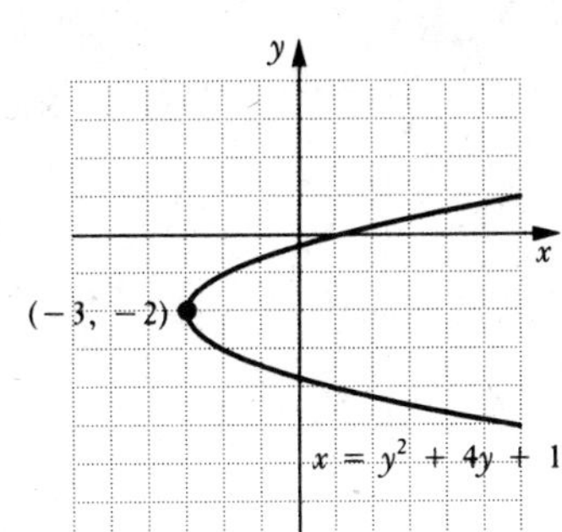

5. Graph: $y = -x^2 + 4x - 5$

We use the formula to find the first coordinate of the vertex:

$$x = -\frac{b}{2a} = -\frac{4}{2(-1)} = 2$$

Then $y = -x^2 + 4x - 5 = -(2)^2 + 4(2) - 5 = -1$.

The vertex is (2,-1); the line of symmetry is $x = 2$. The curve opens downward.

x	y
2	-1
1	-2
3	-2
0	-5
4	-5

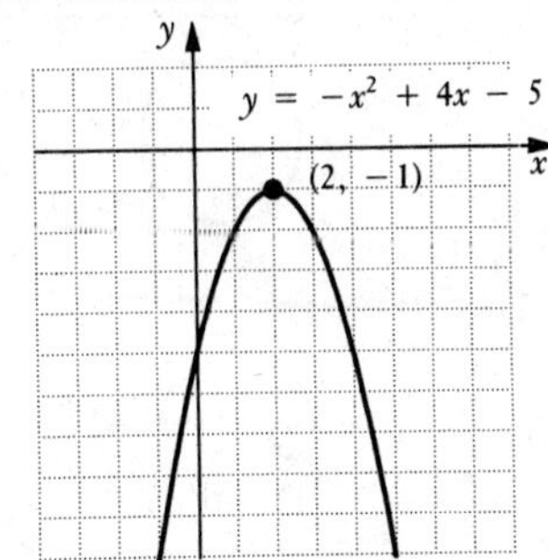

7. Graph: $x = -3y^2 - 6y - 1$

We complete the square.

$x = -3(y^2 + 2y) - 1$
$= -3(y^2 + 2y + 1 - 1) - 1$
$= -3(y^2 + 2y + 1) + 3 - 1$
$= -3(y + 1)^2 + 2$
$= -3[y - (-1)]^2 + 2$

The graph is a parabola. The vertex is (2,-1); the line of symmetry is $y = -1$. The curve opens to the left.

x	y
2	-1
-1	-2
-1	0
-10	-3
-10	1

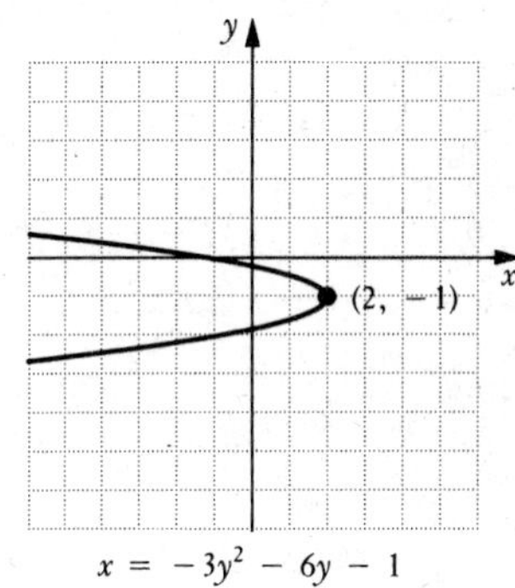

9. $d = \sqrt{(x_1 - x_2)^2 + (y_1 - y_2)^2}$

Let $(x_1,y_1) = (9,5)$ and $(x_2,y_2) = (6,1)$.

$d = \sqrt{(9 - 6)^2 + (5 - 1)^2}$ Substituting
$= \sqrt{(3)^2 + (4)^2}$
$= \sqrt{9 + 16}$
$= \sqrt{25}$
$= 5$

11. $d = \sqrt{(x_1 - x_2)^2 + (y_1 - y_2)^2}$

Let $(x_1,y_1) = (0,-7)$ and $(x_2,y_2) = (3,-4)$.

$d = \sqrt{(0 - 3)^2 + [-7 - (-4)]^2}$ Substituting
$= \sqrt{(-3)^2 + (-3)^2}$
$= \sqrt{9 + 9}$
$= \sqrt{18} \approx 4.243$

13. $d = \sqrt{(x_1 - x_2)^2 + (y_1 - y_2)^2}$

Let $(x_1,y_1) = (2,2)$ and $(x_2,y_2) = (-2,-2)$.

$d = \sqrt{[2 - (-2)]^2 + [2 - (-2)]^2}$ Substituting
$= \sqrt{4^2 + 4^2}$
$= \sqrt{16 + 16}$
$= \sqrt{32} \approx 5.657$

15. $d = \sqrt{(x_1 - x_2)^2 + (y_1 - y_2)^2}$

Let $(x_1,y_1) = (8.6,-3.4)$ and $(x_2,y_2) = (-9.2,-3.4)$.

$d = \sqrt{[8.6 - (9.2)]^2 + [-3.4 - (-3.4)]^2}$ Substituting

$= \sqrt{17.8^2 + 0^2}$

$= \sqrt{17.8^2}$

$= 17.8$

17. $d = \sqrt{(x_1 - x_2)^2 + (y_1 - y_2)^2}$

Let $(x_1,y_1) = (\frac{5}{7},\frac{1}{14})$ and $(x_2,y_2) = (\frac{1}{7},\frac{11}{14})$.

$d = \sqrt{(\frac{5}{7} - \frac{1}{7})^2 + (\frac{1}{14} - \frac{11}{14})^2}$ Substituting

$= \sqrt{(\frac{4}{7})^2 + (-\frac{5}{7})^2}$

$= \sqrt{\frac{16}{49} + \frac{25}{49}}$

$= \sqrt{\frac{41}{49}}$

$= \frac{\sqrt{41}}{7} \approx 0.915$

19. $d = \sqrt{[56 - (-23)]^2 + (-17 - 10)^2}$

$d = \sqrt{79^2 + (-27)^2} = \sqrt{6970} \approx 83.487$

21. $d = \sqrt{(a - 0)^2 + (b - 0)^2}$

$d = \sqrt{a^2 + b^2}$

23. $d = \sqrt{(-\sqrt{7} - \sqrt{2})^2 + [\sqrt{5} - (-\sqrt{3})]^2}$

$d = \sqrt{7 + 2\sqrt{14} + 2 + 5 + 2\sqrt{15} + 3} =$

$\sqrt{17 + 2\sqrt{14} + 2\sqrt{15}} \approx 5.677$

25. $d = \sqrt{[1000 - (-2000)]^2 + (-240 - 580)^2}$

$d = \sqrt{3000^2 + (-820)^2} = \sqrt{9,672,400} =$

$20\sqrt{24,181} \approx 3110.048$

27. $\left(\frac{x_1 + x_2}{2},\frac{y_1 + y_2}{2}\right)$ Midpoint formula

Let $(x_1,y_1) = (-3,6)$ and $(x_2,y_2) = (2,-8)$.

The midpoint is $\left(\frac{-3 + 2}{2},\frac{6 + (-8)}{2}\right)$, or $(-\frac{1}{2},-1)$.

29. $\left(\frac{x_1 + x_2}{2},\frac{y_1 + y_2}{2}\right)$ Midpoint formula

Let $(x_1,y_1) = (8,5)$ and $(x_2,y_2) = (-1,2)$.

The midpoint is $\left(\frac{8 + (-1)}{2},\frac{5 + 2}{2}\right)$, or $(\frac{7}{2},\frac{7}{2})$.

31. $\left(\frac{x_1 + x_2}{2},\frac{y_1 + y_2}{2}\right)$ Midpoint formula

Let $(x_1,y_1) = (-8,-5)$ and $(x_2,y_2) = (6,-1)$.

The midpoint is $\left(\frac{-8 + 6}{2},\frac{-5 + (-1)}{2}\right)$, or $(-1,-3)$.

33. $\left(\frac{x_1 + x_2}{2},\frac{y_1 + y_2}{2}\right)$ Midpoint formula

Let $(x_1,y_1) = (-3.4,8.1)$ and $(x_2,y_2) = (2.9,-8.7)$.

The midpoint is

$\left(\frac{-3.4 + 2.9}{2},\frac{8.1 + (-8.7)}{2}\right)$

$= (\frac{-0.5}{2},\frac{-0.6}{2})$

$= (-0.25,-0.3)$.

35. $\left(\frac{x_1 + x_2}{2},\frac{y_1 + y_2}{2}\right)$ Midpoint formula

Let $(x_1,y_1) = (\frac{1}{6},-\frac{3}{4})$ and $(x_2,y_2) = (-\frac{1}{3},\frac{5}{6})$.

The midpoint is

$\left(\frac{\frac{1}{6} + (-\frac{1}{3})}{2},\frac{-\frac{3}{4} + \frac{5}{6}}{2}\right)$

$= \left(\frac{-\frac{1}{6}}{2},\frac{\frac{1}{12}}{2}\right)$

$= (-\frac{1}{12},\frac{1}{24})$.

37. $\left(\frac{x_1 + x_2}{2},\frac{y_1 + y_2}{2}\right)$ Midpoint formula

Let $(x_1,y_1) = (\sqrt{2},-1)$ and $(x_2,y_2) = (\sqrt{3}, 4)$.

The midpoint is $\left(\frac{\sqrt{2} + \sqrt{3}}{2},\frac{-1 + 4}{2}\right)$, or

$\left(\frac{\sqrt{2} + \sqrt{3}}{2}, \frac{3}{2}\right)$.

39. $(x + 1)^2 + (y + 3)^2 = 4$

$[x - (-1)]^2 + [y - (-3)]^2 = 2^2$ Standard form

The center is $(-1,-3)$, and the radius is 2.

$(x + 1)^2 + (y + 3)^2 = 4$

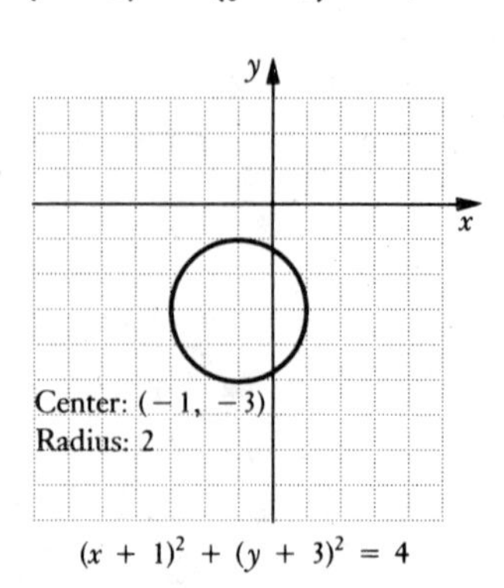

$(x + 1)^2 + (y + 3)^2 = 4$

41. $(x - 3)^2 + y^2 = 2$

$(x - 3)^2 + (y - 0)^2 = (\sqrt{2})^2$ Standard form

The center is (3,0), and the radius is $\sqrt{2}$.

$(x - 3)^2 + y^2 = 2$

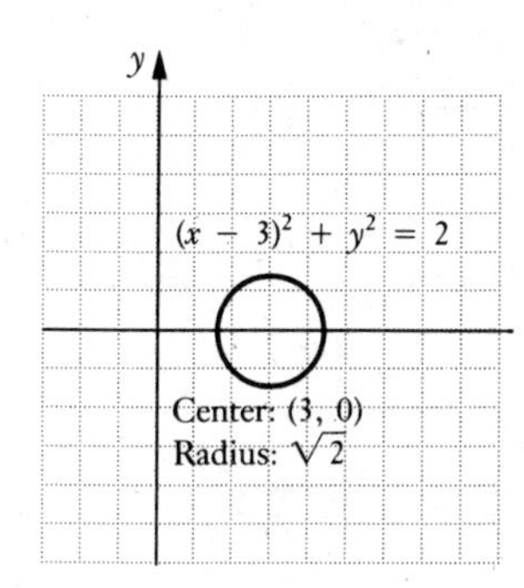

43. $x^2 + y^2 = 25$

$(x - 0)^2 + (y - 0)^2 = 5^2$ Standard form

The center is (0,0), and the radius is 5.

$x^2 + y^2 = 25$

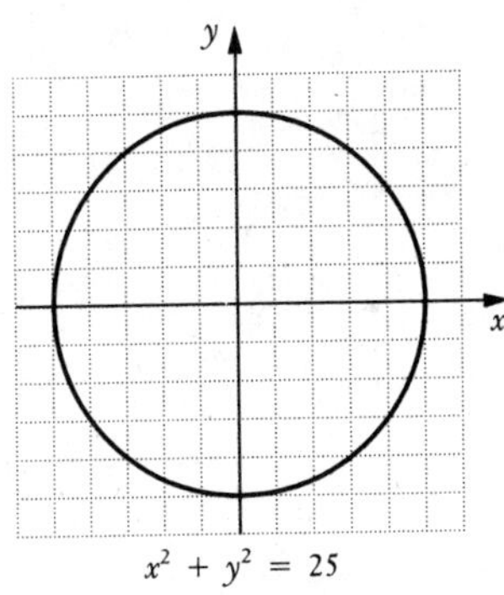

$x^2 + y^2 = 25$
Center: (0, 0)
Radius: 5

45. $(x - a)^2 + (y - b)^2 = r^2$ Standard form

$(x - 0)^2 + (y - 0)^2 = 7^2$ Substituting

$x^2 + y^2 = 49$

47. $(x - a)^2 + (y - b)^2 = r^2$ Standard form

$[x - (-2)]^2 + (y - 7)^2 = (\sqrt{5})^2$ Substituting

$(x + 2)^2 + (y - 7)^2 = 5$

49. $x^2 + y^2 + 8x - 6y - 15 = 0$

$(x^2 + 8x) + (y^2 - 6y) - 15 = 0$ Regrouping

$(x^2+8x+16-16) + (y^2-6y+9-9) - 15 = 0$ Completing the square twice

$(x^2+8x+16) + (y^2-6y+9) - 16 - 9 - 15 = 0$

$(x + 4)^2 + (y - 3)^2 = 40$

$[x - (-4)]^2 + (y - 3)^2 = (\sqrt{40})^2$

$[x - (-4)]^2 + (y - 3)^2 = (2\sqrt{10})^2$

The center is (-4,3), and the radius is $2\sqrt{10}$.

51. $x^2 + y^2 - 8x + 2y + 13 = 0$

$(x^2 - 8x) + (y^2 + 2y) + 13 = 0$ Regrouping

$(x^2-8x+16-16) + (y^2+2y+1-1) + 13 = 0$ Completing the square twice

$(x^2-8x+16) + (y^2+2y+1) - 16 - 1 + 13 = 0$

$(x - 4)^2 + (y + 1)^2 = 4$

$(x - 4)^2 + [y - (-1)]^2 = 2^2$

The center is (4,-1), and the radius is 2.

53. $x^2 + y^2 - 4x = 0$

$(x^2 - 4x) + y^2 = 0$

$(x^2 - 4x + 4 - 4) + y^2 = 0$

$(x^2 - 4x + 4) + y^2 - 4 = 0$

$(x - 2)^2 + y^2 = 4$

$(x - 2)^2 + (y - 0)^2 = 2^2$

The center is (2,0), and the radius is 2.

55. $2x + 3y = 8,$ (1)

$x - 2y = -3$ (2)

We use the elimination method.

$2x + 3y = 8$

$-2x + 4y = 6$ Multiplying (2) by -2

$7y = 14$ Adding

$y = 2$

Substitute 2 for y in either of the original equations and solve for x.

$x - 2y = -3$ (2)

$x - 2\cdot 2 = -3$ Substituting

$x - 4 = -3$

$x = 1$

The solution is (1,2).

57. We first find the length of the radius which is the distance between (0,0) and $(\frac{1}{4}, \frac{\sqrt{31}}{4})$.

$$r = \sqrt{(\frac{1}{4} - 0)^2 + (\frac{\sqrt{31}}{4} - 0)^2}$$

$$= \sqrt{(\frac{1}{4})^2 + (\frac{\sqrt{31}}{4})^2}$$

$$= \sqrt{\frac{1}{16} + \frac{31}{16}}$$

$$= \sqrt{\frac{32}{16}}$$

$$= \sqrt{2}$$

$(x - a)^2 + (y - b)^2 = r^2$ Standard form

$(x - 0)^2 + (y - 0)^2 = (\sqrt{2})^2$ Substituting (0,0) for the center and $\sqrt{2}$ for the radius

$x^2 + y^2 = 2$

59.

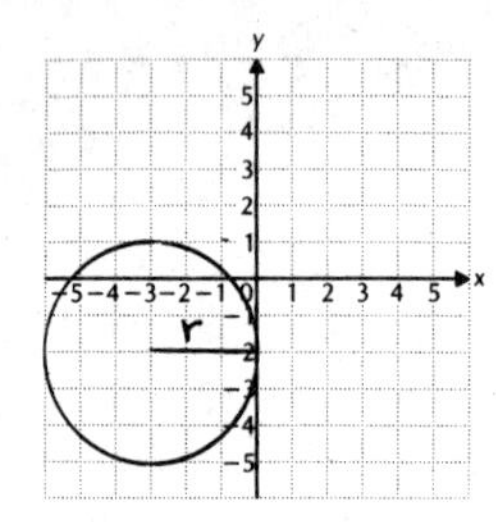

The center is (-3,-2) and the radius is 3.

$(x - a)^2 + (y - b)^2 = r^2$ Standard form

$[x - (-3)]^2 + [y - (-2)]^2 = 3^2$ Substituting

$(x + 3)^2 + (y + 2)^2 = 9$

61. $d = \sqrt{(x_1 - x_2)^2 + (y_1 - y_2)^2}$

Let $(x_1,y_1) = (-1,3k)$ and $(x_2,y_2) = (6,2k)$.

$d = \sqrt{(-1 - 6)^2 + (3k - 2k)^2}$ Substituting

$= \sqrt{(-7)^2 + (k)^2}$

$= \sqrt{49 + k^2}$

63. $d = \sqrt{(x_1 - x_2)^2 + (y_1 - y_2)^2}$

Let $(x_1,y_1) = (6m,-7n)$ and $(x_2,y_2) = (-2m,n)$.

$d = \sqrt{[6m - (-2m)]^2 + (-7n - n)^2}$ Substituting

$= \sqrt{(8m)^2 + (-8n)^2}$

$= \sqrt{64m^2 + 64n^2}$

$= \sqrt{64(m^2 + n^2)}$

$= 8\sqrt{m^2 + n^2}$

65. $d = \sqrt{(x_1 - x_2)^2 + (y_1 - y_2)^2}$

Let $(x_1,y_1) = (-3\sqrt{3},1 - \sqrt{6})$ and $(x_2,y_2) = (\sqrt{3},1 + \sqrt{6})$.

$d = \sqrt{(-3\sqrt{3} - \sqrt{3})^2 + (-\sqrt{6} - \sqrt{6})^2}$

Substituting

$= \sqrt{(-4\sqrt{3})^2 + (-2\sqrt{6})^2}$

$= \sqrt{48 + 24}$

$= \sqrt{72}$

$= \sqrt{36 \cdot 2}$

$= 6\sqrt{2}$

67. The distance between (-8,-5) and (6,1) is

$\sqrt{(-8 - 6)^2 + (-5 - 1)^2} = \sqrt{196 + 36} = \sqrt{232}$.

The distance between (6,1) and (-4,5) is

$\sqrt{[6 - (-4)]^2 + (1 - 5)^2} = \sqrt{100 + 16} = \sqrt{116}$.

The distance between (-4,5) and (-8,-5) is

$\sqrt{[-4 - (-8)]^2 + [5 - (-5)]^2} = \sqrt{16 + 100} = \sqrt{116}$

Since $(\sqrt{116})^2 + (\sqrt{116})^2 = (\sqrt{232})^2$, the points are vertices of a right triangle.

69. $\left(\frac{x_1 + x_2}{2},\frac{y_1 + y_2}{2}\right)$ Midpoint formula

Let $(x_1,y_1) = (2 - \sqrt{3},5\sqrt{2})$ and $(x_2,y_2) = (2 + \sqrt{3},3\sqrt{2})$.

The midpoint is

$$\left(\frac{(2 - \sqrt{3}) + (2 + \sqrt{3})}{2},\frac{5\sqrt{2} + 3\sqrt{2}}{2}\right)$$

$= \left(\frac{4}{2},\frac{8\sqrt{2}}{2}\right)$

$= (2,4\sqrt{2})$

Exercise Set 12.2

1. $\frac{x^2}{4} + \frac{y^2}{1} = 1$

$\frac{x^2}{2^2} + \frac{y^2}{1^2} = 1$

The x-intercepts are (2,0) and (-2,0), and the y-intercepts are (0,1) and (0,-1). We plot these points and connect them with an oval-shaped curve.

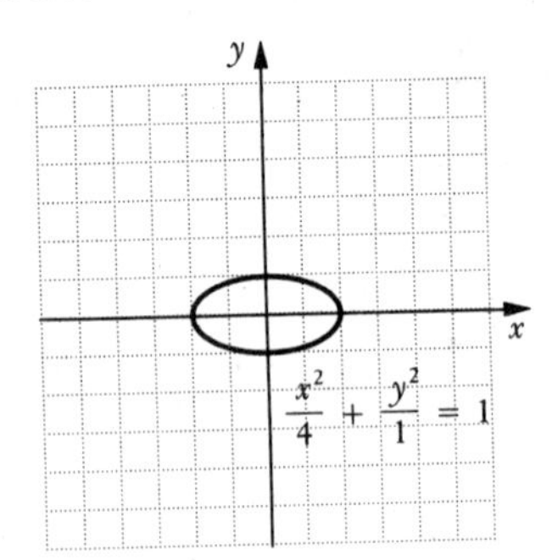

3. $\frac{x^2}{16} + \frac{y^2}{25} = 1$

$\frac{x^2}{4^2} + \frac{y^2}{5^2} = 1$

The x-intercepts are (4,0) and (-4,0), and the y-intercepts are (0,5) and (0,-5). We plot these points and connect them with an oval-shaped curve.

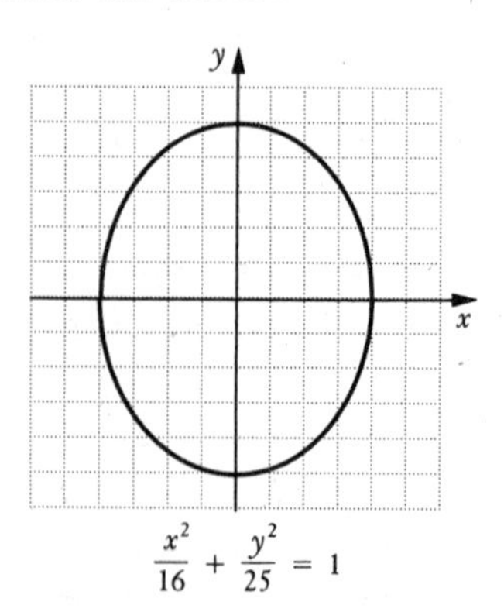

$\frac{x^2}{16} + \frac{y^2}{25} = 1$

5. $4x^2 + 9y^2 = 36$

$\frac{x^2}{9} + \frac{y^2}{4} = 1$ Dividing by 36

$\frac{x^2}{3^2} + \frac{y^2}{2^2} = 1$

The x-intercepts are (3,0) and (-3,0), and the y-intercepts are (0,2) and (0,-2). We plot these points and connect them with an oval-shaped curve.

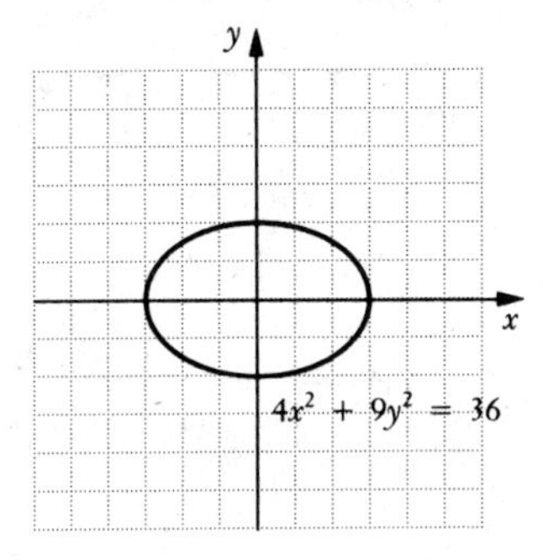

7. $x^2 + 4y^2 = 4$

$\frac{x^2}{4} + \frac{y^2}{1} = 1$ Dividing by 4

$\frac{x^2}{2^2} + \frac{y^2}{1} = 1$

The x-intercepts are (2,0) and (-2,0), and the y-intercepts are (0,1) and (0,-1). We plot these points and connect them with an oval-shaped curve.

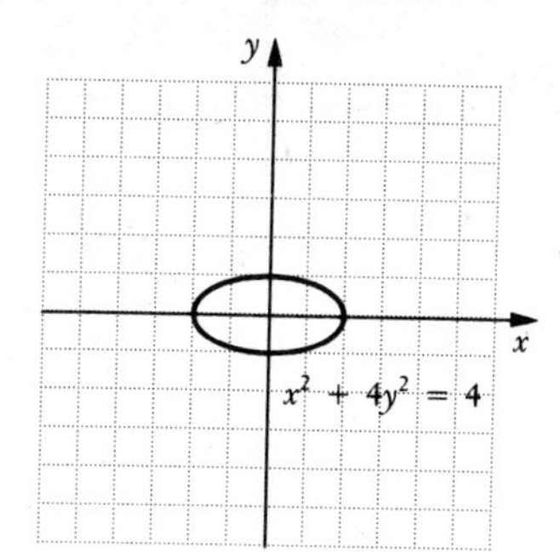

9. $\frac{x^2}{16} - \frac{y^2}{16} = 1$

$\frac{x^2}{4^2} - \frac{y^2}{4^2} = 1$

a) $a = 4$ and $b = 4$, so the asymptotes are $y = \frac{4}{4}x$ and $y = -\frac{4}{4}x$, or $y = x$ and $y = -x$. We sketch them.

b) If we let $y = 0$, we get $x = \pm 4$, so the intercepts are (4,0) and (-4,0).

c) We plot the intercepts and draw smooth curves through them that approach the asymptotes.

9. (continued)

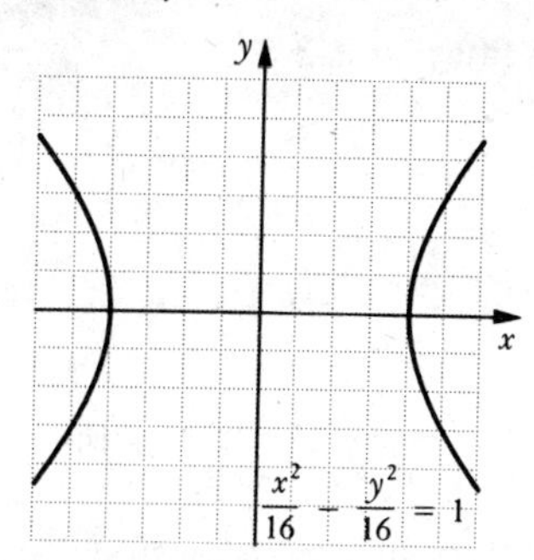

11. $\frac{y^2}{16} - \frac{x^2}{9} = 1$

$\frac{y^2}{4^2} - \frac{x^2}{3^2} = 1$

a) $a = 3$ and $b = 4$, so the asymptotes are $y = \frac{4}{3}x$ and $y = -\frac{4}{3}x$. We sketch them.

b) If we let $x = 0$, we get $y = \pm 4$, so the intercepts are (0,4) and (0,-4).

c) We plot the intercepts and draw smooth curves through them that approach the asymptotes.

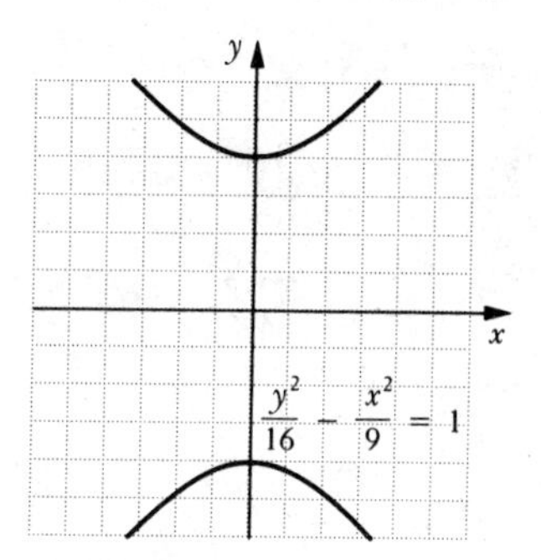

13. $\frac{x^2}{25} - \frac{y^2}{36} = 1$

$\frac{x^2}{5^2} - \frac{y^2}{6^2} = 1$

a) $a = 5$ and $b = 6$, so the asymptotes are $y = \frac{6}{5}x$ and $y = -\frac{6}{5}x$. We sketch them.

b) If we let $y = 0$, we get $x = \pm 5$, so the intercepts are (5,0) and (-5,0).

c) We plot the intercepts and draw smooth curves through them that approach the asymptotes.

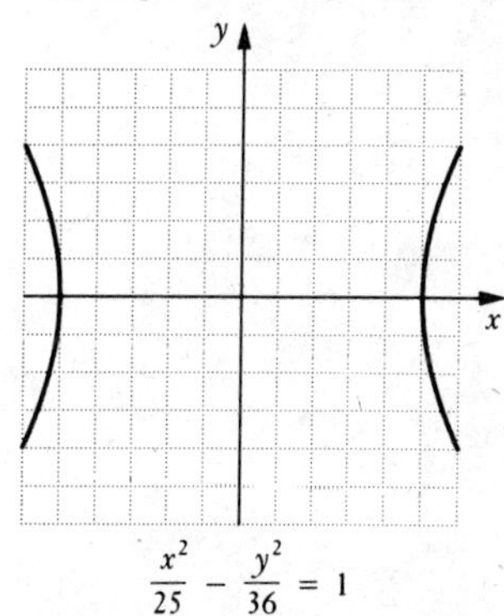

$\frac{x^2}{25} - \frac{y^2}{36} = 1$

15. $x^2 - y^2 = 4$

$\frac{x^2}{4} - \frac{y^2}{4} = 1$ Dividing by 4

$\frac{x^2}{2^2} - \frac{y^2}{2^2} = 1$

a) $a = 2$ and $b = 2$, so the asymptotes are $y = \frac{2}{2}x$ and $y = -\frac{2}{2}x$, or $y = x$ and $y = -x$. We sketch them.

b) If we let $y = 0$, we get $x = \pm 2$, so the intercepts are (2,0) and (-2,0).

c) We plot the intercepts and draw smooth curves through them that approach the asymptotes.

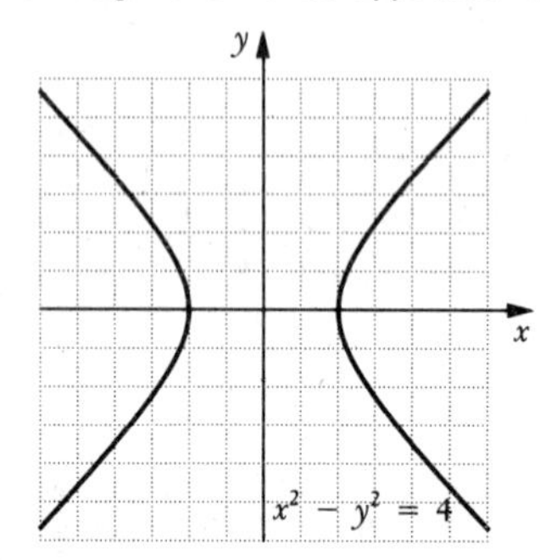

17. $\sqrt[3]{125t^{15}} = \sqrt[3]{5^3 \cdot (t^5)^3} = 5t^5$

19. $\frac{4\sqrt{2} - 5\sqrt{3}}{6\sqrt{3} - 8\sqrt{2}} = \frac{4\sqrt{2} - 5\sqrt{3}}{6\sqrt{3} - 8\sqrt{2}} \cdot \frac{6\sqrt{3} + 8\sqrt{2}}{6\sqrt{3} + 8\sqrt{2}}$

$= \frac{24\sqrt{6} + 32 \cdot 2 - 30 \cdot 3 - 40\sqrt{6}}{36 \cdot 3 - 64 \cdot 2}$

$= \frac{-26 - 16\sqrt{6}}{-20}$

$= \frac{-2(13 + 8\sqrt{6})}{-2 \cdot 10}$

$= \frac{13 + 8\sqrt{6}}{10}$

21. $y = \frac{6}{x}$

We find some solutions, keeping the results in a table.

x	y
1	6
2	3
3	2
6	1
$\frac{1}{2}$	12
$\frac{1}{3}$	18
-1	-6
-2	-3
-3	-2
-6	-1
$-\frac{1}{2}$	-12
$-\frac{1}{3}$	-18

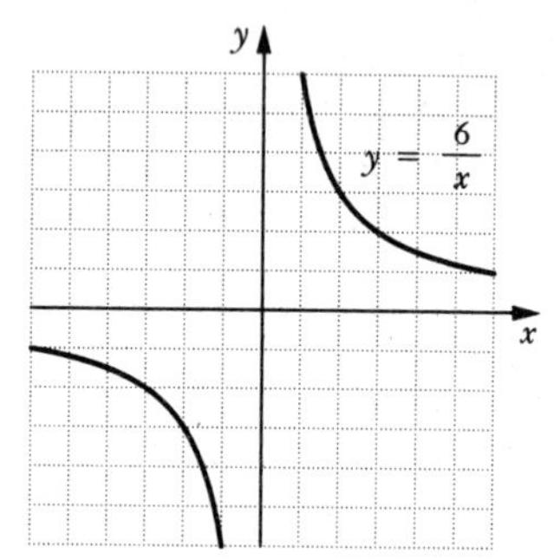

Note that we cannot use 0 for x. The x-axis and the y-axis are the asymptotes.

23. $y = -\frac{1}{x}$

x	y
1	-1
2	$-\frac{1}{2}$
4	$-\frac{1}{4}$
$\frac{1}{2}$	-2
$\frac{1}{4}$	-4
-1	1
-2	$\frac{1}{2}$
-4	$\frac{1}{4}$
$-\frac{1}{2}$	2
$-\frac{1}{4}$	4

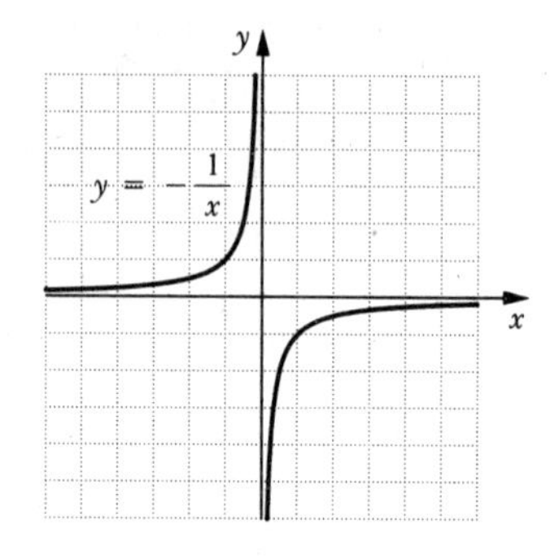

Note that we cannot use 0 for x. The x-axis and the y-axis are the asymptotes.

25.

$16x^2 + y^2 + 96x - 8y + 144 = 0$

$(16x^2 + 96x) + (y^2 - 8y) + 144 = 0$

$16(x^2 + 6x) + (y^2 - 8y) + 144 = 0$

$16(x^2+6x+9-9) + (y^2-8y+16-16) + 144 = 0$

$16(x^2+6x+9) + (y^2-8y+16) - 144 - 16 + 144 = 0$

$16(x + 3)^2 + (y - 4)^2 = 16$

$\frac{(x + 3)^2}{1} + \frac{(y - 4)^2}{16} = 1$

$\frac{[x - (-3)]^2}{1^2} + \frac{(y - 4)^2}{4^2} = 1$

25. (continued)

Center: (-3,4)
Vertices: [1 + (-3),4], [-1 + (-3),4], (-3,4 + 4), (-3,-4 + 4)

or

(-2,4), (-4,4), (-3,8), (-3,0)

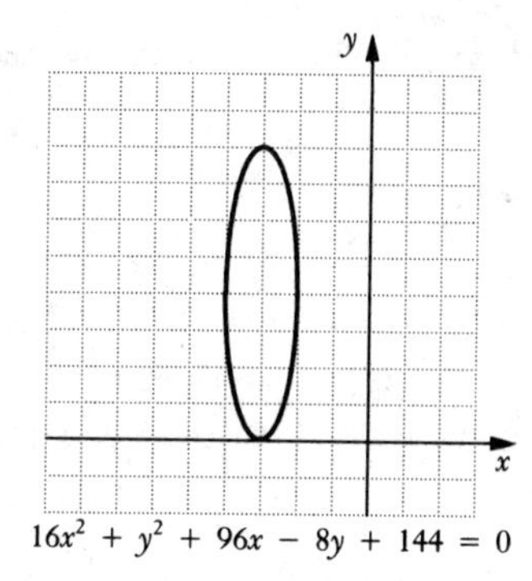

27.
$$x^2 + y^2 - 10x + 8y - 40 = 0$$
$$x^2 - 10x + y^2 + 8y = 40$$
$$(x^2 - 10x + 25) + (y^2 + 8y + 16) = 40 + 25 + 16$$
$$(x - 5)^2 + (y + 4)^2 = 81$$

The graph is a circle.

29. $1 - 3y = 2y^2 - x$
$x = 2y^2 + 3y - 1$

The graph is a parabola.

31.
$$4x^2 + 25y^2 - 8x - 100y + 4 = 0$$
$$4x^2 - 8x + 25y^2 - 100y = -4$$
$$4(x^2 - 2x + 1) + 25(y^2 - 4y + 4) = -4 + 4 + 100$$
$$4(x - 1)^2 + 25(y - 2)^2 = 100$$
$$\frac{(x - 1)^2}{25} + \frac{(y - 2)^2}{4} = 1$$

The graph is an ellipse.

33. $y = ax^2 + bx + c$ Standard form of parabola with line of symmetry parallel to the y-axis.

We substitute values for x and y.

$3 = a(0)^2 + b(0) + c$ Substituting 0 for x and 3 for y

$6 = a(-1)^2 + b(-1) + c$ Substituting -1 for x and 6 for y

$9 = a(2)^2 + b(2) + c$ Substituting 2 for x and 9 for y

or

$3 = c$

$6 = a - b + c$

$9 = 4a + 2b + c$

Next we substitute 3 for c in the second and third equations and solve the resulting system for a and b.

33. (continued)

$6 = a - b + 3$ or $a - b = 3$

$9 = 4a + 2b + 3$ $\quad 4a + 2b = 6$

Solving this system we get a = 2 and b = -1. Thus the equation of the parabola whose line of symmetry is parallel to the y-axis and passes through (0,3), (-1,6), and (2,9) is

$y = 2x^2 - x + 3$

Exercise Set 12.3

1. $x^2 + y^2 = 25,$ (1)
$y - x = 1$ (2)

First solve Eq. (2) for y.

$y = x + 1$ (3)

Then substitute x + 1 for y in Eq. (1) and solve for x.

$$x^2 + y^2 = 25$$
$$x^2 + (x + 1)^2 = 25$$
$$x^2 + x^2 + 2x + 1 = 25$$
$$2x^2 + 2x - 24 = 0$$
$$x^2 + x - 12 = 0 \quad \text{Multiplying by } \frac{1}{2}$$
$$(x + 4)(x - 3) = 0 \quad \text{Factoring}$$

$x + 4 = 0$ or $x - 3 = 0$ Principle of zero products
$x = -4$ or $x = 3$

Now substitute these numbers into Eq. (3) and solve for y.

$y = -4 + 1 = -3$

$y = 3 + 1 = 4$

The pairs (-4,-3) and (3,4) check, so they are the solutions.

3. $4x^2 + 9y^2 = 36,$ (1)
$3y + 2x = 6$ (2)

First solve Eq. (2) for y.

$3y = -2x + 6$

$y = -\frac{2}{3}x + 2$ (3)

Then substitute $-\frac{2}{3}x + 2$ for y in Eq. (1) and solve for x.

$$4x^2 + 9y^2 = 36$$
$$4x^2 + 9\left[-\frac{2}{3}x + 2\right]^2 = 36$$
$$4x^2 + 9\left[\frac{4}{9}x^2 - \frac{8}{3}x + 4\right] = 36$$
$$4x^2 + 4x^2 - 24x + 36 = 36$$
$$8x^2 - 24x = 0$$
$$x^2 - 3x = 0$$
$$x(x - 3) = 0$$

$x = 0$ or $x = 3$

Now substitute these numbers in Eq. (3) and solve for y.

3. (continued)

$y = -\frac{2}{3} \cdot 0 + 2 = 2$

$y = -\frac{2}{3} \cdot 3 + 2 = 0$

The pairs (0,2) and (3,0) check, so they are the solutions.

5. $y^2 = x + 3$, (1)

$2y = x + 4$ (2)

First solve Eq. (2) for x.

$2y - 4 = x$ (3)

Then substitute $2y - 4$ for x in Eq. (1) and solve for y.

$$y^2 = x + 3$$
$$y^2 = (2y - 4) + 3$$
$$y^2 = 2y - 1$$
$$y^2 - 2y + 1 = 0$$
$$(y - 1)(y - 1) = 0$$

$y = 1$ or $y = 1$

Now substitute 1 for y in Eq. (3) and solve for x.

$2 \cdot 1 - 4 = x$

$-2 = x$

The pair (-2,1) checks. It is the solution.

7. $x^2 - xy + 3y^2 = 27$, (1)

$x - y = 2$ (2)

First solve Eq. (2) for y.

$x - 2 = y$ (3)

Then substitute $x - 2$ for y in Eq. (1) and solve for x.

$$x^2 - xy + 3y^2 = 27$$
$$x^2 - x(x - 2) + 3(x - 2)^2 = 27$$
$$x^2 - x^2 + 2x + 3x^2 - 12x + 12 = 27$$
$$3x^2 - 10x - 15 = 0$$

$$x = \frac{-(-10) \pm \sqrt{(-10)^2 - 4(3)(-15)}}{2 \cdot 3}$$
$$= \frac{10 \pm \sqrt{100 + 180}}{6}$$
$$= \frac{10 \pm \sqrt{280}}{6}$$
$$= \frac{10 \pm 2\sqrt{70}}{6}$$
$$= \frac{5 \pm \sqrt{70}}{3}$$

Now substitute these numbers in Eq. (3) and solve for y.

$y = \frac{5 + \sqrt{70}}{3} - 2 = \frac{-1 + \sqrt{70}}{3}$

$y = \frac{5 - \sqrt{70}}{3} - 2 = \frac{-1 - \sqrt{70}}{3}$

The pairs $\left(\frac{5 + \sqrt{70}}{3}, \frac{-1 + \sqrt{70}}{3}\right)$ and $\left(\frac{5 - \sqrt{70}}{3}, \frac{-1 - \sqrt{70}}{3}\right)$ check, so they are the solutions.

9. $x^2 - xy + 3y^2 = 5$, (1)

$x - y = 2$ (2)

First solve Eq. (2) for y.

$x - 2 = y$ (3)

Then substitute $x - 2$ for y in Eq. (1) and solve for x.

$$x^2 - xy + 3y^2 = 5$$
$$x^2 - x(x - 2) + 3(x - 2)^2 = 5$$
$$x^2 - x^2 + 2x + 3x^2 - 12x + 12 = 5$$
$$3x^2 - 10x + 7 = 0$$
$$(3x - 7)(x - 1) = 0$$

$x = \frac{7}{3}$ or $x = 1$

Now substitute these numbers in Eq. (3) and solve for y.

$y = \frac{7}{3} - 2 = \frac{1}{3}$

$y = 1 - 2 = -1$

The pairs $\left(\frac{7}{3}, \frac{1}{3}\right)$ and (1,-1) check, so they are the solutions.

11. $2y^2 + xy = 5$, (1)

$4y + x = 7$ (2)

First solve Eq. (2) for x.

$x = -4y + 7$ (3)

Then substitute $-4y + 7$ for x in Eq. (1) and solve for y.

$$2y^2 + (-4y + 7)y = 5$$
$$2y^2 - 4y^2 + 7y = 5$$
$$0 = 2y^2 - 7y + 5$$
$$0 = (2y - 5)(y - 1)$$

$y = \frac{5}{2}$ or $y = 1$

Now substitute these numbers in Eq. (3) and solve for x.

$x = -4\left(\frac{5}{2}\right) + 7 = -3$

$x = -4(1) + 7 = 3$

The pairs $\left(-3, \frac{5}{2}\right)$ and (3,1) check, so they are the solutions.

13. $2a + b = 1,$ (1)
$b = 4 - a^2$ (2)

Eq. (2) is already solved for b. Substitute $4 - a^2$ for b in Eq. (1) and solve for a.

$2a + 4 - a^2 = 1$
$0 = a^2 - 2a - 3$
$0 = (a - 3)(a + 1)$

$a = 3$ or $a = -1$

Substitute these numbers in Eq. (2) and solve for b.

$b = 4 - 3^2 = -5$
$b = 4 - (-1)^2 = 3$

The pairs (3,-5) and (-1,3) check.

15. $x^2 + y^2 = 5,$ (1)
$x - y = 8$ (2)

First solve Eq. (2) for x.

$x = y + 8$ (3)

Then substitute y + 8 for x in Eq. (1) and solve for y.

$(y + 8)^2 + y^2 = 5$
$y^2 + 16y + 64 + y^2 = 5$
$2y^2 + 16y + 59 = 0$

$y = \frac{-16 \pm \sqrt{(16)^2 - 4(2)(59)}}{2\cdot 2}$

$y = \frac{-16 \pm \sqrt{-216}}{4}$

$y = \frac{-16 \pm 6i\sqrt{6}}{4}$

$y = \frac{-8 \pm 3i\sqrt{6}}{2}$, or $-4 \pm \frac{3}{2}i\sqrt{6}$

Now substitute these numbers in Eq. (3) and solve for x.

$x = -4 + \frac{3}{2}i\sqrt{6} + 8 = 4 + \frac{3}{2}i\sqrt{6}$, or $\frac{8 + 3i\sqrt{6}}{2}$

$x = -4 - \frac{3}{2}i\sqrt{6} + 8 = 4 - \frac{3}{2}i\sqrt{6}$, or $\frac{8 - 3i\sqrt{6}}{2}$

The pairs $\left(4 + \frac{3}{2}i\sqrt{6}, -4 + \frac{3}{2}i\sqrt{6}\right)$ and $\left(4 - \frac{3}{2}i\sqrt{6}, -4 - \frac{3}{2}i\sqrt{6}\right)$, or $\left(\frac{8 + 3i\sqrt{6}}{2}, \frac{-8 + 3i\sqrt{6}}{2}\right)$ and $\left(\frac{8 - 3i\sqrt{6}}{2}, \frac{-8 - 3i\sqrt{6}}{2}\right)$ check.

17. $x^2 + y^2 = 25,$ (1)
$y^2 = x + 5$ (2)

We substitute x + 5 for y^2 in Eq. (1) and solve for x.

$x^2 + y^2 = 25$
$x^2 + (x + 5) = 25$
$x^2 + x - 20 = 0$
$(x + 5)(x - 4) = 0$

$x + 5 = 0$ or $x - 4 = 0$
$x = -5$ or $x = 4$

Next we substitute these numbers for x in either Eq. (1) or Eq. (2) and solve for y. Here we use Eq. (2).

$y^2 = -5 + 5 = 0$ and $y = 0$.
$y^2 = 4 + 5 = 9$ and $y = \pm 3$.

The possible solutions are (-5,0), (4,3), and (4,-3).

Check:

For (-5,0):

$x^2 + y^2 = 25$	
$(-5)^2 + 0^2$	25
25 + 0	
25	

$y^2 = x + 5$	
0^2	-5 + 5
0	0

For (4,3):

$x^2 + y^2 = 25$	
$4^2 + 3^2$	25
16 + 9	
25	

$y^2 = x + 5$	
3^2	4 + 5
9	9

For (4,-3):

$x^2 + y^2 = 25$	
$4^2 + (-3)^2$	25
16 + 9	
25	

$y^2 = x + 5$	
$(-3)^2$	4 + 5
9	9

The solutions are (-5,0), (4,3), and (4,-3).

19. $x^2 + y^2 = 9,$ (1)
$x^2 - y^2 = 9$ (2)

Here we use the addition method.

$x^2 + y^2 = 9$
$x^2 - y^2 = 9$
$2x^2 = 18$ Adding
$x^2 = 9$
$x = \pm 3$

If $x = 3$, $x^2 = 9$, and if $x = -3$, $x^2 = 9$, so substituting 3 or -3 in Eq. (1) give us

$x^2 + y^2 = 9$
$9 + y^2 = 9$
$y^2 = 0$
$y = 0.$

The possible solutions are (3,0) and (-3,0).

19. (continued)

Check:

$x^2 + y^2 = 9$	
$(\pm3)^2 + (0)^2$	9
$9 + 0$	
9	

$x^2 - y^2 = 9$	
$(\pm3)^2 - (0)^2$	9
$9 - 0$	
9	

The solutions are (3,0) and (-3,0).

21. $x^2 + y^2 = 5,$ (1)

$xy = 2$ (2)

First we solve Eq. (2) for y.

$xy = 2$

$y = \frac{2}{x}$

Then we substitute $\frac{2}{x}$ for y in Eq. (1) and solve for x.

$x^2 + y^2 = 5$

$x^2 + \left(\frac{2}{x}\right)^2 = 5$

$x^2 + \frac{4}{x^2} = 5$

$x^4 + 4 = 5x^2$ Multiplying by x^2

$x^4 - 5x^2 + 4 = 0$

$u^2 - 5u + 4 = 0$ Letting $u = x^2$

$(u - 4)(u - 1) = 0$

$u = 4$ or $u = 1$

We now substitute x^2 for u and solve for x.

$x^2 = 4$ or $x^2 = 1$

$x = \pm2$ $x = \pm1$

Since y = 2/x, if x = 2, y = 1; if x = -2, y = -1; if x = 1, y = 2; and if x = -1, y = -2. The pairs (2,1), (-2,-1), (1,2), (-1,-2) check. They are the solutions.

23. $x^2 + y^2 = 13,$ (1)

$xy = 6$ (2)

First we solve Eq. (2) for y.

$xy = 6$

$y = \frac{6}{x}$

Then we substitute $\frac{6}{x}$ for y in Eq. (1) and solve for x.

$x^2 + y^2 = 13$

$x^2 + \left(\frac{6}{x}\right)^2 = 13$

$x^2 + \frac{36}{x^2} = 13$

$x^4 + 36 = 13x^2$ Multiplying by x^2

$x^4 - 13x^2 + 36 = 0$

$u^2 - 13u + 36 = 0$ Letting $u = x^2$

$(u - 9)(u - 4) = 0$

$u = 9$ or $u = 4$

23. (continued)

We now substitute x^2 for u and solve for x.

$x^2 = 9$ or $x^2 = 4$

$x = \pm3$ $x = \pm2$

Since y = 6/x, if x = 3, y = 2; if x = -3, y = -2; if x = 2, y = 3; and if x = -2, y = -3. The pairs (3,2), (-3,-2), (2,3), (-2,-3) check. They are the solutions.

25. $2xy + 3y^2 = 7,$ (1)

$3xy - 2y^2 = 4$ (2)

$6xy + 9y^2 = 21$ Multiplying (1) by 3

$-6xy + 4y^2 = -8$ Multiplying (2) by -2

$13y^2 = 13$

$y^2 = 1$

$y = \pm1$

Substitute for y in Eq. (1) and solve for x.

When y = 1: $2 \cdot x \cdot 1 + 3 \cdot 1^2 = 7$

$2x = 4$

$x = 2$

When y = -1: $2 \cdot x \cdot (-1) + 3(-1)^2 = 7$

$-2x = 4$

$x = -2$

The pairs (2,1) and (-2,-1) check. They are the solutions.

27. $4a^2 - 25b^2 = 0,$ (1)

$2a^2 - 10b^2 = 3b + 4$ (2)

$4a^2 - 25b^2 = 0$

$-4a^2 + 20b^2 = -6b - 8$ Multiplying (2) by -2

$-5b^2 = -6b - 8$

$0 = 5b^2 - 6b - 8$

$0 = (5b + 4)(b - 2)$

$b = -\frac{4}{5}$ or $b = 2$

Substitute for b in Eq. (1) and solve for a.

When $b = -\frac{4}{5}$: $4a^2 - 25\left(-\frac{4}{5}\right)^2 = 0$

$4a^2 = 16$

$a^2 = 4$

$a = \pm2$

When b = 2: $4a^2 - 25(2)^2 = 0$

$4a^2 = 100$

$a^2 = 25$

$a = \pm5$

The pairs $\left(2,-\frac{4}{5}\right)$, $\left(-2,-\frac{4}{5}\right)$, (5,2) and (-5,2) check. They are the solutions.

29. $ab - b^2 = -4,$ (1)
$ab - 2b^2 = -6$ (2)

$$\begin{aligned} ab - b^2 &= -4 \\ -ab + 2b^2 &= 6 \quad \text{Multiplying (2) by -1} \\ \hline b^2 &= 2 \\ b &= \pm\sqrt{2} \end{aligned}$$

Substitute for b in Eq. (1) and solve for a.

When $b = \sqrt{2}$: $a(\sqrt{2}) - (\sqrt{2})^2 = -4$

$$a\sqrt{2} = -2$$

$$a = -\frac{2}{\sqrt{2}} = -\sqrt{2}$$

When $b = -\sqrt{2}$: $a(-\sqrt{2}) - (-\sqrt{2})^2 = -4$

$$-a\sqrt{2} = -2$$

$$a = \frac{-2}{-\sqrt{2}} = \sqrt{2}$$

The pairs $(-\sqrt{2},\sqrt{2})$ and $(\sqrt{2},-\sqrt{2})$ check. They are the solutions.

31. $x^2 + y^2 = 25,$ (1)
$9x^2 + 4y^2 = 36$ (2)

$$\begin{aligned} -4x^2 - 4y^2 &= -100 \quad \text{Multiplying (1) by -4} \\ 9x^2 + 4y^2 &= 36 \\ \hline 5x^2 &= -64 \end{aligned}$$

$$x^2 = -\frac{64}{5}$$

$$x = \pm\sqrt{\frac{-64}{5}} = \pm\frac{8i}{\sqrt{5}}$$

$$x = \pm\frac{8i\sqrt{5}}{5} \quad \text{Rationalizing the denominator}$$

Substituting $\frac{8i\sqrt{5}}{5}$ or $-\frac{8i\sqrt{5}}{5}$ for x in Eq. (1) and solving for y gives us

$$-\frac{64}{5} + y^2 = 25$$

$$y^2 = \frac{189}{5}$$

$$y = \pm\sqrt{\frac{189}{5}} = \pm3\sqrt{\frac{21}{5}}$$

$$y = \pm\frac{3\sqrt{105}}{5}. \quad \text{Rationalizing the denominator}$$

The pairs $\left(\frac{8i\sqrt{5}}{5}, \frac{3\sqrt{105}}{5}\right)$, $\left(-\frac{8i\sqrt{5}}{5}, \frac{3\sqrt{105}}{5}\right)$, $\left(\frac{8i\sqrt{5}}{5}, -\frac{3\sqrt{105}}{5}\right)$, and $\left(-\frac{8i\sqrt{5}}{5}, -\frac{3\sqrt{105}}{5}\right)$ check. They are the solutions.

33. Familiarize. We first make a drawing. We let ℓ and w represent the length and width, respectively.

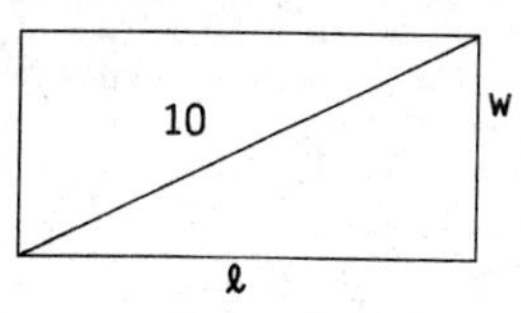

Translate. The perimeter is 28 cm.

$2\ell + 2w = 28$, or $\ell + w = 14$

Using the Pythagorean theorem we have another equation.

$\ell^2 + w^2 = 10^2$, or $\ell^2 + w^2 = 100$

Solve. We solve the system:

$\ell + w = 14,$ (1)
$\ell^2 + w^2 = 100$ (2)

First solve Eq. (1) for w.

$w = 14 - \ell$ (3)

Then substitute $14 - \ell$ for w in Eq. (2) and solve for ℓ.

$$\begin{aligned} \ell^2 + w^2 &= 100 \\ \ell^2 + (14 - \ell)^2 &= 100 \\ \ell^2 + 196 - 28\ell + \ell^2 &= 100 \\ 2\ell^2 - 28\ell + 96 &= 0 \\ \ell^2 - 14\ell + 48 &= 0 \\ (\ell - 8)(\ell - 6) &= 0 \end{aligned}$$

$\ell = 8$ or $\ell = 6$

If $\ell = 8$, then $w = 14 - 8$, or 6. If $\ell = 6$, then $w = 14 - 6$, or 8. Since the length is usually considered to be longer than the width, we have the solution $\ell = 8$ and $w = 6$, or (8,6).

Check. If $\ell = 8$ and $w = 6$, then the perimeter is $2\cdot8 + 2\cdot6$, or 28. The length of a diagonal is $\sqrt{8^2 + 6^2}$, or $\sqrt{100}$, or 10. The numbers check.

State. The length is 8 cm, and the width is 6 cm.

35. Familiarize. We first make a drawing. Let ℓ = the length and w = the width of the rectangle.

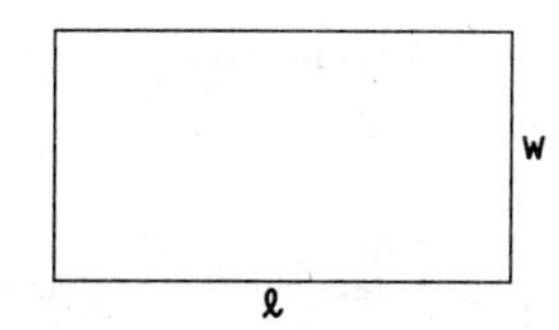

Translate.

Area: $\ell w = 20$

Perimeter: $2\ell + 2w = 18$, or $\ell + w = 9$

Solve. We solve the system:

Solve the second equation for ℓ: $\ell = 9 - w$

Substitute $9 - w$ for ℓ in the first equation and solve for w.

$$\begin{aligned} (9 - w)w &= 20 \\ 9w - w^2 &= 20 \\ 0 &= w^2 - 9w + 20 \\ 0 &= (w - 5)(w - 4) \end{aligned}$$

35. (continued)

$w = 5$ or $w = 4$

If $w = 5$, then $\ell = 9 - w$, or 4. If $w = 4$, then $\ell = 9 - 4$, or 5. Since length is usually considered to be longer than width, we have the solution $\ell = 5$ and $w = 4$, or (5,4).

Check. If $\ell = 5$ and $w = 4$, the area is $5 \cdot 4$, or 20. The perimeter is $2 \cdot 5 + 2 \cdot 4$, or 18. The numbers check.

State. The length is 5 in. and the width is 4 in.

37. Familiarize. We make a drawing of the field. Let ℓ = the length and w = the width.

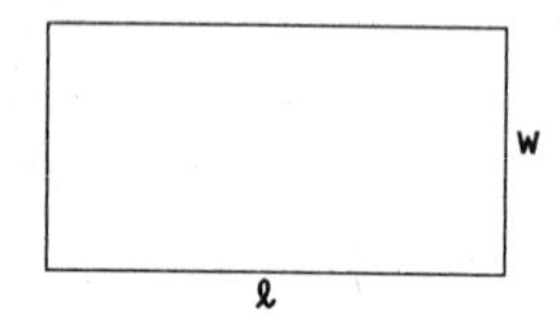

Since it takes 210 yd of fencing to enclose the field, we know that the perimeter is 210 yd.

Translate.

Perimeter: $2\ell + 2w = 210$, or $\ell + w = 105$

Area: $\ell w = 2250$

Solve. We solve the system:

Solve the first equation for ℓ: $\ell = 105 - w$

Substitute $105 - w$ for ℓ in the second equation and solve for w.

$$(105 - w)w = 2250$$
$$105w - w^2 = 2250$$
$$0 = w^2 - 105w + 2250$$
$$0 = (w - 30)(w - 75)$$

$w = 30$ or $w = 75$

If $w = 30$, then $\ell = 105 - 30$, or 75. If $w = 75$, then $\ell = 105 - 75$, or 30. Since length is usually considered to be longer than width, we have the solution $\ell = 75$ and $w = 30$, or (75,30).

Check. If $\ell = 75$ and $w = 30$, the perimeter is $2 \cdot 75 + 2 \cdot 30$, or 210. The area is 75(30), or 2250. The numbers check.

State. The length is 75 yd and the width is 30 yd.

39. Familiarize. We make a drawing and label it. Let x and y represent the lengths of the legs of the triangle.

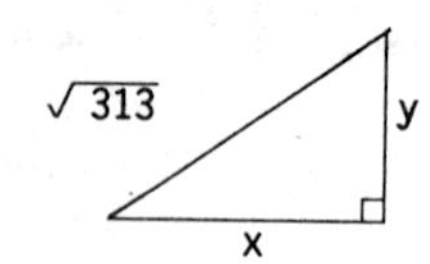

Translate. The product of the lengths of the legs is 156, so we have:

$$xy = 156$$

We use the Pythagorean theorem to get a second equation:

$$x^2 + y^2 = (\sqrt{313})^2, \text{ or } x^2 + y^2 = 313$$

39. (continued)

Solve. We solve the system of equations

$$xy = 156, \quad (1)$$
$$x^2 + y^2 = 313. \quad (2)$$

First solve Eq. (1) for y.

$$xy = 156$$
$$y = \frac{156}{x}$$

Then we substitute $\frac{156}{x}$ for y in Eq. (2) and solve for x.

$$x^2 + y^2 = 313 \quad (2)$$
$$x^2 + \left(\frac{156}{x}\right)^2 = 313$$
$$x^2 + \frac{24,336}{x^2} = 313$$
$$x^4 + 24,336 = 313x^2$$
$$x^4 - 313x^2 + 24,336 = 0$$
$$u^2 - 313u + 24,336 = 0 \quad \text{Letting } u = x^2$$
$$(u - 169)(u - 144) = 0$$

$u = 169$ or $u = 144$

We now substitute x^2 for u and solve for x.

$x^2 = 169$ or $x^2 = 144$

$x = \pm 13$ or $x = \pm 12$

Since $y = 156/x$, if $x = 13$, $y = 12$; if $x = -13$, $y = -12$; if $x = 12$, $y = 13$; and if $x = -12$, $y = -13$. The possible solutions are (13,12), (-13,-12), (12,13), and (-12,-13).

Check. Since negative lengths do not make sense in this problem, we consider only (13,12) and (12,13). Since both possible solutions give the same pair of legs, we only need to check (13,12). If $x = 13$ and $y = 12$, their product is 156. Also, $\sqrt{13^2 + 12^2} = \sqrt{313}$. The numbers check.

State. The lengths of the legs are 13 and 12.

41. Familiarize. We let x = the length of a side of one peanut bed and y = the length of a side of the other peanut bed. Make a drawing.

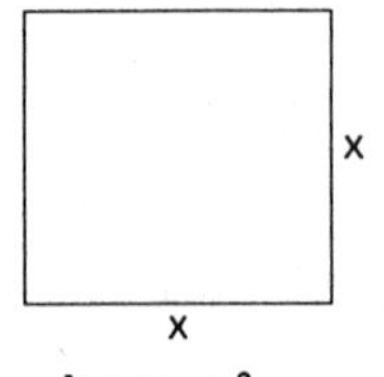

Area: x^2

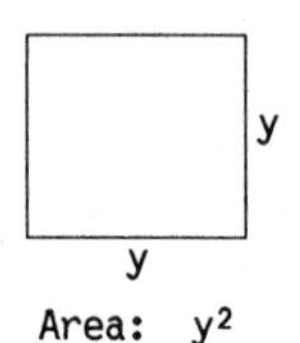

Area: y^2

Translate.

The sum of the areas is 832 ft².

$$x^2 + y^2 = 832$$

The difference of the areas is 320 ft².

$$x^2 - y^2 = 320$$

41. (continued)

Solve. We solve the system of equations.

$$\begin{aligned} x^2 + y^2 &= 832 \\ x^2 - y^2 &= 320 \\ \hline 2x^2 &= 1152 \quad \text{Adding} \\ x^2 &= 576 \\ x &= \pm 24 \end{aligned}$$

Since length cannot be negative, we consider only x = 24. Substitute 24 for x in the first equation and solve for y.

$$24^2 + y^2 = 832$$
$$576 + y^2 = 832$$
$$y^2 = 256$$
$$y = \pm 16$$

Again, we consider only the positive value, 16. The possible solution is (24,16).

Check. The areas of the peanut beds are 24^2, or 576, and 16^2, or 256. The sum of the areas is 576 + 256, or 832. The difference of the areas is 576 - 256, or 320. The values check.

State. The lengths of the beds are 24 ft and 16 ft.

43. Familiarize. We first make a drawing. Let ℓ = the length and w = the width.

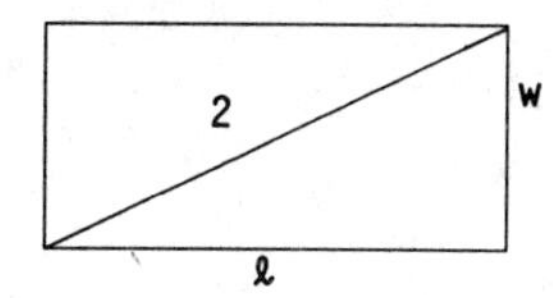

Translate.

Area: $\ell w = \sqrt{3}$ (1)

From the Pythagorean theorem: $\ell^2 + w^2 = 2^2$ (2)

Solve. We solve the system of equations.

We first solve Eq. (1) for w.

$$\ell w = \sqrt{3}$$
$$w = \frac{\sqrt{3}}{\ell}$$

Then we substitute $\frac{\sqrt{3}}{\ell}$ for w in Eq. (2) and solve for ℓ.

$$\ell^2 + \left[\frac{\sqrt{3}}{\ell}\right]^2 = 4$$
$$\ell^2 + \frac{3}{\ell^2} = 4$$
$$\ell^4 + 3 = 4\ell^2$$
$$\ell^4 - 4\ell^2 + 3 = 0$$
$$u^2 - 4u + 3 = 0 \quad \text{Letting } u = \ell^2$$
$$(u - 3)(u - 1) = 0$$
$$u = 3 \quad \text{or} \quad u = 1$$

We now substitute ℓ^2 for u and solve for ℓ.

$$\ell^2 = 3 \quad \text{or} \quad \ell^2 = 1$$
$$\ell = \pm\sqrt{3} \quad \text{or} \quad \ell = \pm 1$$

43. (continued)

Length cannot be negative, so we only need to consider $\ell = \sqrt{3}$ and $\ell = 1$. Since $w = \sqrt{3}/\ell$, if $\ell = \sqrt{3}$, w = 1 and if $\ell = 1$, $w = \sqrt{3}$. Length is usually considered to be longer than width, so we have the solution $\ell = \sqrt{3}$ and w = 1, or $(\sqrt{3},1)$.

Check. If $\ell = \sqrt{3}$ and w = 1, the area is $\sqrt{3}\cdot 1 = \sqrt{3}$. Also $(\sqrt{3})^2 + 1^2 = 3 + 1 = 4 = 2^2$. The numbers check.

State. The length is $\sqrt{3}$ m, and the width is 1 m.

45.
$$3x^2 + 6 = 5x$$
$$3x^2 - 5x + 6 = 0$$
$$x = \frac{-(-5) \pm \sqrt{(-5)^2 - 4\cdot 3\cdot 6}}{2\cdot 3}$$
$$x = \frac{5 \pm \sqrt{-47}}{6}$$
$$x = \frac{5 \pm i\sqrt{47}}{6}$$

47. $\sqrt{48} = \sqrt{16\cdot 3} = 4\sqrt{3}$

49. $\frac{\sqrt{x} - \sqrt{h}}{\sqrt{x} + \sqrt{h}} = \frac{\sqrt{x} - \sqrt{h}}{\sqrt{x} + \sqrt{h}} \cdot \frac{\sqrt{x} + \sqrt{h}}{\sqrt{x} + \sqrt{h}} = \frac{x - h}{x + 2\sqrt{xh} + h}$

51. Familiarize. Let r represent the speed of the boat in still water and t represent the time of the trip upstream. Organize the information in a table.

	Speed	Time	Distance
Upstream	r - 2	t	4
Downstream	r + 2	3 - t	4

Recall that rt = d, or t = d/r.

Translate. From the first line of the table we obtain $t = \frac{4}{r - 2}$. From the second line we obtain $3 - t = \frac{4}{r + 2}$.

Solve. Substitute $\frac{4}{r - 2}$ for t in the second equation and solve for r.

$$3 - \frac{4}{r - 2} = \frac{4}{r + 2}$$
LCM is (r - 2)(r + 2)
$$(r - 2)(r + 2)\left[3 - \frac{4}{r - 2}\right] = (r - 2)(r + 2)\cdot\frac{4}{r + 2}$$
$$3(r - 2)(r + 2) - 4(r + 2) = 4(r - 2)$$
$$3(r^2 - 4) - 4r - 8 = 4r - 8$$
$$3r^2 - 12 - 4r - 8 = 4r - 8$$
$$3r^2 - 8r - 12 = 0$$

51. (continued)

$$r = \frac{-(-8) \pm \sqrt{(-8)^2 - 4(3)(-12)}}{2\cdot 3}$$

$$r = \frac{8 \pm \sqrt{208}}{6} = \frac{8 \pm 4\sqrt{13}}{6}$$

$$r = \frac{4 \pm 2\sqrt{13}}{3}$$

Since negative speed has no meaning in this problem, we consider only the positive square root.

$$r = \frac{4 + 2\sqrt{13}}{3} \approx 3.7$$

Check. Left to the student.

State. The speed of the boat in still water is approximately 3.7 mph.

53. It is helpful to draw a picture.

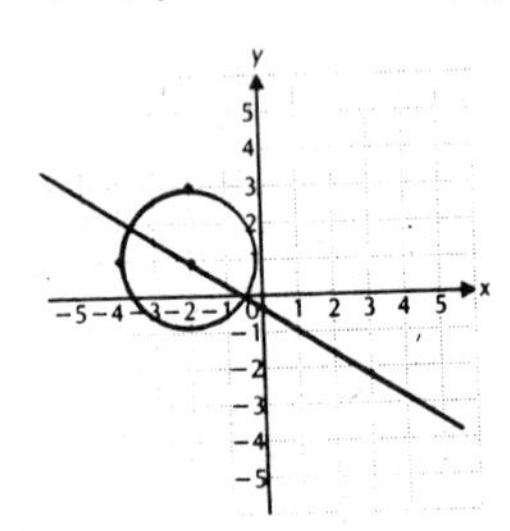

Let (h,k) represent the point on the line $5x + 8y = -2$ which is the center of a circle that passes through the points (-2,3) and (-4,1). The distance between (h,k) and (-2,3) is the same as the distance between (h,k) and (-4,1). This gives us one equation:

$$\sqrt{[h - (-2)]^2 + (k - 3)^2} = \sqrt{[h - (-4)]^2 + (k - 1)^2}$$

$$(h + 2)^2 + (k - 3)^2 = (h + 4)^2 + (k - 1)^2$$

$$h^2 + 4h + 4 + k^2 - 6k + 9 = h^2 + 8h + 16 + k^2 - 2k + 1$$

$$4h - 6k + 13 = 8h - 2k + 17$$

$$-4h - 4k = 4$$

$$h + k = -1$$

We get a second equation by substituting (h,k) in $5x + 8y = -2$.

$$5h + 8k = -2$$

We now solve the following system:

$$h + k = -1$$

$$5h + 8k = -2$$

The solution, which is the center of the circle, is (-2,1).

Next we find the length of the radius. We can find the distance between either (-2,3) or (-4,1) and the center (-2,1). We use (-2,3).

$$r = \sqrt{[-2 - (-2)]^2 + (1 - 3)^2}$$

$$= \sqrt{0^2 + (-2)^2}$$

$$= \sqrt{4}$$

$$= 2$$

53. (continued)

We now have an equation knowing that the center is (-2,1) and the radius is 2.

$$(x - h)^2 + (y - k)^2 = r^2$$

$$[x - (-2)]^2 + (y - 1)^2 = 2^2$$

$$(x + 2)^2 + (y - 1)^2 = 4$$

55. $\frac{x^2}{a^2} + \frac{y^2}{b^2} = 1$ Standard form

Substitute the coordinates of the given points:

$$\frac{2^2}{a^2} + \frac{(-3)^2}{b^2} = 1,$$

$$\frac{1^2}{a^2} + \frac{(\sqrt{13})^2}{b^2} = 1, \text{ or}$$

$$\frac{4}{a^2} + \frac{9}{b^2} = 1, \qquad (1)$$

$$\frac{1}{a^2} + \frac{13}{b^2} = 1 \qquad (2)$$

Solve Eq. (2) for $\frac{1}{a^2}$: $\frac{1}{a^2} = 1 - \frac{13}{b^2}$

$$\frac{1}{a^2} = \frac{b^2 - 13}{b^2} \qquad (3)$$

Substitute $\frac{b^2 - 13}{b^2}$ for $\frac{1}{a^2}$ in Eq. (1) and solve for b^2.

$$4\left(\frac{b^2 - 13}{b^2}\right) + \frac{9}{b^2} = 1$$

$$\frac{4b^2 - 52}{b^2} + \frac{9}{b^2} = 1$$

$$4b^2 - 52 + 9 = b^2$$

$$3b^2 = 43$$

$$b^2 = \frac{43}{3}$$

Substitute $\frac{43}{3}$ for b^2 in Eq. (3) and solve for a^2.

$$\frac{1}{a^2} = \frac{\frac{43}{3} - 13}{\frac{43}{3}} = \frac{\frac{43}{3} - 13}{\frac{43}{3}} \cdot \frac{3}{3}$$

$$\frac{1}{a^2} = \frac{43 - 3\cdot 13}{43} = \frac{43 - 39}{43}$$

$$\frac{1}{a^2} = \frac{4}{43}$$

$$a^2 = \frac{43}{4}$$

The equation of the ellipse is

$$\frac{x^2}{\frac{43}{4}} + \frac{y^2}{\frac{43}{3}} = 1, \text{ or}$$

$$\frac{4x^2}{43} + \frac{3y^2}{43} = 1, \text{ or}$$

$$4x^2 + 3y^2 = 43.$$

57.

$$R = C$$

$$100x - x^2 = 20x + 1500$$

$$0 = x^2 - 80x + 1500$$

$$0 = (x - 30)(x - 50)$$

$x = 30$ or $x = 50$

The break-even points are 30 and 50.

59. $a + b = \frac{5}{6}$, (1)

$\frac{a}{b} + \frac{b}{a} = \frac{13}{6}$ (2)

$b = \frac{5}{6} - a = \frac{5 - 6a}{6}$ Solving (1) for b

$\frac{a}{\frac{5 - 6a}{6}} + \frac{\frac{5 - 6a}{6}}{a} = \frac{13}{6}$ Substituting for b in (2)

$\frac{6a}{5 - 6a} + \frac{5 - 6a}{6a} = \frac{13}{6}$

$36a^2 + 25 - 60a + 36a^2 = 65a - 78a^2$

$150a^2 - 125a + 25 = 0$

$6a^2 - 5a + 1 = 0$

$(3a - 1)(2a - 1) = 0$

$a = \frac{1}{3}$ or $a = \frac{1}{2}$

Substitute for a and solve for b.

When $a = \frac{1}{3}$, $b = \frac{5 - 6\left(\frac{1}{3}\right)}{6} = \frac{1}{2}$.

When $a = \frac{1}{2}$, $b = \frac{5 - 6\left(\frac{1}{2}\right)}{6} = \frac{1}{3}$.

The pairs $\left(\frac{1}{3},\frac{1}{2}\right)$ and $\left(\frac{1}{2},\frac{1}{3}\right)$ check. They are the solutions.

Exercise Set 12.4

1. Each arrow gives an ordered pair. Thus we have the following:

   d = {(New York, Mets), (New York, Yankees), (Los Angeles, Rams), (Los Angeles, Raiders), (Houston, Astros)}

3. Each arrow gives an ordered pair. Thus we have the following:

   f = {(-4,5), (-5,5), (-6,5), (-7,2)}

5. The relation is a function because no two ordered pairs have the same first coordinate and different second coordinates.

7. The relation is not a function because the ordered pairs (20,7) and (20,0) have the same first coordinates and different second coordinates.

9. The domain is the set of all first numbers of ordered pairs in the relation. The domain of c = {9,-5,2,3}.

   The range is the set of all second members of ordered pairs in the relation. The range of c = {1,-2,-1,-9}.

11. Domain of t = {20,21,22,23}

    Range of t = {7,-5,0}

13. $g(x) = -2x - 4$

    $g(-3) = -2(-3) - 4$ Substituting the input, -3

    $= 6 - 4$

    $= 2$

15. $g(x) = -2x - 4$

    $g(\frac{3}{2}) = -2 \cdot \frac{3}{2} - 4$ Substituting the input, $\frac{3}{2}$

    $= -3 - 4$

    $= -7$

17. $h(x) = 3x^2$

    $h(-1) = 3(-1)^2$ Substituting the input, -1

    $= 3 \cdot 1$

    $= 3$

19. $h(x) = 3x^2$

    $h(2) = 3 \cdot 2^2$ Substituting the input, 2

    $= 3 \cdot 4$

    $= 12$

21. $P(x) = x^3 - x$

    $P(2) = 2^3 - 2$ Substituting the input, 2

    $= 8 - 2$

    $= 6$

23. $P(x) = x^3 - x$

    $P(-1) = (-1)^3 - (-1)$ Substituting the input, -1

    $= -1 + 1$

    $= 0$

25. $Q(x) = x^4 - 2x^3 + x^2 - x + 2$

    $Q(-2) = (-2)^4 - 2(-2)^3 + (-2)^2 - (-2) + 2$

    Substituting the input, -2

    $= 16 + 16 + 4 + 2 + 2$

    $= 40$

27. $Q(x) = x^4 - 2x^3 + x^2 - x + 2$

    $Q(0) = 0^4 - 2 \cdot 0^3 + 0^2 - 0 + 2$

    Substituting the input, 0

    $= 0 - 0 + 0 - 0 + 2$

    $= 2$

29. $P(d) = 1 + \frac{d}{33}$

    $P(20) = 1 + \frac{20}{33} = \frac{53}{33}$, or $1\frac{20}{33}$ atm

    $P(30) = 1 + \frac{30}{33} = \frac{63}{33} = \frac{21}{11}$, or $1\frac{10}{11}$ atm

    $P(100) = 1 + \frac{100}{33} = \frac{133}{33}$, or $4\frac{1}{33}$ atm

31. $D(p) = -2.7p + 16.3$

$D(1) = -2.7(1) + 16.3 = -2.7 + 16.3 = 13.6$ million

$D(2) = -2.7(2) + 16.3 = -5.4 + 16.3 = 10.9$ million

$D(3) = -2.7(3) + 16.3 = -8.1 + 16.3 = 8.2$ million

$D(4) = -2.7(4) + 16.3 = -10.8 + 16.3 = 5.5$ million

$D(5) = -2.7(5) + 16.3 = -13.5 + 16.3 = 2.8$ million

33. Graph: $f(x) = -2x - 3$

We find ordered pairs (x,f(x)), plot them, and connect the points.

x	f(x)
0	-3
-2	1
1	-5

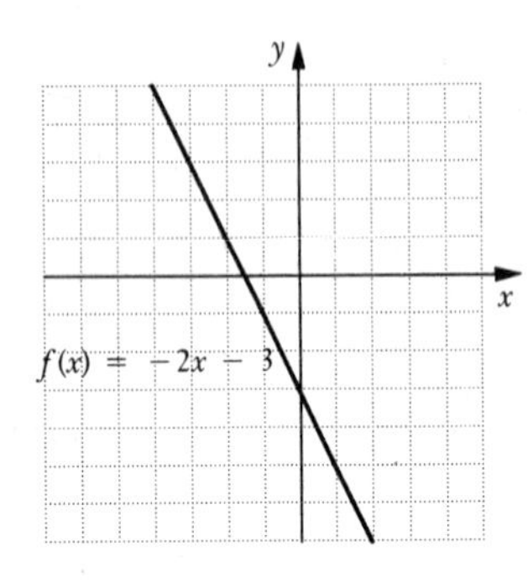

35. Graph: $h(x) = |x|$

We find ordered pairs (x,h(x)), plot them, and connect the points.

x	h(x)
-4	4
-1	1
0	0
2	2
5	5

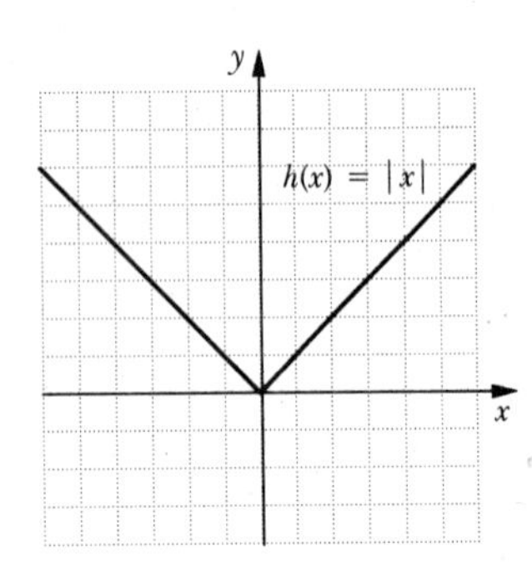

37. Graph: $f(x) = -5$

We find ordered pairs (x,f(x)), plot them, and connect the points.

x	f(x)
-3	-5
0	-5
1	-5
4	-5

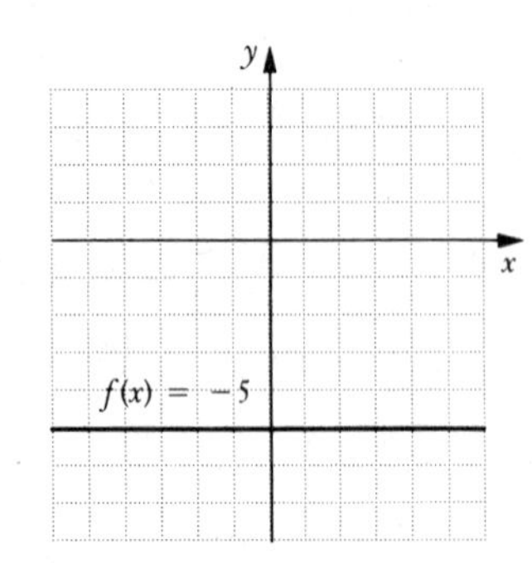

39. $f(x) = x^2 + 2$

We find ordered pairs (x,f(x)), plot them, and connect the points.

x	f(x)
-2	6
-1	3
0	2
1	3
2	6

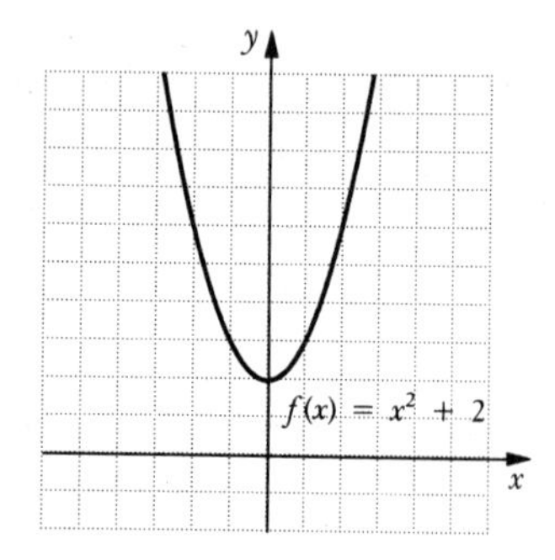

41. $f(x) = 3 - x^2$

We find ordered pairs (x,f(x)), plot them, and connect the points.

x	f(x)
-3	-6
-1	2
0	3
-2	-1
3	-6

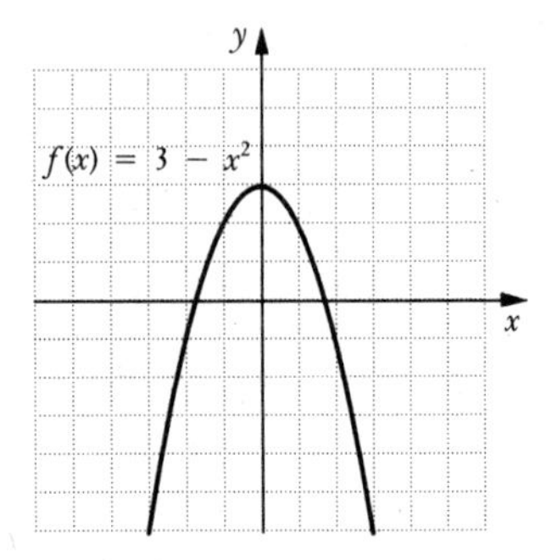

43. $g(x) = -\frac{4}{x}$

We find ordered pairs (x,g(x)), plot them, and connect the points. Note that 0 cannot be an input.

x	g(x)
-4	1
-2	2
-1	4
1	-4
2	-2
4	-1

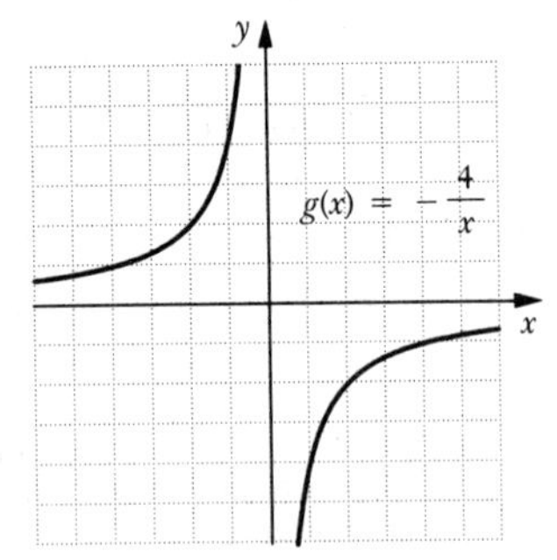

45. $f(x) = 3 - |x|$

We find ordered pairs (x,f(x)), plot them, and connect the points.

x	f(x)
-5	-2
-3	0
-1	2
0	3
2	1
4	-1

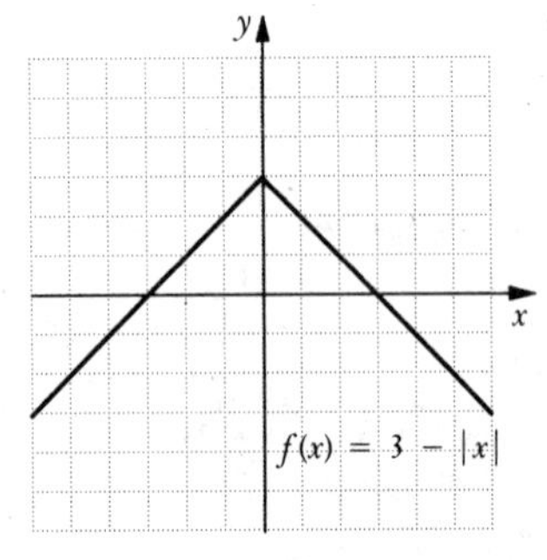

47.

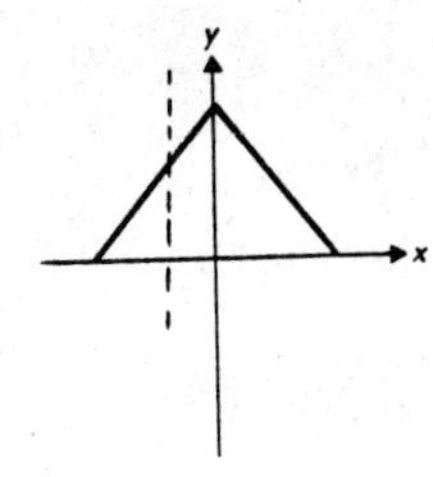

The graph is the graph of a function. It passes the vertical line test. No vertical line intersects the graph more than once.

49.

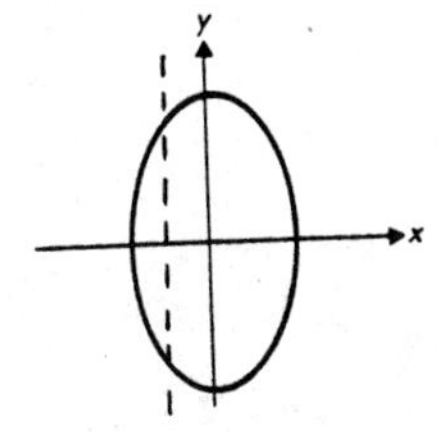

The graph is not the graph of a function. It does not pass the vertical line test. It is possible for a vertical line to intersect the graph more than once.

51. $p(x) = x^3 - x^2 + x - 7$

There are no restrictions on the numbers we can substitute into this formula. The domain is the set of all real numbers.

53. $f(x) = \frac{7}{x^2 - 25}$

This formula is meaningful for all replacements that do not make the denominator 0. To find those replacements that do make the denominator 0, we solve:

$$x^2 - 25 = 0$$
$$(x + 5)(x - 5) = 0$$
$$x = -5 \quad \text{or} \quad x = 5$$

The domain is $\{x | x \neq -5 \text{ and } x \neq 5\}$.

55. $g(x) = \frac{x}{x^2 + 8x + 15}$

We find the replacements that make the denominator 0. Then the formula is meaningful for all other replacements.

$$x^2 + 8x + 15 = 0$$
$$(x + 3)(x + 5) = 0$$
$$x = -3 \quad \text{or} \quad x = -5$$

The domain is $\{x | x \neq -3 \text{ and } x \neq -5\}$.

57. $f(x) = \sqrt{6 + 8x}$

The formula is meaningful for all real numbers x for which the following is true:

$$6 + 8x \geq 0$$
$$8x \geq -6$$
$$x \geq -\frac{6}{8}, \text{ or } -\frac{3}{4}$$

The domain is $\left\{x \middle| x \geq -\frac{3}{4}\right\}$.

59. $g(x) = \frac{4}{5 - 2x}$

We find the replacements that make the denominator 0. Then the formula is meaningful for all other replacements.

$$5 - 2x = 0$$
$$5 = 2x$$
$$\frac{5}{2} = x$$

The domain is $\left\{x \middle| x \neq \frac{5}{2}\right\}$.

61. $f(x) = 3x + 5$

The domain of f = {0,1,2,3}. We find function values by substituting into the formula. The range is the resulting set of function values.

$f(0) = 3 \cdot 0 + 5 = 0 + 5 = 5$

$f(1) = 3 \cdot 1 + 5 = 3 + 5 = 8$

$f(2) = 3 \cdot 2 + 5 = 6 + 5 = 11$

$f(3) = 3 \cdot 3 + 5 = 9 + 5 = 14$

The range of f = {5,8,11,14}.

63. $f(x) = \frac{|x|}{x}$

We find ordered pairs (x,f(x)), plot them, and connect the points. Note that 0 cannot be an input.

x	f(x)
-5	-1
-3	-1
-1	-1
1	1
2	1
4	1

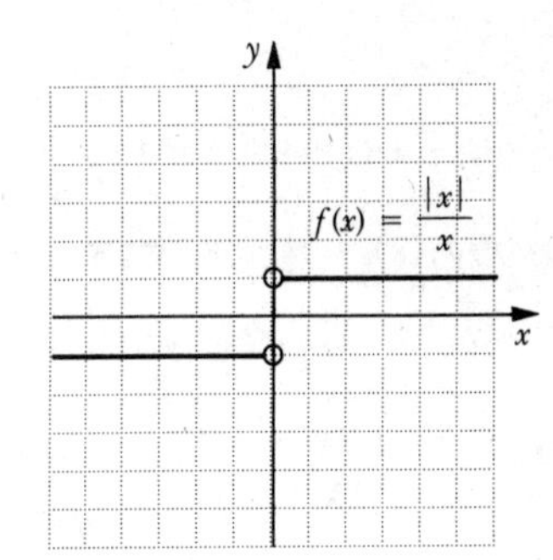

65. $h(x) = |x| - x$

We find ordered pairs (x,h(x)), plot them, and connect the points.

x	h(x)
-3	6
-1	2
0	0
1	0
2	0
4	0

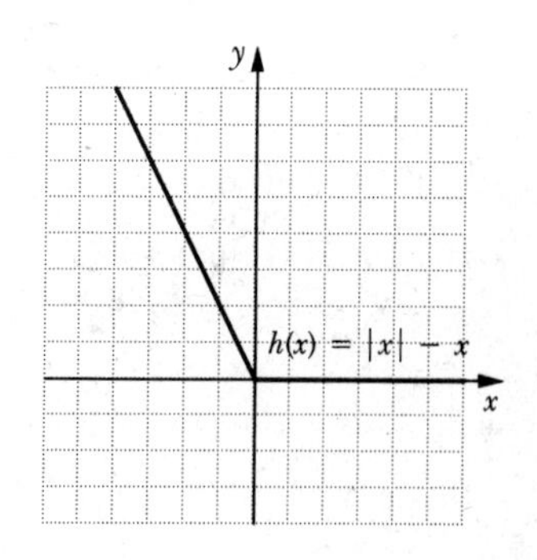

67. $|y| = x$

We choose values for y, find the corresponding x-values, plot the ordered pairs (x,y), or (|y|,y), and connect the points. Then we determine whether the graph passes the vertical line test.

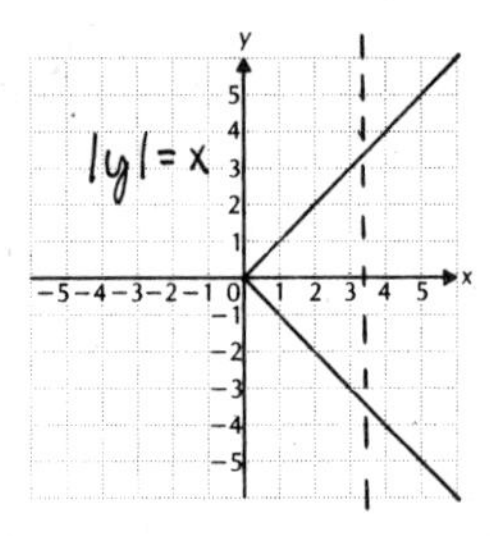

Since the graph does not pass the vertical line test, it is not the graph of a function.

Exercise Set 12.5

1. $f(x) = x + 3$ $g(x) = x - 3$

$(f + g)(x) = f(x) + g(x) = (x + 3) + (x - 3) = 2x$

$(f - g)(x) = f(x) - g(x) = (x + 3) - (x - 3)$
$= x + 3 - x + 3$
$= 6$

$fg(x) = f(x)\cdot g(x) = (x + 3)(x - 3) = x^2 - 9$

$(f/g)(x) = \frac{f(x)}{g(x)} = \frac{x + 3}{x - 3}$

$ff(x) = f(x)\cdot f(x) = (x + 3)(x + 3) = x^2 + 6x + 9$

3. $f(x) = 2x^2 - 3x + 1$ $g(x) = x^3$

$(f + g)(x) = f(x) + g(x) = (2x^2 - 3x + 1) + x^3$
$= x^3 + 2x^2 - 3x + 1$

$(f - g)(x) = f(x) - g(x) = (2x^2 - 3x + 1) - x^3$
$= -x^3 + 2x^2 - 3x + 1$

$fg(x) = f(x)\cdot g(x) = (2x^2 - 3x + 1)(x^3)$
$= 2x^5 - 3x^4 + x^3$

$(f/g)(x) = \frac{f(x)}{g(x)} = \frac{2x^2 - 3x + 1}{x^3}$

$ff(x) = f(x)\cdot f(x) = (2x^2 - 3x + 1)(2x^2 - 3x + 1)$
$= 4x^4 - 6x^3 + 2x^2 - 6x^3 + 9x^2 -$
$3x + 2x^2 - 3x + 1$
$= 4x^4 - 12x^3 + 13x^2 - 6x + 1$

5. $f(x) = -5x^2$ $g(x) = 4x^3$

$(f + g)(x) = f(x) + g(x) = -5x^2 + 4x^3 = 4x^3 - 5x^2$

$(f - g)(x) = f(x) - g(x) = -5x^2 - 4x^3 = -4x^3 - 5x^2$

$fg(x) = f(x)\cdot g(x) = -5x^2\cdot 4x^3 = -20x^5$

5. (continued)

$(f/g)(x) = \frac{f(x)}{g(x)} = \frac{-5x^2}{4x^3} = -\frac{5}{4x}$

$ff(x) = f(x)\cdot f(x) = -5x^2(-5x^2) = 25x^4$

7. $f(x) = 20$ $g(x) = -5$

$(f + g)(x) = f(x) + g(x) = 20 + (-5) = 15$

$(f - g)(x) = f(x) - g(x) = 20 - (-5) = 25$

$fg(x) = f(x)\cdot g(x) = 20(-5) = -100$

$(f/g)(x) = \frac{f(x)}{g(x)} = \frac{20}{-5} = -4$

$ff(x) = f(x)\cdot f(x) = 20\cdot 20 = 400$

9. $f \circ g(x) = f(g(x)) = f(7 - 2x) = 5(7 - 2x) - 8 =$
$35 - 10x - 8 = -10x + 27$

$g \circ f(x) = g(f(x)) = g(5x - 8) = 7 - 2(5x - 8) =$
$7 - 10x + 16 = -10x + 23$

11. $f \circ g(x) = f(g(x)) = f(2x - 1) = 3(2x - 1)^2 + 2 =$
$3(4x^2 - 4x + 1) + 2 =$
$12x^2 - 12x + 3 + 2 = 12x^2 - 12x + 5$

$g \circ f(x) = g(f(x)) = g(3x^2 + 2) = 2(3x^2 + 2) - 1 =$
$6x^2 + 4 - 1 = 6x^2 + 3$

13. $f \circ g(x) = f(g(x)) = f\left(\frac{2}{x}\right) = 4\left(\frac{2}{x}\right)^2 - 1 =$
$4\left(\frac{4}{x^2}\right) - 1 = \frac{16}{x^2} - 1$

$g \circ f(x) = g(f(x)) = g(4x^2 - 1) = \frac{2}{4x^2 - 1}$

15. $f \circ g(x) = f(g(x)) = f(x^2 - 1) = (x^2 - 1)^2 + 1 =$
$x^4 - 2x^2 + 1 + 1 = x^4 - 2x^2 + 2$

$g \circ f(x) = g(f(x)) = g(x^2 + 1) = (x^2 + 1)^2 - 1 =$
$x^4 + 2x^2 + 1 - 1 = x^4 + 2x^2$

17. $h(x) = (5 - 3x)^2$

This is $5 - 3x$ to the 2nd power, so the two most obvious functions are $f(x) = x^2$ and $g(x) = 5 - 3x$.

19. $h(x) = (3x^2 - 7)^5$

This is $3x^2 - 7$ to the 5th power, so the two most obvious functions are $f(x) = x^5$ and $g(x) = 3x^2 - 7$.

21. $h(x) = \frac{1}{x - 1}$

This is the reciprocal of $x - 1$, so the two most obvious functions are $f(x) = \frac{1}{x}$ and $g(x) = x - 1$.

23. $h(x) = \dfrac{1}{\sqrt{7x + 2}}$

This is the reciprocal of the square root of $7x + 2$. Two functions that can be used are $f(x) = \dfrac{1}{\sqrt{x}}$ and $g(x) = 7x + 2$.

25. $h(x) = \dfrac{x^3 + 1}{x^3 - 1}$

Two functions that can be used are $f(x) = \dfrac{x + 1}{x - 1}$ and $g(x) = x^3$.

27. $\sqrt[3]{27x^6y^{12}} = 3x^2y^4$

29. $2x^2 - 7x + 4 = 0$

$a = 2,\ b = -7,\ c = 4$

$$x = \frac{-b \pm \sqrt{b^2 - 4ac}}{2a} = \frac{-(-7) \pm \sqrt{(-7)^2 - 4\cdot2\cdot4}}{2\cdot2}$$

$$x = \frac{7 \pm \sqrt{49 - 32}}{4} = \frac{7 \pm \sqrt{17}}{4}$$

31. $$\frac{f(x + h) - f(x)}{h} = \frac{3(x + h) + 7 - (3x + 7)}{h}$$

$$= \frac{3x + 3h + 7 - 3x - 7}{h}$$

$$= \frac{3h}{h} = 3$$

33. $$\frac{f(x + h) - f(x)}{h} = \frac{-7 - (-7)}{h} = \frac{-7 + 7}{h} = 0$$

35. $$\frac{f(x + h) - f(x)}{h} = \frac{(x + h)^2 - x^2}{h}$$

$$= \frac{x^2 + 2xh + h^2 - x^2}{h}$$

$$= \frac{2xh + h^2}{h} = \frac{h(2x + h)}{h}$$

$$= 2x + h$$

37. $$\frac{f(x + h) - f(x)}{x} = \frac{\frac{1}{x + h} - \frac{1}{x}}{h} = \frac{\frac{x - (x + h)}{x(x + h)}}{h}$$

$$= \frac{\frac{x - x - h}{x(x + h)}}{h} = \frac{\frac{-h}{x(x + h)}}{h}$$

$$= \frac{-h}{x(x + h)} \cdot \frac{1}{h} = \frac{-1}{x(x + h)}, \text{ or}$$

$$-\frac{1}{x(x + h)}$$

39. $$\frac{f(x + h) - f(x)}{h} = \frac{\sqrt{x + h} - \sqrt{x}}{h}$$

$$= \frac{\sqrt{x + h} - \sqrt{x}}{h} \cdot \frac{\sqrt{x + h} + \sqrt{x}}{\sqrt{x + h} + \sqrt{x}}$$

Rationalizing the numerator

$$= \frac{x + h - x}{h(\sqrt{x + h} + \sqrt{x})}$$

$$= \frac{h}{h(\sqrt{x + h} + \sqrt{x})}$$

$$= \frac{1}{\sqrt{x + h} + \sqrt{x}}$$

41. If $C(x)$ is the cost per person, then $x\cdot C(x)$ is the total cost of the bus for x people.

$$x\cdot C(x) = x \cdot \frac{100 + 5x}{x} = 100 + 5x$$

$$40\cdot C(40) = 100 + 5\cdot40 = \$300$$

$$80\cdot C(80) = 100 + 5\cdot80 = \$500$$

Exercise Set 12.6

1.

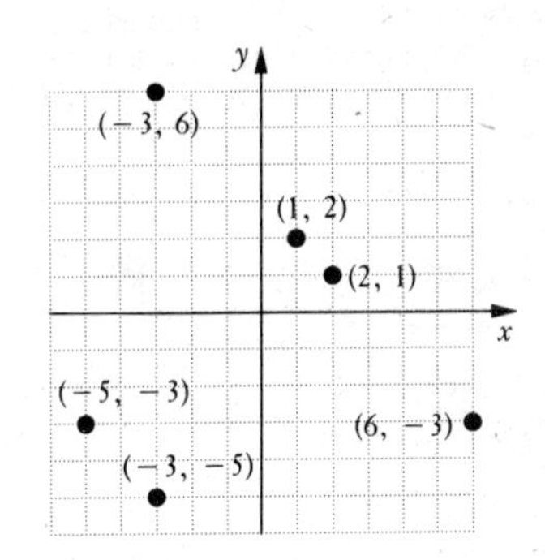

The inverse relation is {(2,1), (-3,6),(-5,-3)}.

3.

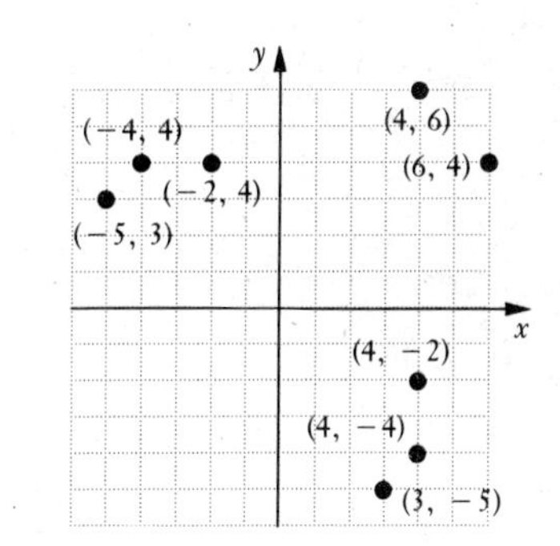

The inverse relation is {(4,6), (4,-2), (-5,3), (-4,4)}.

5. a)

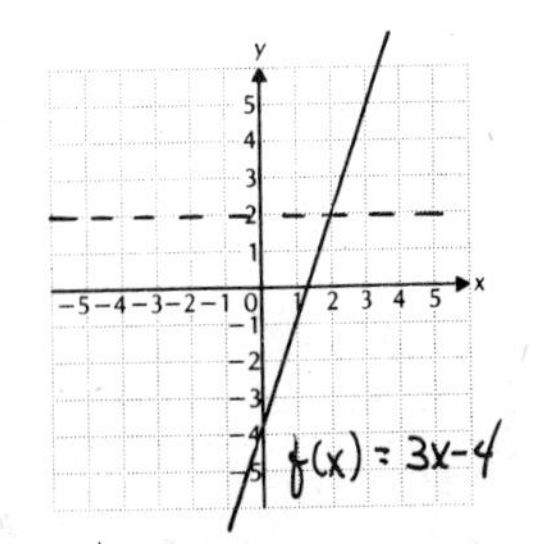

The graph of $f(x) = 3x - 4$ is shown above. It passes the horizontal line test, so it is one-to-one.

b) Replace $f(x)$ by y: $y = 3x - 4$

Interchange x and y: $x = 3y - 4$

Solve for y: $x + 4 = 3y$

$$\frac{x + 4}{3} = y$$

Replace y by $f^{-1}(x)$: $f^{-1}(x) = \dfrac{x + 4}{3}$

7. a)

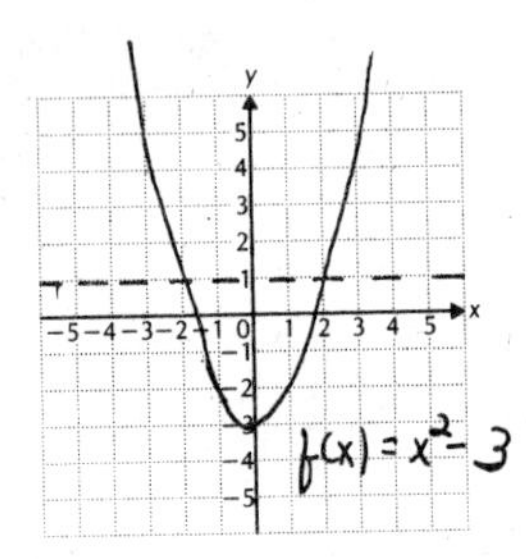

The graph of $f(x) = x^2 - 3$ is shown above. There are many horizontal lines that cross the graph more than once, so the function is not one-to-one.

9. a)

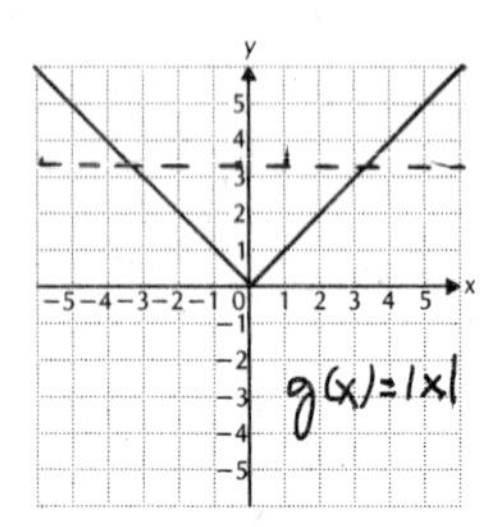

The graph of $g(x) = |x|$ is shown above. There are many horizontal lines that cross the graph more than once, so the function is not one-to-one.

11. a)

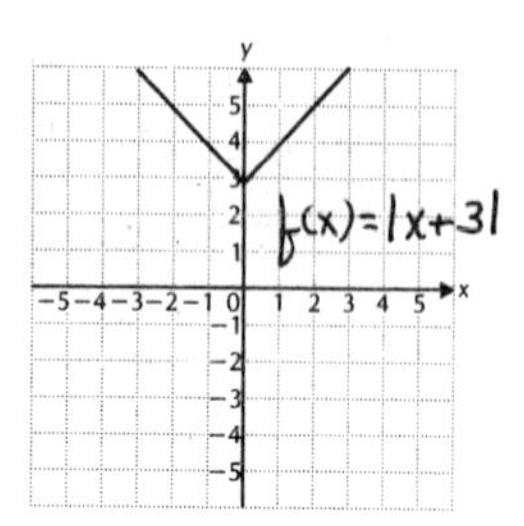

The graph of $f(x) = |x + 3|$ is shown above. There are many horizontal lines that cross the graph more than once, so the function is not one-to-one.

13. a)

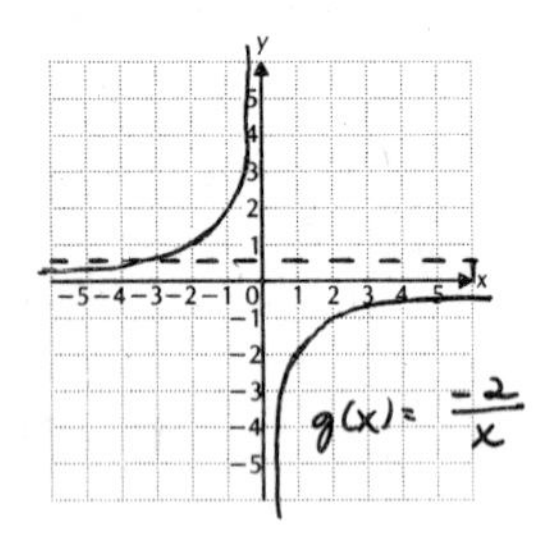

The graph of $g(x) = \frac{-2}{x}$ is shown above. It passes the horizontal line test, so it is one-to-one.

13. (continued)

b) Replace $g(x)$ by y: $y = \frac{-2}{x}$

Interchange x and y: $x = \frac{-2}{y}$

Solve for y: $y = \frac{-2}{x}$

Replace y by $g^{-1}(x)$: $g^{-1}(x) = \frac{-2}{x}$

15. $f(x) = x + 2$

1. Replace $f(x)$ by y: $y = x + 2$
2. Interchange x and y: $x = y + 2$
3. Solve for y: $x - 2 = y$
4. Replace y by $f^{-1}(x)$: $f^{-1}(x) = x - 2$

17. $f(x) = 4 - x$

1. Replace $f(x)$ by y: $y = 4 - x$
2. Interchange x and y: $x = 4 - y$
3. Solve for y: $y = 4 - x$
4. Replace y by $f^{-1}(x)$: $f^{-1}(x) = 4 - x$

19. $g(x) = x - 5$

1. Replace $g(x)$ by y: $y = x - 5$
2. Interchange x and y: $x = y - 5$
3. Solve for y: $x + 5 = y$
4. Replace y by $g^{-1}(x)$: $g^{-1}(x) = x + 5$

21. $f(x) = 3x$

1. Replace $f(x)$ by y: $y = 3x$
2. Interchange x and y: $x = 3y$
3. Solve for y: $\frac{x}{3} = y$
4. Replace y by $f^{-1}(x)$: $f^{-1}(x) = \frac{x}{3}$

23. $g(x) = 3x + 2$

1. Replace $g(x)$ by y: $y = 3x + 2$
2. Interchange x and y: $x = 3y + 2$
3. Solve for y: $x - 2 = 3y$

   $\frac{x - 2}{3} = y$
4. Replace y by $g^{-1}(x)$: $g^{-1}(x) = \frac{x - 2}{3}$

25. $h(x) = \frac{2}{x + 5}$

1. Replace h(x) by y: $y = \frac{2}{x + 5}$
2. Interchange x and y: $x = \frac{2}{y + 5}$
3. Solve for y: $x(y + 5) = 2$

$$y + 5 = \frac{2}{x}$$

$$y = \frac{2}{x} - 5$$

4. Replace y by $h^{-1}(x)$: $h^{-1}(x) = \frac{2}{x} - 5$

27. $f(x) = \frac{2x + 1}{5x + 3}$

1. Replace f(x) by y: $y = \frac{2x + 1}{5x + 3}$
2. Interchange x and y: $x = \frac{2y + 1}{5y + 3}$
3. Solve for y: $5xy + 3x = 2y + 1$

$$5xy - 2y = 1 - 3x$$

$$y(5x - 2) = 1 - 3x$$

$$y = \frac{1 - 3x}{5x - 2}$$

4. Replace y by $f^{-1}(x)$: $f^{-1}(x) = \frac{1 - 3x}{5x - 2}$

29. $g(x) = \frac{x - 3}{x + 4}$

1. Replace g(x) by y: $y = \frac{x - 3}{x + 4}$
2. Interchange x and y: $x = \frac{y - 3}{y + 4}$
3. Solve for y: $xy + 4x = y - 3$

$$4x + 3 = y - xy$$

$$4x + 3 = y(1 - x)$$

$$\frac{4x + 3}{1 - x} = y$$

4. Replace y by $g^{-1}(x)$: $g^{-1}(x) = \frac{4x + 3}{1 - x}$

31. $f(x) = x^3 - 1$

1. Replace f(x) by y: $y = x^3 - 1$
2. Interchange x and y: $x = y^3 - 1$
3. Solve for y: $x + 1 = y^3$

$$\sqrt[3]{x + 1} = y$$

4. Replace y by $f^{-1}(x)$: $f^{-1}(x) = \sqrt[3]{x + 1}$

33. $G(x) = (x - 2)^3$

1. Replace G(x) by y: $y = (x - 2)^3$
2. Interchange x and y: $x = (y - 2)^3$
3. Solve for y: $\sqrt[3]{x} = y - 2$

$$\sqrt[3]{x} + 2 = y$$

4. Replace y by $G^{-1}(x)$: $G^{-1}(x) = \sqrt[3]{x} + 2$

35. $f(x) = \sqrt[3]{x}$

1. Replace f(x) by y: $y = \sqrt[3]{x}$
2. Interchange x and y: $x = \sqrt[3]{y}$
3. Solve for y: $x^3 = y$
4. Replace y by $f^{-1}(x)$: $f^{-1}(x) = x^3$

37. Relation: $y = -\frac{1}{2}x + 2$

Inverse: $x = -\frac{1}{2}y + 2$

We graph the relation and its inverse:

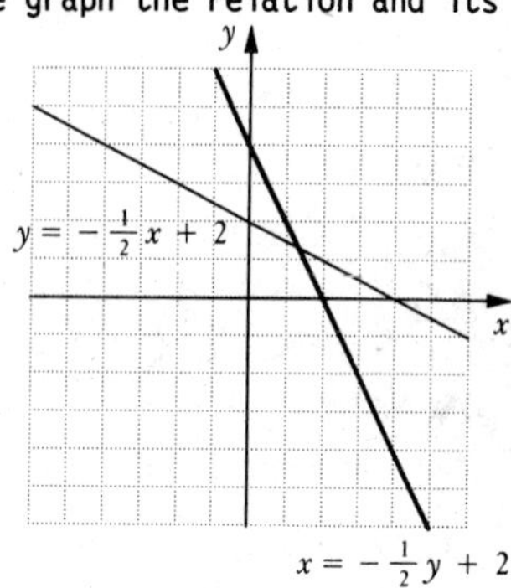

39. Relation: $y = x^2 - 3$

Inverse: $x = y^2 - 3$

We graph the relation and its inverse:

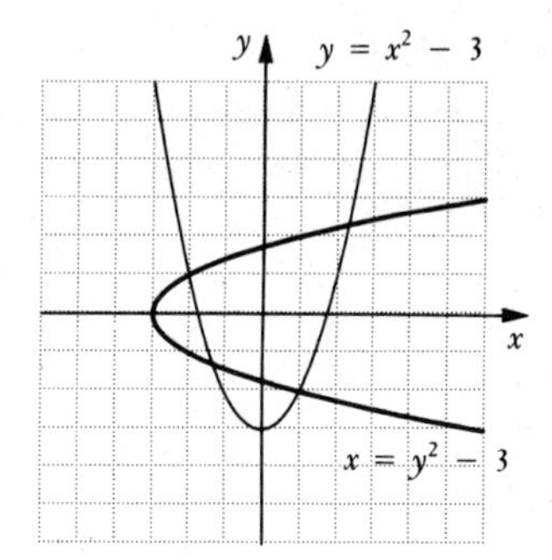

41. We first graph $f(x) = \frac{1}{2}x - 3$. The graph of f^{-1} can be obtained by reflecting the graph of f across the line y = x.

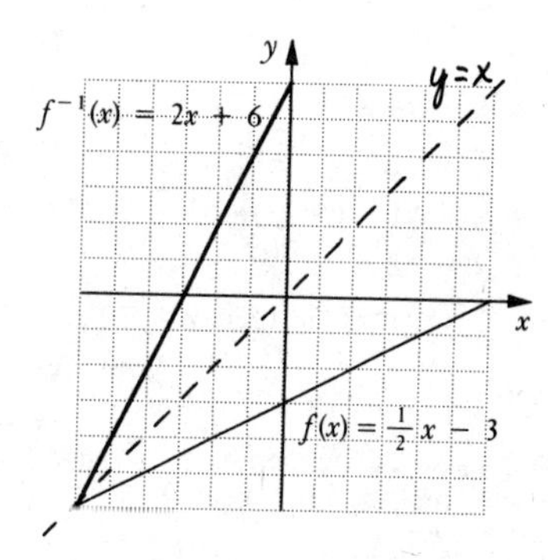

43. We first graph $f(x) = x^3$. The graph of f^{-1} can be obtained by reflecting the graph of f across the line y = x.

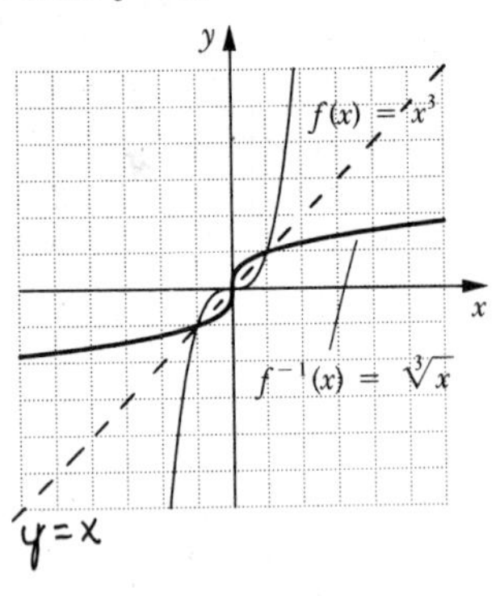

45. a) $f(8) = 8 + 32 = 40$

Size 40 in France corresponds to size 8 in the U.S.

$f(10) = 10 + 32 = 42$

Size 42 in France corresponds to size 10 in the U.S.

$f(14) = 14 + 32 = 46$

Size 46 in France corresponds to size 14 in the U.S.

$f(18) = 18 + 32 = 50$

Size 50 in France corresponds to size 18 in the U.S.

b) The graph of $f(x) = x + 32$ is shown below.

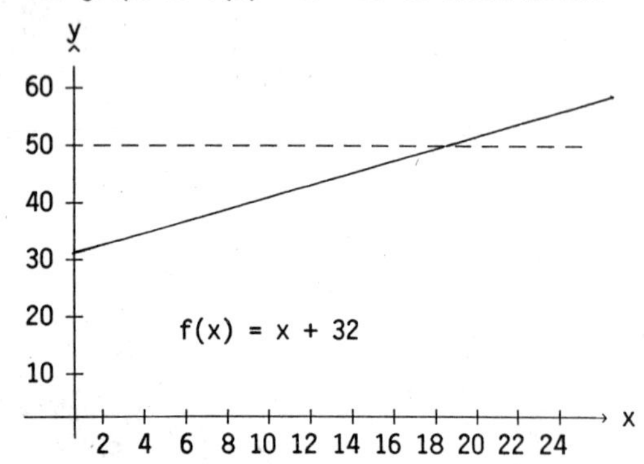

It passes the horizontal line test, so the function is one-to-one and, hence, has an inverse that is a function. We now find a formula for the inverse.

Replace f(x) by y: $y = x + 32$

Interchange x and y: $x = y + 32$

Solve for y: $x - 32 = y$

Replace y by $f^{-1}(x)$: $f^{-1}(x) = x - 32$

c) $f^{-1}(40) = 40 - 32 = 8$

Size 8 in the U.S. corresponds to size 40 in France.

$f^{-1}(42) = 42 - 32 = 10$

Size 10 in the U.S. corresponds to size 42 in France.

45. (continued)

$f^{-1}(46) = 46 - 32 = 14$

Size 14 in the U.S. corresponds to size 46 in France.

$f^{-1}(50) = 50 - 32 = 18$

Size 18 in the U.S. corresponds to size 50 in France.

47. The graph of $f(x) = 4$ is shown below. Since the horizontal line y = 4 crosses the graph in more than one place, the function does not have an inverse that is a function.

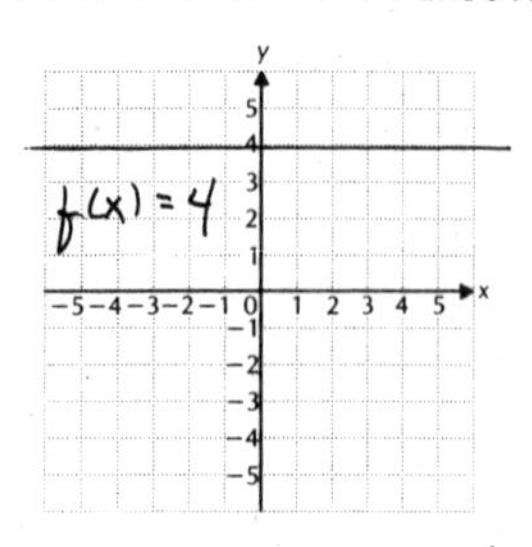

49. $C(x) = \dfrac{100 + 5x}{x}$

1. Replace C(x) by y: $y = \dfrac{100 + 5x}{x}$

2. Interchange x and y: $x = \dfrac{100 + 5y}{y}$

3. Solve for y:

$$xy = 100 + 5y$$
$$xy - 5y = 100$$
$$y(x - 5) = 100$$
$$y = \frac{100}{x - 5}$$

4. Replace y by $C^{-1}(x)$: $C^{-1}(x) = \dfrac{100}{x - 5}$

$C^{-1}(x)$ gives the number of people in the group, where x is the cost per person (in dollars) of chartering a bus.

51. $f \circ f^{-1}(x) = f(f^{-1}(x)) = f(3x - 7) = \dfrac{(3x - 7) + 7}{3} = \dfrac{3x}{3} = x$

$f^{-1} \circ f(x) = f^{-1}(f(x)) = 3\left(\dfrac{x + 7}{3}\right) - 7 = x + 7 - 7 = x$

53. $f \circ f^{-1}(x) = f(f^{-1}(x)) = (\sqrt[3]{x + 5})^3 - 5 = x + 5 - 5 = x$

$f^{-1} \circ f(x) = f^{-1}(f(x)) = \sqrt[3]{(x^3 - 5) + 5} = \sqrt[3]{x^3} = x$

55. $f \circ g(x) = f(g(x)) = \sqrt[5]{x^5} = x$

$g \circ f(x) = g(f(x)) = (\sqrt[5]{x})^5 = x$

f and g are inverses of each other.

57. $$f \circ g(x) = f(g(x)) = \frac{2\left(\frac{7x - 4}{3x + 2}\right) - 3}{4\left(\frac{7x - 4}{3x + 2}\right) + 7}$$

$$= \frac{\frac{14x - 8}{3x + 2} - 3}{\frac{28x - 16}{3x + 2} + 7}$$

$$= \frac{\frac{14x - 8}{3x + 2} - 3}{\frac{28x - 16}{3x + 2} + 7} \cdot \frac{3x + 2}{3x + 2}$$

$$= \frac{14x - 8 - 3(3x + 2)}{28x - 16 + 7(3x + 2)}$$

$$= \frac{14x - 8 - 9x - 6}{28x - 16 + 21x + 14}$$

$$= \frac{5x - 14}{49x - 2}$$

Since $f \circ g(x) \neq x$, f and g are not inverses of each other.

Exercise Set 13.1

1. Graph: $y = 2^x$

We compute some function values, thinking of y as f(x), and keep the results in a table.

$f(0) = 2^0 = 1$

$f(1) = 2^1 = 2$

$f(2) = 2^2 = 4$

$f(-1) = 2^{-1} = \frac{1}{2^1} = \frac{1}{2}$

$f(-2) = 2^{-2} = \frac{1}{2^2} = \frac{1}{4}$

x	y, or f(x)
0	1
1	2
2	4
-1	$\frac{1}{2}$
-2	$\frac{1}{4}$

Next we plot these points and connect them with a smooth curve.

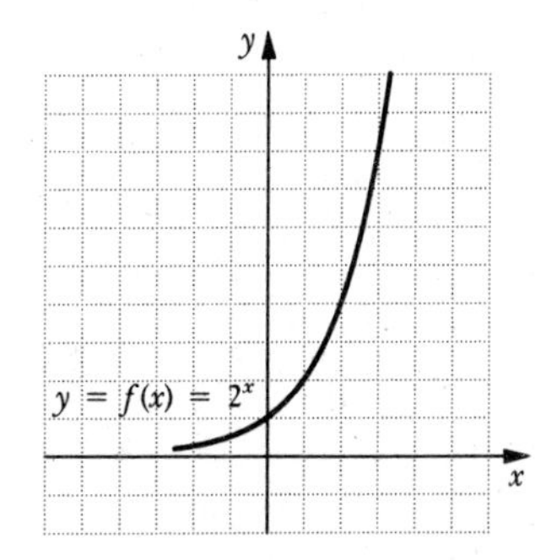

3. Graph: $y = 5^x$

We compute some function values, thinking of y as f(x), and keep the results in a table.

$f(0) = 5^0 = 1$

$f(1) = 5^1 = 5$

$f(2) = 5^2 = 25$

$f(-1) = 5^{-1} = \frac{1}{5^1} = \frac{1}{5}$

$f(-2) = 5^{-2} = \frac{1}{5^2} = \frac{1}{25}$

x	y, or f(x)
0	1
1	5
2	25
-1	$\frac{1}{5}$
-2	$\frac{1}{25}$

Next we plot these points and connect them with a smooth curve.

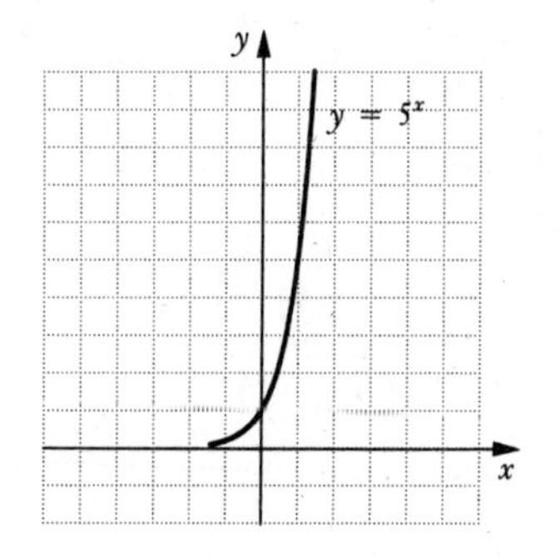

5. Graph: $y = 2^{x+1}$

We compute some function values, thinking of y as f(x), and keep the results in a table.

$f(0) = 2^{0+1} = 2^1 = 2$

$f(-1) = 2^{-1+1} = 2^0 = 1$

$f(-2) = 2^{-2+1} = 2^{-1} = \frac{1}{2^1} = \frac{1}{2}$

$f(-3) = 2^{-3+1} = 2^{-2} = \frac{1}{2^2} = \frac{1}{4}$

$f(1) = 2^{1+1} = 2^2 = 4$

$f(2) = 2^{2+1} = 2^3 = 8$

x	y, or f(x)
0	2
-1	1
-2	$\frac{1}{2}$
-3	$\frac{1}{4}$
1	4
2	8

Next we plot these points and connect them with a smooth curve.

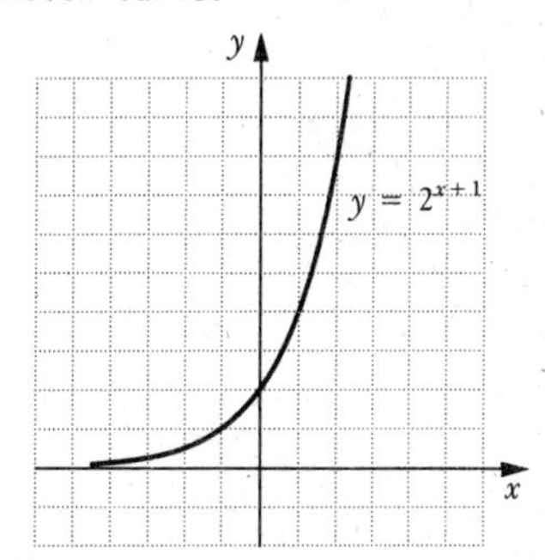

7. Graph: $y = 3^{x-2}$

We compute some function values, thinking of y as f(x), and keep the results in a table.

$f(0) = 3^{0-2} = 3^{-2} = \frac{1}{3^2} = \frac{1}{9}$

$f(1) = 3^{1-2} = 3^{-1} = \frac{1}{3^1} = \frac{1}{3}$

$f(2) = 3^{2-2} = 3^0 = 1$

$f(3) = 3^{3-2} = 3^1 = 3$

$f(4) = 3^{4-2} = 3^2 = 9$

$f(-1) = 3^{-1-2} = 3^{-3} = \frac{1}{3^3} = \frac{1}{27}$

$f(-2) = 3^{-2-2} = 3^{-4} = \frac{1}{3^4} = \frac{1}{81}$

x	y, or f(x)
0	$\frac{1}{9}$
1	$\frac{1}{3}$
2	1
3	3
4	9
-1	$\frac{1}{27}$
-2	$\frac{1}{81}$

Next we plot these points and connect them with a smooth curve.

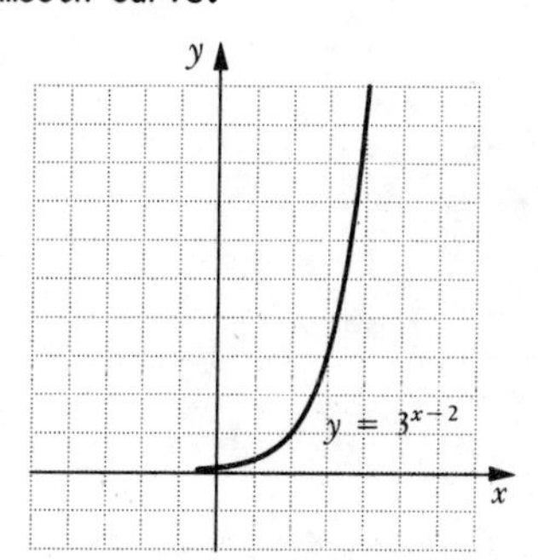

9. Graph: $y = 2^x - 3$

We construct a table of values, thinking of y as f(x). Then we plot the points and connect them with a smooth curve.

$f(0) = 2^0 - 3 = 1 - 3 = -2$

$f(1) = 2^1 - 3 = 2 - 3 = -1$

$f(2) = 2^2 - 3 = 4 - 3 = 1$

$f(3) = 2^3 - 3 = 8 - 3 = 5$

$f(-1) = 2^{-1} - 3 = \frac{1}{2} - 3 = -\frac{5}{2}$

$f(-2) = 2^{-2} - 3 = \frac{1}{4} - 3 = -\frac{11}{4}$

x	y, or f(x)
0	-2
1	-1
2	1
3	5
-1	$-\frac{5}{2}$
-2	$-\frac{11}{4}$

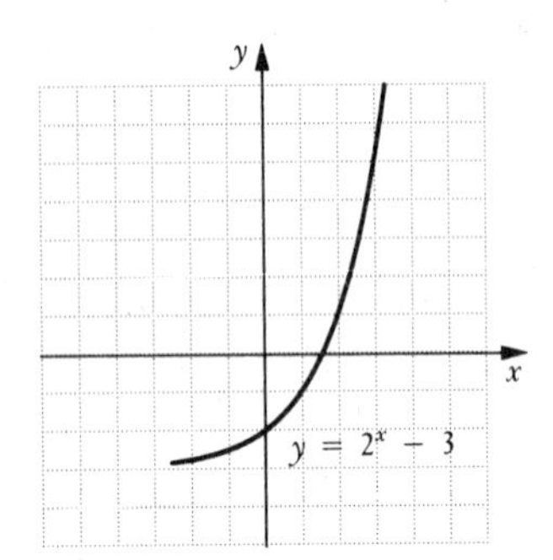

11. Graph: $y = 5^{x+3}$

We construct a table of values, thinking of y as f(x). Then we plot the points and connect them with a smooth curve.

$f(0) = 5^{0+3} = 5^3 = 125$

$f(-1) = 5^{-1+3} = 5^2 = 25$

$f(-2) = 5^{-2+3} = 5^1 = 5$

$f(-3) = 5^{-3+3} = 5^0 = 1$

$f(-4) = 5^{-4+3} = 5^{-1} = \frac{1}{5}$

$f(-5) = 5^{-5+3} = 5^{-2} = \frac{1}{25}$

x	y, or f(x)
0	125
-1	25
-2	5
-3	1
-4	$\frac{1}{5}$
-5	$\frac{1}{25}$

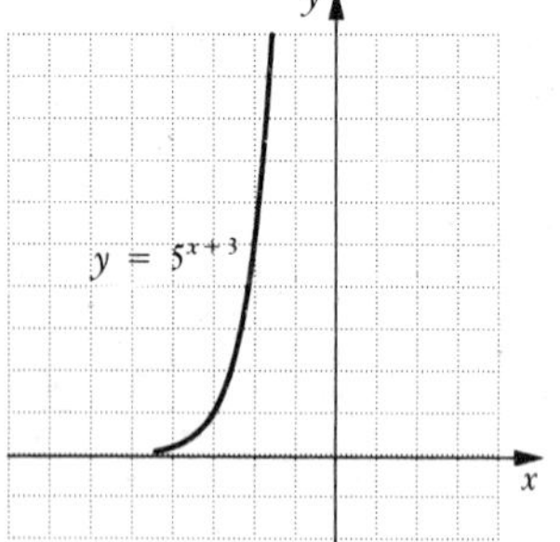

13. Graph: $y = \left(\frac{1}{2}\right)^x$

We construct a table of values, thinking of y as f(x). Then we plot the points and connect them with a smooth curve.

$f(0) = \left(\frac{1}{2}\right)^0 = 1$

$f(1) = \left(\frac{1}{2}\right)^1 = \frac{1}{2}$

$f(2) = \left(\frac{1}{2}\right)^2 = \frac{1}{4}$

$f(3) = \left(\frac{1}{2}\right)^3 = \frac{1}{8}$

$f(-1) = \left(\frac{1}{2}\right)^{-1} = \frac{1}{\left(\frac{1}{2}\right)^1} = \frac{1}{\frac{1}{2}} = 2$

$f(-2) = \left(\frac{1}{2}\right)^{-2} = \frac{1}{\left(\frac{1}{2}\right)^2} = \frac{1}{\frac{1}{4}} = 4$

$f(-3) = \left(\frac{1}{2}\right)^{-3} = \frac{1}{\left(\frac{1}{2}\right)^3} = \frac{1}{\frac{1}{8}} = 8$

x	y, or f(x)
0	1
1	$\frac{1}{2}$
2	$\frac{1}{4}$
3	$\frac{1}{8}$
-1	2
-2	4
-3	8

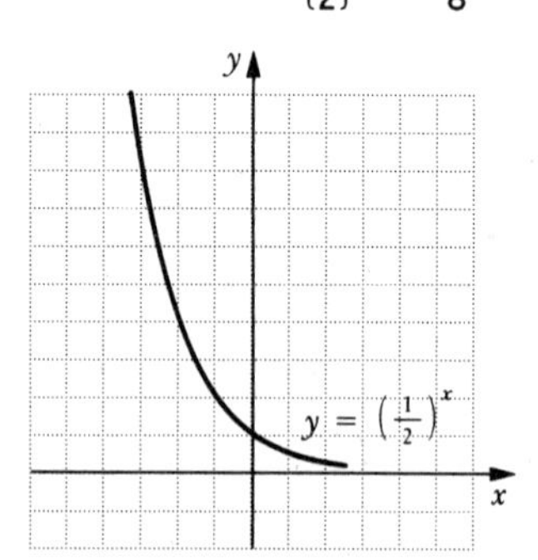

15. Graph: $y = \left(\frac{1}{5}\right)^x$

We construct a table of values, thinking of y as f(x). Then we plot the points and connect them with a smooth curve.

$f(0) = \left(\frac{1}{5}\right)^0 = 1$

$f(1) = \left(\frac{1}{5}\right)^1 = \frac{1}{5}$

$f(2) = \left(\frac{1}{5}\right)^2 = \frac{1}{25}$

$f(-1) = \left(\frac{1}{5}\right)^{-1} = \frac{1}{\frac{1}{5}} = 5$

$f(-2) = \left(\frac{1}{5}\right)^{-2} = \frac{1}{\frac{1}{25}} = 25$

x	y, or f(x)
0	1
1	$\frac{1}{5}$
2	$\frac{1}{25}$
-1	5
-2	25

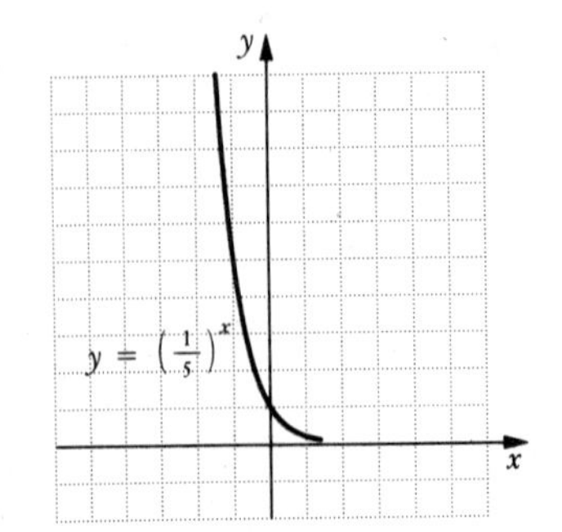

17. Graph: $y = 2^{2x-1}$

We construct a table of values, thinking of y as f(x). Then we plot the points and connect them with a smooth curve.

$f(0) = 2^{2\cdot 0-1} = 2^{-1} = \frac{1}{2}$

$f(1) = 2^{2\cdot 1-1} = 2^{1} = 2$

$f(2) = 2^{2\cdot 2-1} = 2^{3} = 8$

$f(-1) = 2^{2(-1)-1} = 2^{-3} = \frac{1}{8}$

$f(-2) = 2^{2(-2)-1} = 2^{-5} = \frac{1}{32}$

x	y, or f(x)
0	$\frac{1}{2}$
1	2
2	8
-1	$\frac{1}{8}$
-2	$\frac{1}{32}$

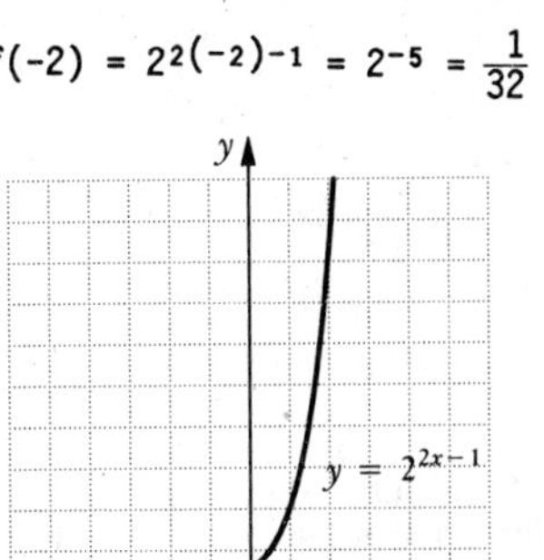

19. Graph: $x = 2^y$

We can find ordered pairs by choosing values for y and then computing values for x.

For y = 0, $x = 2^0 = 1$.

For y = 1, $x = 2^1 = 2$.

For y = 2, $x = 2^2 = 4$.

For y = 3, $x = 2^3 = 8$.

For y = -1, $x = 2^{-1} = \frac{1}{2^1} = \frac{1}{2}$.

For y = -2, $x = 2^{-2} = \frac{1}{2^2} = \frac{1}{4}$.

For y = -3, $x = 2^{-3} = \frac{1}{2^3} = \frac{1}{8}$.

x	y
1	0
2	1
4	2
8	3
$\frac{1}{2}$	-1
$\frac{1}{4}$	-2
$\frac{1}{8}$	-3

(1) Choose values for y.

(2) Compute values for x.

We plot these points and connect them with a smooth curve.

19. (continued)

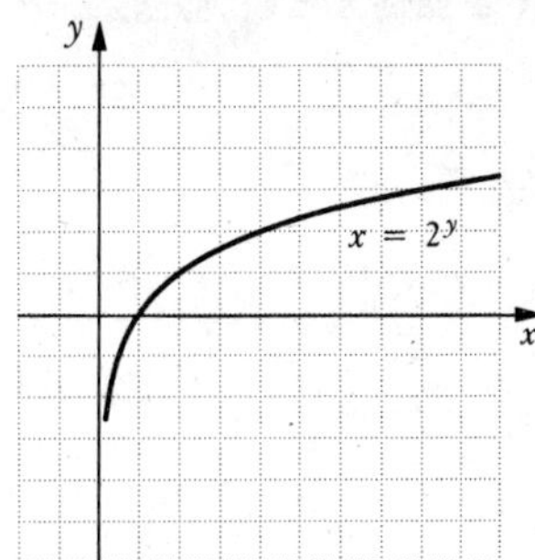

21. Graph: $x = \left(\frac{1}{2}\right)^y$

We can find ordered pairs by choosing values for y and then computing values for x. Then we plot these points and connect them with a smooth curve.

For y = 0, $x = \left(\frac{1}{2}\right)^0 = 1$.

For y = 1, $x = \left(\frac{1}{2}\right)^1 = \frac{1}{2}$.

For y = 2, $x = \left(\frac{1}{2}\right)^2 = \frac{1}{4}$.

For y = 3, $x = \left(\frac{1}{2}\right)^3 = \frac{1}{8}$.

For y = -1, $x = \left(\frac{1}{2}\right)^{-1} = \frac{1}{\frac{1}{2}} = 2$.

For y = -2, $x = \left(\frac{1}{2}\right)^{-2} = \frac{1}{\frac{1}{4}} = 4$.

For y = -3, $x = \left(\frac{1}{2}\right)^{-3} = \frac{1}{\frac{1}{8}} = 8$.

x	y
1	0
$\frac{1}{2}$	1
$\frac{1}{4}$	2
$\frac{1}{8}$	3
2	-1
4	-2
8	-3

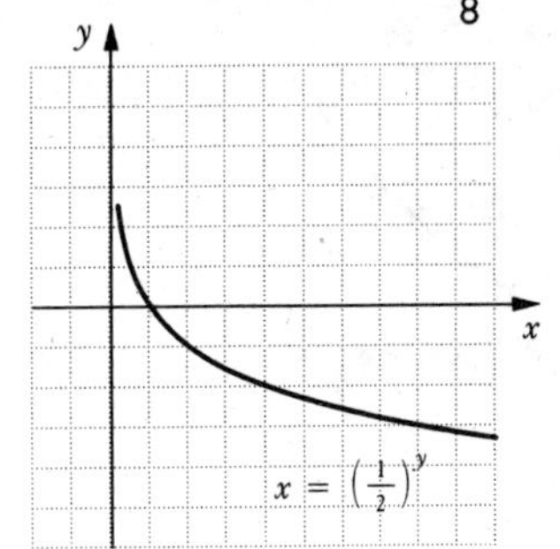

23. Graph: $x = 5^y$

We can find ordered pairs by choosing values for y and then computing values for x. Then we plot these points and connect them with a smooth curve.

For y = 0, $x = 5^0 = 1$.

For y = 1, $x = 5^1 = 5$.

For y = 2, $x = 5^2 = 25$.

For y = -1, $x = 5^{-1} = \frac{1}{5}$.

For y = -2, $x = 5^{-2} = \frac{1}{25}$.

x	y
1	0
5	1
25	2
$\frac{1}{5}$	-1
$\frac{1}{25}$	-2

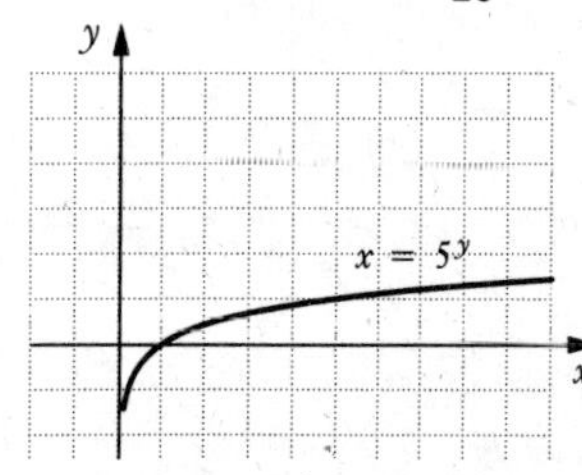

25. Graph $y = 2^x$ (see Exercise 1) and $x = 2^y$ (see Exercise 19) using the same set of axes.

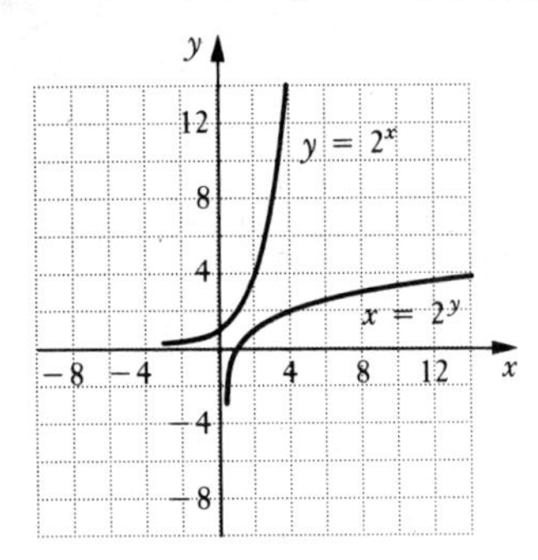

27. a) Use the formula $A = P(1 + r)^t$. Substitute \$50,000 for P and 8%, or 0.08, for r:

$A(t) = \$50,000(1 + 0.08)^t = \$50,000(1.08)^t$

b) Substitute for t and compute.

$A(0) = \$50,000(1.08)^0 = \$50,000(1) = \$50,000$

$A(4) = \$50,000(1.08)^4 \approx \$50,000(1.36048896) \approx \$68,024.45$

$A(8) = \$50,000(1.08)^8 \approx \$50,000(1.85093021) \approx \$92,546.51$

$A(10) = \$50,000(1.08)^{10} \approx \$50,000(2.158924997) \approx \$107,946.25$

c) We use the function values computed in part (b) with others, if we wish, and draw the graph. Note that the axes are scaled differently because of the large function values.

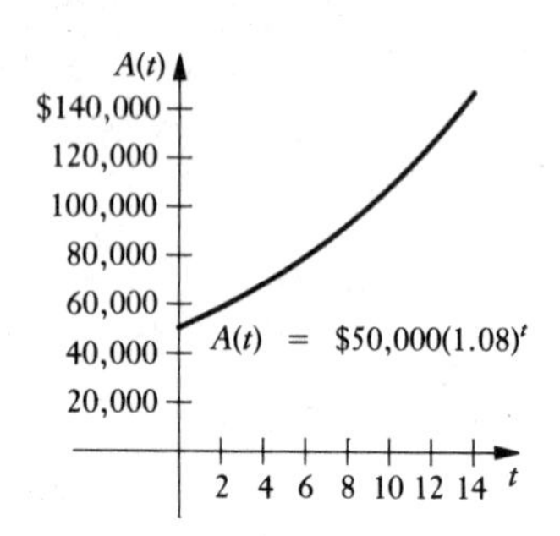

29. a) $V(0) = \$5200(0.8)^0 = \$5200;$

$V(1) = \$5200(0.8)^1 = \$4160;$

$V(2) = \$5200(0.8)^2 = \$3328;$

$V(5) = \$5200(0.8)^5 \approx \$1703.94;$

$V(10) = \$5200(0.8)^{10} \approx \$558.35$

b) We use the function values computed in part (a) and others, if we wish, and draw the graph. Note that the axes are scaled differently because of the large function values.

29. (continued)

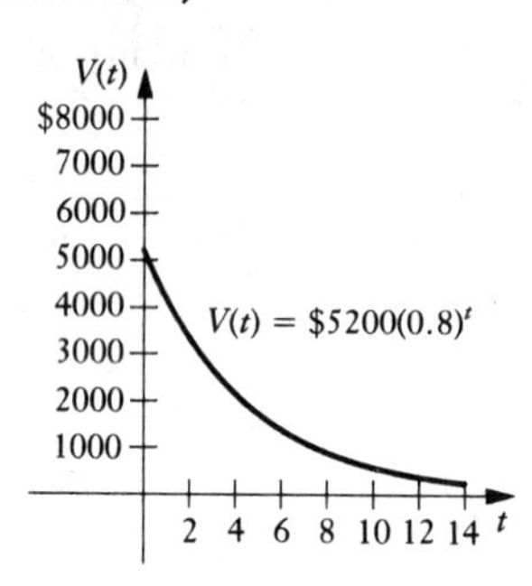

31. $x^{-5} \cdot x^3 = x^{-5+3} = x^{-2} = \frac{1}{x^2}$

33. $\frac{x^{-3}}{x^4} = x^{-3-4} = x^{-7} = \frac{1}{x^7}$

35. Graph: $y = 2^x + 2^{-x}$

Construct a table of values, thinking of y as f(x). Then plot these points and connect them with a curve.

$f(0) = 2^0 + 2^{-0} = 1 + 1 = 2$

$f(1) = 2^1 + 2^{-1} = 2 + \frac{1}{2} = 2\frac{1}{2}$

$f(2) = 2^2 + 2^{-2} = 4 + \frac{1}{4} = 4\frac{1}{4}$

$f(3) = 2^3 + 2^{-3} = 8 + \frac{1}{8} = 8\frac{1}{8}$

$f(-1) = 2^{-1} + 2^{-(-1)} = \frac{1}{2} + 2 = 2\frac{1}{2}$

$f(-2) = 2^{-2} + 2^{-(-2)} = \frac{1}{4} + 4 = 4\frac{1}{4}$

$f(-3) = 2^{-3} + 2^{-(-3)} = \frac{1}{8} + 8 = 8\frac{1}{8}$

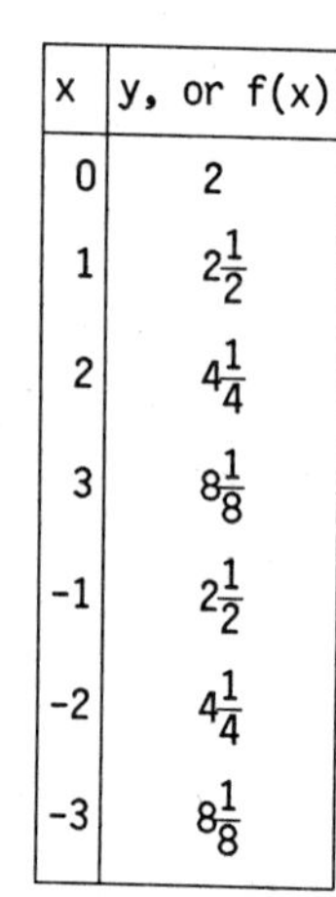

x	y, or f(x)
0	2
1	$2\frac{1}{2}$
2	$4\frac{1}{4}$
3	$8\frac{1}{8}$
-1	$2\frac{1}{2}$
-2	$4\frac{1}{4}$
-3	$8\frac{1}{8}$

37. Graph: $y = 3^x + 3^{-x}$

We construct a table of values, thinking of y as f(x). Then plot these points and connect them with a curve.

$f(0) = 3^0 + 3^{-0} = 1 + 1 = 2$

$f(1) = 3^1 + 3^{-1} = 3 + \frac{1}{3} = 3\frac{1}{3}$

$f(2) = 3^2 + 3^{-2} = 9 + \frac{1}{9} = 9\frac{1}{9}$

$f(-1) = 3^{-1} + 3^{-(-1)} = \frac{1}{3} + 3 = 3\frac{1}{3}$

$f(-2) = 3^{-2} + 3^{-(-2)} = \frac{1}{9} + 9 = 9\frac{1}{9}$

x	y, or f(x)
0	2
1	$3\frac{1}{3}$
2	$9\frac{1}{9}$
-1	$3\frac{1}{3}$
-2	$9\frac{1}{9}$

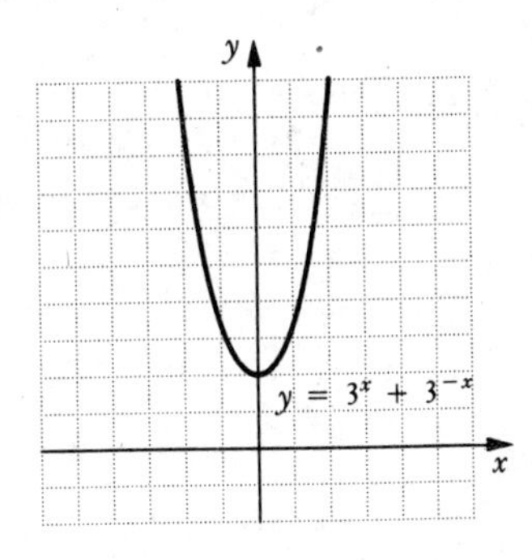

39. Graph: $y = |2^{x^2} - 1|$

We construct a table of values, thinking of y as f(x). Then we plot these points and connect them with a curve.

$f(0) = |2^{0^2} - 1| = |1 - 1| = 0$

$f(1) = |2^{1^2} - 1| = |2 - 1| = 1$

$f(2) = |2^{2^2} - 1| = |16 - 1| = 15$

$f(-1) = |2^{(-1)^2} - 1| = |2 - 1| = 1$

$f(-2) = |2^{(-2)^2} - 1| = |16 - 1| = 15$

x	y, or f(x)
0	0
1	1
2	15
-1	1
-2	15

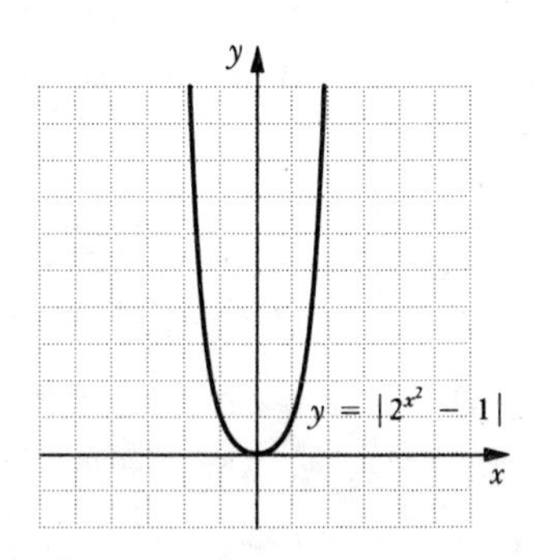

Exercise Set 13.2

1. Graph: $y = \log_2 x$

The equation $y = \log_2 x$ is equivalent to $2^y = x$. We can find ordered pairs by choosing values for y and computing the corresponding x-values.

For $y = 0$, $x = 2^0 = 1$.

For $y = 1$, $x = 2^1 = 2$.

For $y = 2$, $x = 2^2 = 4$.

For $y = 3$, $x = 2^3 = 8$.

For $y = -1$, $x = 2^{-1} = \frac{1}{2}$.

For $y = -2$, $x = 2^{-2} = \frac{1}{4}$.

x, or 2^y	y
1	0
2	1
4	2
8	3
$\frac{1}{2}$	-1
$\frac{1}{4}$	-2

(1) Select y.
(2) Compute x.

We plot the set of ordered pairs and connect the points with a smooth curve.

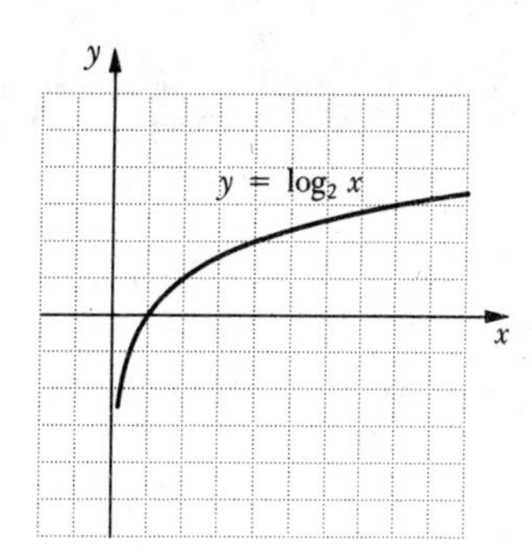

3. Graph: $y = \log_6 x$

The equation $y = \log_6 x$ is equivalent to $6^y = x$. We can find ordered pairs by choosing values for y and computing the corresponding x-values.

For $y = 0$, $x = 6^0 = 1$.

For $y = 1$, $x = 6^1 = 6$.

For $y = 2$, $x = 6^2 = 36$.

For $y = -1$, $x = 6^{-1} = \frac{1}{6}$.

For $y = -2$, $x = 6^{-2} = \frac{1}{36}$.

x, or 6^y	y
1	0
6	1
36	2
$\frac{1}{6}$	-1
$\frac{1}{36}$	-2

We plot the set of ordered pairs and connect the points with a smooth curve.

3. (continued)

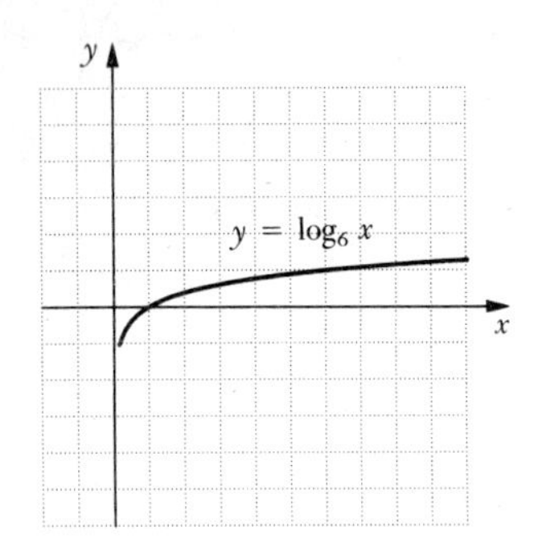

5. Graph $f(x) = 3^x$ (see Exercise Set 10.1, Exercise 2) and $f^{-1}(x) = \log_3 x$ on the same set of axes. We can obtain the graph of f^{-1} by reflecting the graph of f across the line y = x.

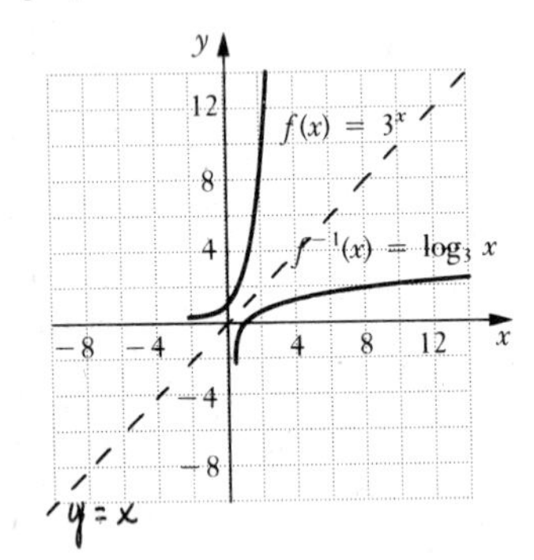

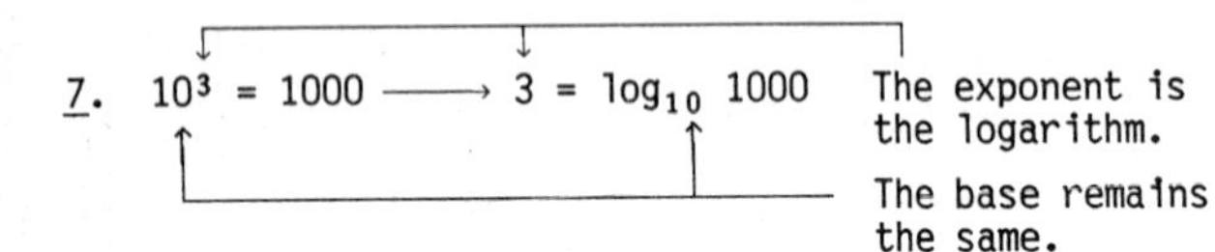

7. $10^3 = 1000 \longrightarrow 3 = \log_{10} 1000$ The exponent is the logarithm. The base remains the same.

9. $5^{-3} = \frac{1}{125} \longrightarrow -3 = \log_5 \frac{1}{125}$ The exponent is the logarithm. The base remains the same.

11. $8^{1/3} = 2 \longrightarrow \frac{1}{3} = \log_8 2$

13. $10^{0.3010} = 2 \longrightarrow 0.3010 = \log_{10} 2$

15. $e^2 = t \longrightarrow 2 = \log_e t$

17. $Q^t = x \longrightarrow t = \log_Q x$

19. $e^2 = 7.3891 \longrightarrow 2 = \log_e 7.3891$

21. $e^{-2} = 0.1353 \longrightarrow -2 = \log_e 0.1353$

23. $t = \log_3 8 \longrightarrow 3^t = 8$ The logarithm is the exponent. The base remains the same.

25. $\log_5 25 = 2 \longrightarrow 5^2 = 25$ The logarithm is the exponent. The base remains the same.

27. $\log_{10} 0.1 = -1 \longrightarrow 10^{-1} = 0.1$

29. $\log_{10} 7 = 0.845 \longrightarrow 10^{0.845} = 7$

31. $\log_e 20 = 2.9957 \longrightarrow e^{2.9957} = 20$

33. $\log_t Q = k \longrightarrow t^k = Q$

35. $\log_3 x = 2$

$3^2 = x$ Converting to an exponential equation

$9 = x$ Computing 3^2

37. $\log_x 16 = 2$

$x^2 = 16$ Converting to an exponential equation

$x = 4$ or $x = -4$ Principle of square roots

$\log_4 16 = 2$ because $4^2 = 16$. Thus, 4 is a solution. Since all logarithm bases must be positive, $\log_{-4} 16$ is not defined and -4 is not a solution.

39. $\log_2 16 = x$

$2^x = 16$ Converting to an exponential equation

$2^x = 2^4$

$x = 4$ The exponents are the same.

41. $\log_3 27 = x$

$3^x = 27$ Converting to an exponential equation

$3^x = 3^3$

$x = 3$ The exponents are the same.

43. $\log_x 13 = 1$

$x^1 = 13$ Converting to an exponential equation

$x = 13$ Simplifying x^1

45. $\log_6 x = 0$

$6^0 = x$ Converting to an exponential equation

$1 = x$ Computing 6^0

47. $\log_2 x = -1$

$2^{-1} = x$ Converting to an exponential equation

$\frac{1}{2} = x$ Simplifying

49. $\log_8 x = \frac{1}{3}$

$8^{1/3} = x$

$2 = x$

51. Let $\log_{10} 100 = x$.
Then $10^x = 100$
$10^x = 10^2$
$x = 2$.
Thus, $\log_{10} 100 = 2$.

53. Let $\log_{10} 0.1 = x$.
Then $10^x = 0.1 = \frac{1}{10}$
$10^x = 10^{-1}$
$x = -1$.
Thus, $\log_{10} 0.1 = -1$.

55. Let $\log_{10} 1 = x$.
Then $10^x = 1$
$10^x = 10^0 \quad (10^0 = 1)$
$x = 0$.
Thus, $\log_{10} 1 = 0$.

57. Let $\log_5 625 = x$.
Then $5^x = 625$
$5^x = 5^4$
$x = 4$.
Thus, $\log_5 625 = 4$.

59. Think of the meaning of $\log_7 49$. It is the exponent to which you raise 7 to get 49. That exponent is 2. Therefore, $\log_7 49 = 2$.

61. Think of the meaning of $\log_2 8$. It is the exponent to which you raise 2 to get 8. That exponent is 3. Therefore, $\log_2 8 = 3$.

63. Let $\log_5 \frac{1}{25} = x$.
Then $5^x = \frac{1}{25}$
$5^x = 5^{-2}$
$x = -2$.
Thus, $\log_5 \frac{1}{25} = -2$.

65. Let $\log_3 1 = x$.
Then $3^x = 1$
$3^x = 3^0 \quad (3^0 = 1)$
$x = 0$.
Thus, $\log_3 1 = 0$.

67. Let $\log_e e = x$.
Then $e^x = e$
$e^x = e^1$
$x = 1$.
Thus, $\log_e e = 1$.

69. Let $\log_{27} 9 = x$.
Then $27^x = 9$
$(3^3)^x = 3^2$
$3^{3x} = 3^2$
$3x = 2$
$x = \frac{2}{3}$.
Thus, $\log_{27} 9 = \frac{2}{3}$.

71. Graph: $y = \left(\frac{3}{2}\right)^x$

x	y, or $\left(\frac{3}{2}\right)^x$
0	1
1	$\frac{3}{2}$
2	$\frac{9}{4}$
3	$\frac{27}{8}$
-1	$\frac{2}{3}$
-2	$\frac{4}{9}$

Graph: $y = \log_{3/2} x$, or $x = \left(\frac{3}{2}\right)^y$

x, or $\left(\frac{3}{2}\right)^y$	y
1	0
$\frac{3}{2}$	1
$\frac{9}{4}$	2
$\frac{27}{8}$	3
$\frac{2}{3}$	-1
$\frac{4}{9}$	-2

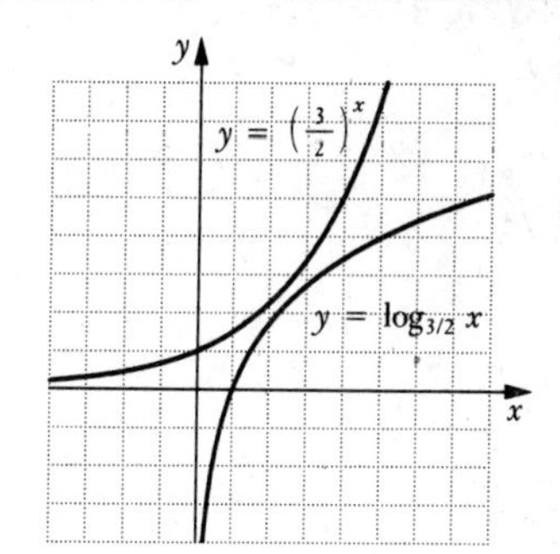

73. Graph: $y = \log_3 |x + 1|$

x	y
0	0
2	1
8	2
-2	0
-4	1
-9	2

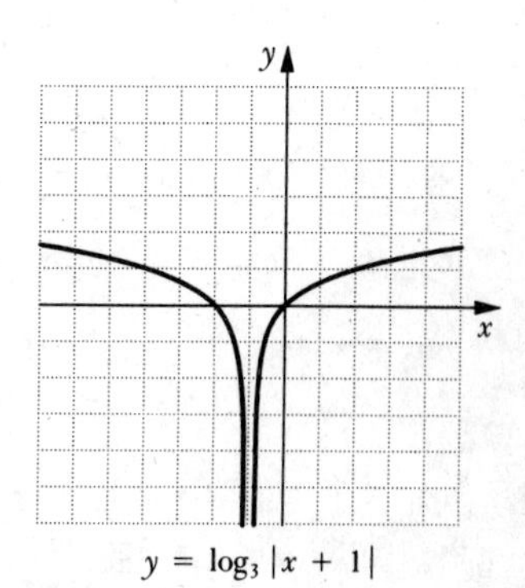

75. $\log_{125} x = \frac{2}{3}$

$125^{2/3} = x$

$(5^3)^{2/3} = x$

$5^2 = x$

$25 = x$

77. $\log_8 (2x + 1) = -1$

$8^{-1} = 2x + 1$

$\frac{1}{8} = 2x + 1$

$1 = 16x + 8$ Multiplying by 8

$-7 = 16x$

$-\frac{7}{16} = x$

79. Let $\log_{1/4} \frac{1}{64} = x$.

Then $\left(\frac{1}{4}\right)^x = \frac{1}{64}$

$\left(\frac{1}{4}\right)^x = \left(\frac{1}{4}\right)^3$

$x = 3.$

Thus, $\log_{1/4} \frac{1}{64} = 3$.

81. $\log_{10} (\log_4 (\log_3 81))$

$= \log_{10} (\log_4 4)$ $(\log_3 81 = 4)$

$= \log_{10} 1$ $(\log_4 4 = 1)$

$= 0$

83. Let $\log_{1/5} 25 = x$.

Then $\left(\frac{1}{5}\right)^x = 25$

$(5^{-1})^x = 25$

$5^{-x} = 5^2$

$-x = 2$

$x = -2$

Exercise Set 13.3

1. $\log_2 (32 \cdot 8) = \log_2 32 + \log_2 8$ Property 1

3. $\log_4 (64 \cdot 16) = \log_4 64 + \log_4 16$ Property 1

5. $\log_c Bx = \log_c B + \log_c x$ Property 1

7. $\log_a 6 + \log_a 70 = \log_a (6 \cdot 70)$ Property 1

$= \log_a 420$

9. $\log_c K + \log_c y = \log_c K \cdot y$ Property 1

$= \log_c Ky$

11. $\log_a x^3 = 3 \log_a x$ Property 2

13. $\log_c y^6 = 6 \log_c y$ Property 2

15. $\log_b C^{-3} = -3 \log_b C$ Property 2

17. $\log_a \frac{67}{5} = \log_a 67 - \log_a 5$ Property 3

19. $\log_b \frac{3}{4} = \log_b 3 - \log_b 4$ Property 3

21. $\log_a 15 - \log_a 7 = \log_a \frac{15}{7}$ Property 3

23. $\log_a x^2y^3z$

$= \log_a x^2 + \log_a y^3 + \log_a z$ Property 1

$= 2 \log_a x + 3 \log_a y + \log_a z$ Property 2

25. $\log_b \frac{xy^2}{z^3}$

$= \log_b xy^2 - \log_b z^3$ Property 3

$= \log_b x + \log_b y^2 - \log_b z^3$ Property 1

$= \log_b x + 2 \log_b y - 3 \log_b z$ Property 2

27. $\log_c \sqrt[3]{\frac{x^4}{y^3z^2}}$

$= \log_c \left(\frac{x^4}{y^3z^2}\right)^{1/3}$

$= \frac{1}{3} \log_c \frac{x^4}{y^3z^2}$ Property 2

$= \frac{1}{3} (\log_c x^4 - \log_c y^3z^2)$ Property 3

$= \frac{1}{3} [\log_c x^4 - (\log_c y^3 + \log_c z^2)]$ Property 1

$= \frac{1}{3} (\log_c x^4 - \log_c y^3 - \log_c z^2)$ Removing parentheses

$= \frac{1}{3}(4 \log_c x - 3 \log_c y - 2 \log_c z)$ Property 2

$= \frac{4}{3} \log_c x - \log_c y - \frac{2}{3} \log_c z$

29. $\log_a \sqrt[4]{\frac{m^8n^{12}}{a^3b^5}}$

$= \log_a \left(\frac{m^8n^{12}}{a^3b^5}\right)^{1/4}$

$= \frac{1}{4} \log_a \frac{m^8n^{12}}{a^3b^5}$ Property 2

$= \frac{1}{4} (\log_a m^8n^{12} - \log_a a^3b^5)$ Property 3

$= \frac{1}{4} [\log_a m^8 + \log_a n^{12} - (\log_a a^3 + \log_a b^5)]$ Property 1

$= \frac{1}{4} (\log_a m^8 + \log_a n^{12} - \log_a a^3 - \log_a b^5)$ Removing parentheses

$= \frac{1}{4} (\log_a m^8 + \log_a n^{12} - 3 - \log_a b^5)$ Property 4

$= \frac{1}{4} (8 \log_a m + 12 \log_a n - 3 - 5 \log_a b)$ Property 2

$= 2 \log_a m + 3 \log_a n - \frac{3}{4} - \frac{5}{4} \log_a b$

31. $\frac{2}{3}\log_a x - \frac{1}{2}\log_a y$

$= \log_a x^{2/3} - \log_a y^{1/2}$ Property 2

$= \log_a \sqrt[3]{x^2} - \log_a \sqrt{y}$

$= \log_a \frac{\sqrt[3]{x^2}}{\sqrt{y}}$ Property 3

$= \log_a \frac{\sqrt[3]{x^2}\sqrt{y}}{y}$ Multiplying by $\frac{\sqrt{y}}{\sqrt{y}}$

33. $\log_a 2x + 3(\log_a x - \log_a y)$

$= \log_a 2x + 3\log_a x - 3\log_a y$

$= \log_a 2x + \log_a x^3 - \log_a y^3$ Property 2

$= \log_a 2x^4 - \log_a y^3$ Property 1

$= \log_a \frac{2x^4}{y^3}$ Property 3

35. $\log_a \frac{a}{\sqrt{x}} - \log_a \sqrt{ax}$

$= \log_a ax^{-1/2} - \log_a a^{1/2}x^{1/2}$

$= \log_a \frac{ax^{-1/2}}{a^{1/2}x^{1/2}}$ Property 3

$= \log_a \frac{a^{1/2}}{x}$

$= \log_a \frac{\sqrt{a}}{x}$

37. $\log_b 15 = \log_b (3\cdot 5)$

$= \log_b 3 + \log_b 5$ Property 1

$= 1.099 + 1.609$

$= 2.708$

39. $\log_b \frac{5}{3} = \log_b 5 - \log_b 3$ Property 3

$= 1.609 - 1.099$

$= 0.51$

41. $\log_b \frac{1}{5} = \log_b 1 - \log_b 5$ Property 3

$= 0 - 1.609$ ($\log_b 1 = 0$)

$= -1.609$

43. $\log_b \sqrt{b^3} = \log_b b^{3/2} = \frac{3}{2}$ Property 4

45. $\log_b 5b = \log_b 5 + \log_b b$ Property 1

$= 1.609 + 1$ ($\log_b b = 1$)

$= 2.609$

47. $\log_t t^9 = 9$ Property 4

49. $\log_e e^m = m$ Property 4

51. $\log_3 3^4 = x$

$4 = x$ Property 4

53. $\log_e e^x = -7$

$x = -7$ Property 4

55. $i^{29} = i^{28}\cdot i = (i^4)^7\cdot i = 1^7\cdot i = 1\cdot i = i$

57. $\frac{2+i}{2-i} = \frac{2+i}{2-i}\cdot\frac{2+i}{2+i} = \frac{4+4i+i^2}{4-i^2} = \frac{4+4i-1}{4-(-1)} =$

$\frac{3+4i}{5} = \frac{3}{5} + \frac{4}{5}i$

59. $\log_a (x^8 - y^8) - \log_a (x^2 + y^2)$

$= \log_a \frac{x^8 - y^8}{x^2 + y^2}$ Property 3

$= \log_a \frac{(x^4 + y^4)(x^2 + y^2)(x + y)(x - y)}{x^2 + y^2}$ Factoring

$= \log_a [(x^4 + y^4)(x^2 - y^2)]$ Simplifying

$= \log_a (x^6 - x^4y^2 + x^2y^4 - y^6)$ Multiplying

61. $\log_a \sqrt{1 - s^2}$

$= \log_a (1 - s^2)^{1/2}$

$= \frac{1}{2}\log_a (1 - s^2)$

$= \frac{1}{2}\log_a [(1 - s)(1 + s)]$

$= \frac{1}{2}\log_a (1 - s) + \frac{1}{2}\log_a (1 + s)$

63. False. For example, let a = 10, P = 100, and Q = 10.

$\frac{\log 100}{\log 10} = \frac{2}{1} = 2$, but

$\log \frac{100}{10} = \log 10 = 1$.

65. True, by Property 1

67. False. For example, let a = 2, P = 1, and Q = 1.

$\log_2 (1 + 1) = \log_2 2 = 1$, but

$\log_2 1 + \log_2 1 = 0 + 0 = 0$.

Exercise Set 13.4

1. 0.3010

3. 0.9031

5. 0.8021

7. 1.7952

9. 2.6405

11. 4.1271

13. -0.2441

15. -1.2840

17. -2.2069

19. 1000

21. 501.1872

23. 3.0001

25. 0.2841

27. 0.0011

29. 0.6931

31. 2.0794

33. 4.1271

35. -5.0832

37. 36.7890

39. 0.0023

41. 1.0057

43. 1.190×10^{10}

45. 8.1490

47. -3.3496

49. 1637.9488

51. 7.6331

53. We will use common logarithms for the conversion. Let a = 10, b = 6, and M = 100 and substitute into the change-of-base formula.

$$\log_b M = \frac{\log_a M}{\log_a b}$$

$$\log_6 100 = \frac{\log_{10} 100}{\log_{10} 6} \approx \frac{2}{0.7782} \approx 2.5702$$

55. We will use common logarithms for the conversion. Let a = 10, b = 2, and M = 10 and substitute in the change-of-base formula.

$$\log_2 10 = \frac{\log_{10} 10}{\log_{10} 2} \approx \frac{1}{0.3010} \approx 3.3219$$

57. We will use natural logarithms for the conversion. Let a = e, b = 200, and M = 30 and substitute in the change-of-base formula.

$$\log_{200} 30 = \frac{\ln 30}{\ln 200} \approx \frac{3.4012}{5.2983} \approx 0.6419$$

59. We will use natural logarithms for the conversion. Let a = e, b = 0.5, and M = 5 and substitute in the change-of-base formula.

$$\log_{0.5} 5 = \frac{\ln 5}{\ln 0.5} \approx \frac{1.6094}{-0.6931} \approx -2.3219$$

61. We will use common logarithms for the conversion. Let a = 10, b = 2, and M = 0.2 and substitute in the change-of-base formula.

$$\log_2 0.2 = \frac{\log_{10} 0.2}{\log_{10} 2} \approx \frac{-0.6990}{0.3010} \approx -2.3219$$

63. We will use natural logarithms for the conversion. Let a = e, b = π, and M = 58 and substitute in the change-of-base formula.

$$\log_\pi 58 = \frac{\ln 58}{\ln \pi} \approx \frac{4.0604}{1.1447} \approx 3.5471$$

65. $ax^2 - b = 0$

$ax^2 = b$

$x^2 = \frac{b}{a}$

$x = \pm\sqrt{\frac{b}{a}}$

The solution is $\pm\sqrt{\frac{b}{a}}$.

67. $x^{1/2} - 6x^{1/4} + 8 = 0$

Let $u = x^{1/4}$.

$u^2 - 6u + 8 = 0$ Substituting

$(u - 4)(u - 2) = 0$

$u = 4$ or $u = 2$

$x^{1/4} = 4$ or $x^{1/4} = 2$

$x = 256$ or $x = 16$ Raising both sides to the fourth power

Both numbers check. The solutions are 256 and 16.

69. Use the change-of-base formula with a = e and b = 10. We obtain

$$\log M = \frac{\ln M}{\ln 10}, \text{ or } 0.4343 \ln M.$$

71.

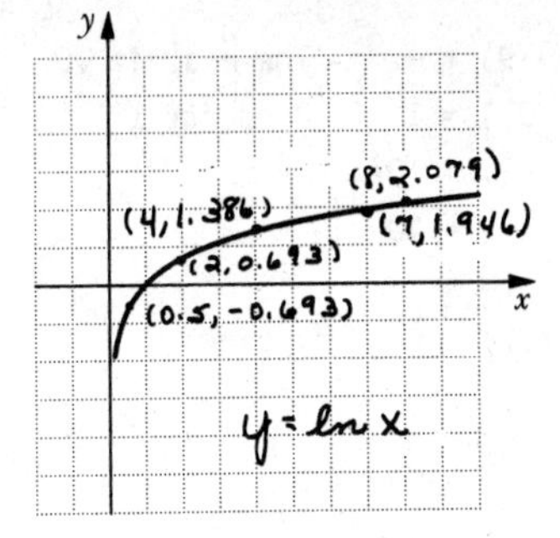

73. $\dfrac{\log_2 47}{\log_2 16} = \log_{16} 47$ Change of base formula

≈ 1.3886

75. $\dfrac{4.31}{\ln x} = \dfrac{28}{3.01}$

$\dfrac{4.31(3.01)}{28} = \ln x$

$0.463325 = \ln x$

$1.5893 \approx x$

Exercise Set 13.5

1. $2^x = 8$

$2^x = 2^3$

$x = 3$ The exponents are the same.

3. $4^x = 256$

$4^x = 4^4$

$x = 4$ The exponents are the same.

5. $2^{2x} = 32$

$2^{2x} = 2^5$

$2x = 5$

$x = \dfrac{5}{2}$

7. $3^{5x} = 27$

$3^{5x} = 3^3$

$5x = 3$

$x = \dfrac{3}{5}$

9. $2^x = 9$

$\log 2^x = \log 9$ Taking the common logarithm on both sides

$x \log 2 = \log 9$ Property 2

$x = \dfrac{\log 9}{\log 2}$ Solving for x

$x \approx 3.170$ Using a calculator

11. $2^x = 10$

$\log 2^x = \log 10$ Taking the common logarithm on both sides

$x \log 2 = \log 10$ Property 2

$x = \dfrac{\log 10}{\log 2}$ Solving for x

$x \approx 3.322$ Using a calculator

13. $5^{4x-7} = 125$

$5^{4x-7} = 5^3$

$4x - 7 = 3$ The exponents are the same.

$4x = 10$

$x = \dfrac{10}{4}$, or $\dfrac{5}{2}$

15. $3^{x^2} \cdot 3^{4x} = \dfrac{1}{27}$

$3^{x^2+4x} = 3^{-3}$

$x^2 + 4x = -3$

$x^2 + 4x + 3 = 0$

$(x + 3)(x + 1) = 0$

$x = -3$ or $x = -1$

17. $4^x = 7$

$\log 4^x = \log 7$

$x \log 4 = \log 7$

$x = \dfrac{\log 7}{\log 4}$

$x \approx \dfrac{0.8451}{0.6021}$

$x \approx 1.404$

19. $e^t = 100$

$\ln e^t = \ln 100$ Taking ln on both sides

$t = \ln 100$ Property 4

$t \approx 4.605$ Using a calculator

21. $e^{-t} = 0.1$

$\ln e^{-t} = \ln 0.1$ Taking ln on both sides

$-t = \ln 0.1$ Property 4

$-t \approx -2.303$

$t \approx 2.303$

23. $e^{-0.02t} = 0.06$

$\ln e^{-0.02t} = \ln 0.06$ Taking ln on both sides

$-0.02t = \ln 0.06$ Property 4

$t = \dfrac{\ln 0.06}{-0.02}$

$t \approx \dfrac{-2.8134}{-0.02}$

$t \approx 140.671$

25.
$$2^x = 3^{x-1}$$
$$\log 2^x = \log 3^{x-1}$$
$$x \log 2 = (x - 1) \log 3$$
$$x \log 2 = x \log 3 - \log 3$$
$$\log 3 = x \log 3 - x \log 2$$
$$\log 3 = x(\log 3 - \log 2)$$
$$\frac{\log 3}{\log 3 - \log 2} = x$$
$$\frac{0.4771}{0.4771 - 0.3010} \approx x$$
$$-2.710 \approx x$$

27.
$$(2.8)^x = 41$$
$$\log (2.8)^x = \log 41$$
$$x \log 2.8 = \log 41$$
$$x = \frac{\log 41}{\log 2.8}$$
$$x \approx \frac{1.6128}{0.4472}$$
$$x \approx 3.607$$

29. $\log_3 x = 3$

$x = 3^3$ Writing an equivalent exponential expression

$x = 27$

31. $\log_2 x = -3$

$x = 2^{-3}$ Writing an equivalent exponential expression

$x = \frac{1}{8}$

33. $\log x = 1$ The base is 10.

$x = 10^1$

$x = 10$

35. $\log x = -2$ The base is 10.

$x = 10^{-2}$

$x = \frac{1}{100}$

37. $\ell n\ x = 2$

$x = e^2 \approx 7.389$

39. $\ell n\ x = -1$

$x = e^{-1}$

$x = \frac{1}{e} \approx 0.368$

41.
$$\log_5 (2x - 7) = 3$$
$$2x - 7 = 5^3$$
$$2x - 7 = 125$$
$$2x = 132$$
$$x = 66$$

The answer checks. The solution is 66.

43. $\log x + \log (x - 9) = 1$ The base is 10.

$\log_{10} [x(x - 9)] = 1$ Property 1

$$x(x - 9) = 10^1$$
$$x^2 - 9x = 10$$
$$x^2 - 9x - 10 = 0$$
$$(x - 10)(x + 1) = 0$$

$x = 10$ or $x = -1$

Check: For 10:

$\log x + \log (x - 9) = 1$	
$\log 10 + \log (10 - 9)$	1
$\log 10 + \log 1$	
$1 + 0$	
1	TRUE

For -1:

$\log x + \log (x - 9) = 1$		
$\log (-1) + \log (-1 - 9)$	1	FALSE

The number -1 does not check, because negative numbers do not have logarithms. The solution is 10.

45. $\log x - \log (x + 3) = -1$ The base is 10.

$\log_{10} \frac{x}{x + 3} = -1$ Property 3

$$\frac{x}{x + 3} = 10^{-1}$$
$$\frac{x}{x + 3} = \frac{1}{10}$$
$$10x = x + 3$$
$$9x = 3$$
$$x = \frac{1}{3}$$

The answer checks. The solution is $\frac{1}{3}$.

47. $\log_2 (x + 1) + \log_2 (x - 1) = 3$

$\log_2 [(x + 1)(x - 1)] = 3$ Property 1

$$(x + 1)(x - 1) = 2^3$$
$$x^2 - 1 = 8$$
$$x^2 = 9$$
$$x = \pm 3$$

The number 3 checks, but -3 does not. The solution is 3.

49. $\log_4 (x + 6) - \log_4 x = 2$

$\log_4 \frac{x + 6}{x} = 2$ Property 3

$$\frac{x + 6}{x} = 4^2$$
$$\frac{x + 6}{x} = 16$$
$$x + 6 = 16x$$
$$6 = 15x$$
$$\frac{2}{5} = x$$

The answer checks. The solution is $\frac{2}{5}$.

51. $\log_4 (x + 3) + \log_4 (x - 3) = 2$

$\log_4 [(x + 3)(x - 3)] = 2$ Property 1

$(x + 3)(x - 3) = 4^2$

$x^2 - 9 = 16$

$x^2 = 25$

$x = \pm 5$

The number 5 checks, but -5 does not. The solution is 5.

53. $\log_3 (2x - 6) - \log_3 (x + 4) = 2$

$\log_3 \frac{2x - 6}{x + 4} = 2$ Property 3

$\frac{2x - 6}{x + 4} = 3^2$

$\frac{2x - 6}{x + 4} = 9$

$2x - 6 = 9x + 36$

Multiplying by $(x + 4)$

$-42 = 7x$

$-6 = x$

Check:

$\log_3 (2x - 6) - \log_3 (x + 4)$	$= 2$
$\log_3 [2(-6) - 6] - \log_3 (-6 + 4)$	2
$\log_3 (-18) - \log_3 (-2)$	
	FALSE

The number -6 does not check, because negative numbers do not have logarithms. There is no solution.

55. $x^2 + y^2 = 25$ (1)

$y - x = 1$ (2)

First solve equation (2) for y.

$y - x = 1$ (2)

$y = x + 1$

Then substitute x + 1 for y in equation (1) and solve for x.

$x^2 + y^2 = 25$ (1)

$x^2 + (x + 1)^2 = 25$ Substituting

$x^2 + x^2 + 2x + 1 = 25$

$2x^2 + 2x - 24 = 0$

$x^2 + x - 12 = 0$

$(x + 4)(x - 3) = 0$

$x + 4 = 0$ or $x - 3 = 0$

$x = -4$ or $x = 3$

We now substitute these numbers for x in equation (2) and solve for y.

$y = x + 1$ (2)

When $x = -4$, $y = -4 + 1$, or -3.

When $x = 3$, $y = 3 + 1$, or 4.

The ordered pairs (-4,-3) and (3,4) check and are the solutions.

57. $2x^2 + 1 = y^2$ or $2x^2 - y^2 = -1$ (1)

$2y^2 + x^2 = 22$ $x^2 + 2y^2 = 22$ (2)

Solve the system using the elimination method.

$4x^2 - 2y^2 = -2$ Multiplying by 2

$x^2 + 2y^2 = 22$

$5x^2 = 20$ Adding

$x^2 = 4$

$x = \pm 2$

Substitute these values for x in either of the original equations and solve for y. Here we use equation (1).

$2x^2 + 1 = y^2$ (1)

When $x = 2$, $y^2 = 2 \cdot 2^2 + 1$, or 9. Thus $y = \pm 3$.

When $x = -2$, $y^2 = 2(-2)^2 + 1$, or 9. Thus $y = \pm 3$.

The ordered pairs (2,3), (2,-3), (-2,3), and (-2,-3) check and are the solutions.

59. $8^x = 16^{3x+9}$

$(2^3)^x = (2^4)^{3x+9}$

$2^{3x} = 2^{12x+36}$

$3x = 12x + 36$

$-36 = 9x$

$-4 = x$

61. $\log_6 (\log_2 x) = 0$

$\log_2 x = 6^0$

$\log_2 x = 1$

$x = 2^1$

$x = 2$

63. $\log_5 \sqrt{x^2 - 9} = 1$

$\sqrt{x^2 - 9} = 5^1$

$x^2 - 9 = 25$ Squaring both sides

$x^2 = 34$

$x = \pm\sqrt{34}$

Both numbers check. The solutions are $\pm\sqrt{34}$.

65. $\log (\log x) = 5$ The base is 10.

$\log x = 10^5$

$\log x = 100,000$

$x = 10^{100,000}$

The number checks. The solution is $10^{100,000}$.

67. $\log x^2 = (\log x)^2$

$2 \log x = (\log x)^2$

$0 = (\log x)^2 - 2 \log x$

Let $u = \log x$.

$0 = u^2 - 2u$

$0 = u(u - 2)$

$u = 0$ or $u = 2$

$\log x = 0$ or $\log x = 2$

$x = 10^0$ or $x = 10^2$

$x = 1$ or $x = 100$

Both numbers check. The solutions are 1 and 100.

69. $\log_a a^{x^2+4x} = 21$

$x^2 + 4x = 21$ Property 4

$x^2 + 4x - 21 = 0$

$(x + 7)(x - 3) = 0$

$x = -7$ or $x = 3$

Both numbers check. The solutions are -7 and 3.

71. $3^{2x} - 8 \cdot 3^x + 15 = 0$

Let $u = 3^x$ and substitute.

$u^2 - 8u + 15 = 0$

$(u - 5)(u - 3) = 0$

$u = 5$ or $u = 3$

$3^x = 5$ or $3^x = 3$ Substituting 3^x for u

$\log 3^x = \log 5$ or $3^x = 3^1$

$x \log 3 = \log 5$ or $x = 1$

$x = \frac{\log 5}{\log 3}$ or $x = 1$

Both numbers check. Note that we can also express $\frac{\log 5}{\log 3}$ as $\log_3 5$ using the change of base formula.

73. $\log_5 125 = 3$ and $\log_{125} 5 = \frac{1}{3}$, so

$x = (\log_{125} 5)^{\log_5 125}$ is equivalent to

$x = \left(\frac{1}{3}\right)^3 = \frac{1}{27}$. Then $\log_3 x = \log_3 \frac{1}{27} = -3$.

Exercise Set 13.6

1. a) We set $A(t) = \$450,000$ and solve for t:

$450,000 = 50,000(1.06)^t$

$\frac{450,000}{50,000} = (1.06)^t$

$9 = (1.06)^t$

$\log 9 = \log (1.06)^t$ Taking the common logarithm on both sides

$\log 9 = t \log 1.06$ Property 2

$t = \frac{\log 9}{\log 1.06} \approx \frac{0.95424}{0.02531} \approx 37.7$

It will take about 37.7 years for the \$50,000 to grow to \$450,000.

b) We set $A(t) = \$100,000$ and solve for t:

$100,000 = 50,000(1.06)^t$

$2 = (1.06)^t$

$\log 2 = \log (1.06)^t$

Taking the common logarithm on both sides

$\log 2 = t \log 1.06$ Property 2

$t = \frac{\log 2}{\log 1.06} \approx \frac{0.30103}{0.02531} \approx 11.9$

The doubling time is about 11.9 years.

3. a) We set $N(t) = 60,000$ and solve for t:

$60,000 = 250,000\left(\frac{1}{4}\right)^t$

$\frac{60,000}{250,000} = \left(\frac{1}{4}\right)^t$

$0.24 = (0.25)^t$ $\left(\frac{1}{4} = 0.25\right)$

$\log 0.24 = \log (0.25)^t$

$\log 0.24 = t \log 0.25$

$t = \frac{\log 0.24}{\log 0.25} \approx \frac{-0.61979}{-0.60206} \approx 1.0$

After about 1 year 60,000 cans will still be in use.

b) We set $N(t) = 10$ and solve for t.

$10 = 250,000\left(\frac{1}{4}\right)^t$

$\frac{10}{250,000} = \left(\frac{1}{4}\right)^t$

$0.00004 = (0.25)^t$

$\log 0.00004 = \log (0.25)^t$

$\log 0.00004 = t \log 0.25$

$t = \frac{\log 0.00004}{\log 0.25} \approx \frac{-4.39794}{-0.60206} \approx 7.3$

After about 7.3 years only 10 cans will still be in use.

5. a) One billion is 1000 millions, so we set $N(t) = 1000$ and solve for t:

$$1000 = 7.5(6)^{0.5t}$$
$$\frac{1000}{7.5} = (6)^{0.5t}$$
$$\log \frac{1000}{7.5} = \log(6)^{0.5t}$$
$$\log 1000 - \log 7.5 = 0.5t \log 6$$
$$t = \frac{\log 1000 - \log 7.5}{0.5 \log 6}$$
$$t \approx \frac{3 - 0.87506}{0.5(0.77815)} \approx 5.5$$

After about 5.5 years, one billion compact discs will be sold in a year.

b) When $t = 0$, $N(t) = 7.5(6)^{0.5(0)} = 7.5(6)^0 = 7.5(1) = 7.5$. Twice this initial number is 15, so we set $N(t) = 15$ and solve for t:

$$15 = 7.5(6)^{0.5t}$$
$$2 = (6)^{0.5t}$$
$$\log 2 = \log(6)^{0.5t}$$
$$\log 2 = 0.5t \log 6$$
$$t = \frac{\log 2}{0.5 \log 6} \approx \frac{0.30103}{0.5(0.77815)} \approx 0.8$$

The doubling time is about 0.8 year.

7. We substitute 175 for P in the function for walking speed, since P is in thousands.

$$R(P) = 0.37 \ln P + 0.05$$
$$R(175) = 0.37 \ln 175 + 0.05$$
$$\approx 0.37(5.1648) + 0.05 \quad \text{Finding } \ln 175 \text{ on a calculator}$$
$$\approx 2.0 \text{ ft/sec}$$

9. We substitute 50.4 for P in the function, since P is in thousands.

$$R(50.4) = 0.37 \ln 50.4 + 0.05$$
$$\approx 0.37(3.9200) + 0.05$$
$$\approx 1.5 \text{ ft/sec}$$

11. a) The equation $P(t) = P_0e^{kt}$ can be used to model population growth. At $t = 0$ (1987), the population was 5.0 billion. We substitute 5 for P_0 and 2.8%, or 0.028, for k to obtain the exponential growth function:

$P(t) = 5e^{0.028t}$, where t is the number of years after 1987.

b) In 1996, $t = 1996 - 1987$, or 9. To find the population in 1996, we substitute 9 for t:

$$P(9) = 5e^{0.028(9)}$$
$$= 5e^{0.252}$$
$$\approx 5(1.2866)$$
$$\approx 6.4$$

We can predict that the population will be about 6.4 billion in 1996.

11. (continued)

In 2000, $t = 2000 - 1987$, or 13. To find the population in 2000, we substitute 13 for t:

$$P(13) = 5e^{0.028(13)}$$
$$= 5e^{0.364}$$
$$\approx 5(1.4391)$$
$$\approx 7.2$$

We can predict that the population will be about 7.2 billion in 2000.

c) We set $P(t) = 6$ and solve for t:

$$6 = 5e^{0.028t}$$
$$1.2 = e^{0.028t}$$
$$\ln 1.2 = \ln e^{0.028t}$$
$$\ln 1.2 = 0.028t \quad \text{Property 4}$$
$$\frac{\ln 1.2}{0.028} = t$$
$$\frac{0.1823}{0.028} \approx t$$
$$7 \approx t$$

The population will be 6.0 billion about 7 years after 1987, or in 1994.

13. a) We can use the function $C(t) = C_0 e^{kt}$ as a model. At $t = 0(1962)$, the cost was 5¢. We substitute 0.05 for C_0 and 9.7%, or 0.097, for k:

$C(t) = 5 e^{0.097t}$, where t is the number of years after 1962.

b) In 2010, $t = 2010 - 1962$, or 48. We substitute 48 for t:

$$C(48) = 5 e^{0.097(48)}$$
$$= 5 e^{4.656}$$
$$\approx 5(105.2144)$$
$$\approx 526$$

In 2010 a Hershey bar will cost about 526¢, or \$5.26.

c) We set $C(t) = 500$(\$5 = 500¢) and solve for t:

$$500 = 5 e^{0.097t}$$
$$100 = e^{0.097t}$$
$$\ln 100 = \ln e^{0.097t}$$
$$\ln 100 = 0.097t$$
$$\frac{\ln 100}{0.097} = t$$
$$\frac{4.6052}{0.097} \approx t$$
$$47 \approx t$$

A Hershey bar will cost \$5 about 47 years after 1962, or in 2009.

13. (continued)

d) To find the doubling time, we set $C(t) = 2(5¢)$, or 10¢ and solve for t:

$$10 = 5\,e^{0.097t}$$
$$2 = e^{0.097t}$$
$$\ln 2 = \ln e^{0.097t}$$
$$\ln 2 = 0.097t$$
$$\frac{\ln 2}{0.097} = t$$
$$\frac{0.6931}{0.097} \approx t$$
$$7.1 \approx t$$

The doubling time is about 7.1 years.

15. a) The exponential growth function is $P(t) = P_0\,e^{kt}$. Substituting 52 for P_0, we have

$$P(t) = 52\,e^{kt},$$

where t is the number of years after 1970.

To find k we can use the fact that at t = 8 (1978), P = 66. We substitute and solve for k:

$$66 = 52\,e^{k(8)}$$
$$\frac{66}{52} = e^{8k}$$
$$\frac{33}{26} = e^{8k}$$
$$\ln \frac{33}{26} = \ln e^{8k}$$
$$\ln \frac{33}{26} = 8k$$
$$\frac{\ln \frac{33}{26}}{8} = k$$
$$\frac{0.2384}{8} \approx k$$
$$0.03 \approx k$$

The exponential growth function is

$$P(t) = 52\,e^{0.03t}.$$

b) In 1994, t = 1994 - 1970, or 24. We find P(24):

$$P(24) = 52\,e^{0.03(24)} \approx 107$$

The cost of a cone in 1994 will be about 107¢, or $1.07.

c) We substitute 2(52), or 104 for P and solve for t:

$$104 = 52\,e^{0.03t}$$
$$2 = e^{0.03t}$$
$$\ln 2 = \ln e^{0.03t}$$
$$\ln 2 = 0.03t$$
$$\frac{\ln 2}{0.03} = t$$
$$23.1 \approx t$$

The cost of a cone will be twice that of 1970 after about 23.1 yr.

15. (continued)

d) We substitute 300 ($3 = 300¢) for P and solve for t:

$$300 = 52\,e^{0.03t}$$
$$\frac{300}{52} = e^{0.03t}$$
$$\frac{75}{13} = e^{0.03t}$$
$$\ln \frac{75}{13} = \ln e^{0.03t}$$
$$\ln \frac{75}{13} = 0.03t$$
$$\frac{\ln \frac{75}{13}}{0.03} = t$$
$$58 \approx t$$

The cost of a cone will be $3 about 58 years after 1970, or in 2028.

17. a) The exponential growth function is $P(t) = P_0\,e^{kt}$. Since $P_0 = 84{,}000$, we have $P(t) = 84{,}000\,e^{kt}$, where t is the number of years after 1947. To find k we use the fact that at t = 40 (1987), the value of the painting was $53,900,000. We substitute and solve for k:

$$53{,}900{,}000 = 84{,}000\,e^{k(40)}$$
$$\frac{1925}{3} = e^{40k}$$
$$\ln \frac{1925}{3} = \ln e^{40k}$$
$$\ln \frac{1925}{3} = 40k$$
$$\frac{\ln \frac{1925}{3}}{40} = k$$
$$\frac{6.4641}{40} \approx k$$
$$0.1616 \approx k$$

The exponential growth rate is about 0.1616, or 16.16%. The exponential growth function is $P(t) = 84{,}000\,e^{0.1616t}$, where t is the number of years after 1947.

b) We find P(50):

$$P(50) = 84{,}000\,e^{0.1616(50)} =$$
$$84{,}000\,e^{8.08} \approx$$
$$84{,}000(3229.2332) \approx \$271{,}000{,}000$$

c) We substitute 2(84,000), or 168,000 for P and solve for t:

$$168{,}000 = 84{,}000\,e^{0.1616t}$$
$$2 = e^{0.1616t}$$
$$\ln 2 = \ln e^{0.1616t}$$
$$\ln 2 = 0.1616t$$
$$\frac{\ln 2}{0.1616} = t$$
$$\frac{0.6931}{0.1616} \approx t$$
$$4.3 \approx t$$

The doubling time is about 4.3 yr.

17. (continued)

d) We substitute 1,000,000,000 for P and solve for t:

$$1{,}000{,}000{,}000 = 84{,}000\, e^{0.1616t}$$

$$\frac{250{,}000}{21} = e^{0.1616t}$$

$$\ln \frac{250{,}000}{21} = \ln e^{0.1616t}$$

$$\ln \frac{250{,}000}{21} = 0.1616t$$

$$\frac{\ln \frac{250{,}000}{21}}{0.1616} = t$$

$$\frac{9.3847}{0.1616} \approx t$$

$$58 \approx t$$

The value of the painting will be \$1 billion after about 58 years.

19. We will use the function derived in Example 6:

$$P(t) = P_0\, e^{-0.00012t}$$

If the tusk has lost 20% of its carbon-14 from an initial amount P_0, then 80% (P_0) is the amount present. To find the age of the tusk t, we substitute 80% (P_0), or $0.8P_0$, for P(t) in the function above and solve for t.

$$0.8P_0 = P_0\, e^{-0.00012t}$$

$$0.8 = e^{-0.00012t}$$

$$\ln 0.8 = \ln e^{-0.00012t}$$

$$-0.2231 \approx -0.00012t$$

$$t \approx \frac{-0.2231}{-0.00012} \approx 1860$$

The tusk is about 1860 years old.

21. a) V(0) = \$28,000

b) $V(2) = \$28{,}000\, e^{-2} \approx \$3789.39$